Numerical Python

Scientific Computing and Data Science
Applications with Numpy, SciPy
and Matplotlib

Third Edition

Robert Johansson

Apress®

Numerical Python: Scientific Computing and Data Science Applications with Numpy, SciPy and Matplotlib

Robert Johansson
Urayasu-shi, Chiba, Japan

ISBN-13 (pbk): 979-8-8688-0412-0   ISBN-13 (electronic): 979-8-8688-0413-7
https://doi.org/10.1007/979-8-8688-0413-7

Managing Director, Apress Media LLC: Welmoed Spahr
Acquisitions Editor: Celestin Suresh John
Development Editor: James Markham
Coordinating Editor: Gryffin Winkler
Copyeditor: Kim Burton

Cover designed by eStudioCalamar

Cover image by author

Distributed to the book trade worldwide by Apress Media, LLC, 1 New York Plaza, New York, NY 10004, U.S.A. Phone 1-800-SPRINGER, fax (201) 348-4505, e-mail orders-ny@springer-sbm.com, or visit www.springeronline.com. Apress Media, LLC is a California LLC and the sole member (owner) is Springer Science + Business Media Finance Inc (SSBM Finance Inc). SSBM Finance Inc is a **Delaware** corporation.

For information on translations, please e-mail booktranslations@springernature.com; for reprint, paperback, or audio rights, please e-mail bookpermissions@springernature.com.

Apress titles may be purchased in bulk for academic, corporate, or promotional use. eBook versions and licenses are also available for most titles. For more information, reference our Print and eBook Bulk Sales web page at http://www.apress.com/bulk-sales.

Any source code or other supplementary material referenced by the author in this book is available to readers on GitHub (https://github.com/Apress). For more detailed information, please visit https://www.apress.com/gp/services/source-code.

If disposing of this product, please recycle the paper

To Mika, Erika, and Mia

Table of Contents

About the Author

Robert Johansson is an experienced Python programmer and computational scientist with a Ph.D. in Theoretical Physics from Chalmers University of Technology, Sweden. He has worked with scientific computing in academia and industry for over 15 years and participated in open source and proprietary research and development projects. His open-source contributions include work on QuTiP, a popular Python framework for simulating the dynamics of quantum systems, and he has also contributed to several other popular Python libraries in the scientific computing landscape. Robert is passionate about scientific computing and software development, teaching and communicating best practices for combining these fields with optimal outcomes: novel, reproducible, extensible, and impactful computational results.

About the Technical Reviewer

Shovon Sengupta is a distinguished data science expert specializing in advanced predictive analytics, machine learning, deep learning, and reinforcement learning. As the principal data scientist at the AI Center of Excellence for Fidelity Investment in the United States, Shovon is at the forefront of driving innovative initiatives that leverage artificial intelligence (specifically Generative AI) to solve complex business challenges. Shovon holds a US patent in automated predictive call routing using reinforcement learning.

He has also authored a few courses in the realm of machine learning. He has also presented at various international conferences on machine learning, time series forecasting, and building trustworthy artificial intelligence. His primary research is concentrated on deep reinforcement learning, deep learning, natural language processing, knowledge graphs, causality analysis, and time series analysis.

Shovon is also a PhD scholar specializing in applying machine learning algorithms in finance. His primary research interests include deep reinforcement learning, natural language processing, knowledge graphs, causality analysis, and time series analysis. His dedication to advancing the field of data science is evident in his continuous pursuit of knowledge and innovation.

For more details about Shovon's work, please check out his LinkedIn page at www.linkedin.com/in/shovon-sengupta-272aa917/.

Introduction

Scientific and numerical computing is a booming field in research, engineering, and analytics. The revolution in the computer industry over the last several decades has provided new and powerful tools for computational practitioners. This has enabled computational undertakings of previously unprecedented scale and complexity. Entire fields and industries have sprung up as a result. This development is ongoing, creating new opportunities as hardware, software, and algorithms keep improving. Ultimately, the enabling technology for this movement is the powerful computing hardware developed in recent decades. However, for a computational practitioner, the software environment used for computational work is as important as, if not more important than, the hardware on which the computations are carried out. This book is about one popular and fast-growing environment for numerical computing: the Python programming language and its vibrant ecosystem of libraries and extensions for computational work.

Computing is an interdisciplinary activity that requires experience and expertise in both theoretical and practical subjects: a firm understanding of mathematics and scientific thinking is a fundamental requirement for effective computational work. Equally important is solid training in computer programming and computer science. The role of this book is to bridge these two subjects by introducing how scientific computing can be done using the Python programming language and the computing environment that has appeared around this language. In this book, the reader is assumed to have some previous training in mathematics and numerical methods and basic knowledge of Python programming. The book's focus is to give a practical introduction to computational problem-solving with Python. Brief introductions to the theory of the covered topics are provided in each chapter to introduce notation and remind readers of the basic methods and algorithms. However, this book is not a self-consistent treatment of numerical methods. To assist readers who have yet to become familiar with some of the topics of this book, references for further reading are given at the end of each chapter. Likewise, readers without experience in Python programming will find it helpful to read this book with a book that focuses on the Python programming language itself.

How This Book Is Organized

The first chapter in this book introduces general principles for scientific computing and the main development environments available for computing in Python: the focus is on IPython and its interactive Python prompt, the excellent Jupyter Notebook application, and the Spyder IDE.

Chapter 2 introduces the NumPy library and generally discusses array-based computing and its virtues. Chapter 3 turns attention to symbolic computing—which in many respects complements array-based computing—using the SymPy library. Chapter 4 covers plotting and visualization using the Matplotlib library. These three chapters provide the basic computational tools used for domain-specific problems throughout the rest of the book: numerics, symbolics, and visualization.

Chapter 5 focuses on equation solving and explores numerical and symbolic methods, using the SciPy and SymPy libraries. Chapter 6 explores optimization, which is a natural extension of equation solving. It mainly works with the SciPy library and briefly with the cvxopt library. Chapter 7 deals with interpolation, another basic mathematical method with many applications, and important roles in higher-level algorithms and methods. Chapter 8 covers numerical and symbolic integration. Chapters 5 to 8 cover core computational techniques that are pervasive in all types of computational work. Most of the methods from these chapters are found in the SciPy library.

Chapter 9 covers ordinary differential equations. Chapter 10 is a detour into sparse matrices and graph methods, which helps prepare the field for the following chapter. Chapter 11 discusses partial differential equations, which conceptually are closely related to ordinary differential equations but require a different set of techniques that necessitates the introduction of sparse matrices, the topic of Chapter 10.

Chapter 12 changes direction and begins exploring data analysis and statistics. It introduces the Pandas library and its excellent data analysis framework. Chapter 13 covers basic statistical analysis and methods from the SciPy stats package. Chapter 14 moves on to statistical modeling using the statsmodels library. In Chapter 15, the theme of statistics and data analysis is continued with a discussion of machine learning using the scikit-learn library. Chapter 16 wraps up the statistics-related chapters with a discussion of Bayesian statistics and the PyMC library. Chapters 12 through 16 introduce the broad field of statistics and data analytics, which has been developing rapidly within and outside the scientific Python community in recent years.

Chapter 17 briefly returns to a core subject in scientific computing: signal processing. Chapter 18 discusses data input and output, and several methods for reading and writing numerical data to files, which is a basic topic required for most types of computational work. Chapter 19 introduces two methods for speeding up Python code using the Numba and Cython libraries.

The Appendix covers the installation of the software used in this book. This book uses the conda package manager to install the required software (mostly Python libraries). Conda can also be used to create virtual and isolated Python environments, which is an important topic for creating stable and reproducible computational environments. The Appendix also discusses how to work with such environments using the conda package manager.

Source Code Listings

Each chapter in this book has an accompanying Jupyter Notebook that contains the chapter's source code listings. These notebooks and the data files required to run them can be downloaded by visiting the book's GitHub page at `https://github.com/Apress/Numerical-Python-3rd-ed`.

CHAPTER 1

■ ■ ■

Introduction to Computing with Python

This book is about using Python for numerical computing. Python is a high-level, general-purpose interpreted programming language widely used in scientific computing and engineering. As a general-purpose language, Python was not specifically designed for numerical computing, but many of its characteristics make it well-suited for this task. First and foremost, Python is well known for its clean and easy-to-read code syntax. Good code readability improves maintainability, reduces bugs, and leads to better applications overall. It also enables rapid code development, since readability and expressiveness are essential in exploratory and interactive computing, where fast turnaround for testing various ideas and models is important.

In computational problem-solving, it is important to consider algorithms' performance and implementations. It is natural to strive for efficient high-performance code, and optimal performance is crucial for many computational problems. In such cases, it may be necessary to use a low-level program language, such as C or Fortran, to obtain the best performance out of the hardware that runs the code. However, it is not always the case that optimal runtime performance should be the highest priority. It is also important to consider the development time required to solve a problem in a programming language or an environment. While the best possible runtime performance can be achieved in a low-level programming language, working in a high-level language such as Python reduces the development time and often results in more flexible and extensible code.

These conflicting objectives present a trade-off between high performance but long development time and lower performance but shorter development time. Figure 1-1 shows a schematic visualization of this concept. When choosing a computational environment for solving a particular problem, it is important to consider this trade-off and to decide whether person-hours spent on the development or CPU-hours spent on running the computations are more valuable. It is worth noting that CPU-hours are cheap and getting even cheaper, but person-hours are expensive. Your own time is a very valuable resource. This makes a strong case for minimizing development time rather than the computation runtime by using a high-level programming language and environment such as Python and its scientific computing libraries.

© Robert Johansson 2024
R. Johansson, *Numerical Python*, https://doi.org/10.1007/979-8-8688-0413-7_1

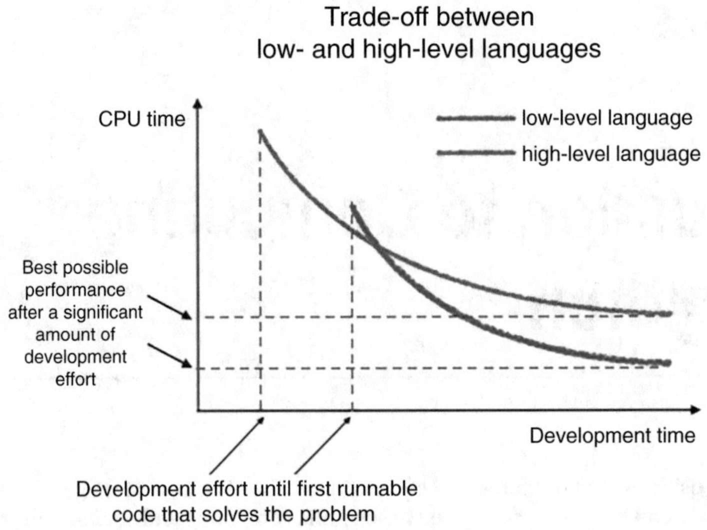

Figure 1-1. *The trade-off between low- and high-level programming languages. While a low-level language typically gives the best performance when a significant amount of development time is invested in the implementation of a solution to a problem, the development time required to obtain a first runnable code that solves the problem is typically shorter in a high-level language such as Python*

A solution that partially avoids the trade-off between high- and low-level languages is to use a multilanguage model, where a high-level language is used to interface libraries and software packages written in low-level languages. In a high-level scientific computing environment, this type of interoperability with software packages written in low-level languages (e.g., Fortran, C, or C++) is an important requirement. Python excels at this type of integration, and as a result, Python has become a popular "glue language" used as an interface for setting up and controlling computations that use code written in low-level programming languages for time-consuming number crunching. This is an important reason why Python is a popular language for numerical computing. The multilanguage model enables rapid code development in a high-level language while retaining most of the performance of low-level languages.

Due to the multilanguage model, scientific and technical computing with Python involves much more than just the Python language itself. In fact, the Python language is only a piece of an entire ecosystem of software and solutions that provide a complete environment for scientific and technical computing. This ecosystem includes development tools and interactive programming environments, such as Spyder and IPython, which are designed particularly with scientific computing in mind. It also includes a vast collection of Python packages for scientific computing. This ecosystem of scientifically oriented libraries ranges from generic core libraries—such as NumPy, SciPy, and Matplotlib—to more specific libraries for problem domains. Another crucial layer in the scientific Python stack exists below the various Python modules. Many scientific Python libraries interface with low-level, high-performance scientific software packages, such as optimized LAPACK and BLAS libraries[1] for low-level vector, matrix, and linear algebra routines or other specialized libraries for specific computational tasks. These libraries are typically implemented in a compiled low-level language and can be highly optimized and efficient. Without the foundation that such libraries provide, scientific computing with Python would not be practical. Figure 1-2 is an overview of the various layers of the software stack for computing with Python.

---

[1] For example, MKL, the Math Kernel Library from Intel at `https://software.intel.com/en-us/intel-mkl`; openBLAS at `www.openblas.net`; or ATLAS, the Automatically Tuned Linear Algebra Software at `http://math-atlas.sourceforge.net`

Figure 1-2. *An overview of the components and layers in the scientific computing environment for Python, from a user's perspective from top to bottom. Users typically only interact with the top three layers, but the bottom layer constitutes a very important part of the software stack*

■ **Tip**   The SciPy organization (`www.scipy.org`) provides a centralized resource for information about the core packages in the scientific Python ecosystem, lists of additional specialized packages, and documentation and tutorials. It is a valuable resource when working with scientific and technical computing in Python. Another great resource is the Numeric and Scientific page on the official Python Wiki (`http://wiki.python.org/moin/NumericAndScientific`).

Besides the technical reasons why Python provides a good environment for computational work, it is also significant that it and its scientific computing libraries are free and open source. This eliminates economic constraints on when and how applications developed with the environment can be deployed and distributed by its users. Equally significant, it makes it possible for a dedicated user to obtain complete insight into how the language and the domain-specific packages are implemented and what methods are used. For academic work where transparency and reproducibility are hallmarks, this is increasingly recognized as an important requirement of software used in research. For commercial use, it provides freedom on how the environment is used and integrated into products and how such solutions are distributed to customers. All users benefit from the relief of not paying license fees, which may otherwise inhibit deployments on large computing environments, such as clusters and cloud computing platforms.

The social component of the scientific computing ecosystem for Python is another important aspect of its success. Vibrant user communities have emerged around the core packages and many domain-specific projects. Project-specific mailing lists, Stack Overflow groups, and issue trackers (e.g., on GitHub, `www.github.com`) are typically very active and provide forums for discussing problems and obtaining help, as well as a way of getting involved in developing these tools. The Python computing community also organizes yearly conferences and meet-ups at many venues around the world, such as the SciPy (`http://conference.scipy.org`) and PyData (`http://pydata.org`) conference series.

Environments for Computing with Python

Several different environments are suitable for working with Python for scientific and technical computing. This diversity has both advantages and disadvantages compared to a single endorsed environment that is common in proprietary computing products: diversity provides flexibility and dynamism that lends itself to specialization for particular uses, but on the other hand, it can also be confusing for new users, and it can be more complicated to set up a full productive environment. Here, I give an orientation of common environments for scientific computing so that their benefits can be weighed against each other and an informed decision can be reached regarding which one to use in different situations and for different purposes. The following are the three environments discussed in this chapter.

- The Python interpreter or the IPython console run code interactively. Together with a text editor for writing code, this provides a lightweight development environment.

- The Jupyter Notebook is a web application in which Python code can be written and executed through a web browser. This environment is great for numerical computing, analysis, and problem-solving because it allows us to collect the code, the output produced by the code, related technical documentation, and the analysis and interpretation all in one document.

- The Spyder Integrated Development Environment can write and interactively run Python code. An IDE like Spyder is a great tool for developing libraries and reusable Python modules.

These environments have justified uses, and it is largely a matter of personal preference for which one to use. However, I recommend exploring the Jupyter Notebook environment, because it is highly suitable for interactive and exploratory computing and data analysis, where data, code, documentation, and results are tightly connected. For the development of Python modules and packages, I recommend using the Spyder IDE because of its integration with code analysis tools and the Python debugger.

Python and the rest of the software stack required for scientific computing with Python can be installed and configured in many ways, and in general, the installation details also vary from system to system. The Appendix goes through one popular cross-platform method to install the tools and libraries required for this book.

Python

The Python programming language and the standard implementation of the Python interpreter are frequently updated and made available through new releases.[2] Currently, the active version of Python available for production use is the Python 3 series; this book requires Python 3.8 or greater. Note that at the time of writing, versions prior to Python 3.8 have already passed end-of-life, meaning they will no longer receive important bug fixes and security updates. Should you encounter any such legacy Python environment, it is therefore recommended that you upgrade the Python interpreter to a newer version.

Interpreter

The standard way to execute Python code is to run the program directly through the Python interpreter. On most systems, the Python interpreter is invoked using the python command. When a Python source file is passed as an argument to this command, the Python code in the file is executed.

---

[2] The Python language and the default Python interpreter are managed and maintained by the Python Software Foundation (www.python.org).

```
$ python hello.py
Hello from Python!
```

Here, the hello.py file contains a single line.

```
print("Hello from Python!")
```

To see which version of Python is installed, we can invoke the python command with the --version argument.

```
$ python --version
Python 3.11.4
```

It is common to have more than one version of Python installed on the same system. Each version of Python maintains its own set of libraries and provides its own interpreter command (so each Python environment can have different libraries installed). On many systems, specific versions of the Python interpreter are available through commands such as python3.11. It is also possible to set up *virtual* Python environments independent of the system-provided environments, which has many advantages. I strongly recommend becoming familiar with this way of working with Python. Appendix A describes setting up and working with these kinds of environments.

In addition to executing Python script files, a Python interpreter can be used as an interactive console (also known as a REPL (read–evaluate–print–loop)). Entering python at the command prompt (without any Python files as arguments) launches the Python interpreter in an interactive mode. When doing so, you are presented with a prompt.

```
$ python
Python 3.11.4 (main, Jul  5 2023, 08:41:25) [Clang 14.0.6 ] on darwin
Type "help", "copyright", "credits" or "license" for more information.
>>>
```

From here, Python code can be entered, and for each statement, the interpreter evaluates the code and prints the result to the screen. The Python interpreter itself already provides a very useful environment for interactively exploring Python code, especially since the release of Python 3.4, which includes basic facilities such as a command history and basic autocompletion.

IPython Console

Although the interactive command-line interface provided by the standard Python interpreter has been greatly improved in the Python interpreter itself, it is still, in certain aspects, rudimentary, and it does not provide a complete environment for interactive computing. IPython[3] is an enhanced command-line REPL environment for Python, with additional interactive and exploratory computing features. For example, IPython provides improved command history browsing (also between sessions), an input and output caching system, improved auto-completion, more verbose and helpful exception tracebacks, and more. IPython is now much more than an enhanced Python command-line interface, which is explored in more detail later in this chapter and throughout the book. For instance, under the hood, IPython is a client-server application that separates the front end (user interface) from the back end (kernel) and executes the Python code. This allows multiple types of user interfaces to communicate and work with the same kernel, and a user-interface application can connect multiple kernels using IPython's framework for parallel computing.

---

[3] See the IPython project web page, http://ipython.org, for more information and its official documentation.

Running the ipython command launches the IPython command prompt.

```
$ ipython
Python 3.11.4 (main, Jul  5 2023, 08:41:25) [Clang 14.0.6 ]
Type 'copyright', 'credits' or 'license' for more information
IPython 8.12.2 -- An enhanced Interactive Python. Type '?' for help
In [1]:
```

■ **Caution**   Each IPython installation corresponds to a specific version of Python. If several versions of Python are available on your system, you may also have several versions of IPython. On many systems, IPython for Python 3 is invoked with the ipython3 command, although the exact setup varies from system to system. Note that here, the "3" refers to the Python version, which differs from the version of IPython itself (at the time of writing it is 8.12.2).

The following sections briefly overview some of the IPython features that are most relevant to interactive computing. It is worth noting that IPython is used in many different contexts in scientific computing with Python, for example, as a kernel in the Jupyter Notebook application and in the Spyder IDE, which is covered in more detail later in this chapter. It is time well spent to get familiar with the tricks and techniques that IPython offers to improve your productivity when working with interactive computing.

Input and Output Caching

In the IPython console, the input prompt is denoted as In [1]: and the corresponding output is denoted as Out [1]:, where the numbers within the square brackets are incremented for each new input and output. These inputs and outputs are called *cells* in IPython. The input and the output of previous cells can later be accessed through the In and Out variables that IPython automatically creates. The In and Out variables are a list and a dictionary, respectively, that can be indexed with a cell number. For instance, consider the following IPython session.

```
In [1]: 3 * 3
Out[1]: 9
In [2]: In[1]
Out[2]: '3 * 3'
In [3]: Out[1]
Out[3]: 9
In [4]: In
Out[4]: ['', '3 * 3', 'In[1]', 'Out[1]', 'In']
In [5]: Out
Out[5]: {1: 9, 2: '3 * 3', 3: 9, 4: ['', '3 * 3', 'In[1]', 'Out[1]', 'In', 'Out']}
```

Here, the first input was 3 * 3, and the result was 9, which later is available as In[1] and Out[1]. A single underscore _ is a shorthand notation for referring to the most recent output, and a double underscore __ refers to the output that preceded the most recent output. Input and output caching is often useful in interactive and exploratory computing since the result of a computation can be accessed even if it was not explicitly assigned to a variable.

Note that when a cell is executed, the value of the last statement in an input cell is, by default, displayed in the corresponding output cell unless the statement is an assignment or if the value is Python null value None. The output can be suppressed by ending the statement with a semicolon.

```
In [6]: 1 + 2
Out[6]: 3
In [7]: 1 + 2;    # output suppressed by the semicolon
In [8]: x = 1     # no output for assignments
In [9]: x = 2; x  # these are two statements. The value of 'x' is shown in
                  # the output
Out[9]: 2
```

Autocompletion and Object Introspection

In IPython, pressing the TAB key activates autocompletion, which displays a list of symbols (variables, functions, classes, etc.) with names that are valid completions of what has already been typed. The autocompletion in IPython is contextual, and it looks for matching variables and functions in the current namespace or among the attributes and methods of a class when invoked after the name of a class instance. For example, os.<TAB> produces a list of the variables, functions, and classes in the os module, and pressing TAB after typing os.w results in a list of symbols in the os module that starts with w.

```
In [10]: import os
In [11]: os.w<TAB>
os.wait   os.wait3   os.wait4   os.waitpid   os.walk   os.write   os.writev
```

This feature is called *object introspection*, a powerful tool for interactively exploring the properties of Python objects. Object introspection works on modules, classes, attributes, methods, functions, and arguments.

Documentation

Object introspection is convenient for exploring the API of a module, such as its member classes and functions, and together with the documentation strings or *docstrings* that are commonly provided in Python code, it provides a built-in dynamic reference manual for almost any Python module that is installed and can be imported. A Python object followed by a question mark displays the documentation string for the object. This is similar to the Python function help. An object can also be followed by two question marks, in this case, IPython tries to display more detailed documentation, including the Python source code, if available. For example, to display help for the cos function in the math library.

```
In [12]: import math
In [13]: math.cos?
Signature: math.cos(x, /)
Docstring: Return the cosine of x (measured in radians).
Type:      builtin_function_or_method
```

Docstrings can be specified for Python modules, functions, classes, and their attributes and methods. A well-documented module includes full API documentation in the code itself. From a developer's point of view, it is convenient to document a code together with the implementation. This encourages writing and maintaining documentation, and Python modules tend to be well-documented.

Interaction with the System Shell

IPython also provides extensions to the Python language that make interacting with the underlying system convenient. Anything that follows an exclamation mark is evaluated using the system shell (such as bash shell). For example, on a Unix-like system, such as Linux or macOS, listing files in the current directory can be done using the following.

```
In[14]: !ls
file1.py      file2.py      file3.py
```

In Microsoft Windows, the equivalent command would be !dir. This method for interacting with the operating system is a powerful feature that makes it easy to navigate the file system and use the IPython console as a system shell. The output generated by a command following an exclamation mark can easily be captured in a Python variable. For example, a file listing produced by !ls can be stored in a Python list using the following.

```
In[15]: files = !ls
In[16]: len(files)
3
In[17] : files
['file1.py', 'file2.py', 'file3.py']
```

Likewise, we can pass the values of Python variables to shell commands by prefixing the variable name with a $ sign.

```
In[18]: file = "file1.py"
In[19]: !ls -l $file
-rw-r--r--  1 rob   staff 131 Oct 22 16:38 file1.py
```

This two-way communication with the IPython console and the system shell can be very convenient, for example, when processing data files.

IPython Extensions

IPython provides extension commands that are called *magic functions* in IPython terminology. These commands all start with one or two % signs.[4] A single % sign is used for one-line commands, and two % signs are used for commands that operate on cells (multiple lines). For a complete list of available extension commands, type %lsmagic, and the documentation for each command can be obtained by typing the magic command followed by a question mark.

```
In[20]: %lsmagic?
Docstring: List currently available magic functions.
File:          /usr/local/lib/python3.6/site-packages/IPython/core/magics/basic.py
```

---

[4] When %automagic is activated (type %automagic at the IPython prompt to toggle this feature), the % sign that precedes the IPython commands can be omitted, unless there is a name conflict with a Python variable or function. However, for clarity, the % signs are explicitly shown here.

File System Navigation

In addition to the interaction with the system shell described in the previous section, IPython provides commands for navigating and exploring the file system. These commands are familiar to Unix shell users: %ls (list files), %pwd (return current working directory), %cd (change working directory), %cp (copy file), %less (show the content of a file in the pager), and %%writefile filename (write content of a cell to the file filename). Note that autocomplete in IPython also works with the files in the current working directory, which makes IPython as convenient to explore the file system as the system shell. It is worth noting that these IPython commands are system-independent and can be used on both Unix-like operating systems and Windows.

Running Scripts from the IPython Console

The %run command is an important and useful extension, perhaps one of the most important features of the IPython console. This command can execute an external Python source code file within an interactive IPython session. Keeping a session active between multiple runs of a script makes it possible to explore the variables and functions defined in a script interactively after the execution of the script has finished. To demonstrate this functionality, consider a script file fib.py that contains the following code.

```python
def fib(n):
    """
    Return a list of the first n Fibonacci numbers.
    """
    f0, f1 = 0, 1
    f = [1] * n
    for i in range(1, n):
        f[i] = f0 + f1
        f0, f1 = f1, f[i]
    return f

print(fib(10))
```

It defines a function that generates a sequence of n Fibonacci numbers and prints the result for $n = 10$ to the standard output. It can be run from the system terminal using the standard Python interpreter.

```
$ python fib.py
[1, 1, 2, 3, 5, 8, 13, 21, 34, 55]
```

It can also be run from an interactive IPython session, which produces the same output but also adds the symbols defined in the file to the local namespace so that the fib function is available in the interactive session after the %run command has been issued.

```
In [21]: %run fib.py
Out[22]: [1, 1, 2, 3, 5, 8, 13, 21, 34, 55]
In [23]: %who
fib
In [23]: fib(6)
Out[23]: [1, 1, 2, 3, 5, 8]
```

The preceding example also used the %who command, which lists all defined symbols (variables and functions).[5] The %whos command is similar, but also gives more detailed information about the type and value of each symbol, when applicable.

Debugger

IPython includes a handy debugger mode, which can be invoked postmortem after a Python exception (error) has been raised. After the traceback of an unintercepted exception has been printed to the IPython console, it is possible to step directly into the Python debugger using the IPython command %debug. This possibility can eliminate the need to rerun the program from the beginning using the debugger or after employing the common debugging method of sprinkling print statements into the code. If the exception is unexpected and happens late in a time-consuming computation, this can be a big time-saver.

To see how the %debug command can be used, consider the following incorrect invocation of the fib function defined earlier. It is incorrect because a float is passed to the function while the function is implemented, assuming that the argument passed to it is an integer. On line 7 the code runs into a type error, and the Python interpreter raises an exception of TypeError. IPython catches the exception and prints a useful traceback of the call sequence on the console. If we are clueless about why the code on line 7 contains an error, entering the debugger by typing %debug in the IPython console could be useful. We then get access to the local namespace at the source of the exception, which can allow us to explore in more detail why the exception was raised.

```
In [24]: fib(1.0)
---------------------------------------------------------------------------
TypeError                                 Traceback (most recent call last)
<ipython-input-24-874ca58a3dfb> in <module>()
  ----> 1 fib.fib(1.0)
/Users/rob/code/fib.py in fib(n)
      5      """
      6      f0, f1 = 0, 1
  ----> 7      f = [1] * n
      8      for i in range(1, n):
      9          f[n] = f0 + f1
TypeError: can't multiply sequence by non-int of type 'float'
In [25]: %debug
> /Users/rob/code/fib.py(7)fib()
      6      f0, f1 = 0, 1
  ----> 7      f = [1] * n
      8      for i in range(1, n):
ipdb> print(n)
1.0
```

■ Tip Type a question mark at the debugger prompt to show a help menu that lists available commands.

```
ipdb> ?
```

More information about the Python debugger and its features is available in the Python Standard Library documentation: http://docs.python.org/3/library/pdb.html.

---

[5] The Python function dir provides a similar feature.

Reset

Resetting the namespace of an IPython session is often useful to ensure that a program is run in a pristine environment, uncluttered by existing variables and functions. The %reset command provides this functionality (use the -f flag to force the reset). Using this command can often eliminate the need for otherwise common exit-restart cycles of the console. Although it is necessary to reimport modules after the %reset command has been used, it is important to know that even if the modules have changed since the last import, a new import after a %reset does not import the new module but rather reenable a cached version of the module from the previous import. When developing Python modules, this is usually not the desired behavior. In that case, a reimport of a previously imported (and since updated) module can often be achieved by using the reload function from IPython.lib.deepreload. However, this method does not always work, as some libraries run code at import time that is only intended to run once. In this case, the only option might be to terminate and restart the IPython interpreter.

Timing and Profiling Code

The %timeit and %time commands provide simple benchmarking facilities useful when looking for bottlenecks and attempting to optimize code. The %timeit command runs a Python statement several times and estimates the runtime (use %%timeit to do the same for a multiline cell). The exact number of times the statement is run is determined heuristically unless explicitly set using the -n and -r flags. See %timeit? for details. The %timeit command does not return the resulting value of the expression. If the result of the computation is required, the %time or %%time (for a multiline cell) commands can be used instead, but %time and %%time only run the statement once and give a less accurate estimate of the average runtime.

The following example demonstrates a typical usage of the %timeit and %time commands.

```
In [26]: %timeit fib(100)
100000 loops, best of 3: 16.9 µs per loop
In [27]: result = %time fib(100)
CPU times: user 33 µs, sys: 0 ns, total: 33 µs
Wall time: 48.2
```

While the %timeit and %time commands are useful for measuring the elapsed runtime of a computation, they do not give detailed information about what part of the computation takes more time. Such analyses require a more sophisticated code profiler, such as the one provided by the Python standard library module cProfile.[6] The Python profiler is accessible in IPython through the %prun (for statements) and %run commands with the -p flag (for running external script files). The output from the profiler is rather verbose and can be customized using optional flags to the %prun and %run -p commands (see %prun? for a detailed description of the available options).

As an example, consider a function that simulates N random walkers, each taking M steps, and then calculates the furthest distance from the starting point achieved by any of the random walkers.

```
In [28]: import numpy as np
In [29]: def random_walker_max_distance(M, N):
    ...:     """
    ...:     Simulate N random walkers taking M steps, and return the largest
    ...:     Distance from the starting point achieved by any of the random
    ...:     walkers.
```

---

[6] Which can, for example, be used with the standard Python interpreter to profile scripts by running python -m cProfile script.py

```
...:      """
...:      trajectories = [np.random.randn(M).cumsum() for _ in range(N)]
...:      return np.max(np.abs(trajectories))
```

Calling this function using the profiler with %prun results in the following output, which includes information about how many times each function was called and a breakdown of the total and cumulative time spent in each function. From this information, we can conclude that in this simple example, the calls to the np.random.randn function consume the bulk of the elapsed computation time.

```
In [30]: %prun random_walker_max_distance(400, 10000)

        20011 function calls in 0.285 seconds

Ordered by: internal time

ncalls  tottime  percall  cumtime  percall  filename:lineno(function)
10000    0.181    0.000    0.181    0.000  {method 'randn' of
                                           'mtrand.RandomState' objects}
10000    0.053    0.000    0.053    0.000  {method 'cumsum' of
                                           'numpy.ndarray' objects}
    1    0.020    0.020    0.277    0.277  2615584822.py:3(
                                           random_walker_max_distance)
    1    0.019    0.019    0.253    0.253  2615584822.py:8(<listcomp>)
    1    0.008    0.008    0.285    0.285  <string>:1(<module>)
    1    0.004    0.004    0.004    0.004  {method 'reduce' of
                                           'numpy.ufunc' objects}
    1    0.000    0.000    0.285    0.285  {built-in method builtins.exec}
    1    0.000    0.000    0.004    0.004  fromnumeric.py:71(_wrapreduction)
    1    0.000    0.000    0.004    0.004  fromnumeric.py:2692(max)
    1    0.000    0.000    0.000    0.000  fromnumeric.py:72(<dictcomp>)
    1    0.000    0.000    0.000    0.000  fromnumeric.py:2687(_max_dispatcher)
    1    0.000    0.000    0.000    0.000  {method 'disable' of
                                           '_lsprof.Profiler' objects}
    1    0.000    0.000    0.000    0.000  {method 'items' of 'dict' objects}
```

Interpreter and Text Editor as Development Environment

In principle, the Python or the IPython interpreter and a good text editor are all required for a fully productive Python development environment. This simple setup is, in fact, the preferred development environment for many experienced programmers. However, the following sections look at the Jupyter Notebook and Spyder's integrated development environment. These environments provide richer features that improve productivity when working with interactive and exploratory computing applications.

Jupyter

The Jupyter project[7] is a spin-off from the IPython project that includes the Python independent frontends—most notably the notebook application, which is discussed in more detail in the following section—and the communication framework that enables the separation of the frontend from the computational backends,

---

[7] For more information about Jupyter, see http://jupyter.org.

known as *kernels*. Prior to the creation of the Jupyter project, the notebook application and its underlying framework were a part of the IPython project. However, because the notebook frontend is language agnostic (it can also be used with many other languages, such as R and Julia), it was spun off a separate project to better cater to the wider computational community and avoid a perceived bias toward Python. Now, the remaining role of IPython is to focus on Python-specific applications, such as the interactive Python console, and to provide a Python kernel for the Jupyter environment.

In the Jupyter framework, the front end can be connected to multiple computational backend kernels, for example, for different programming languages, versions of Python, or for different Python environments. The kernel maintains the state of the interpreter. It performs the actual computations, while the front end manages how code is entered and organized and how the results of calculations are visualized to the user.

This section discusses the Jupyter QtConsole and Notebook frontends. It briefly introduces some of their rich display and interactivity features and the workflow organization that the notebook provides. The Jupyter Notebook is the Python environment for computational work that I generally recommend in this book, and the code listings in the rest of this book are understood to be read as if they are cells in a notebook.

The Jupyter QtConsole

The Jupyter QtConsole is an enhanced console application that can substitute for the standard IPython console. The QtConsole is launched by passing the `qtconsole` argument to the `jupyter` command.

```
$ jupyter qtconsole
```

This opens a new IPython application in a console that can display rich media objects such as images, figures, and mathematical equations. The Jupyter QtConsole also provides a menu-based mechanism for displaying autocompletion results, and it shows docstrings for functions in a pop-up window when typing the opening parenthesis of a function or a method call. A screenshot of the Jupyter Qtconsole is shown in Figure 1-3.

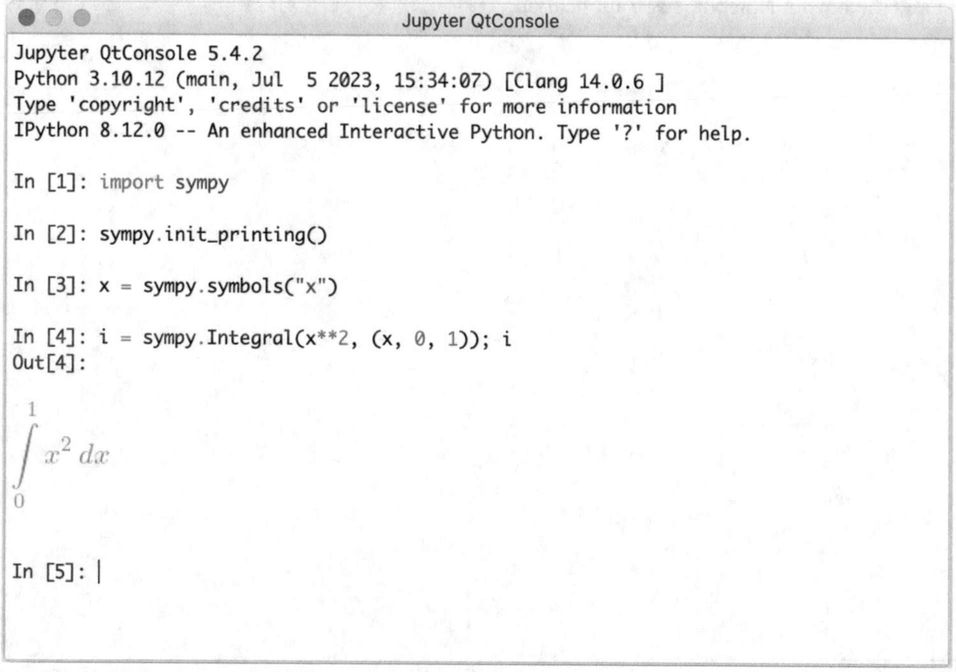

Figure 1-3. *A screenshot of the Jupyter QtConsole application*

The Jupyter Notebook

In addition to the interactive console, Jupyter also provides a web-based notebook application that has made it famous. The notebook offers many advantages over a traditional development environment when working with data analysis and computational problem-solving. In particular, the notebook environment allows us to write and run code, display the output produced by the code, and document and interpret the code and the results—all in one document. This means the entire analysis workflow is captured in one file, which can be saved, restored, and reused later. In contrast, when working with a text editor or an IDE, the code, the corresponding data files and figures, and the documentation are spread out over multiple files in the file system, and it takes a significant effort and discipline to keep such a workflow organized.

The Jupyter Notebook features a rich display system that can show media such as equations, figures, and videos as embedded objects in the notebook. Creating user interface (UI) elements with HTML and JavaScript is possible using Jupyter's widget system. These widgets can be used in interactive applications that connect the web application with Python code executed in the IPython kernel (on the server side). These and many other features of the Jupyter Notebook make it a great environment for interactive and literate computing, as we will see in examples throughout this book.

To launch the Jupyter Notebook environment, the `notebook` argument is passed to the `jupyter` command-line application.

```
$ jupyter notebook
```

This launches a notebook kernel and a web application that, by default, serves up a web server on port 8888 on localhost, which is accessed using the local address `http://localhost:8888/` in a web browser.[8] By default, running `jupyter notebook` opens a web page in the default web browser. The application lists all notebooks that are available in the directory from where the Jupyter Notebook was launched, as well as a simple directory browser that can be used to navigate subdirectories relative to the location where the notebook server was launched and to open notebooks from therein. Figure 1-4 shows a screenshot of a web browser and the Jupyter Notebook web application.

---

[8] This web application is by default only accessible locally from the system where the notebook application was launched.

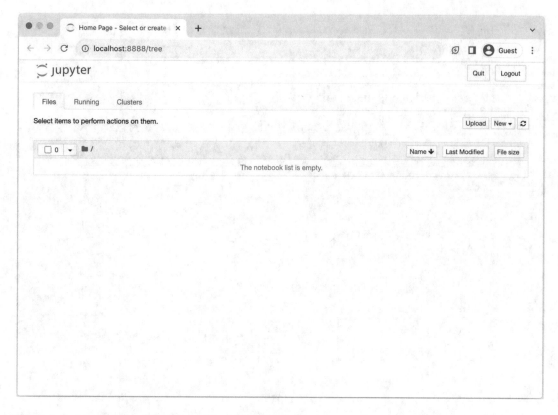

Figure 1-4. *A screenshot of the Jupyter Notebook web application*

Clicking the New button creates a new notebook and opens it in a new page in the browser (see Figure 1-5). A newly created notebook is named Untitled, or Untitled1, for example, depending on the availability of unused filenames. A notebook can be renamed by clicking the title field on the top of the notebook page. The Jupyter Notebook files are stored in JSON format using the ipynb filename extension. A Jupyter Notebook file is not pure Python code. When necessary, the Python code in a notebook can easily be extracted using either File ➤ Download as ➤ Python or the Jupyter utility nbconvert (see in the following section).

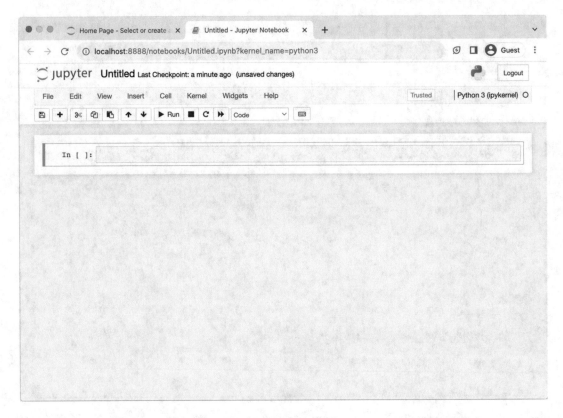

Figure 1-5. *A newly created and empty Jupyter Notebook*

Jupyter Lab

Jupyter Lab is an alternative development environment from the Jupyter project. The application can be launched by running `jupyter lab` from the command line. It combines the Jupyter Notebook interface with a file browser, text editor, shell, and IPython consoles in a web-based IDE-like environment; see Figure 1-6.

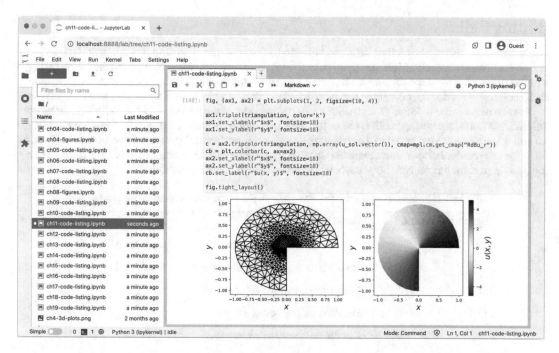

Figure 1-6. *The Jupyter Lab interface includes a file browser (left) and multitab notebook editor (right). The notebook displayed here shows code and output from an example in Chapter 11*

The Jupyter Lab environment consolidates the many advantages of the notebook environment and the strengths of traditional IDEs. Access to shell consoles and text editors all within the same web frontend is also convenient when working on a Jupyter server that runs on a remote system, such as a computing cluster or in the cloud.

Cell Types

The main content of a notebook, below the menu bar and the toolbar, is organized as input and output cells. The cells can be of several types, and the selected cell type can be changed using the cell-type drop-down menu in the toolbar (which initially displays "Code"). The following describes the most important types.

- **Code**: A code cell can contain an arbitrary amount of multiline Python code. Pressing Shift-Enter sends the code in the cell to the kernel process, where the kernel evaluates it using the Python interpreter. The result is sent back to the browser and displayed in the corresponding output cell.

- **Markdown**: The content of a Markdown cell can contain marked-up plain text, which is interpreted using the Markdown language and HTML. A Markdown cell can also contain LaTeX formatted equations, which are rendered in the notebook using the JavaScript-based LaTeX engine MathJax.

- **Raw**: A raw text cell is displayed without any processing.

Editing Cells

Using the menu bar and the toolbar, cells can be added, removed, moved up and down, cut and pasted, and so on. These functions are also mapped to keyboard shortcuts, which are convenient and time-saving when working with Jupyter Notebooks. The notebook uses a two-mode input interface, with an edit mode and a command mode. The edit mode can be entered by clicking a cell or by pressing the Enter key on the keyboard when a cell is in focus. Once in edit mode, the content of the input cell can be edited. Leaving the edit mode is done by pressing the ESC key or by using Shift-Enter to execute the cell. In command mode, the up and down arrows can be used to move focus between cells, and many keyboard shortcuts are mapped to the basic cell manipulation actions available through the toolbar and the menu bar. Table 1-1 summarizes the most important Jupyter Notebook keyboard shortcuts for the command mode.

Table 1-1. *A Summary of Keyboard Shortcuts in the Jupyter Notebook Command Mode*

Keyboard Shortcut	Description
b	Create a new cell *below* the currently selected cell.
a	Create a new cell *above* the currently selected cell.
d-d	Delete the currently selected cell.
1 to 6	Heading cell for levels 1 to 6.
x	Cut the currently selected cell.
c	Copy the currently selected cell.
v	Paste the cell from the clipboard.
m	Convert a cell to a Markdown cell.
y	Convert a cell to a code cell.
Up	Select the previous cell.
Down	Select the next cell.
Enter	Enter edit mode.
Escape	Exit edit mode.
Shift-Enter	Run the cell.
0-0	Restart the kernel.
i-i	Interrupt an executing cell.
s	Save the notebook.

While a notebook cell is being executed, the input prompt number is represented with an asterisk, In[*], and an indicator in the upper right corner of the page signals that the IPython kernel is busy. The execution of a cell can be interrupted using the menu option Kernel ➤ Interrupt or by typing i-i in the command mode (i.e., press the i key twice in a row).

Markdown Cells

One of the key features of the Jupyter Notebook is that code cells and output cells can be complemented with documentation contained in text cells. Text input cells are called *Markdown cells*. The input text is interpreted and reformatted using the Markdown markup language. The Markdown language is designed to be a lightweight typesetting system that allows text with simple markup rules to be converted to HTML and other formats for richer display. The markup rules are designed to be user-friendly and readable as is in plain-text format. For example, a piece of text can be made italics by surrounding it with asterisks, *text*. We can make it bold by surrounding it with double asterisks, **text**. Markdown also supports creating enumerated and bulleted lists, tables, and hyper-references. Jupyter supports an extension to Markdown that allows mathematical expressions to be typeset in LaTeX using the JavaScript LaTeX library MathJax. Generously documenting the code and the resulting output using Markdown cells and the many rich display options they provide is a good way to take advantage of Jupyter's offers. Table 1-2 introduces basic Markdown and equation formatting features that can be used in a Jupyter Notebook Markdown cell.

Table 1-2. *Summary of Markdown Syntax for Jupyter Notebook Markdown Cells*

Function	Syntax by Example
Italics	`*text*`
Bold	`**text**`
Strike-through	`~~text~~`
Fixed-width font	`` `text` ``
URL	`[URL text](http://www.example.com)`
New paragraph	Separate the text of two paragraphs with an empty line.

Verbatim	Lines that start with four blank spaces are displayed as is, without any further processing, using a fixed-width font. This is useful for code-like text segments.
	`␣␣␣␣def func(x):`
	`␣␣␣␣    return x ** 2`

Table	`| A | B | C |`
	`+---+---+---+`
	`| 1 | 2 | 3 |`
	`+---+---+---+`
	`| 4 | 5 | 6 |`

Horizontal line	A line containing three dashes is rendered as a horizontal line separator: `---`

Heading	`# Level 1 heading`
	`## Level 2 heading`
	`### Level 3 heading`
	`...`

Block quote	Lines that start with a ">" are rendered as a block quote.
	`> Text here is indented and offset`
	`> from the main text body.`

(continued)

Table 1-2. (*continued*)

Function	Syntax by Example

Unordered list	```* Item one```
	```* Item two```
	```* Item three```

Ordered list	```1. Item one```
	```2. Item two```
	```3. Item three```

Image	```![Alternative text](image-file.png)```[9]
	or
	```![Alternative text](http://www.example.com/image.png)```

Inline LaTeX equation	```$\LaTeX$```
Displayed LaTeX equation (centered and on a new line)	```$$\LaTeX$$``` or ```\begin{env}...\end{env}``` where env can be a LaTeX environment such as ```equation```, ```eqnarray```, ```align```, etc.

Markdown cells can also contain HTML code, and the Jupyter Notebook interface displays it as rendered HTML. It is a very powerful feature for the Jupyter Notebook. Its disadvantage is that such HTML code cannot be converted to other formats, such as PDF, using the `nbconvert` tool (see later section in this chapter). Therefore, it is generally better to use Markdown formatting when possible and resort to HTML only when necessary.

More information about MathJax and Markdown is available on the projects' web pages at `www.mathjax.com` and `http://daringfireball.net/projects/markdown`, respectively.

Rich Output Display

The result produced by the last statement in a notebook cell is normally displayed in the corresponding output cell, just like in the standard Python interpreter or the IPython console. The default output cell formatting is a string representation of the object generated, for example, by the `__repr__` method. However, the notebook environment enables a much richer output formatting, as it, in principle, allows displaying arbitrary HTML in the output cell area. The `IPython.display` module provides several classes and functions that make it easy to programmatically render formatted output in a notebook. For example, the `Image` class provides a way to display images from the local file system or online resources in a notebook, as shown in Figure 1-7. Other useful classes from the same module are `HTML`, for rendering HTML code, and `Math`, for rendering LaTeX expressions. The `display` function can explicitly request an object to be rendered and displayed in the output area.

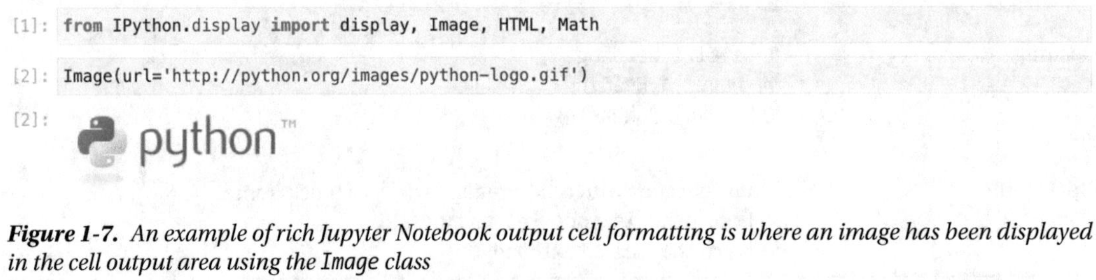

```
[1]: from IPython.display import display, Image, HTML, Math
```

```
[2]: Image(url='http://python.org/images/python-logo.gif')
```

[2]:

Figure 1-7. *An example of rich Jupyter Notebook output cell formatting is where an image has been displayed in the cell output area using the Image class*

---

[9] The path/filename is relative to the notebook directory.

An example of how HTML code can be rendered in the notebook using the HTML class is shown in Figure 1-8. Here, we first construct a string containing HTML code for a table with version information for a list of Python libraries. This HTML code is then rendered in the output cell area by creating an instance of the HTML class. Since this statement is the last (and only) statement in the corresponding input cell, Jupyter renders the representation of this object in the output cell area.

```
[3]: import scipy, numpy, matplotlib
     modules = [numpy, matplotlib, scipy]
     row = "<tr> <td>%s</td> <td>%s</td> </tr>"
     rows = "\n".join([row % (module.__name__, module.__version__) for module in modules])
     s = "<table> <tr><th>Library</th><th>Version</th> </tr> %s</table>" % rows
```

```
[4]: s
```

```
[4]: '<table> <tr><th>Library</th><th>Version</th> </tr> <tr> <td>numpy</td> <td>1.25.2</td> </tr>\n<tr> <t
     d>matplotlib</td> <td>3.7.1</td> </tr>\n<tr> <td>scipy</td> <td>1.11.1</td> </tr></table>'
```

```
[5]: HTML(s)
```

[5]:

Library	Version
numpy	1.25.2
matplotlib	3.7.1
scipy	1.11.1

Figure 1-8. *Example of rich Jupyter Notebook output cell formatting, where an HTML table containing module version information has been rendered and displayed using the HTML class*

For an object to be displayed in an HTML formatted representation, all we need to do is to add a method called _repr_hmtl_ to the class definition. For example, we can easily implement your own primitive version of the HTML class and use it to render the same HTML code as in the previous example, as demonstrated in Figure 1-9.

```
[6]: class HTMLDisplayer(object):
         def __init__(self, code):
             self.code = code

         def _repr_html_(self):
             return self.code
```

```
[7]: HTMLDisplayer(s)
```

[7]:

Library	Version
numpy	1.25.2
matplotlib	3.7.1
scipy	1.11.1

Figure 1-9. *Example of how to render HTML code in the Jupyter Notebook, using a class that implements the _repr_hmtl_ method*

Jupyter supports a large number of representations in addition to the _repr_hmtl_ shown in the preceding text, for example, _repr_png_, _repr_svg_, and _repr_latex_, to mention a few. The former two can generate and display graphics in the notebook output cell, as used by, for example, the Matplotlib library (see the following interactive example and Chapter 4). The Math class, which uses the _repr_latex_ method, can render mathematical formulas in the Jupyter Notebook. This is often useful in scientific and technical applications. Examples of how formulas can be rendered using the Math class and the _repr_latex_ method are shown in Figure 1-10.

```
[8]: Math(r'\hat{H} = -\frac{1}{2}\epsilon \hat{\sigma}_z-\frac{1}{2}\delta \hat{\sigma}_x')
```

[8]: $\hat{H} = -\dfrac{1}{2}\epsilon\hat{\sigma}_z - \dfrac{1}{2}\delta\hat{\sigma}_x$

```
[9]: class QubitHamiltonian(object):
         def __init__(self, epsilon, delta):
             self.epsilon = epsilon
             self.delta = delta

         def _repr_latex_(self):
             return "$\hat{H} = -%.2f\hat{\sigma}_z-%.2f\hat{\sigma}_x$" % \
                 (self.epsilon/2, self.delta/2)
```

```
[10]: QubitHamiltonian(0.5, 0.25)
```

[10]: $\hat{H} = -0.25\hat{\sigma}_z - 0.12\hat{\sigma}_x$

Figure 1-10. *An example of how a LaTeX formula is rendered using the* Math *class and how the* _repr_latex_ *method can be used to generate a LaTeX formatted representation of an object*

Using the various representation methods recognized by Jupyter or the convenience classes in the IPython.display module, we have great flexibility in shaping how results are visualized in the Jupyter Notebook. However, the possibilities do not stop there: an exciting feature of the Jupyter Notebook is that interactive applications, with two-way communication between the frontend and the backend kernel, can be created using, for example, a library of widgets (UI components) or directly with JavaScript and HTML. For example, using the interact function from the ipywidgets library, we can very easily create an interactive graph that takes an input parameter that is determined from a UI slider, as shown in Figure 1-11.

```
[11]: import matplotlib.pyplot as plt
      import numpy as np
      from scipy import stats

      def f(mu):
          X = stats.norm(loc=mu, scale=np.sqrt(mu))
          N = stats.poisson(mu)
          x = np.linspace(0, X.ppf(0.999))
          n = np.arange(0, x[-1])

          fig, ax = plt.subplots()
          ax.plot(x, X.pdf(x), color='black', lw=2, label="Normal($\mu=%d, \sigma^2=%d$)" % (mu, mu))
          ax.bar(n, N.pmf(n), align='edge', label=r"Poisson($\lambda=%d$)" % mu)
          ax.set_ylim(0, X.pdf(x).max() * 1.25)
          ax.legend(loc=2, ncol=2)
          plt.close(fig)
          return fig
```

```
[12]: from ipywidgets import interact
      import ipywidgets as widgets
```

```
[13]: interact(f, mu=widgets.FloatSlider(min=1.0, max=20.0, step=1.0));
```

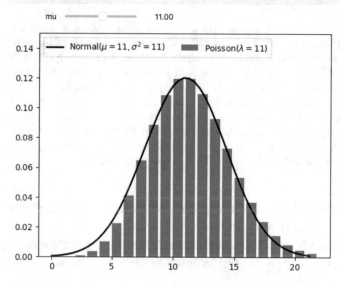

Figure 1-11. *An example of an interactive application created using the IPython widget* `interact`. *The interact widget provides a slider UI element that allows the value of an input parameter to be changed. When the slider is dragged, the provided function is reevaluated, which, in this case, renders a new graph*

The example in Figure 1-11 plots the distribution functions for the Normal distribution and the Poisson distribution, where the distributions' mean and variance are taken as input from the UI object created by the `interact` function. Moving the slider back and forth shows how the Normal and the Poisson distributions (with equal variance) approach each other as the distribution mean increases and how they behave very differently for small mean values. Interactive graphs like this are a great tool for building intuition and exploring computation problems, and the Jupyter Notebook is a fantastic enabler for this kind of investigation.[10]

---

[10] For more information about how to create interactive applications using Jupyter and IPython widgets, see the documentation for the `ipywidgets` library `https://ipywidgets.readthedocs.io/en/latest`.

nbconvert

Jupyter Notebooks can be converted to many different read-only formats using the nbconvert application, which can be invoked by calling nbconvert from the command line. Supported formats include, among others, PDF and HTML. Converting Jupyter Notebooks to PDF or HTML is useful when sharing notebooks with colleagues or publishing them online when the reader does not necessarily need to run the code but primarily views the results contained in the notebooks.

HTML

In the notebook web application, the menu option File ➤ Download as ➤ HTML can generate an HTML document representing a static view of a notebook. An HTML document can also be generated from the command prompt using the nbconvert application. For example, a notebook called Notebook.ipynb can be converted to HTML using the following command.

```
$ nbconvert --to html Notebook.ipynb
```

This generates an HTML page that is self-contained in terms of style sheets and JavaScript resources (which are loaded from public CDN servers), and it can be published as is online. However, image resources using Markdown or HTML tags are not included and must be distributed with the resulting HTML file.

For public online publishing of Jupyter Notebooks, the Jupyter project provides a convenient web service called nbviewer, available at http://nbviewer.org. By feeding it a URL to a public notebook file, the nbviewer application automatically converts the notebook to HTML and displays the result. One of the many benefits of this method of publishing Jupyter Notebooks is that the notebook author only needs to maintain one file—the notebook file itself—and when it is updated and uploaded to its online location, the static view of the notebook provided by nbviewer is automatically updated as well. However, it requires publishing the source notebook at a publicly accessible URL, so it can only be used for public sharing.

■ **Tip** The Jupyter project maintains a Wiki page that indexes many interesting Jupyter Notebooks published online at http://github.com/jupyter/jupyter/wiki#a-gallery-of-interesting-Jupyter-Notebooks. These notebooks demonstrate many of IPython's and Jupyter's more advanced features. They can be a great resource for learning more about Jupyter Notebooks as well as the many topics covered by those notebooks.

PDF

Converting a Jupyter Notebook to PDF format requires first converting the notebook to LaTeX and then compiling the LaTeX document to PDF format. To do the LaTeX to PDF conversion, a LaTeX environment must be available on the system (see Appendix A for pointers on how to install these tools). The nbconvert application can do both the notebook-to-LaTeX and the LaTeX-to-PDF conversions in one go, using the --to pdf argument (the --to latex argument can be used to obtain the intermediate LaTeX source).

```
$ nbconvert --to pdf Notebook.ipynb
```

The style of the resulting document can be specified using the --template name argument, where built-in templates include base, article, and report (these templates can be found in the nbconvert/templates/latex directory where Jupyter is installed). Extending one of the existing templates makes it easy to customize the appearance of the generated document. For example, in LaTeX it is common to include additional information about the document that is not available in Jupyter Notebooks, such as a document title (if different from the notebook filename) and the author of the document. This information can be added to a LaTeX document generated by the nbconvert application by creating a custom template. For example, the following template extends the built-in template article and overrides the title and author blocks.

```
((*- extends 'article.tplx' -*))
((* block title *)) \title{Document title} ((* endblock title *))
((* block author *)) \author{Author's Name} ((* endblock author *))
```

Assuming that this template is stored in a file called custom_template.tplx, the following command can be used to convert a notebook to PDF format using this customized template.

```
$ nbconvert --to pdf --template custom_template.tplx Notebook.ipynb
```

The result is LaTeX and PDF documents where the title and author fields are set as requested in the template.

Python

A Jupyter Notebook in its JSON-based file format can be converted to Python code using the nbconvert application and the python format.

```
$ nbconvert --to python Notebook.ipynb
```

This generates the Notebook.py file, which only contains executable Python code (or, if IPython extensions were used in the notebook, a file that is executable with ipython). The non-code content of the notebook is also included in the resulting Python code file in the form of comments that do not prevent the file from being interpreted by the Python interpreter. Converting a notebook to Python code is useful, for example, when using the Jupyter Notebooks to develop functions and classes that need to be imported into other Python files or notebooks.

Spyder: An Integrated Development Environment

An integrated development environment is an enhanced text editor with features such as code execution, documentation, and debugging from within the editor. Many free and commercial IDE environments have good support for Python-based projects. Spyder[11] is an excellent free IDE that is particularly well suited for computing and data analysis using Python. The rest of this section focuses on Spyder and further explores its features. However, there are also many other suitable IDEs. For example, Eclipse[12] is a popular and powerful multilanguage IDE, and the PyDev[13] extension to Eclipse provides a good Python environment. PyCharm[14] is another powerful Python IDE that has gained significant popularity among Python developers recently, and Visual Studio Code[15] from Microsoft is yet another great option. For readers with previous experience with any of these tools, they could be a productive and familiar environment for computational work.

---

[11] https://spyder-ide.org
[12] http://www.eclipse.org
[13] http://pydev.org
[14] http://www.jetbrains.com/pycharm
[15] https://code.visualstudio.com

However, the Spyder IDE was specifically created for Python programming, particularly for scientific computing with Python. As such, it has features useful for interactive and exploratory computing, most notably, integration with the IPython console directly in the IDE. The Spyder user interface consists of several optional panes, which can be arranged in different ways within the IDE application. The following are the most important panes.

- Source code editor
- Consoles for the Python and the IPython interpreters and the system shell
- Object inspector, for showing documentation for Python objects
- Variable explorer
- File explorer
- Command history
- Profiler

Each pane can be configured to be shown or hidden, depending on the user's preferences and needs, using the View ➤ Panes menu. Furthermore, panes can be organized together in tabbed groups. In the default layout, three pane groups are displayed. The left pane group contains the source code editor. The top-right pane group contains the variable explorer, the file explorer, and the object inspector. The bottom right pane group contains Python and IPython consoles.

Running the spyder command at the shell prompt launches the Spyder IDE. Figure 1-12 shows a screenshot of the default layout of the Spyder application.

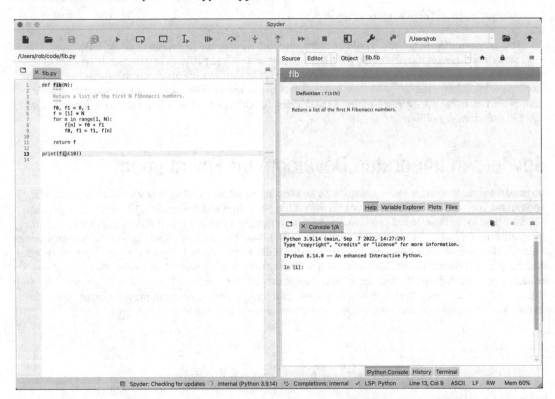

Figure 1-12. *A screenshot of the Spyder IDE application. The code editor is in the left panel, the top-right panel shows the object inspector (help viewer), and the bottom-right panel shows an IPython console*

Source Code Editor

The source code editor in Spyder supports code highlighting, intelligent autocompletion, working with multiple open files simultaneously, parenthesis matching, indentation guidance, and many other features that one would expect from a modern source code editor. The added benefit of using an IDE is that code in the editor can be run—as a whole (shortcut F5) or a selection (shortcut F9)—in attached Python or IPython consoles with persistent sessions between successive runs.

In addition, the Spyder editor has very useful support for static code checking with pylint,[16] pyflakes,[17] and pep8,[18] which are external tools that analyze Python source code and report errors such as undefined symbols, syntax errors, coding style violations, and more. Such warnings and errors are shown line-by-line as a yellow triangle with an exclamation mark in the editor's left margin, next to the line number. Static code checking is extremely important in Python programming. Since Python is an interpreted and lazily evaluated language, simple bugs like undefined symbols may not be discovered until the offending code line reaches runtime. For rarely used code paths, sometimes such bugs can be very hard to discover. Real-time static code checking and coding style checks in the Spyder editor can be activated and deactivated in the preference windows (Spyder ➤ Preferences in macOS, and Tools ➤ Preferences in Linux and Windows). In the Completion and Linting section, I recommend checking "Enable code style linting" in the "Code style and formatting" tab.

■ **Tip** The Python language is versatile, and equivalent Python source code can be written in various styles and manners. However, a Python coding style standard, PEP8, has been implemented to encourage a uniform appearance of Python code. I strongly recommend studying the PEP8 coding style standard and complying with it in your code. The PEP8 is described at `www.python.org/dev/peps/pep-0008`.

Consoles in Spyder

The integrated Python and IPython consoles can be used to execute a file edited in the text editor window or run interactively typed Python code. When executing Python source code files from the editor, the namespace variables created in the script are retained in the IPython or Python session in the console. It is an important feature that makes Spyder an interactive computing environment, in addition to a traditional IDE application, since it allows exploring the values of variables after a script has finished executing. Spyder supports having multiple Python and IPython consoles opened simultaneously, and, for example, a new IPython console can be launched through the menu option Consoles ➤ Open an IPython console. When running a script from the editor, by pressing F5 or the run button in the toolbar, the script is run in the most recently activated console by default. This makes it possible to maintain different consoles with independent namespaces for different scripts or projects.

Use the %reset command and the reload function to clear a namespace and reload updated modules when possible. If that is insufficient, it is possible to restart the IPython kernel corresponding to an IPython console or the Python interpreter via the drop-down menu for the top-right icon in the console panel. Finally, a value feature for audit trail and logging is that IPython console sessions can be exported as an HTML file by right-clicking the console window and selecting Save as HTML/XML in the pop-up menu.

---

[16] `http://www.pylint.org`
[17] `http://github.com/PyCQA/pyflakes`
[18] `http://pep8.readthedocs.org`

Object Inspector

The object inspector (Help pane) is a great aid when writing Python code. It can display richly formatted documentation strings for objects defined in source code created with the editor and for symbols defined in library modules installed on the system. The object text field at the top of the object inspector panel can be used to type the name of a module, function, or class to display the documentation string. Modules and symbols do not need to be imported into the local namespace to be able to display their docstrings using the object inspector. The documentation for an object in the editor or the console can also be opened in the object inspector by selecting the object with the cursor and using the shortcut Ctrl-i (Cmd-i in macOS). It is even possible to automatically display docstrings for callable objects when its opening left parenthesis is typed. This gives an immediate reminder of the arguments and their order for the callable object, which can be a great productivity booster. To activate this feature, navigate to the Help page in the Preferences window and check the boxes in the "Automatic connections" section.

Summary

This chapter introduced the Python environment for scientific and technical computing. This environment is, in fact, an ecosystem of libraries and tools for computing, which includes not only Python software but everything from low-level number-crunching libraries up to graphical user interface applications and web applications. In this multilanguage ecosystem, Python is the language that ties it all together into a coherent and productive environment for computing. IPython is a core component of Python's computing environment, and we briefly surveyed some of its most important features before covering the higher-level user environments provided by the Jupyter Notebook and the Spyder IDE. These are the tools in which most exploratory and interactive computing is carried out. The rest of this book focuses on computing using Python libraries, assuming that we are working within one of the environments provided by IPython, the Jupyter Notebook, or Spyder.

Further Reading

The Jupyter Notebook is a particularly rich platform for interactive computing, and it is also a very actively developed software. One of the most recent developments within the Jupyter Notebook is its widget system, which consists of user-interface components that can be used to create interactive interfaces within the browser that is displaying the notebook. This book briefly touches upon Jupyter widgets, but it is a very interesting and rapidly developing part of the Jupyter project, and I do recommend exploring their potential applications for interactive computing. The Jupyter Notebook widgets are well documented at `https://ipywidgets.readthedocs.io/en/latest`. There are also two interesting books by Cyrille Rossant on this topic that I highly recommend: *Learning IPython for Interactive Computing and Data Visualization* (Packt, 2013) and IPython Interactive Computing and Visualization Cookbook (Packt, 2014).

CHAPTER 2

∎∎∎

Vectors, Matrices, and Multidimensional Arrays

Vectors, matrices, and arrays of higher dimensions are essential tools in numerical computing. They enable us to represent and manipulate data efficiently, particularly when computations need to be repeated for a set of input values. By formulating computations in terms of array operations, we can perform vectorized[1] computing, eliminating the need for many explicit loops over array elements by applying batch operations to the array data. Vectorized computing results in concise and more maintainable code, allowing the implementation of array operations to be delegated to more efficient low-level libraries. As a result, vectorized computations can be significantly faster than sequential element-by-element computations. This is particularly important in interpreted languages like Python, where looping over arrays element-by-element incurs significant performance overhead.

In Python's scientific computing environment, the NumPy library provides efficient data structures for working with arrays. The core of NumPy is implemented in C and provides efficient functions for manipulating and processing arrays. At first glance, NumPy arrays may resemble Python's list data structure. But an important difference is that while Python lists are generic containers of objects, NumPy arrays are homogenous and typed arrays of fixed size. Homogenous means that all elements in the array have the same data type. Fixed-size means that an array cannot be resized (without creating a new array). For these and other reasons, operations and functions acting on NumPy arrays can be much more efficient than those using Python lists. Along with the data structures for arrays, NumPy offers a large collection of fundamental operators and functions that work on these data structures. It also includes submodules that implement higher-level algorithms such as linear algebra and fast Fourier transform.

This chapter first looks at the basic data structure for arrays in NumPy, including several ways to create them. Next, let's look at operations for manipulating arrays and doing computations with arrays. NumPy's multidimensional data arrays are a foundation for nearly all numerical libraries for Python. Spending time on getting familiar with NumPy and developing an understanding of how NumPy works is therefore important.

---

[1] Many modern processors provide instructions that operate on arrays. These are also known as vectorized operations, but here vectorized refers to high-level array-based operations, regardless of how they are implemented at the processor level.

© Robert Johansson 2024
R. Johansson, *Numerical Python*, https://doi.org/10.1007/979-8-8688-0413-7_2

■ **NumPy** The NumPy library provides data structures for representing a rich variety of arrays and methods and functions for operating on such arrays. NumPy provides the numerical backend for nearly every scientific or technical library for Python. It is a very important part of the scientific Python ecosystem. At the time of writing, the latest version of NumPy is 1.24.2. More information about NumPy is available at `www.numpy.org`.

Importing the Modules

To use the NumPy library, we need to import it into the program. By convention, the numpy module is imported under the alias np, like so.

```
In [1]: import numpy as np
```

After this, we can access functions and classes in the numpy module using the np namespace. Throughout this book, assume that the NumPy module is imported this way.

The NumPy Array Object

The core of the NumPy library is the data structures for representing multidimensional arrays of homogeneous data. Homogeneous refers to all elements in an array having the same data type.[2] The main data structure for multidimensional arrays in NumPy is the ndarray class. In addition to the data stored in the array, this data structure also contains important metadata, such as shape, size, data type, and other attributes. Table 2-1 presents a detailed description of these attributes. A full list of attributes with descriptions is available in the ndarray docstring, which can be accessed by calling help(np.ndarray) in the Python interpreter or np.ndarray? in an IPython console.

Table 2-1. *Basic Attributes of the ndarray Class*

Attribute	Description
shape	A tuple that contains the number of elements (i.e., the length) for each dimension (axis) of the array
size	The total number of elements in the array
ndim	The number of dimensions (axes)
nbytes	The number of bytes used to store the data
dtype	The data type of the elements in the array

The following example demonstrates how these attributes are accessed for an instance data of the ndarray class.

```
In [2]: data = np.array([[1, 2], [3, 4], [5, 6]])
```

---

[2] This does not necessarily need to be the case for Python lists, which therefore can be heterogenous.

```
In [3]: type(data)
Out[3]: numpy.ndarray
In [4]: data
Out[4]: array([[1, 2],
               [3, 4],
               [5, 6]])
In [5]: data.ndim
Out[5]: 2
In [6]: data.shape
Out[6]: (3, 2)
In [7]: data.size
Out[7]: 6
In [8]: data.dtype
Out[8]: dtype('int64')
In [9]: data.nbytes
Out[9]: 48
```

Here the ndarray instance data is created from a nested Python list using the np.array function. More ways to create ndarray instances from data and rules of various kinds are introduced later in this chapter. In the preceding example, the data is a two-dimensional array (data.ndim) of shape 3 × 2, as indicated by data. shape, and in total, it contains six elements (data.size) of type int64 (data.dtype), which amounts to a total size of 48 bytes (data.nbytes).

Data Types

In the previous section, we encountered the dtype attribute of the ndarray object. This attribute describes the data type of the elements within the array. As the array is homogeneous, all elements have the same data type. The basic numerical data types supported in NumPy are shown in Table 2-2. Nonnumerical data types, such as strings, objects, and user-defined compound types, are also supported.

Table 2-2. *Basic Numerical Data Types Available in NumPy*

dtype	Variants	Description
int	int8, int16, int32, int64	Integers
uint	uint8, uint16, uint32, uint64	Unsigned (nonnegative) integers
bool	bool	Boolean (True or False)
float	float16, float32, float64, float128	Floating-point numbers
complex	complex64, complex128, complex256	Complex-valued floating-point numbers

For numerical work, the most important data types are int (for integers), float (for floating-point numbers), and complex (for complex floating-point numbers). Each of these data types comes in different sizes, such as int32 for 32-bit integers, int64 for 64-bit integers, and so on. This offers more fine-grained control over data types than the standard Python types, which only provide one type for integers and one type for floats. When working with data types, it's typically not essential to explicitly select the bit size. However, it's often necessary to choose whether to use arrays of integers, floating-point numbers, or complex values explicitly.

The following example demonstrates how to use the dtype attribute to generate arrays of integer-, float-, and complex-valued elements.

31

```
In [10]: np.array([1, 2, 3], dtype=int)
Out[10]: array([1, 2, 3])
In [11]: np.array([1, 2, 3], dtype=float)
Out[11]: array([ 1.,  2.,  3.])
In [12]: np.array([1, 2, 3], dtype=complex)
Out[12]: array([ 1.+0.j,  2.+0.j,  3.+0.j])
```

Once a NumPy array is created, its dtype cannot be changed other than by creating a new copy with type-casted array values. Typecasting an array is straightforward and can be done using the np.array function.

```
In [13]: data = np.array([1, 2, 3], dtype=float)
In [14]: data
Out[14]: array([ 1.,  2.,  3.])
In [15]: data.dtype
Out[15]: dtype('float64')
In [16]: data = np.array(data, dtype=int)
In [17]: data.dtype
Out[17]: dtype('int64')
In [18]: data
Out[18]: array([1, 2, 3])
```

Or, it can be done using the astype method of the ndarray class.

```
In [19]: data = np.array([1, 2, 3], dtype=float)
In [20]: data
Out[20]: array([ 1.,  2.,  3.])
In [21]: data.astype(int)
Out[21]: array([1, 2, 3])
```

When computing with NumPy arrays, the data type might get converted from one type to another, if required by the operation. For example, when adding float-valued and complex-valued arrays, the result is a complex-valued array.

```
In [22]: d1 = np.array([1, 2, 3], dtype=float)
In [23]: d2 = np.array([1, 2, 3], dtype=complex)
In [24]: d1 + d2
Out[24]: array([ 2.+0.j,  4.+0.j,  6.+0.j])
In [25]: (d1 + d2).dtype
Out[25]: dtype('complex128')
```

In some cases, depending on the application and its requirements, it is essential to create arrays with data types appropriately set to, for example, int or complex. The default type is float. Consider the following example.

```
In [26]: np.sqrt(np.array([-1, 0, 1]))
Out[26]: RuntimeWarning: invalid value encountered in sqrt
```

```
              array([ nan,    0.,    1.])
In [27]: np.sqrt(np.array([-1, 0, 1], dtype=complex))
Out[27]: array([ 0.+1.j,   0.+0.j,   1.+0.j])
```

Here, using the np.sqrt function to compute the square root of each element in an array gives different results depending on the data type of the array. The square root of −1 results in the imaginary unit (denoted as 1j in Python) only when the data type of the array is complex.

Real and Imaginary Parts

Regardless of the value of the dtype attribute, all NumPy array instances have the real and imag attributes for extracting the real and imaginary parts of the array, respectively.

```
In [28]: data = np.array([1, 2, 3], dtype=complex)
In [29]: data
Out[29]: array([ 1.+0.j,   2.+0.j,   3.+0.j])
In [30]: data.real
Out[30]: array([ 1.,   2.,   3.])
In [31]: data.imag
Out[31]: array([ 0.,   0.,   0.])
```

The same functionality is also provided by the functions np.real and np.imag, which also can be applied to other array-like objects, such as Python lists. Note that Python supports complex numbers, and the imag and real attributes are also available for Python scalars.

Order of Array Data in Memory

Multidimensional arrays are stored as contiguous data in memory. There is freedom of choice in arranging the array elements in this memory segment. Consider a two-dimensional array containing rows and columns: one possible way to store this array as a consecutive sequence of values is to store the rows after each other, and another equally valid approach is to store the columns one after another. The former is called row-major format, and the latter is column-major format. Whether to use row-major or column-major is a matter of conventions. For example, row-major format is used in the C programming language, and Fortran uses the column-major format. A NumPy array can be specified to be stored in row-major format, using the keyword argument order= 'C', and column-major format, using the keyword argument order= 'F', when the array is created or reshaped. The default format is row-major. The 'C' or 'F' ordering of NumPy array is particularly relevant when NumPy arrays are used in interfaces with software written in C and Fortran, which is often required when working with numerical computing with Python.

Row-major and column-major ordering are special cases of strategies for mapping the index used to address an element to the offset for the element in the array's memory segment. In general, the NumPy array attribute ndarray.strides defines exactly how this mapping is done. The strides attribute is a tuple of the same length as the number of axes (dimensions) of the array. Each value in strides is the factor by which the index for the corresponding axis is multiplied when calculating the memory offset (in bytes) for a given index expression.

For example, consider a C-order array A with shape (2, 3), which corresponds to a two-dimensional array with two and three elements along the first and the second dimensions, respectively. If the data type is int32, then each element uses 4 bytes, and the total memory buffer for the array, therefore, uses $2 \times 3 \times 4 = 24$ bytes. The strides attribute of this array is $(4 \times 3, 4 \times 1) = (12, 4)$ because each increment of m in A[n, m] increases the memory offset with one item, or 4 bytes. Likewise, each increment of n increases the memory offset with three items or 12 bytes (because the second dimension of the array has length 3). If, on the other hand, the same array was stored in 'F' order, the strides would instead be (4, 8). Using strides to describe the mapping of array index to array memory offset is clever because it can be used to describe different mapping strategies, and many common operations on arrays, such as the transpose, can be implemented by simply changing the strides attribute, which can eliminate the need for moving data around in the memory. Operations that only require changing the strides attribute result in new ndarray objects that refer to the same data as the original array. Such arrays are called views. For efficiency, NumPy strives to create views rather than copies when applying operations on arrays. This is generally a good thing, but it is important to be aware that some array operations result in views rather than new independent arrays because modifying their data also modifies the data of the original array. Several examples of this behavior are presented later in this chapter.

Creating Arrays

The previous section looked at NumPy's basic data structure for representing arrays, the ndarray class and the basic attributes of this class. This section focuses on functions from the NumPy library that can create ndarray instances.

Arrays can be generated in several ways, depending on their properties and the applications they are used for. For example, as we saw in the previous section, one way to initialize an ndarray instance is to use the np.array function on a Python list, which, for example, can be explicitly defined. However, this method is limited to small arrays. In many situations, it is necessary to generate arrays with elements that follow some given rule, such as filled with constant values, increasing integers, uniformly spaced numbers, random numbers, and so forth. In other cases, we might need to create arrays from data stored in a file. The requirements are many and varied, and the NumPy library provides a comprehensive set of functions for creating arrays in a variety of ways. This section looks at many of these functions in more detail. For a complete list, see the NumPy reference manual or the available docstrings by typing help(np) or using the autocompletion np.<TAB>. A summary of frequently used array-generating functions is given in Table 2-3.

Table 2-3. *Summary of NumPy Functions for Generating Arrays*

Function Name	Type of Array
np.array	Create an array for which the elements are given by an array-like object, which, for example, can be a (nested) Python list, a tuple, an iterable sequence, or another ndarray instance.
np.zeros	Create an array with the specified dimensions and data type that is filled with zeros.
np.ones	Create an array with the specified dimensions and data type that is filled with ones.
np.diag	Create a diagonal array with specified values along the diagonal and zeros elsewhere.
np.arange	Create an array with evenly spaced values between the specified start, end, and increment values.
np.linspace	Create an array with evenly spaced values between specified start and end values, using a specified number of elements.
np.logspace	Create an array with values that are logarithmically spaced between the given start and end values.
np.meshgrid	Generate coordinate matrices (and higher-dimensional coordinate arrays) from one-dimensional coordinate vectors.
np.fromfunction	Create an array and fills it with values specified by a given function, which is evaluated for each combination of indices for the given array size.
np.fromfile	Create an array with the data from a binary (or text) file. NumPy also provides a corresponding function np.tofile with which NumPy arrays can be stored to disk and later read back using np.fromfile.
np.genfromtxt, np.loadtxt	Create an array from data read from a text file, for example, a comma-separated value (CSV) file. The np.genfromtxt function also supports data files with missing values.
np.random.rand	Generate an array with random numbers that are uniformly distributed between 0 and 1. Other types of distributions are also available in the np.random module.

Arrays Created from Lists and Other Array-Like Objects

Using the np.array function, NumPy arrays can be constructed from explicit Python lists, iterable expressions, and other array-like objects (such as other ndarray instances). For example, to create a one-dimensional array from a Python list, we simply pass the Python list as an argument to the np.array function.

```
In [32]: np.array([1, 2, 3, 4])
Out[32]: array([ 1,   2,   3, 4])
In [33]: data.ndim
Out[33]: 1
In [34]: data.shape
Out[34]: (4,)
```

We can use a nested Python list to create a two-dimensional array with the same data as in the previous example.

```
In [35]: np.array([[1, 2], [3, 4]])
Out[35]: array([[1,  2],
                [3, 4]])
In [36]: data.ndim
Out[36]: 2
In [37]: data.shape
Out[37]: (2, 2)
```

Arrays Filled with Constant Values

The functions np.zeros and np.ones create and return arrays filled with zeros and ones, respectively. As first argument, they take an integer or a tuple that describes the number of elements along each dimension of the array. For example, to create a 2 × 3 array filled with zeros, and an array of length 4 filled with ones, we can use the following.

```
In [38]: np.zeros((2, 3))
Out[38]: array([[ 0.,  0.,  0.],
                [ 0.,  0.,  0.]])
In [39]: np.ones(4)
Out[39]: array([ 1.,  1.,  1.,  1.])
```

Like other array-generating functions, the np.zeros and np.ones functions also accept an optional keyword argument that specifies the data type for the elements in the array. By default, the data type is float64, and it can be changed to the required type by explicitly specifying the dtype argument.

```
In [40]: data = np.ones(4)
In [41]: data.dtype
Out[41]: dtype('float64')
In [42]: data = np.ones(4, dtype=np.int64)
In [43]: data.dtype
Out[43]: dtype('int64')
```

An array filled with an arbitrary constant value can be generated by creating an array filled with ones and then multiplying the array with the desired fill value. However, NumPy also provides the np.full function that does this in one step. The following two ways of constructing arrays with ten elements, which are initialized to the numerical value 5.4 in this example, produce the same results. However, np.full is slightly more efficient since it avoids multiplication.

```
In [44]: x1 = 5.4 * np.ones(10)
In [45]: x2 = np.full(10, 5.4)
```

An existing array can also be filled with constant values using the np.fill function, which takes an array and a value as arguments and sets all elements in the array to the given value. The following two methods to create an array give the same results.

```
In [46]: x1 = np.empty(5)
In [47]: x1.fill(3.0)
In [48]: x1
```

```
Out[48]: array([ 3.,   3.,   3.,   3.,   3.])
In  [49]: x2 = np.full(5, 3.0)
In  [50]: x2
Out[50]: array([ 3.,   3.,   3.,   3.,   3.])
```

This last example also used the np.empty function, which generates an array with uninitialized values, of the given size. This function should only be used when the initialization of all elements can be guaranteed by other means, such as an explicit loop over the array elements or another explicit assignment. This function is described in more detail later in this chapter.

Arrays Filled with Incremental Sequences

In numerical computing, it is very common to require arrays with evenly spaced values between a starting value and an ending value. NumPy provides two similar functions to create such arrays: np.arange and np.linspace. Both functions take three arguments, where the first two arguments are the start and end values. The third argument of np.arange is the increment, while for np.linspace, it is the total number of points in the array.

For example, we could use either of the following to generate arrays with values between 0 and 10, with increment 1.

```
In  [51]: np.arange(0.0, 11, 1)
Out[51]: array([ 0.,   1.,   2.,   3.,   4.,   5.,   6.,   7.,   8.,   9. ,  10.])
In  [52]: np.linspace(0, 10, 11)
Out[52]: array([ 0.,   1.,   2.,   3.,   4.,   5.,   6.,   7.,   8.,   9.,  10.])
```

However, note that np.arange does not include the end value (11), while by default np.linspace does (although this behavior can be changed using the optional endpoint keyword argument). Whether to use np.arange or np.linspace is mostly a matter of personal preference, but it is generally recommended to use np.linspace whenever the increment is a noninteger.

Arrays Filled with Logarithmic Sequences

The np.logspace function is similar to np.linspace, but the increments between the elements in the array are logarithmically distributed, and the first two arguments for the start and end values, are the powers of the optional base keyword argument (which defaults to 10). For example, we can use the following to generate an array with logarithmically distributed values between 1 and 100.

```
In  [53]: np.logspace(0, 2, 5)   # 5 data points between 10**0=1 to 10**2=100
Out[53]: array([ 1. ,  3.16227766,  10. ,  31.6227766 ,  100.])
```

Meshgrid Arrays

Multidimensional coordinate grids can be generated using the np.meshgrid function. Given two one-dimensional coordinate arrays (i.e., arrays containing a set of coordinates along a given dimension), we can generate two-dimensional coordinate arrays using the np.meshgrid function. An illustration of this is given in the following example.

```
In  [54]: x = np.array([-1, 0, 1])
In  [55]: y = np.array([-2, 0, 2])
In  [56]: X, Y = np.meshgrid(x, y)
```

```
In [57]: X
Out[57]: array([[-1,  0,  1],
                [-1,  0,  1],
                [-1,  0,  1]])
In [58]: Y
Out[58]: array([[-2, -2, -2],
                [ 0,  0,  0],
                [ 2,  2,  2]])
```

A common use for two-dimensional coordinate arrays, like X and Y in this example, is to evaluate functions over two variables x and y. This can be used, for example, when plotting functions as colormap plots and contour plots. To evaluate the expression $(x + y)^2$ at all combinations of values from the x and y arrays in the preceding section, we can use the two-dimensional coordinate arrays X and Y.

```
In [59]: Z = (X + Y) ** 2
In [60]: Z
Out[60]: array([[9, 4, 1],
                [1, 0, 1],
                [1, 4, 9]])
```

It is also possible to generate higher-dimensional coordinate arrays by passing more arrays as an argument to the np.meshgrid function. Alternatively, the functions np.mgrid and np.ogrid can also be used to generate coordinate arrays using a slightly different syntax based on indexing and slice objects. See their docstrings or the NumPy documentation for details.

Creating Uninitialized Arrays

To create an array of specific size and data type, but without initializing the elements in the array to any particular values, we can use the np.empty function. The advantage of using this function, instead of, for example, np.zeros, which creates an array initialized with zero-valued elements, is that we can avoid the initiation step. If all elements are guaranteed to be initialized later in the code, this can save a little bit of time, especially when working with large arrays. To illustrate the use of the np.empty function, consider the following example.

```
In [61]: np.empty(3, dtype=float)
Out[61]: array([  1.28822975e-231,   1.28822975e-231,   2.13677905e-314])
```

This generated a new array with three elements of type float. There is no guarantee that the elements have any particular values, and the actual values vary from time to time. For this reason, it is important that all values are explicitly assigned before the array is used; otherwise, unpredictable errors are likely to arise. Often the np.zeros function is a safer alternative to np.empty, and if the performance gain is not essential, it is better to use np.zeros, to minimize the likelihood of subtle and hard-to-reproduce bugs due to uninitialized values in the array returned by np.empty.

Creating Arrays with Properties of Other Arrays

It is often necessary to create new arrays that share properties, such as shape and data type, with another array. NumPy provides a family of functions for this purpose: np.ones_like, np.zeros_like, np.full_like, and np.empty_like. A typical use case is a function that takes arrays of unspecified type and size as arguments and requires working arrays of the same size and type. For example, a boilerplate example of this situation is given in the following function.

```
def f(x):
    y = np.ones_like(x)
    # compute with x and y
    return y
```

At the first line of the body of this function, a new array y is created using np.ones_like, which results in an array of the same size and data type as x and filled with ones.

Creating Matrix Arrays

Matrices, or two-dimensional arrays, are an important case in numerical computing. One of the useful features of NumPy is its ability to generate common matrices. The np.identity function, for example, creates a square matrix with ones on the diagonal and zeros in all other positions.

```
In [62]: np.identity(4)
Out[62]: array([[ 1.,  0.,  0.,  0.],
                [ 0.,  1.,  0.,  0.],
                [ 0.,  0.,  1.,  0.],
                [ 0.,  0.,  0.,  1.]])
```

The similar function np.eye generates matrices with ones on a diagonal (optionally offset). This is illustrated in the following example, which produces matrices with nonzero diagonals above and below the diagonal, respectively.

```
In [63]: np.eye(3, k=1)
Out[63]: array([[ 0.,  1.,  0.],
                [ 0.,  0.,  1.],
                [ 0.,  0.,  0.]])
In [64]: np.eye(3, k=-1)
Out[64]: array([[ 0.,  0.,  0.],
                [ 1.,  0.,  0.],
                [ 0.,  1.,  0.]])
```

To construct a matrix with an arbitrary one-dimensional array on the diagonal, we can use the np. diag function (which also takes the optional keyword argument k to specify an offset from the diagonal), as demonstrated in the following.

```
In [65]: np.diag(np.arange(0, 20, 5))
Out[65]: array([[0,  0,  0,  0],
                [0,  5,  0,  0],
                [0,  0, 10,  0],
                [0,  0,  0, 15]])
```

This gave a third argument to the np.arange function, which specifies the step size in the enumeration of elements in the array returned by the function. The resulting array, therefore, contains the values [0, 5, 10, 15], which are inserted on the diagonal of a two-dimensional matrix by the np.diag function.

Indexing and Slicing

Elements and subarrays of NumPy arrays are accessed using the standard square bracket notation that is also used with Python lists. Within the square bracket, a variety of different index formats are used for different types of element selection. In general, the expression within the bracket is a tuple, where each item in the tuple is a specification of which elements to select from each axis (dimension) of the array.

One-Dimensional Arrays

Along a single axis, integers are used to select single elements, and slices are used to select ranges and sequences of elements. Positive integers are used to index elements from the beginning of the array (index starts at 0), and negative integers are used to index elements from the end of the array, where the last element is indexed with –1, the second to last element with –2, and so on.

Slices are specified using the : notation that is also used for Python lists. In this notation, a range of elements can be selected using an expression like m:n, which selects elements starting with m and ending with $n - 1$ (note that the nth element is not included). The slice m:n can also be written more explicitly as m : n : 1, where the number 1 specifies that every element between m and n should be selected. To select every second element between m and n, use m : n : 2, and to select every p element, use m : n : p, and so on. If p is negative, elements are returned in reversed order starting from m to $n + 1$ (which implies that m has to be larger than n in this case). Table 2-4 summarizes indexing and slicing operations for NumPy arrays.

Table 2-4. *Examples of Array Indexing and Slicing Expressions*

Expression	Description
a[m]	Select the element at index m, where m is an integer (start counting form 0).
a[-m]	Select the nth element from the end of the list, where m is an integer. The last element in the list is addressed as –1, the second to last element as –2, and so on.
a[m:n]	Select elements with index starting at m and ending at $n - 1$ (m and n are integers).
a[:]	Select all elements in the given axis.
a[:n]	Select elements starting with index 0 and going up to index $n - 1$ (integer).
a[m:]	Select elements starting with index m (integer) and going up to the last element in the array.
a[m:n:p]	Select elements with index m through n (exclusive), with increment p.
a[::-1]	Select all the elements, in reverse order.

The following examples demonstrate index and slicing operations for NumPy arrays. To begin with, consider an array with a single axis (dimension) that contains a sequence of integers between 0 and 10.

```
In [66]: a = np.arange(0, 11)
In [67]: a
Out[67]: array([ 0,  1,  2,  3,  4,  5,  6,  7,  8,  9, 10])
```

Note that the end value 11 is not included in the array. To select specific elements from this array, for example, the first, the last, and the fifth element, we can use integer indexing.

```
In [68]: a[0]  # the first element
Out[68]: 0
In [69]: a[-1] # the last element
```

```
Out[69]: 10
In [70]: a[4]   # the fifth element, at index 4
Out[70]: 4
```

To select a range of elements, say from the second to the second-to-last element, selecting every element and every second element, respectively, we can use index slices.

```
In [71]: a[1:-1]
Out[71]: array([1, 2, 3, 4, 5, 6, 7, 8, 9])
In [72]: a[1:-1:2]
Out[72]: array([1, 3, 5, 7, 9])
```

To select the first five and the last five elements from an array, we can use the slices :5 and –5:, since if m or n is omitted in m:n, the defaults are the beginning and the end of the array, respectively.

```
In [73]: a[:5]
Out[73]: array([0, 1, 2, 3, 4])
In [74]: a[-5:]
Out[74]: array([6, 7, 8, 9, 10])
```

To reverse the array and select only every second value, we can use the slice : :-2, as shown in the following example.

```
In [75]: a[::-2]
Out[75]: array([10,  8,  6,  4,  2,  0])
```

Multidimensional Arrays

With multidimensional arrays, element selections like those introduced in the previous section can be applied on each axis (dimension). The result is a reduced array where each element matches the given selection rules. As a specific example, consider the following two-dimensional array.

```
In [76]: f = lambda m, n: n + 10 * m
In [77]: A = np.fromfunction(f, (6, 6), dtype=int)
In [78]: A
Out[78]: array([[ 0,  1,  2,  3,  4,  5],
                [10, 11, 12, 13, 14, 15],
                [20, 21, 22, 23, 24, 25],
                [30, 31, 32, 33, 34, 35],
                [40, 41, 42, 43, 44, 45],
                [50, 51, 52, 53, 54, 55]])
```

We can extract columns and rows from this two-dimensional array using a combination of slice and integer indexing.

```
In [79]: A[:, 1]   # the second column
Out[79]: array([ 1, 11, 21, 31, 41, 51])
In [80]: A[1, :]   # the second row
Out[80]: array([10, 11, 12, 13, 14, 15])
```

By applying a slice on each of the array axes, we can extract subarrays (submatrices in this two-dimensional example).

```
In [81]: A[:3, :3]  # upper half diagonal block matrix
Out[81]: array([[ 0,  1,  2],
                [10, 11, 12],
                [20, 21, 22]])
In [82]: A[3:, :3]  # lower left off-diagonal block matrix
Out[82]: array([[30, 31, 32],
                [40, 41, 42],
                [50, 51, 52]])
```

With element spacing other than 1, submatrices made up from nonconsecutive elements can be extracted.

```
In [83]: A[::2, ::2]  # every second element starting from 0, 0
Out[83]: array([[ 0,  2,  4],
                [20, 22, 24],
                [40, 42, 44]])
In [84]: A[1::2, 1::3]  # every second and third element starting from 1, 1
Out[84]: array([[11, 14],
                [31, 34],
                [51, 54]])
```

This ability to extract subsets of data from a multidimensional array is a simple but very powerful feature with many applications in data processing.

Views

Subarrays extracted from arrays using slice operations are alternative *views* of the same underlying array data. More specifically, they are arrays that refer to the same data in the memory as the original array, but with a different `strides` configuration. When elements in a view are assigned new values, the original array's values are updated. The following is an example.

```
In [85]: B = A[1:5, 1:5]
In [86]: B
Out[86]: array([[11, 12, 13, 14],
                [21, 22, 23, 24],
                [31, 32, 33, 34],
                [41, 42, 43, 44]])
In [87]: B[:, :] = 0
In [88]: A
Out[88]: array([[ 0,  1,  2,  3,  4,  5],
                [10,  0,  0,  0,  0, 15],
                [20,  0,  0,  0,  0, 25],
                [30,  0,  0,  0,  0, 35],
                [40,  0,  0,  0,  0, 45],
                [50, 51, 52, 53, 54, 55]])
```

In assigning new values to the elements in array B, which is created from array A, we also modify the values in A (since both arrays refer to the same data in the memory). The fact that extracting subarrays results in views rather than new independent arrays eliminates the need for copying data and improves performance. When a copy rather than a view is needed, the view can be copied explicitly using the copy method of the ndarray instance.

```
In [89]: C = B[1:3, 1:3].copy()
In [90]: C
Out[90]: array([[0, 0],
                [0, 0]])
In [91]: C[:, :] = 1  # this does not affect B since C is a copy of the view B[1:3, 1:3]
In [92]: C
Out[92]: array([[1, 1],
                [1, 1]])
In [93]: B
Out[93]: array([[0, 0, 0, 0],
                [0, 0, 0, 0],
                [0, 0, 0, 0],
                [0, 0, 0, 0]])
```

In addition to the copy attribute of the ndarray class, an array can be copied using the np.copy function or, equivalently, using the np.array function with the copy=True keyword argument.

Fancy Indexing and Boolean-Valued Indexing

The previous section looked at indexing NumPy arrays with integers and slices to extract individual elements or ranges of elements. NumPy provides another convenient method to index arrays, called fancy indexing. With fancy indexing, an array can be indexed with another NumPy array, a Python list, or a sequence of integers whose values select elements in the indexed array. To clarify this concept, consider the following example: we first create a NumPy array with 11 floating-point numbers and then index the array with another NumPy array (and Python list) to extract element numbers 0, 2, and 4 from the original array.

```
In [94]: A = np.linspace(0, 1, 11)
Out[94]: array([ 0. ,  0.1,  0.2,  0.3,  0.4,  0.5,  0.6,  0.7,  0.8,  0.9,  1. ])
In [95]: A[np.array([0, 2, 4])]
Out[95]: array([ 0. ,  0.2,  0.4])
In [96]: A[[0, 2, 4]]  # The same thing can be accomplished by indexing with a Python list
Out[96]: array([ 0. ,  0.2,  0.4])
```

This indexing method can be used along each axis (dimension) of a multidimensional NumPy array. It requires that the elements in the array or list used for indexing are integers.

Another variant of indexing NumPy arrays is to use Boolean-valued index arrays. In this case, each element (with values True or False) indicates whether or not to select the element from the list with the corresponding index. That is, if element n in the indexing array of Boolean values is True, then element n is selected from the indexed array. If the value is False, then element n is not selected. This indexing method is handy when filtering out elements from an array. For example, to select all the elements from the array A (as defined in the preceding section) that exceed the value 0.5, we can use the following combination of the comparison operator applied to a NumPy array and indexing using a Boolean-valued array.

```
In [97]: A > 0.5
Out[97]: array([False, False, False, False, False, False, True, True, True, True, True],
dtype=bool)
In [98]: A[A > 0.5]
Out[98]: array([ 0.6,  0.7,  0.8,  0.9,  1. ])
```

Unlike arrays created using slices, the arrays returned using fancy indexing and Boolean-valued indexing are not views but new independent arrays. Nonetheless, it is possible to assign values to elements selected using fancy indexing.

```
In [99]: A = np.arange(10)
In [100]: indices = [2, 4, 6]
In [101]: B = A[indices]
In [102]: B[0] = -1  # this does not affect A
In [103]: A
Out[103]: array([0, 1, 2, 3, 4, 5, 6, 7, 8, 9])
In [104]: A[indices] = -1  # this alters A
In [105]: A
Out[105]: array([ 0,  1, -1,  3, -1,  5, -1,  7,  8,  9])
```

And likewise for Boolean-valued indexing.

```
In [106]: A = np.arange(10)
In [107]: B = A[A > 5]
In [108]: B[0] = -1  # this does not affect A
In [109]: A
Out[109]: array([0, 1, 2, 3, 4, 5, 6, 7, 8, 9])
In [110]: A[A > 5] = -1  # this alters A
In [111]: A
Out[111]: array([ 0,  1,  2,  3,  4,  5, -1, -1, -1, -1])
```

A visual summary of different methods to index NumPy arrays is given in Figure 2-1. Note that each type of indexing discussed here can be independently applied to each dimension of an array.

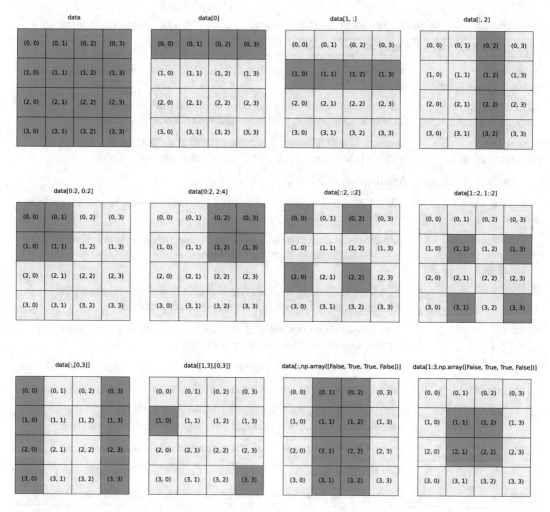

Figure 2-1. *A visual summary of indexing methods for NumPy arrays. These diagrams represent NumPy arrays of shape (4, 4), and the highlighted elements are selected using the indexing expression shown above the block representations of the arrays*

Reshaping and Resizing

When working with data in array form, it is often useful to rearrange arrays and alter the way they are interpreted. For example, an $N \times N$ matrix array could be rearranged into a vector of length N^2, or a set of one-dimensional arrays could be concatenated together or stacked next to each other to form a matrix. NumPy provides a rich set of functions for this type of manipulation. Table 2-5 summarizes a selection of these functions.

Table 2-5. *Summary of NumPy Functions for Manipulating the Dimensions and the Shape of Arrays*

Function/Method	Description
`np.reshape`, `np.ndarray.reshape`	Reshape an N-dimensional array. The total number of elements must remain the same.
`np.ndarray.flatten`	Create a copy of an N-dimensional array and reinterprets it as a one-dimensional array (i.e., all dimensions are collapsed into one).
`np.ravel`, `np.ndarray.ravel`	Create a view (if possible, otherwise a copy) of an N-dimensional array in which it is interpreted as a one-dimensional array.
`np.squeeze`	Remove axes with length 1.
`np.expand_dims`, `np.newaxis`	Add a new axis (dimension) of length 1 to an array, where `np.newaxis` is used with array indexing.
`np.transpose`, `np.ndarray.transpose`, `np.ndarray.T`	Transpose the array. The transpose operation corresponds to reversing (or, more generally, permuting) the axes of the array.
`np.hstack`	Stack a list of arrays horizontally (along axis 1): for example, given a list of column vectors, it appends the columns to form a matrix.
`np.vstack`	Stack a list of arrays vertically (along axis 0): for example, given a list of row vectors, it appends the rows to form a matrix.
`np.dstack`	Stack arrays depth-wise (along axis 2).
`np.concatenate`	Create a new array by appending arrays after each other along a given axis.
`np.resize`	Resize an array. Create a new copy of the original array, with the requested size. If necessary, the original array is repeated to fill up the new array.
`np.append`	Append an element to an array. Create a new copy of the array.
`np.insert`	Insert a new element at a given position. Create a new copy of the array.
`np.delete`	Delete an element at a given position. Create a new copy of the array.

Reshaping an array does not require modifying the underlying array data; it only changes how the data is interpreted by redefining the array's `strides` attribute. An example of this type of operation is a 2×2 array (matrix) that is reinterpreted as a 1×4 array (vector). In NumPy, the `np.reshape` function or the `ndarray` class `reshape` method can reconfigure how the underlying data is interpreted. It takes an array and the new shape of the array as arguments.

```
In [112]: data = np.array([[1, 2], [3, 4]])
In [113]: np.reshape(data, (1, 4))
Out[113]: array([[1, 2, 3, 4]])
In [114]: data.reshape(4)
Out[114]: array([1, 2, 3, 4])
```

The requested new array shape must match the number of elements in the original size. However, the number of axes (dimensions) does not need to be conserved, as illustrated in the previous example, whereas in the first case, the new array has dimension 2 and shape (1, 4). In contrast, in the second case, the new array has dimension 1 and shape (4,). This example also demonstrates two ways of invoking the reshape operation: the np.reshape function and the ndarray reshape method. Note that reshaping an array produces a view of the array, and if an independent copy of the array is needed, the view has to be copied explicitly (e.g., using np.copy).

The np.ravel (and its corresponding ndarray method) is a special case of reshape, which collapses all dimensions of an array and returns a flattened one-dimensional array with a length corresponding to the total number of elements in the original array. The ndarray method flatten performs the same function but returns a copy instead of a view.

```
In [115]: data = np.array([[1, 2], [3, 4]])
In [116]: data
Out[116]: array([[1, 2],
                  [3, 4]])
In [117]: data.flatten()
Out[117]: array([ 1,  2,  3,  4])
In [118]: data.flatten().shape
Out[118]: (4,)
```

While np.ravel and np.flatten collapse the axes of an array into a one-dimensional array, it is also possible to introduce new axes into an array, either by using np.reshape or, when adding new empty axes, using indexing notation and the np.newaxis keyword at the place of a new axis. In the following example, the array data has one axis, so it should normally be indexed with a tuple with one element. However, if it is indexed with a tuple with more than one element, and if the extra indices in the tuple have the value np.newaxis, the corresponding new axes are added.

```
In [119]: data = np.arange(0, 5)
In [120]: column = data[:, np.newaxis]
In [121]: column
Out[121]: array([[0],
                  [1],
                  [2],
                  [3],
                  [4]])
In [122]: row = data[np.newaxis, :]
In [123]: row
Out[123]: array([[0, 1, 2, 3, 4]])
```

The np.expand_dims function can also be used to add new dimensions to an array, and in the preceding example, the expression data[:, np.newaxis] is equivalent to np.expand_dims(data, axis=1), and data[np.newaxis, :] is equivalent to np.expand_dims(data, axis=0). Here the axis argument specifies the location relative to the existing axes where the new axis is to be inserted.

Up to now, we have seen methods to rearrange arrays in ways that do not affect the underlying data. Earlier in this chapter, we learned how to extract subarrays using various indexing techniques. In addition to reshaping and selecting subarrays, it is often necessary to merge arrays into bigger arrays, for example, when joining separately computed or measured data series into a higher-dimensional array, such as a matrix. For this task, NumPy provides the functions np.vstack, for vertical stacking of, for example, rows into a matrix, and np.hstack for horizontal stacking of, for example, columns into a matrix. The np.concatenate function provides similar functionality, but it takes a keyword argument axis that specifies the axis along which the arrays will be concatenated.

The shape of the arrays passed to np.hstack, np.vstack, and np.concatenate is important to achieve the desired type of array joining. For example, consider the following case: we have one-dimensional arrays of data and want to stack them vertically to obtain a matrix where the rows are made up of one-dimensional arrays. We can use np.vstack to achieve this.

```
In [124]: data = np.arange(5)
In [125]: data
Out[125]: array([0, 1, 2, 3, 4])
In [126]: np.vstack((data, data, data))
Out[126]: array([[0, 1, 2, 3, 4],
                  [0, 1, 2, 3, 4],
                  [0, 1, 2, 3, 4]])
```

If we instead want to stack the arrays horizontally to obtain a matrix where the arrays are the column vectors, we might first attempt something similar using np.hstack.

```
In [127]: data = np.arange(5)
In [128]: data
Out[128]: array([0, 1, 2, 3, 4])
In [129]: np.hstack((data, data, data))
Out[129]: array([0, 1, 2, 3, 4, 0, 1, 2, 3, 4, 0, 1, 2, 3, 4])
```

This stacks the arrays horizontally, but not in the way intended here. To make np.hstack treat the input arrays as columns and stack them accordingly, we need to make the input arrays two-dimensional arrays of shape (1, 5) rather than one-dimensional arrays of shape (5,). As discussed earlier, we can insert a new axis by indexing with np.newaxis.

```
In [130]: data = data[:, np.newaxis]
In [131]: np.hstack((data, data, data))
Out[131]: array([[0, 0, 0],
                  [1, 1, 1],
                  [2, 2, 2],
                  [3, 3, 3],
                  [4, 4, 4]])
```

The behavior of the functions for horizontal and vertical stacking, as well as concatenating arrays using np.concatenate, is clearest when the stacked arrays have the same number of dimensions as the final array and when the input arrays are stacked along an axis for which they have length 1.

The number of elements in a NumPy array cannot be changed once the array has been created. For example, to insert, append, and remove elements from a NumPy array using the np.append, np.insert, and np.delete functions, a new array must be created and the data copied to it. It may sometimes be tempting to use these functions to grow or shrink the size of a NumPy array. But, due to the overhead of creating new arrays and copying the data, pre-allocating arrays with sizes is usually a good idea to pre-allocate them with sizes such that they do not need to be resized later.

Vectorized Expressions

The purpose of storing numerical data in arrays is to be able to process the data with concise vectorized expressions that represent batch operations that are applied to all elements in the arrays. Efficient use of vectorized expressions eliminates the need for many explicit for loops. This approach makes the code more concise, easier to maintain, and performs better. NumPy implements functions and vectorized operations corresponding to most fundamental mathematical functions and operators. Many of these functions and operations act on arrays on an elementwise basis, and binary operations require all arrays in an expression to be of compatible size. The meaning of compatible size is normally that the variables in an expression represent either scalars or arrays of the same size and shape. More generally, a binary operation involving two arrays is well defined if the arrays can be *broadcasted* into the same shape and size.

In an operation between a scalar and an array, broadcasting refers to the scalar being distributed and the operation applied to each element in the array. When an expression contains arrays of unequal sizes, the operations may still be well defined if the smaller of the array can be broadcasted ("effectively expanded") to match the larger array according to NumPy's broadcasting rule: an array can be broadcasted over another array if their axes on a one-by-one basis either have the same length or if either of them has length 1. If the number of axes of the two arrays is not equal, the array with fewer axes is padded with new axes of length 1 from the left until the numbers of dimensions of the two arrays agree.

Two simple examples that illustrate array broadcasting are shown in Figure 2-2: a 3 × 3 matrix is added to a 1 × 3 row vector and a 3 × 1 column vector, respectively, and in both cases the result is a 3 × 3 matrix. However, the elements in the two resulting matrices are different, because the way the elements of the row and column vectors are broadcasted to the shape of the larger array is different depending on the shape of the arrays, according to NumPy's broadcasting rule.

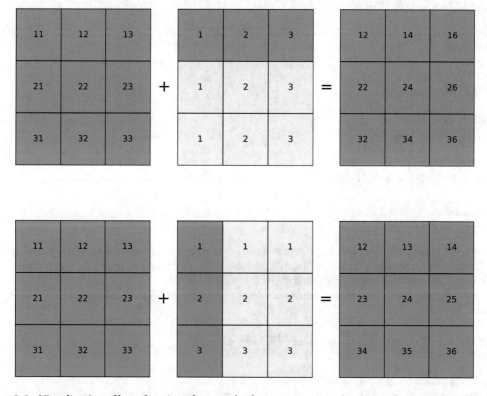

Figure 2-2. *Visualization of broadcasting of row and column vectors into the shape of a matrix. The highlighted elements represent true elements of the arrays, while the light blue elements describe the broadcasting of the elements of the array of smaller size*

Arithmetic Operations

The standard arithmetic operations with NumPy arrays perform elementwise operations. Consider, for example, the addition, subtraction, multiplication, and division of equal-sized arrays.

```
In [132]: x = np.array([[1, 2], [3, 4]])
In [133]: y = np.array([[5, 6], [7, 8]])
In [134]: x + y
Out[134]: array([[ 6,  8],
                 [10, 12]])
In [135]: y - x
Out[135]: array([[4, 4],
                 [4, 4]])
In [136]: x * y
Out[136]: array([[ 5, 12],
                 [21, 32]])
In [137]: y / x
Out[137]: array([[ 5.        ,  3.        ],
                 [ 2.33333333,  2.        ]])
```

In operations between scalars and arrays, the scalar value is applied to each element in the array, as we could expect.

```
In [138]: x * 2
Out[138]: array([[2, 4],
                 [6, 8]])
In [139]: 2 ** x
Out[139]: array([[ 2,  4],
                 [ 8, 16]])
In [140]: y / 2
Out[140]: array([[ 2.5,  3. ],
                 [ 3.5,  4. ]])
In [141]: (y / 2).dtype
Out[141]: dtype('float64')
```

Note that the dtype of the resulting array for an expression can be promoted if the computation requires it, as shown in the preceding example with the division between an integer array and an integer scalar, which in that case resulted in an array with a dtype that is np.float64.

If an arithmetic operation is performed on arrays with incompatible sizes or shapes, a ValueError exception is raised.

```
In [142]: x = np.array([1, 2, 3, 4]).reshape(2, 2)
In [143]: z = np.array([1, 2, 3, 4])
In [144]: x / z
---------------------------------------------------------------------
ValueError                                Traceback (most recent call last)
<ipython-input-144-b88ced08eb6a> in <module>()
----> 1 x / z
ValueError: operands could not be broadcast together with shapes (2,2) (4,)
```

Here the array x has shape $(2, 2)$ and the array z has shape $(4,)$, which cannot be broadcasted into a form that is compatible with $(2, 2)$. If, on the other hand, z has shape $(2,)$, $(2, 1)$, or $(1, 2)$, then it can be broadcasted to the shape $(2, 2)$ by effectively repeating the array z along the axis with length 1. Let's first consider an example with an array z of shape $(1, 2)$, where the first axis (axis 0) has length 1.

```
In [145]: z = np.array([[2, 4]])
In [146]: z.shape
Out[146]: (1, 2)
```

Dividing the array x with array z is equivalent to dividing x with an array zz that is constructed by repeating (here using np.concatenate) the row vector z to obtain an array zz that has the same dimensions as x.

```
In [147]: x / z
Out[147]: array([[ 0.5,  0.5],
                 [ 1.5,  1. ]])
In [148]: zz = np.concatenate([z, z], axis=0)
In [149]: zz
Out[149]: array([[2, 4],
                 [2, 4]])
In [150]: x / zz
Out[150]: array([[ 0.5,  0.5],
                 [ 1.5,  1. ]])
```

Let's also consider the example in which the array z has shape $(2, 1)$ and where the second axis (axis 1) has length 1.

```
In [151]: z = np.array([[2], [4]])
In [152]: z.shape
Out[152]: (2, 1)
```

In this case, dividing x with z is equivalent to dividing x with an array zz that is constructed by repeating the column vector z until a matrix with the same dimension as x is obtained.

```
In [153]: x / z
Out[153]: array([[ 0.5 ,  1.  ],
                 [ 0.75,  1.  ]])
In [154]: zz = np.concatenate([z, z], axis=1)
In [155]: zz
Out[155]: array([[2, 2],
                 [4, 4]])
In [156]: x / zz
Out[156]: array([[ 0.5 ,  1.  ],
                 [ 0.75,  1.  ]])
```

In summary, these examples show how arrays with shapes $(1, 2)$ and $(2, 1)$ are broadcasted to the shape $(2, 2)$ of the array x when the operation x / z is performed. In both cases, the result of the operation x / z is the same as first repeating the smaller array z along its axis of length 1 to obtain a new array zz with the same shape as x and then performing the equal-sized array operation x / zz. However, the implementation of the broadcasting does not explicitly perform this expansion and the corresponding memory copies. But thinking of the array broadcasting in these terms can be helpful.

A summary of the operators for arithmetic operations with NumPy arrays is given in Table 2-6. These operators use the standard symbols used in Python. The result of an arithmetic operation with one or two arrays is a new independent array with its own data in the memory. Evaluating complicated arithmetic expressions might trigger many memory allocation and copy operations. Working with large arrays can lead to a large memory footprint and negatively impact performance. In such cases, in-place operation (see Table 2-6) can reduce the memory footprint and improve performance. As an example of in-place operators, consider the following two statements with the same effect.

```
In [157]: x = x + y
In [158]: x += y
```

Table 2-6. *Operators for Elementwise Arithmetic Operation on NumPy Arrays*

Operator	Operation
+, +=	Addition
-, -=	Subtraction
*, *=	Multiplication
/, /=	Division
//, //=	Integer division
**, **=	Exponentiation

The two expressions have the same effect, but in the first case, x is reassigned to a new array, while in the second case, the values of array x are updated in place. Extensive use of in-place operators tends to impair code readability, and in-place operators should be used only when necessary.

Elementwise Functions

In addition to arithmetic expressions using operators, NumPy provides vectorized functions for elementwise evaluation of many elementary mathematical functions and operations. Table 2-7 gives a summary of elementary mathematical functions in NumPy.[3] Each of these functions takes a single array (of arbitrary dimension) as input and returns a new array of the same shape, where for each element, the function has been applied to the corresponding element in the input array. The data type of the output array is not necessarily the same as that of the input array.

---

[3] Note that this is not a complete list of the available elementwise functions in NumPy. See the NumPy reference documentations for comprehensive lists.

Table 2-7. *Selection of NumPy Functions for Elementwise Elementary Mathematical Functions*

NumPy Function	Description
np.cos, np.sin, np.tan	Trigonometric functions
np.arccos, np.arcsin, np.arctan	Inverse trigonometric functions
np.cosh, np.sinh, np.tanh	Hyperbolic trigonometric functions
np.arccosh, np.arcsinh, np.arctanh	Inverse hyperbolic trigonometric functions
np.sqrt	Square root
np.exp	Exponential
np.log, np.log2, np.log10	Logarithms of base e, 2, and 10, respectively

For example, the np.sin function (which takes only one argument) is used to compute the sine function for all values in the array.

```
In [159]: x = np.linspace(-1, 1, 11)
In [160]: x
Out[160]: array([-1. , -0.8, -0.6, -0.4, -0.2,  0. ,  0.2,  0.4,  0.6,  0.8,  1.])
In [161]: y = np.sin(np.pi * x)
In [162]: np.round(y, decimals=4)
Out[162]: array([-0., -0.5878, -0.9511, -0.9511, -0.5878, 0., 0.5878, 0.9511, 0.9511,
0.5878, 0.])
```

The np.pi constant and the np.round function were used to round the values of y to four decimals. Like the np.sin function, many elementary math functions take one input array and produce one output array. In contrast, many mathematical operator functions (see Table 2-8) operate on two input arrays and return one array.

```
In [163]: np.add(np.sin(x) ** 2, np.cos(x) ** 2)
Out[163]: array([ 1.,  1.,  1.,  1.,  1.,  1.,  1.,  1.,  1.,  1.,  1.])
In [164]: np.sin(x) ** 2 + np.cos(x) ** 2
Out[164]: array([ 1.,  1.,  1.,  1.,  1.,  1.,  1.,  1.,  1.,  1.,  1.])
```

Table 2-8. *Summary of NumPy Functions for Elementwise Mathematical Operations*

NumPy Function	Description
np.add, np.subtract, np.multiply, np.divide	Addition, subtraction, multiplication, and division of two NumPy arrays
np.power	Raises first input argument to the power of the second input argument (applied elementwise)
np.remainder	The remainder of the division
np.reciprocal	The reciprocal (inverse) of each element
np.real, np.imag, np.conj	The real part, imaginary part, and the complex conjugate of the elements in the input arrays
np.sign, np.abs	The sign and the absolute value
np.floor, np.ceil, np.rint	Converts to integer values
np.round	Rounds to a given number of decimals

Note that in this example, np.add and the + operator are equivalent; for normal use, the operator should be used.

Occasionally, it is necessary to define new functions that operate on NumPy arrays on an element-by-element basis. A good way to implement such functions is to express it in terms of existing NumPy operators and expressions. But when this is not possible, the np.vectorize function can be a convenient tool. This function takes a nonvectorized function and returns a vectorized function. For example, consider the following implementation of the Heaviside step function, which works for scalar input.

```
In [165]: def heaviside(x):
   ...:        return 1 if x > 0 else 0
In [166]: heaviside(-1)
Out[166]: 0
In [167]: heaviside(1.5)
Out[167]: 1
```

However, unfortunately, this function does not work for NumPy array input.

```
In [168]: x = np.linspace(-5, 5, 11)
In [169]: heaviside(x)
...
ValueError: The truth value of an array with more than one element is ambiguous. Use a.any()
or a.all()
```

Using np.vectorize the scalar Heaviside function can be converted into a vectorized function that works with NumPy arrays as input.

```
In [170]: heaviside = np.vectorize(heaviside)
In [171]: heaviside(x)
Out[171]: array([0, 0, 0, 0, 0, 0, 1, 1, 1, 1, 1])
```

Although the function returned by np.vectorize works with arrays, it will be relatively slow since the original function must be called for each element in the array. There are much better ways to implement this function using arithmetic with Boolean-valued arrays, as discussed later in this chapter.

```
In [172]: def heaviside(x):
    ...:        return 1.0 * (x > 0)
```

Nonetheless, np.vectorize can often be a quick and convenient way to vectorize a function written for scalar input.

In addition to NumPy's functions for elementary mathematical functions, as summarized in Table 2-7, there are numerous functions in NumPy for mathematical operations. A summary of a selection of these functions is given in Table 2-8.

Aggregate Functions

NumPy provides another set of functions for calculating aggregates for NumPy arrays, which take an array as input and, by default, return a scalar as output. For example, statistics such as averages, standard deviations, and variances of the values in the input array and functions for calculating the sum and the product of elements in an array are all aggregate functions.

A summary of aggregate functions is given in Table 2-9. All of these functions are also available as methods in the ndarray class. For example, np.mean(data) and data.mean() in the following example are equivalent.

```
In [173]: data = np.random.normal(size=(15,15))
In [174]: np.mean(data)
Out[174]: -0.032423651106794522
In [175]: data.mean()
Out[175]: -0.032423651106794522
```

Table 2-9. *NumPy Functions for Calculating Aggregates of NumPy Arrays*

NumPy Function	Description
np.mean	The average of all values in the array
np.std	Standard deviation
np.var	Variance
np.sum	The sum of all elements
np.prod	The product of all elements
np.cumsum	The cumulative sum of all elements
np.cumprod	The cumulative product of all elements
np.min, np.max	The minimum/maximum value in an array
np.argmin, np.argmax	The index of the minimum/maximum value in an array
np.all	Returns True if all elements in the argument array are nonzero
np.any	Returns True if any of the elements in the argument array is nonzero

By default, the functions in Table 2-9 aggregate over the entire input array. Using the `axis` keyword argument with these functions, and their corresponding `ndarray` methods, it is possible to control which axis in the array aggregation is carried out over. The axis argument can be an integer, which specifies the axis to aggregate values over. In many cases, the axis argument can also be a tuple of integers, which specifies multiple axes to aggregate over. The following example demonstrates how calling the aggregate function `np.sum` on the array of shape (5, 10, 15) reduces the dimensionality of the array depending on the values of the `axis` argument.

```
In [176]: data = np.random.normal(size=(5, 10, 15))
In [177]: data.sum(axis=0).shape
Out[177]: (10, 15)
In [178]: data.sum(axis=(0, 2)).shape
Out[178]: (10,)
In [179]: data.sum()
Out[179]: -31.983793284860798
```

A visual illustration of how aggregation over all elements, over the first axis, and over the second axis of a 3 × 3 array is shown in Figure 2-3. In this example, the `data` array is filled with integers between 1 and 9.

```
In [180]: data = np.arange(1,10).reshape(3,3)
In [181]: data
Out[181]: array([[1, 2, 3],
                 [4, 5, 6],
                 [7, 8, 9]])
```

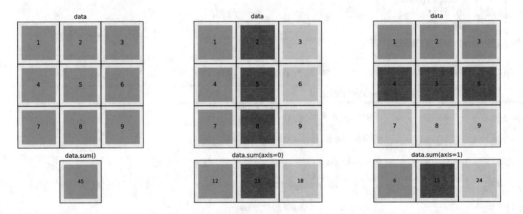

Figure 2-3. *Illustration of array aggregation functions along all axes (left), the first axis (center), and the second axis (right) of a two-dimensional array of shape 3 × 3*

We compute the aggregate sum of the entire array, over axis 0, and over axis 1, respectively.

```
In [182]: data.sum()
Out[182]: 45
In [183]: data.sum(axis=0)
Out[183]: array([12, 15, 18])
In [184]: data.sum(axis=1)
Out[184]: array([ 6, 15, 24])
```

Boolean Arrays and Conditional Expressions

When computing with NumPy arrays, there is often a need to compare elements in different arrays and perform conditional computations based on the results of such comparisons. Like with arithmetic operators, NumPy arrays can be used with the usual comparison operators, for example, >, <, >=, <=, ==, and !=, and the comparisons are made on an element-by-element basis. The broadcasting rules also apply to comparison operators. If two operators have compatible shapes and sizes, the result of the comparison is a new array with Boolean values (with dtype as np.bool) that gives the result of the comparison for each element.

```
In [185]: a = np.array([1, 2, 3, 4])
In [186]: b = np.array([4, 3, 2, 1])
In [187]: a < b
Out[187]: array([ True,  True, False, False], dtype=bool)
```

To use the result of a comparison between arrays in, for example, an if statement, we need to aggregate the Boolean values of the resulting arrays in some suitable fashion, to obtain a single True or False value. A common use is to apply the np.all or np.any aggregation functions, depending on the situation.

```
In [188]: np.all(a < b)
Out[188]: False
In [189]: np.any(a < b)
Out[189]: True
In [190]: if np.all(a < b):
    ...:     print("All elements in a are smaller than their corresponding element in b")
    ...: elif np.any(a < b):
    ...:     print("Some elements in a are smaller than their corresponding element in b")
    ...: else:
    ...:     print("All elements in b are smaller than their corresponding element in a")
Some elements in a are smaller than their corresponding element in b
```

However, the advantage of Boolean-valued arrays is that they often make it possible to avoid conditional if statements altogether. Using Boolean-valued arrays in arithmetic expressions makes writing conditional computations in vectorized form possible. When appearing in an arithmetic expression with a scalar number, or another NumPy array with a numerical data type, a Boolean array is converted to a numerical-valued array with values 0 and 1 in place of False and True, respectively.

```
In [191]: x = np.array([-2, -1, 0, 1, 2])
In [192]: x > 0
Out[192]: array([False, False, False,  True,  True], dtype=bool)
In [193]: 1 * (x > 0)
Out[193]: array([0, 0, 0, 1, 1])
In [194]: x * (x > 0)
Out[194]: array([0, 0, 0, 1, 2])
```

This is a useful property for conditional computing, such as defining piecewise functions. For example, if we need to define a function describing a pulse of a given height, width, and position, we can implement this function by multiplying the height (a scalar variable) with two Boolean-valued arrays for the spatial extension of the pulse.

```
In [195]: def pulse(x, position, height, width):
     ...:        return height * (x >= position) * (x <= (position + width))
In [196]: x = np.linspace(-5, 5, 11)
In [197]: pulse(x, position=-2, height=1, width=5)
Out[197]: array([0, 0, 0, 1, 1, 1, 1, 1, 1, 0, 0])
In [198]: pulse(x, position=1, height=1, width=5)
Out[198]: array([0, 0, 0, 0, 0, 0, 1, 1, 1, 1, 1])
```

In this example, the expression (x >= position) * (x <= (position + width)) is a multiplication of two Boolean-valued arrays. For this case, the multiplication operator acts as an elementwise AND operator. The pulse function could also be implemented using NumPy's np.logical_and function for elementwise AND operations.

```
In [199]: def pulse(x, position, height, width):
     ...:        return height * np.logical_and(x >= position, x <= (position + width))
```

There are functions for other logical operations, such as NOT, OR, and XOR, and functions for selectively picking values from different arrays depending on a given np.where condition, a list of np.select conditions, and an array of np.choose indices. Table 2-10 summarizes such functions, and the following examples demonstrate their basic usage. The np.where function selects elements from two arrays (second and third arguments), given a Boolean-valued array condition (the first argument). For elements where the condition is True, the corresponding values from the array given as second argument are selected. If the condition is False, values from the array given as third argument are selected.

```
In [200]: x = np.linspace(-4, 4, 9)
In [201]: np.where(x < 0, x**2, x**3)
Out[201]: array([ 16.,   9.,   4.,   1.,   0.,   1.,   8.,  27.,  64.])
```

Table 2-10. *NumPy Functions for Conditional and Logical Expressions*

Function	Description
np.where	Chooses values from two arrays depending on the value of a condition array
np.choose	Chooses values from a list of arrays depending on the values of a given index array
np.select	Chooses values from a list of arrays depending on a list of conditions
np.nonzero	Returns an array with indices of nonzero elements
np.logical_and	Performs an elementwise AND operation
np.logical_or, np.logical_xor	Elementwise OR/XOR operations
np.logical_not	Elementwise NOT operation (inverting)

The np.select function works similarly, but instead of a Boolean-valued condition array, it expects a list of Boolean-valued condition arrays and a corresponding list of value arrays.

```
In [202]: np.select([x < -1, x < 2, x >= 2],
     ...:           [x**2 , x**3 , x**4])
Out[202]: array([ 16.,   9.,   4.,  -1.,   0.,   1.,  16.,  81., 256.])
```

The np.choose takes as a first argument a list or an array with indices that determine from which array in a given list of arrays an element is picked.

```
In [203]: np.choose([0, 0, 0, 1, 1, 1, 2, 2, 2],
     ...:          [x**2, x**3, x**4])
Out[203]: array([ 16.,    9.,    4.,   -1.,    0.,    1.,   16.,   81.,  256.])
```

The np.nonzero function returns a tuple of indices that can be used to index the array (e.g., the one that the condition was based on). This has the same results as indexing the array directly with abs(x) > 2. But it uses fancy indexing with the indices returned by np.nonzero rather than Boolean-valued array indexing.

```
In [204]: np.nonzero(abs(x) > 2)
Out[204]: (array([0, 1, 7, 8]),)
In [205]: x[np.nonzero(abs(x) > 2)]
Out[205]: array([-4., -3.,  3.,  4.])
In [206]: x[abs(x) > 2]
Out[206]: array([-4., -3.,  3.,  4.])
```

Set Operations

The Python language provides a convenient *set* data structure for managing unordered collections of unique objects. The NumPy array class ndarray can also describe such sets, and NumPy contains functions for operating on sets stored as NumPy arrays. These functions are summarized in Table 2-11. Using NumPy arrays to describe and operate on sets allows expressing certain operations in vectorized form. For example, testing if the values in a NumPy array are included in a set can be done using the np.in1d function, which tests for the existence of each element of its first argument in the array passed as the second argument. To see how this works, consider the following example: first, to ensure that a NumPy array is a proper set, we can use the np.unique function, which returns a new array with unique values.

```
In [207]: a = np.unique([1, 2, 3, 3])
In [208]: b = np.unique([2, 3, 4, 4, 5, 6, 5])
In [209]: np.in1d(a, b)
Out[209]: array([False,  True,  True], dtype=bool)
```

Table 2-11. *NumPy Functions for Operating on Sets*

Function	Description
np.unique	Creates a new array with unique elements, where each value only appears once
np.in1d	Tests for the existence of an array of elements in another array
np.intersect1d	Returns an array with elements that are contained in two given arrays
np.setdiff1d	Returns an array with elements that are contained in one, but not the other, of two given arrays
np.union1d	Returns an array with elements that are contained in either, or both, of two given arrays

Here, the existence of each element in a in the set b was tested, and the result is a Boolean-valued array. Note that we can use the in keyword to test for the existence of single elements in a set represented as a NumPy array.

```
In [210]: 1 in a
Out[210]: True
In [211]: 1 in b
Out[211]: False
```

To test if a is a subset of b, we can use the np.in1d, as in the previous example, together with the aggregation function np.all (or the corresponding ndarray method).

```
In [212]: np.all(np.in1d(a, b))
Out[212]: False
```

The fundamental set operations union (the set of elements included in either or both sets), intersection (elements included in both sets), and difference (elements included in one of the sets but not the other) is provided by np.union1d, np.intersect1d, and np.setdiff1d, respectively.

```
In [213]: np.union1d(a, b)
Out[213]: array([1, 2, 3, 4, 5, 6])
In [214]: np.intersect1d(a, b)
Out[214]: array([2, 3])
In [215]: np.setdiff1d(a, b)
Out[215]: array([1])
In [216]: np.setdiff1d(b, a)
Out[216]: array([4, 5, 6])
```

Operations on Arrays

In addition to elementwise and aggregation functions, some operations act on arrays as a whole and produce a transformed array of the same size. An example of this type of operation is the transpose, which flips the order of the axes of an array. For the special case of a two-dimensional array (i.e., a matrix, the transpose simply exchanges rows and columns).

```
In [217]: data = np.arange(9).reshape(3, 3)
In [218]: data
Out[218]: array([[0, 1, 2],
                 [3, 4, 5],
                 [6, 7, 8]])
In [219]: np.transpose(data)
Out[219]: array([[0, 3, 6],
                 [1, 4, 7],
                 [2, 5, 8]])
```

The transpose function np.transpose also exists as a method in ndarray and as the special method name ndarray.T. For an arbitrary N-dimensional array, the transpose operation reverses all the axes, as seen from the following example (note that the shape attribute is used here to display the number of values along each axis of the array).

```
In [220]: data = np.random.randn(1, 2, 3, 4, 5)
In [221]: data.shape
Out[221]: (1, 2, 3, 4, 5)
In [222]: data.T.shape
Out[222]: (5, 4, 3, 2, 1)
```

The np.fliplr (flip left-right) and np.flipud (flip up-down) functions perform operations that are similar to the transpose: they reshuffle the elements of an array so that the elements in rows (np.fliplr) or columns (np.flipud) are reversed, and the shape of the output array is the same as the input. The np.rot90 function rotates the elements in the first two axes in an array by 90 degrees, and like the transpose function, it can change the shape of the array. Table 2-12 summarizes NumPy functions for common array operations.

Table 2-12. *Summary of NumPy Functions for Array Operations*

Function	Description
np.transpose, np.ndarray.transpose, np.ndarray.T	Transpose (reverse axes) an array.
np.fliplr/np.flipud	Reverse the elements in each row/column.
np.rot90	Rotate the elements along the first two axes by 90 degrees.
np.sort, np.ndarray.sort	Sort an array's elements along a specified axis (which defaults to the last axis of the array). The np.ndarray method sort performs the sorting in place, modifying the input array.

Matrix and Vector Operations

So far, I have discussed general N-dimensional arrays. One of the main applications of such arrays is to represent the mathematical concepts of vectors, matrices, and tensors. In this case, we also frequently need to calculate vector and matrix operations such as scalar (inner) products, dot (matrix) products, and tensor (outer) products. Table 2-13 summarizes NumPy's functions for matrix operations.

Table 2-13. *Summary of NumPy Functions for Matrix Operations*

NumPy Function	Description
np.dot	Matrix multiplication (dot product) between two given arrays representing vectors, arrays, or tensors
np.inner	Scalar multiplication (inner product) between two arrays representing vectors
np.cross	The cross product between two arrays that represent vectors
np.tensordot	Dot product along specified axes of multidimensional arrays
np.outer	Outer product (tensor product of vectors) between two arrays representing vectors
np.kron	Kronecker product (tensor product of matrices) between arrays representing matrices and higher-dimensional arrays
np.einsum	Evaluates Einstein's summation convention for multidimensional arrays

In NumPy, the * operator is used for elementwise multiplication. For two two-dimensional arrays A and B, the expression A * B does not compute a matrix product (in contrast to many other computing environments). Originally, there was no operator for denoting matrix multiplication in Python. Instead, the NumPy function np.dot has been used for this purpose, together with the corresponding method in the ndarray class. However, the @ operator has recently been introduced for matrix multiplication[4], and while it is supported by NumPy, this operator is still not widely used. It is, therefore, still important to be familiar with the explicit methods in NumPy for matrix multiplication.

To compute the product of two matrices, A and B, of size $N \times M$ and $M \times P$, which results in a matrix of size $N \times P$, we can use np.dot.

```
In [223]: A = np.arange(1, 7).reshape(2, 3)
In [224]: A
Out[224]: array([[1, 2, 3],
                 [4, 5, 6]])
In [225]: B = np.arange(1, 7).reshape(3, 2)
In [226]: B
Out[226]: array([[1, 2],
                 [3, 4],
                 [5, 6]])
In [227]: np.dot(A, B)  # or equivalently: A @ B
Out[227]: array([[22, 28],
                 [49, 64]])
In [228]: np.dot(B, A)  # or equivalently: B @ A
Out[228]: array([[ 9, 12, 15],
                 [19, 26, 33],
                 [29, 40, 51]])
```

The np.dot function can also be used for matrix-vector multiplication (i.e., multiplication of a two-dimensional array, which represents a matrix, with a one-dimensional array representing a vector). The following is an example.

```
In [229]: A = np.arange(9).reshape(3, 3)
In [230]: A
Out[230]: array([[0, 1, 2],
                 [3, 4, 5],
                 [6, 7, 8]])
In [231]: x = np.arange(3)
In [232]: x
Out[232]: array([0, 1, 2])
In [233]: np.dot(A, x)
Out[233]: array([5, 14, 23])
```

In this example, x can be either a two-dimensional array of shape $(1, 3)$ or a onedimensional array with shape $(3,)$. In addition to the np.dot function, there is a corresponding dot method in ndarray, which is used in the following example.

```
In [234]: A.dot(x)
Out[234]: array([5, 14, 23])
```

---

[4] As of Python 3.5 the @ symbol has been adopted for denoting matrix multiplication. See http://legacy.python.org/dev/peps/pep-0465 for details.

Unfortunately, nontrivial matrix multiplication expressions can often become complex and hard to read when using either np.dot or np.ndarray.dot. For example, even a relatively simple matrix expression like the one for a similarity transform, $A' = BAB^{-1}$, must be represented with relatively cryptic nested expressions,[5] such as

```
In [235]: A = np.random.rand(3,3)
In [236]: B = np.random.rand(3,3)
In [237]: Ap = np.dot(B, np.dot(A, np.linalg.inv(B)))
   or
In [238]: Ap = B.dot(A.dot(np.linalg.inv(B)))
```

The @ operator was introduced partly to address this type of readability problem. Also, since long before this operator was available, NumPy has provided an alternative to ndarray data structure named matrix, for which expressions like A * B are implemented as matrix multiplication. It also provides some convenient special attributes, like matrix.I for the inverse matrix and matrix.H for the complex conjugate transpose of a matrix. With instances of this matrix class, we can use a vastly more readable expression.

```
In [239]: A = np.matrix(A)
In [240]: B = np.matrix(B)
In [241]: Ap = B * A * B.I
```

This may seem like a practical compromise, but unfortunately, using the matrix class has a few disadvantages, and its use is often discouraged. The main objection against using matrix is that an expression like A * B is context-dependent: that is, it is not immediately clear if A * B denotes elementwise or matrix multiplication because it depends on the type of A and B, creating another code-readability problem. This can be a particularly relevant issue if A and B are user-supplied arguments to a function, in which case it would be necessary to cast all input arrays explicitly to matrix instances, using, for example, np.asmatrix or the np.matrix function (since there would be no guarantee that the user calls the function with arguments of type matrix rather than ndarray). The np.asmatrix function creates a view of the original array in the form of an np.matrix instance. This does not add much to computational costs, but explicitly casting arrays back and forth between ndarray and matrix does offset much of the benefits of the improved readability of matrix expressions. A related issue is that some functions that operate on arrays and matrices might not respect the type of the input and may return an ndarray even though it was called with an input argument of type matrix. This way, a matrix of type matrix might be unintentionally converted to ndarray, which would change the behavior of expressions like A * B. This behavior is not likely to occur when using NumPy's array and matrix functions, but it is likely to happen when using functions from other packages. However, despite all the arguments for not using matrix matrices too extensively, I believe that using matrix class instances for complicated matrix expressions is important. In these cases, it might be a good idea to explicitly cast arrays to matrices before the computation and explicitly cast the result back to the ndarray type, following the pattern.

```
In [242]: A = np.asmatrix(A)
In [243]: B = np.asmatrix(B)
In [244]: Ap = B * A * B.I
In [245]: Ap = np.asarray(Ap)
```

---

[5] With the new infix matrix multiplication operator, this same expression can be expressed as the considerably more readable: Ap = B @ A @ np.linalg.inv(B).

The inner product (scalar product) between two arrays representing vectors can be computed using the np.inner function.

```
In [246]: np.inner(x, x)
Out[246]: 5
```

Or, equivalently, use np.dot.

```
In [247]: np.dot(x, x)
Out[247]: 5
```

The main difference is that np.inner expects two input arguments with the same dimension, while np.dot can take input vectors of shape $1 \times N$ and $N \times 1$, respectively.

```
In [248]: y = x[:, np.newaxis]
In [249]: y
Out[249]: array([[0],
                 [1],
                 [2]])
In [250]: np.dot(y.T, y)
Out[250]: array([[5]])
```

While the inner product maps two vectors to a scalar, the outer product performs the complementary operation of mapping two vectors to a matrix.

```
In [251]: x = np.array([1, 2, 3])
In [252]: np.outer(x, x)
Out[252]: array([[1, 2, 3],
                 [2, 4, 6],
                 [3, 6, 9]])
```

The outer product can also be calculated using the Kronecker product using the np.kron function, which, in contrast to np.outer, produces an output array of shape (M*P, N*Q) if the input arrays have shapes (M, N) and (P, Q), respectively. Thus, for two one-dimensional arrays of length M and P, the resulting array has shape (M*P,).

```
In [253]: np.kron(x, x)
Out[253]: array([1, 2, 3, 2, 4, 6, 3, 6, 9])
```

To obtain the result that corresponds to np.outer(x, x), the input array x must be expanded to shape (N, 1) and (1, N), in the first and second argument to np.kron, respectively.

```
In [254]: np.kron(x[:, np.newaxis], x[np.newaxis, :])
Out[254]: array([[1, 2, 3],
                 [2, 4, 6],
                 [3, 6, 9]])
```

In general, while the np.outer function is primarily intended for vectors as input, the np.kron function can be used for computing tensor products of arrays of arbitrary dimension (but both inputs must have the same number of axes). For example, we can use the following to compute the tensor product of two 2 × 2 matrices.

```
In [255]: np.kron(np.ones((2,2)), np.identity(2))
Out[255]: array([[ 1.,  0.,  1.,  0.],
                 [ 0.,  1.,  0.,  1.],
                 [ 1.,  0.,  1.,  0.],
                 [ 0.,  1.,  0.,  1.]])
In [256]: np.kron(np.identity(2), np.ones((2,2)))
Out[256]: array([[ 1.,  1.,  0.,  0.],
                 [ 1.,  1.,  0.,  0.],
                 [ 0.,  0.,  1.,  1.],
                 [ 0.,  0.,  1.,  1.]])
```

When working with multidimensional arrays, it is often possible to express common array operations concisely using Einstein's summation convention, in which an implicit summation is assumed over each index that occurs multiple times in an expression. For example, the scalar product between two vectors x and y is compactly expressed as $x_n y_n$, and the matrix multiplication of two matrices, A and B, is expressed as $A_{mk} B_{kn}$. NumPy provides the np.einsum function for carrying out Einstein summations. Its first argument is an index expression, followed by an arbitrary number of arrays included in the expression. The index expression is a string with comma-separated indices, where each comma separates the indices of each array. Each array can have any number of indices. For example, the scalar product expression $x_n y_n$ can be evaluated with np.einsum using the index expression "n,n", that is using np.einsum("n,n", x, y).

```
In [257]: x = np.array([1, 2, 3, 4])
In [258]: y = np.array([5, 6, 7, 8])
In [259]: np.einsum("n,n", x, y)
Out[259]: 70
In [260]: np.inner(x, y)
Out[260]: 70
```

Similarly, the matrix multiplication $A_{mk} B_{kn}$ can be evaluated using np.einsum and the index expression "mk,kn".

```
In [261]: A = np.arange(9).reshape(3, 3)
In [262]: B = A.T
In [263]: np.einsum("mk,kn", A, B)
Out[263]: array([[  5,  14,  23],
                 [ 14,  50,  86],
                 [ 23,  86, 149]])
In [264]: np.alltrue(np.einsum("mk,kn", A, B) == np.dot(A, B))
Out[264]: True
```

The Einstein summation convention can be particularly convenient when dealing with multidimensional arrays since the index expression defining the operation makes it explicit which operation is carried out and which axes it is performed. An equivalent computation using, for example, np.tensordot might require giving the axes along which the dot product will be evaluated.

Summary

This chapter provided a brief introduction to array-based programming with the NumPy library that can serve as a reference for the upcoming chapters in this book. NumPy is a core library for computing with Python that provides a foundation for nearly all computational libraries for Python. To be familiar with the NumPy library and its patterns of application is a fundamental skill for using Python for scientific and technical computing. I started with introducing NumPy's data structure for N-dimensional arrays—the ndarray object—and continued by discussing functions for creating and manipulating arrays, including indexing and slicing for extracting elements from arrays. I also discussed functions and operators for performing computations with ndarray objects, emphasizing vectorized expressions and operators for efficient computation with arrays. Throughout the rest of this book, we will see examples of higher-level libraries for specific fields in scientific computing that use the array framework provided by NumPy.

Further Reading

The NumPy library is the topic of several books, including the *Guide to NumPy*, by creator Travis E. Oliphant, available at http://web.mit.edu/dvp/Public/numpybook.pdf, and a series of books by Ivan Idris: *NumPy Beginner's Guide* (Packt, 2015), *NumPy Cookbook* (Packt, 2012), and *Learning NumPy Array* (Packt, 2014). NumPy is also covered in fair detail in *Python for Data Analysis* by Wes McKinney (O'Reilly, 2013).

CHAPTER 3

■ ■ ■

Symbolic Computing

Symbolic computing is an entirely different paradigm than the numerical array-based computing introduced in the previous chapter. In symbolic computing software, also known as a computer algebra system (CAS), representations of mathematical objects and expressions are manipulated and transformed analytically. Symbolic computing is mainly about using computers to automate analytical computations that can be done by hand with pen and paper. However, by automating the bookkeeping and the manipulations of mathematical expressions using a computer algebra system, it is possible to take analytical computing much further than can realistically be done by hand. Symbolic computing is a great tool for checking and debugging analytical calculations done by hand, but more importantly, it enables carrying out analytical analysis that may not otherwise be possible.

Analytical and symbolic computing is a key part of the scientific and technical computing landscape. Even for problems that can only be solved numerically (which is common because analytical methods are not feasible in many practical problems), pushing the limits for what can be done analytically can make a big difference before resorting to numerical techniques. It can, for example, reduce the complexity or size of the numerical problem that finally needs to be solved. In other words, instead of tackling a problem in its original form directly using numerical methods, it may be possible to use analytical methods to simplify the problem first.

In the scientific Python environment, the main library for symbolic computing is SymPy (Symbolic Python). SymPy is written in Python and provides tools for a wide range of analytical and symbolic problems. This chapter explores how SymPy can be used for symbolic computing with Python.

■ **SymPy** The Symbolic Python (SymPy) library aims to provide a full-featured computer algebra system. However, SymPy is primarily a library rather than a full environment. This makes SymPy well-suited for integration in applications and computations using other Python libraries. At the time of writing, the latest version is 1.12. More information about SymPy is available at `www.sympy.org` and `https://github.com/sympy/sympy/wiki/Faq`.

Importing SymPy

The SymPy library provides the Python module named `sympy`. It is common to import all symbols from this module when working with SymPy, using `from sympy import *`. But in the interest of clarity and to avoid namespace conflicts between functions and variables from SymPy and other packages such NumPy and

R. Johansson, *Numerical Python*, https://doi.org/10.1007/979-8-8688-0413-7_3

SciPy (see later chapters), here we import the library in its entirety as sympy. For the rest of this book, assume that SymPy is imported this way.

```
In [1]: import sympy
In [2]: sympy.init_printing()
```

This also called the sympy.init_printing function, which configures SymPy's printing system to display nicely formatted renditions of mathematical expressions, as shown in examples later in this chapter. In the Jupyter Notebook, this sets up printing so that the MathJax JavaScript library renders SymPy expressions, and the results are displayed on the browser page of the notebook.

For convenience and readability of the example codes in this chapter, assume that the following frequently used symbols are explicitly imported from SymPy into the local namespace.

```
In [3]: from sympy import I, pi, oo
```

■ **Caution**   NumPy and SymPy, as well as many other libraries, provide many functions and variables with the same name. But these symbols are rarely interchangeable. For example, numpy.pi is a numerical approximation of the mathematical symbol π, while sympy.pi is a symbolic representation of π. It is important not to mix them up and use, for instance, numpy.pi in place of sympy.pi when doing symbolic computations, or vice versa. The same holds for many fundamental mathematical functions, such as numpy.sin vs. sympy.sin. Therefore, it is important to consistently use namespaces when using more than one package in computing with Python.

Symbols

A core functionality in SymPy is to represent mathematical symbols as Python objects. In the SymPy library, the sympy.Symbol class can be used for this purpose. A sympy.Symbol instance has a name and set of attributes describing its properties and methods for querying those properties and for operating on the symbol object. A symbol by itself is not of much practical use, but symbols are used as nodes in trees to represent algebraic expressions (see next section). Among the first steps in analyzing a problem with SymPy is to create symbols for the various mathematical variables and quantities required to describe the problem.

The symbol name is a string, which optionally can contain LaTeX-like markup to make the symbol name display well in, for example, IPython's and Jupyter's rich display system. The name of a sympy.Symbol object is set when it is created. Symbols can be created in a few different ways in SymPy, for example, using sympy.Symbol, sympy.symbols, and sympy.var. Normally, it is desirable to associate SymPy symbols with Python variables with the same name or a name that closely corresponds to the symbol name. For example, to create a symbol named x and bind it to the Python variable with the same name, we can use the constructor of the Symbol class and pass a string containing the symbol name as the first argument.

```
In [4]: x = sympy.Symbol("x")
```

The variable x now represents an abstract mathematical symbol x, of which very little information is known by default. At this point, x could represent, for example, a real number, an integer, a complex number, a function, and many other possibilities. In many cases, it is sufficient to represent a mathematical symbol with this abstract, unspecified Symbol object. Yet, sometimes, it is necessary to give the SymPy library more hints about exactly what type of symbol a Symbol object represents. This may help SymPy to manipulate or simplify analytical expressions more efficiently. We can add various assumptions that narrow down the possible properties of a symbol by adding optional keyword arguments to the symbol-creating functions, such as Symbol. Table 3-1 summarizes frequently used assumptions associated with a Symbol class instance. For example, if we have a mathematical variable *y* known as a real number, we can use the real=True keyword argument when creating the corresponding symbol instance. We can verify that SymPy recognizes that the symbol is real by using the is_real attribute of the Symbol class.

```
In [5]: y = sympy.Symbol("y", real=True)
In [6]: y.is_real
Out[6]: True
```

If, on the other hand, we were to use is_real to query the previously defined symbol x, which was not explicitly specified as real and can represent both real and nonreal variables, we get None as a result.

```
In [7]: x.is_real is None
Out[7]: True
```

Note that the is_real returns True if the symbol is known to be real, False if the symbol is known not to be real, and None if it is not known if the symbol is real. Other attributes (see Table 3-1) for querying assumptions on Symbol objects work in the same way. Consider the following example demonstrating a symbol for which the is_real attribute is False.

```
In [8]: sympy.Symbol("z", imaginary=True).is_real
Out[8]: False
```

Table 3-1. *Selected Assumptions and Their Corresponding Keyword Arguments for Symbol Objects**

Assumption Keyword Arguments	Attributes	Description
real, imaginary	is_real, is_imaginary	Specify that a symbol represents a real or imaginary number.
positive, negative	is_positive, is_negative	Specify that a symbol is positive or negative.
Integer	is_integer	The symbol represents an integer.
odd, even	is_odd, is_even	The symbol represents an odd or even integer.
Prime	is_prime	The symbol is a prime number and also an integer.
finite, infinite	is_finite, is_infinite	The symbol represents a quantity that is finite or infinite.

**For a complete list, see the docstring for sympy.Symbol.*

Among the assumptions in Table 3-1, the most important ones to explicitly specify when creating new symbols are real and positive. When applicable, adding these assumptions to symbols can frequently help SymPy simplify various expressions further than otherwise. Consider the following simple example.

```
In  [9]: x = sympy.Symbol("x")
In [10]: y = sympy.Symbol("y", positive=True)
In [11]: sympy.sqrt(x ** 2)
```

```
Out[11]: √x²
In [12]: sympy.sqrt(y ** 2)
Out[12]: y
```

This created two symbols, x and y, and computed the square root of the square of that symbol using the SymPy function `sympy.sqrt`. If nothing is known about the symbol in the computation, then no simplification can be done. If, on the other hand, the symbol is known to be representing a positive number, then $\sqrt{y^2} = y$ and SymPy correctly recognizes this in the latter example.

When working with mathematical symbols that represent integers rather than real numbers, it is also useful to explicitly specify this when creating the corresponding SymPy symbols, using, for example, the `integer=True`, `even=True,` or `odd=True`, if applicable. This may also allow SymPy to analytically simplify certain expressions and function evaluations, such as in the following example.

```
In [13]: n1 = sympy.Symbol("n")
In [13]: n2 = sympy.Symbol("n", integer=True)
In [13]: n3 = sympy.Symbol("n", odd=True)
In [14]: sympy.cos(n1 * pi)
Out[14]: cos(πn)
In [15]: sympy.cos(n2 * pi)
Out[15]: (−1)ⁿ
In [16]: sympy.cos(n3 * pi)
Out[16]: −1
```

To formulate a nontrivial mathematical problem, it is often necessary to define a large number of symbols. Using `sympy.Symbol` to specify each symbol one by one may become tedious, and for convenience, SymPy provides a `sympy.symbols` function for creating multiple symbols in one function call. This function takes a comma-separated string of symbol names and an arbitrary set of keyword arguments (which apply to all the symbols), and it returns a tuple of newly created symbols. Using Python's tuple unpacking syntax and a call to `sympy.symbols` is a convenient way to create symbols.

```
In [17]: a, b, c = sympy.symbols("a, b, c", negative=True)
In [18]: d, e, f = sympy.symbols("d, e, f", positive=True)
```

Numbers

The purpose of creating Python objects for mathematical symbols is to use them to represent mathematical expressions. To be able to do this, we also need to represent other mathematical objects, such as numbers, functions, and constants. This section looks at SymPy's classes for representing number objects. These classes have many methods and attributes shared with instances of `sympy.Symbol`, which allows us to treat symbols and numbers on equal footing when representing expressions.

For example, in the previous section, we saw that `sympy.Symbol` instances have attributes for querying properties of symbol objects, such as `is_real`. We need to be able to use the same attributes for all types of objects, including for example numbers such as integers and floating-point numbers, when manipulating symbolic expressions in SymPy. For this reason, we cannot directly use the built-in Python objects for integers, `int`, floating-point numbers, `float`, and so on. Instead, SymPy provides the `sympy.Integer` and `sympy.Float` classes for representing integers and floating-point numbers within the SymPy framework. This distinction is important to be aware of when working with SymPy, but fortunately, we rarely need to be concerned with creating objects of type `sympy.Integer` and `sympy.Float` to represent specific numbers, since SymPy automatically promotes Python numbers to instances of these classes when they occur in

SymPy expressions. However, to demonstrate this difference between Python's built-in number types and the corresponding types in SymPy, the following example explicitly creates instances of sympy.Integer and sympy.Float and use some of their attributes to query their properties.

```
In [19]: i = sympy.Integer(19)
In [20]: type(i)
Out[20]: sympy.core.numbers.Integer
In [21]: i.is_Integer, i.is_real, i.is_odd
Out[21]: (True, True, True)
In [22]: f = sympy.Float(2.3)
In [23]: type(f)
Out[23]: sympy.core.numbers.Float
In [24]: f.is_Integer, f.is_real, f.is_odd
Out[24]: (False, True, False)
```

■ **Tip**   We can cast instances of sympy.Integer and sympy.Float back to Python built-in types using the standard type casting int(i) and float(f).

To create a SymPy representation of a number, or in general, an arbitrary expression, we can also use the sympy.sympify function. This function takes a wide range of inputs and derives a SymPy-compatible expression, and it eliminates the need for specifying explicitly what types of objects are to be created. For the simple case of number input, we can use the following.

```
In [25]: i, f = sympy.sympify(19), sympy.sympify(2.3)
In [26]: type(i), type(f)
Out[26]: (sympy.core.numbers.Integer, sympy.core.numbers.Float)
```

Integer

The previous section used the Integer class to represent integers. It's worth pointing out that there is a difference between a Symbol instance with the assumption integer=True and an instance of Integer. While the Symbol with integer=True represents some integer, the Integer instance represents a specific integer. For both cases, the is_integer attribute is True, but there is also an is_Integer attribute (note the capital I), which is only True for Integer instances. In general, attributes with names in the form is_Name indicate if the object is of type Name, and attributes with names in the form is_name indicate if the object is known to satisfy the condition name. Thus, there is also an is_Symbol attribute that is True for Symbol instances.

```
In [27]: n = sympy.Symbol("n", integer=True)
In [28]: n.is_integer, n.is_Integer, n.is_positive, n.is_Symbol
Out[28]: (True, False, None, True)
In [29]: i = sympy.Integer(19)
In [30]: i.is_integer, i.is_Integer, i.is_positive, i.is_Symbol
Out[30]: (True, True, True, False)
```

Integers in SymPy are of arbitrary precision, allowing for a limitless range without fixed lower or upper bounds. This sets them apart from representing integers with a specific bit size, such as in NumPy. Consequently, SymPy enables working with extremely large numbers, as demonstrated in the following examples.

```
In [31]: i ** 50
Out[31]: 8663234049605954426644038200675212212900743262211018069459689001
In [32]: sympy.factorial(100)
Out[32]: 93326215443944152681699238856266700490715968264381621468592963895217599993229
9156089414639761565182862536979208272237582511852109168640000000000000000000000000000
```

Float

We have also already encountered the type sympy.Float in the previous sections. Like Integer, Float is arbitrary precision, unlike Python's built-in float type and the float types in NumPy. This means that a Float can represent a real number with an arbitrary number of decimals. When a Float instance is created using its constructor, there are two arguments: the first argument is a Python float or a string representing a floating-point number, and the second (optional) argument is the precision (number of significant decimal digits) of the Float object. For example, it is well known that the real number 0.3 cannot be represented exactly as a regular fixed bit-size floating-point number, and when printing 0.3 to 20 significant digits, it is displayed as 0.299999999999999988977698. The SymPy Float object can represent the real number 0.3 without the limitations of floating-point numbers.

```
In [33]: "%.25f" % 0.3  # Create a string representation with 25 decimals
Out[33]: '0.2999999999999999888977698'
In [34]: sympy.Float(0.3, 25)
Out[34]: 0.2999999999999999888977698
In [35]: sympy.Float('0.3', 25)
Out[35]: 0.3
```

However, note that to correctly represent 0.3 as a Float object, it is necessary to initialize it from a string '0.3 rather than the Python float 0.3, which already contains a floating-point error.

Rational

A rational number is a fraction p/q of two integers, the numerator p and the denominator q. SymPy represents this type of number using the sympy.Rational class. Rational numbers can be created explicitly using sympy.Rational and the numerator and denominator as arguments.

```
In [36]: sympy.Rational(11, 13)
```
$$\text{Out[36]: } \frac{11}{13}$$

Or they can be a result of a simplification carried out by SymPy. In either case, arithmetic operations between rational and integers remain rational.

```
In [37]: r1 = sympy.Rational(2, 3)
In [38]: r2 = sympy.Rational(4, 5)
In [39]: r1 * r2
```
$$\text{Out[39]: } \frac{8}{15}$$

```
In [40]: r1 / r2
```
$$\text{Out[40]: } \frac{5}{6}$$

Constants and Special Symbols

SymPy provides predefined symbols for mathematical constants and special objects, such as the imaginary unit i and infinity. These are summarized in Table 3-2, together with their corresponding symbols in SymPy. Note that the imaginary unit is written as I in SymPy.

Table 3-2. *Selected Mathematical Constants and Special Symbols and Their Corresponding Symbols in SymPy*

Mathematical Symbol	SymPy Symbol	Description
π	sympy.pi	Ratio of the circumference to the diameter of a circle
E	sympy.E	The base of the natural logarithm, $e = \exp(1)$
γ	sympy.EulerGamma	Euler's constant
I	sympy.I	The imaginary unit
∞	sympy.oo	Infinity

Functions

In SymPy, objects that represent functions can be created with sympy.Function. Like Symbol, the Function object takes a name as the first argument. SymPy distinguishes between defined and undefined functions and between applied and unapplied functions. Creating a function with Function results in an undefined (abstract) and unapplied function, which has a name but cannot be evaluated because its expression, or body, is not defined. Such a function can represent an arbitrary function of arbitrary numbers of input variables since it also has not yet been applied to any particular symbols. An unapplied function can be applied to a set of input symbols that represent the domain of the function by calling the function instance with those symbols as arguments.[1] The result is still an unevaluated function but one that has been applied to the specified input variables and, therefore, has a set of dependent variables. As an example of these concepts, consider the following code listing where we create an undefined function f, which we apply to the symbol x, and another g function, which we directly apply to the set of symbols x, y, z.

```
In [41]: x, y, z = sympy.symbols("x, y, z")
In [42]: f = sympy.Function("f")
In [43]: type(f)
Out[43]: sympy.core.function.UndefinedFunction
In [44]: f(x)
Out[44]: f(x)
In [45]: g = sympy.Function("g")(x, y, z)
In [46]: g
```

---

[1] Here it is important to keep in mind the distinction between a Python function, or callable Python object such as sympy.Function, and the symbolic function that a sympy.Function class instance represents.

```
Out[46]: g(x,y,z)
In [47]: g.free_symbols
Out[47]: {x,y,z}
```

This used the `free_symbols` property, which returns a set of unique symbols contained in an expression (in this case, the applied undefined g function), to demonstrate that an applied function is associated with a specific set of input symbols. This will be important later in this chapter, for example, when considering derivatives of abstract functions. One important application of undefined functions is for specifying differential equations or, in other words, when an equation for the function is known, but the function itself is unknown.

In contrast to undefined functions, a defined function has a specific implementation and can be numerically evaluated for all valid input parameters. It is possible to define this type of function, for example, by subclassing `sympy.Function`, but in most cases, it is sufficient to use the mathematical functions provided by SymPy. There are built-in functions for many standard mathematical functions in the global SymPy namespace. See the module documentation for `sympy.functions.elementary`, `sympy.functions.combinatorial`, and `sympy.functions.special` and their submodules for comprehensive lists of the numerous available functions, using the Python `help` function. For example, the SymPy function for the sine function is available as `sympy.sin` (with our import convention). Note that this is not a function in the Python sense of the word (it is, in fact, a subclass of `sympy.Function`), and it represents an unevaluated sine function that can be applied to a numerical value, a symbol, or an expression.

```
In [48]: sympy.sin
Out[48]: sympy.functions.elementary.trigonometric.sin
In [49]: sympy.sin(x)
Out[49]: sin(x)
In [50]: sympy.sin(pi * 1.5)
Out[50]: -1
```

When applied to an abstract symbol, such as x, the `sin` function remains unevaluated. But when possible, it is evaluated to a numerical value, for example, when applied to a number or, in some cases, when applied to expressions with certain properties, as in the following example.

```
In [51]: n = sympy.Symbol("n", integer=True)
In [52]: sympy.sin(pi * n)
Out[52]: 0
```

A third type of function in SymPy is lambda functions, or anonymous functions, which do not have names associated with them but have a specific function body that can be evaluated. Lambda functions can be created with `sympy.Lambda`.

```
In [53]: h = sympy.Lambda(x, x**2)
In [54]: h
Out[54]: (x ↦ x²)
In [55]: h(5)
Out[55]: 25
In [56]: h(1 + x)
Out[56]: (1 + x)²
```

Expressions

The symbols introduced in the previous sections are the fundamental building blocks required to express mathematical expressions. In SymPy, mathematical expressions are represented as trees, where leaves are symbols and nodes are class instances that represent mathematical operations. These classes include Add, Mul, and Pow for basic arithmetic operators and Sum, Product, Integral, and Derivative for analytical mathematical operations. In addition, there are many other classes for mathematical operations, which are demonstrated in examples later in this chapter.

Consider, for example, the mathematical expression $1+2x^2+3x^3$. To represent this in SymPy, we only need to create the symbol x and write the expression as Python code.

```
In [54]: x = sympy.Symbol("x")
In [55]: expr = 1 + 2 * x**2 + 3 * x**3
In [56]: expr
Out[56]: 3x³ + 2x² + 1
```

Here expr is an instance of Add, with the subexpressions 1, 2*x**2, and 3*x**3. The entire expression tree for expr is visualized in Figure 3-1. Note that we do not need to explicitly construct the expression tree, since it is automatically built up from the expression with symbols and operators. Nevertheless, to understand how SymPy works, it is important to know how expressions are represented.

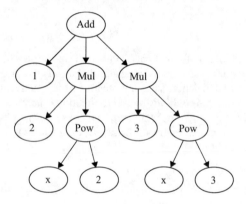

Figure 3-1. *Visualization of the expression tree for 1 + 2*x**2 + 3*x**3*

The expression tree can be traversed explicitly using the args attribute, which all SymPy operations and symbols provide. For an operator, the args attribute is a tuple of subexpressions combined with the rule implemented by the operator class. For symbols, the args attribute is an empty tuple, which signifies that it is a leaf in the expression tree. The following example demonstrates how the expression tree can be explicitly accessed.

```
In [57]: expr.args
Out[57]: (1,2x²,3x³)
In [58]: expr.args[1]
Out[58]: 2x²
In [59]: expr.args[1].args[1]
Out[59]: x²
```

```
In [60]: expr.args[1].args[1].args[0]
Out[60]: x
In [61]: expr.args[1].args[1].args[0].args
Out[61]: ()
```

In the basic use of SymPy, explicitly manipulating expression trees is rarely necessary. But when the methods for manipulating expressions that are introduced in the following section are not sufficient, it is useful to be able to implement functions of your own that traverse and manipulate the expression tree using the args attribute.

Manipulating Expressions

Manipulating expression trees is one of the main jobs for SymPy, and numerous functions are provided for different types of transformations. The general idea is that expression trees can be transformed between mathematically equivalent forms using simplification and rewrite functions. These functions generally do not change the expressions passed to the functions but rather create a new expression corresponding to the modified expression. Expressions in SymPy should thus be considered immutable objects (that cannot or should not be changed in place). All the functions in this section treat SymPy expressions as immutable objects and return new expression trees rather than modify expressions in place.

Simplification

The most desirable manipulation of a mathematical expression is to simplify it. This is perhaps also the most ambiguous operation since it is nontrivial to determine algorithmically if one expression appears simpler than another to a human being, and in general, it is not obvious which methods should be employed to arrive at a simpler expression. Nonetheless, black-box simplification is an important part of any CAS, and SymPy includes the sympy.simplify function, which attempts to simplify a given expression using a variety of methods and approaches. As illustrated in the following example, the simplification function can also be invoked through the simplify method.

```
In [67]: expr = 2 * (x**2 - x) - x * (x + 1)
In [68]: expr
Out[68]: 2x² - x(x+1)-2x
In [69]: sympy.simplify(expr)
Out[69]: x(x-3)
In [70]: expr.simplify()
Out[70]: x(x-3)
In [71]: expr
Out[71]: 2x² - x(x+1)-2x
```

Note that here, both sympy.simplify(expr) and expr.simplify() return new expression trees, and leave the expression expr untouched, as mentioned earlier. In this example, the expression expr can be simplified by expanding the products, canceling terms, and then factoring the expression again. In general, sympy.simplify attempts a variety of different strategies and also simplifies, for example, trigonometric and power expressions. This is exemplified in the following example.

```
In [72]: expr = 2 * sympy.cos(x) * sympy.sin(x)
In [73]: expr
Out[73]: 2 sin(x)cos(x)
```

```
In [74]: sympy.simplify(expr)
Out[74]: sin(2x)
```

The following is another example.

```
In [75]: expr = sympy.exp(x) * sympy.exp(y)
In [76]: expr
Out[76]: exp(x)exp(y)
In [77]: sympy.simplify(expr)
Out[77]: exp(x+y)
```

Each specific type of simplification can also be carried out with more specialized functions, such as `sympy.trigsimp` and `sympy.powsimp`, for trigonometric and power simplifications, respectively. These functions only perform the simplification that their names indicate and leave other parts of an expression in their original form. A summary of simplification functions is given in Table 3-3. When the exact simplification steps are known, it is generally better to rely on the more specific simplification functions since their actions are more well-defined and less likely to change in future versions of SymPy. The `sympy.simplify` function, on the other hand, relies on heuristic approaches that may change in the future and, as a consequence, produce different results for a particular input expression.

Table 3-3. *Summary of Selected SymPy Functions for Simplifying Expressions*

Function	Description
sympy.simplify	Attempt various methods and approaches to obtain a simpler form of a given expression.
sympy.trigsimp	Attempt to simplify an expression using trigonometric identities.
sympy.powsimp	Attempt to simplify an expression using laws of powers.
sympy.compsimp	Simplify combinatorial expressions.
sympy.ratsimp	Simplify an expression by writing on a common denominator.

Expand

When the black-box simplification provided by `sympy.simplify` does not produce satisfying results, it is often possible to make progress by manually guiding SymPy using more specific algebraic operations. An important tool in this process is to expand expression in various ways. The `sympy.expand` function performs a variety of expansions depending on the values of optional keyword arguments. By default, the function distributes products over additions into a fully expanded expression. For example, a product of the type $(x + 1)(x + 2)$ can be expanded to $x^2 + 3x + 2$ using the following.

```
In [78]: expr = (x + 1) * (x + 2)
In [79]: sympy.expand(expr)
Out[79]: x² + 3x + 2
```

Some of the available keyword arguments are `mul=True` for expanding products (as in the preceding example) and `trig=True` for trigonometric expansions.

```
In [80]: sympy.sin(x + y).expand(trig=True)
Out[80]: sin(x)cos(y) + sin(y)cos(x)
```

log=True is for expanding logarithms.

```
In [81]: a, b = sympy.symbols("a, b", positive=True)
In [82]: sympy.log(a * b).expand(log=True)
Out[82]: log(a) + log(b)
```

complex=True is for separating real and imaginary parts of an expression.

```
In [83]: sympy.exp(I*a + b).expand(complex=True)
Out[83]: ie^b sin(a) + e^b cos(a)
```

And power_base=True and power_exp=True for expanding the base and the exponent of a power expression, respectively.

```
In [84]: sympy.expand((a * b)**x, power_base=True)
Out[84]: a^x b^x
In [85]: sympy.exp((a-b)*x).expand(power_exp=True)
Out[85]: e^{ax} e^{-bx}
```

Calling the sympy.expand function with these keyword arguments set to True is equivalent to calling the more specific functions sympy.expand_mul, sympy.expand_trig, sympy.expand_log, sympy.expand_complex, sympy.expand_power_base, and sympy.expand_power_exp, respectively, but an advantage of the sympy.expand function is that several types of expansions can be performed in a single function call.

Factor, Collect, and Combine

A common use pattern for the sympy.expand function is to expand an expression, let SymPy cancel terms or factors, and then factor or combine the expression again. The sympy.factor function attempts to factor an expression as far as possible and is in some sense the opposite to sympy.expand with mul=True. It can be used to factor algebraic expressions, such as the following.

```
In [86]: sympy.factor(x**2 - 1)
Out[86]: (x - 1)(x + 1)
In [87]: sympy.factor(x * sympy.cos(y) + sympy.sin(z) * x)
Out[87]: x(sin(x) + cos(y))
```

The inverse of the other types of expansions in the previous section can be carried out using sympy.trigsimp, sympy.powsimp, and sympy.logcombine, as shown in the following example.

```
In [90]: sympy.logcombine(sympy.log(a) - sympy.log(b))
```
$$\text{Out}[90]: \log\left(\frac{a}{b}\right)$$

When working with mathematical expressions, fine-grained control over factoring is often necessary. The SymPy function sympy.collect factors terms that contain a given symbol or list of symbols. For example, $x + y + xyz$ cannot be completely factorized, but we can partially factor terms containing x or y.

```
In [89]: expr = x + y + x * y * z
In [90]: expr.collect(x)
Out[90]: x(yz + 1) + y
In [91]: expr.collect(y)
Out[91]: x + y(xz + 1)
```

By passing a list of symbols or expressions to the `sympy.collect` function, or to the corresponding `collect` method, we can collect multiple symbols in one function call. Also, when using the `collect` method, which returns the new expression, it is possible to chain multiple methods calls in the following way.

```
In [93]: expr = sympy.cos(x + y) + sympy.sin(x - y)
In [94]: expr.expand(trig=True).collect(
   ...:        [sympy.cos(x), sympy.sin(x)]
   ...: ).collect(
   ...:        sympy.cos(y) - sympy.sin(y)
   ...: )
Out[95]: (sin(x) + cos(x))(-sin(y) + cos(y))
```

Apart, Together, and Cancel

The final type of mathematical simplification to consider here is the rewriting of fractions. The functions `sympy.apart` and `sympy.together`, which rewrite a fraction as a partial fraction and combine partial fractions to a single fraction, can be used in the following way.

```
In [95]: sympy.apart(1/(x**2 + 3*x + 2), x)
```

$$\text{Out[95]:} \quad -\frac{1}{x+2} + \frac{1}{x+1}$$

```
In [96]: sympy.together(1 / (y * x + y) + 1 / (1+x))
```

$$\text{Out[96]:} \quad \frac{y+1}{y(x+1)}$$

```
In [97]: sympy.cancel(y / (y * x + y))
```

$$\text{Out[97]:} \quad \frac{1}{x+1}$$

The first example used `sympy.apart` to rewrite the expression $(x^2+3x+2)^{-1}$ as the partial fraction $-\frac{1}{x+2} + \frac{1}{x+1}$ and used `sympy.together` to combine the sum of fractions $1/(yx+y) + 1/(1+x)$ into an expression in the form of a single fraction. This example also used the `sympy.cancel` function to cancel shared factors between numerator and the denominator in the expression $y/(yx+y)$.

Substitutions

The previous sections have been concerned with rewriting expressions using various mathematical identities. Another frequently used form of manipulation of mathematical expressions is the substitution of symbols or subexpressions within an expression. For example, we may want to perform a variable substitution and replace the variable x with y or replace a symbol with another expression. In SymPy, there are two methods for carrying out substitutions: `subs` and `replace`. Usually, `subs` is the most suitable alternative, but in some cases `replace` provides a more powerful tool, which, for example, can make replacements based on wildcard expressions (see docstring for `sympy.Symbol.replace` for details).

In the most basic use of subs, the method is invoked from an expression, and the symbol or expression that is to be replaced (x) is given as the first argument, and the new symbol or the expression (y) is given as the second argument. The result is that all occurrences of x in the expression are replaced with y.

```
In [98]: (x + y).subs(x, y)
Out[98]: 2y
In [99]: sympy.sin(x * sympy.exp(x)).subs(x, y)
Out[99]: sin(ye^y)
```

Instead of chaining multiple subs calls when several substitutions are required, we can alternatively pass a dictionary that maps old symbols or expressions to new symbols or expressions as the first and only argument to subs.

```
In [100]: sympy.sin(x * z).subs({z: sympy.exp(y), x: y, sympy.sin: sympy.cos})
Out[100]: cos(ye^y)
```

A common application of the subs method is to substitute numerical values instead of symbols for numerical evaluation (see the following section for more details). A convenient way of doing this is to define a dictionary that translates the symbols to numerical values and pass this dictionary as the argument to the subs method. For example, consider the following.

```
In [101]: expr = x * y + z**2 *x
In [102]: values = {x: 1.25, y: 0.4, z: 3.2}
In [103]: expr.subs(values)
Out[103]: 13.3
```

Numerical Evaluation

Even when working with symbolic mathematics, it is almost invariably sooner or later required to evaluate the symbolic expressions numerically, for example, when producing plots or concrete numerical results. A SymPy expression can be evaluated using either the sympy.N function or the evalf method of SymPy expression instances.

```
In [104]: sympy.N(1 + pi)
Out[104]: 4.14159265358979
In [105]: sympy.N(pi, 50)
Out[105]: 3.1415926535897932384626433832795028841971693993751
In [106]: (x + 1/pi).evalf(10)
Out[106]: x + 0.3183098862
```

Both sympy.N and the evalf method take an optional argument that specifies the number of significant digits to which the expression is to be evaluated, as shown in the previous example where SymPy's arbitrary precision float capabilities were leveraged to evaluate the value of π up to 50 digits.

When we need to evaluate an expression numerically for a range of input values, we could loop over the values and perform successive evalf calls, as shown in the following example.

```
In [114]: expr = sympy.sin(pi * x * sympy.exp(x))
In [115]: [expr.subs(x, xx).evalf(3) for xx in range(0, 10)]
Out[115]: [0,0.774,0.642,0.722,0.944,0.205,0.974,0.977,-0.870,-0.695]
```

However, this method is rather slow, and SymPy provides a more efficient method for doing this operation using the `sympy.lambdify` function. This function takes a set of free symbols and an expression as arguments and generates a function that efficiently evaluates the numerical value of the expression. The produced function takes the same number of arguments as the number of free symbols passed as the first argument to `sympy.lambdify`.

```
In [109]: expr_func = sympy.lambdify(x, expr)
In [110]: expr_func(1.0)
Out[110]: 0.773942685266709
```

Note that the `expr_func` function expects numerical (scalar) values as arguments, so we cannot, for example, pass a symbol as an argument to this function; it is strictly for numerical evaluation. The `expr_func` created in the previous example is a scalar function and is not directly compatible with vectorized input in the form of NumPy arrays, as discussed in Chapter 2. However, SymPy is also able to generate functions that are NumPy-array aware: passing 'numpy' as the optional third argument to `sympy.lambdify` creates a vectorized function that accepts NumPy arrays as input. This is often an efficient way to numerically evaluate symbolic expressions[2] for a large number of input parameters. The following code exemplifies how the SymPy expression `expr` is converted into a NumPy-array aware vectorized function that can be efficiently evaluated:

```
In [111]: expr_func = sympy.lambdify(x, expr, 'numpy')
In [112]: import numpy as np
In [113]: xvalues = np.arange(0, 10)
In [114]: expr_func(xvalues)
Out[114]: array([ 0.        ,  0.77394269,  0.64198244,  0.72163867,  0.94361635,
                  0.20523391,  0.97398794,  0.97734066, -0.87034418, -0.69512687])
```

This method for generating data from SymPy expressions is useful for plotting and many other data-oriented applications.

Calculus

So far, we have looked at how to represent mathematical expressions in SymPy and how to perform basic simplification and transformation of such expressions. With this framework in place, we are now ready to explore symbolic calculus, or analysis, which is a cornerstone in applied mathematics and has many applications throughout science and engineering. The central concept in calculus is the change of functions as input variables are varied, as quantified with derivatives and differentials, and accumulations of functions over ranges of input, as quantified by integrals. This section looks at how to compute derivatives and integrals of functions in SymPy.

---

[2] See also the `ufuncity` from the `sympy.utilities.autowrap` module and the `aesara_function` from the `sympy.printing.aesaracode` module. These provide similar functionality as `sympy.lambdify`, but use different computational backends.

Derivatives

A function's derivative describes its change rate at a given point. In SymPy, we can calculate the derivative of a function using `sympy.diff` or the `diff` method of SymPy expression instances. The argument to these functions is a symbol, or several symbols, with respect to which the function or the expression is to be derived. To represent the first-order derivative of an abstract function $f(x)$ with respect to x, we can do the following.

```
In [119]: f = sympy.Function('f')(x)
In [120]: sympy.diff(f, x)              # equivalent to f.diff(x)
```
$$\text{Out[120]: } \frac{d}{dx}f(x)$$

To represent higher-order derivatives, we must repeat the symbol x in the argument list in the call to `sympy.diff` or specify an integer as an argument following a symbol, which defines the number of times the expression should be derived with respect to that symbol.

```
In [117]: sympy.diff(f, x, x)
```
$$\text{Out[117]: } \frac{d^2}{dx^2}f(x)$$

```
In [118]: sympy.diff(f, x, 3)    # equivalent to sympy.diff(f, x, x, x)
```
$$\text{Out[118]: } \frac{d^3}{dx^3}f(x)$$

This method is readily extended to multivariate functions.

```
In [119]: g = sympy.Function('g')(x, y)
In [120]: g.diff(x, y)            # equivalent to sympy.diff(g, x, y)
```
$$\text{Out[120]: } \frac{\partial^2}{\partial x \partial y}g(x,y)$$

```
In [121]: g.diff(x, 3, y, 2)     # equivalent to sympy.diff(g, x, x, x, y, y)
```
$$\text{Out[121]: } \frac{\partial^5}{\partial x^3 \partial y^2}g(x,y)$$

These examples so far only involve formal derivatives of undefined functions. Naturally, we can also evaluate the derivatives of defined functions and expressions, which result in new expressions that correspond to the evaluated derivatives. For example, `sympy.diff` lets us easily evaluate derivatives of arbitrary mathematical expressions, such as polynomials.

```
In [122]: expr = x**4 + x**3 + x**2 + x + 1
In [123]: expr.diff(x)
Out[123]: 4x³ + 3x² + 2x+1
In [124]: expr.diff(x, x)
Out[124]: 2(6x² + 3x + 1)
In [125]: expr = (x + 1)**3 * y ** 2 * (z - 1)
In [126]: expr.diff(x, y, z)
Out[126]: 6y(x + 1)²
```

We can also easily evaluate trigonometric and other more complicated mathematical expressions.

```
In [127]: expr = sympy.sin(x * y) * sympy.cos(x / 2)
In [128]: expr.diff(x)
```

Out[128]: $y\cos\left(\dfrac{x}{2}\right)\cos(xy) - \dfrac{1}{2}\sin\left(\dfrac{x}{2}\right)\sin(xy)$

```
In [129]: expr = sympy.functions.special.polynomials.hermite(x, 0)
In [130]: expr.diff(x).doit()
```

Out[130]: $\dfrac{2^x \sqrt{\pi}\ \text{polygamma}\left(0, -\dfrac{x}{2} + \dfrac{1}{2}\right)}{2\Gamma\left(-\dfrac{x}{2} + \dfrac{1}{2}\right)} + \dfrac{2^x \sqrt{\pi}\ \log(2)}{\Gamma\left(-\dfrac{x}{2} + \dfrac{1}{2}\right)}$

Derivatives are usually relatively easy to compute, and `sympy.diff` should be able to evaluate the derivative of most standard mathematical functions defined in SymPy.

Note that in these examples, calling `sympy.diff` on an expression directly results in a new expression. If we want to symbolically represent the derivative of a definite expression, we can create an instance of the `sympy.Derivative` class, passing the expression as the first argument, followed by the symbols with respect to which the derivative is to be computed.

```
In [131]: d = sympy.Derivative(sympy.exp(sympy.cos(x)), x)
In [132]: d
```

Out[132]: $\dfrac{d}{dx} e^{\cos(x)}$

This formal representation of a derivative can then be evaluated by calling the `doit` method on the `sympy.Derivative` instance.

```
In [133]: d.doit()
Out[133]: -e^{cos(x)} sin(x)
```

This pattern of delayed evaluation is reoccurring throughout SymPy, and full control of when a formal expression is evaluated to a specific result is useful in many situations, particularly with expressions that can be simplified or manipulated while represented as a formal expression rather than after it has been evaluated.

Integrals

SymPy evaluates integrals using the `sympy.integrate` function, and formal integrals can be represented using `sympy.Integral` (which, as in the case with `sympy.Derivative`, can be explicitly evaluated by calling the `doit` method). Integrals come in two basic forms: definite and indefinite, where a definite integral has specified integration limits and can be interpreted as an area or volume. In contrast, an indefinite integral does not have integration limits and denotes the antiderivative (inverse of the derivative of a function). SymPy handles both indefinite and definite integrals using the `sympy.integrate` function.

If the sympy.integrate function is called with only an expression as an argument, the indefinite integral is computed. On the other hand, a definite integral is computed if the sympy.integrate function additionally is passed a tuple in the form (x, a, b), where x is the integration variable and a and b are the integration limits. For a single-variable $f(x)$ function, the indefinite and definite integrals are computed using the following.

```
In [135]: a, b, x, y = sympy.symbols("a, b, x, y")
     ...: f = sympy.Function("f")(x)
In [136]: sympy.integrate(f)
Out[136]: ∫f(x)dx
In [137]: sympy.integrate(f, (x, a, b))
```

$$\text{Out[137]:} \quad \int_{a}^{b} f(x)dx$$

When these methods are applied to explicit functions, the integrals are evaluated accordingly.

```
In [138]: sympy.integrate(sympy.sin(x))
Out[138]: -cos(x)
In [139]: sympy.integrate(sympy.sin(x), (x, a, b))
Out[139]: cos(a) - cos(b)
```

Definite integrals can also include limits that extend from negative infinity or to positive infinite, using SymPy's symbol for infinity oo.

```
In [139]: sympy.integrate(sympy.exp(-x**2), (x, 0, oo))
```

$$\text{Out[139]:} \quad \frac{\sqrt{\pi}}{2}$$

```
In [140]: a, b, c = sympy.symbols("a, b, c", positive=True)
In [141]: sympy.integrate(a * sympy.exp(-((x-b)/c)**2), (x, -oo, oo))
```

$$\text{Out[141]:} \quad \sqrt{\pi}ac$$

Computing integrals symbolically is generally a difficult problem, and SymPy cannot give symbolic results for any integral we can come up with. When SymPy fails to evaluate an integral, an instance of sympy.Integral, representing the formal integral, is returned instead.

```
In [142]: sympy.integrate(sympy.sin(x * sympy.cos(x)))
Out[142]: ∫sin(x cos(x))dx
```

Multivariable expressions can also be integrated with sympy.integrate. In an indefinite integral of a multivariable expression, the integration variable must be specified explicitly.

```
In [140]: expr = sympy.sin(x*sympy.exp(y))
In [141]: sympy.integrate(expr, x)
Out[141]: -e-ycos(xe^y)
In [142]: expr = (x + y)**2
In [143]: sympy.integrate(expr, x)
```

$$\text{Out[143]:} \quad \frac{x^3}{3} + x^2 y + xy^2$$

We can carry out multiple integrations by passing more than one symbol, or multiple tuples that contain symbols and their integration limits.

```
In [144]: sympy.integrate(expr, x, y)
Out[144]:
In [145]: sympy.integrate(expr, (x, 0, 1), (y, 0, 1))
```

Out[145]: $\dfrac{7}{6}$

Series

Series expansions are an important tool in many disciplines in computing. With a series expansion, an arbitrary function can be written as a polynomial, with coefficients given by the function's derivatives at the point around which the series expansion is made. The nth-order approximation of the function is obtained by truncating the series expansion at some order n. In SymPy, the series expansion of a function or an expression can be computed using the function `sympy.series` or the `series` method available in SymPy expression instances. The first argument to `sympy.series` is a function or expression to be expanded, followed by a symbol with respect to which the expansion is to be computed (it can be omitted for single-variable expressions and functions). In addition, it is also possible to request a particular point around which the series expansions are to be performed (using the x0 keyword argument, with default x0=0), specifying the order of the expansion (using the n keyword argument, with default n=6) and specifying the direction from which the series is computed (i.e., from below or above x0 using the `dir` keyword argument, which defaults to dir='+').

For an undefined $f(x)$ function, the expansion up to the sixth order around x0=0 is computed as follows.

```
In [147]: x, y = sympy.symbols("x, y")
In [148]: f = sympy.Function("f")(x)
In [149]: sympy.series(f, x)
```

Out[149]:
$$f(0) + x \frac{d}{dx} f(x) \Big|_{x=0} + \frac{x^2}{2} \frac{d^2}{dx^2} f(x) \Big|_{x=0} + \frac{x^3}{6} \frac{d^3}{dx^3} f(x) \Big|_{x=0}$$
$$+ \frac{x^4}{24} \frac{d^4}{dx^4} f(x) \Big|_{x=0} + \frac{x^5}{120} \frac{d^5}{dx^5} f(x) \Big|_{x=0} + \mathcal{O}\left(x^6\right)$$

To change the point around which the function is expanded, we specify the x0 argument as in the following example.

```
In [147]: x0 = sympy.Symbol("{x_0}")
In [151]: f.series(x, x0, n=2)
```

Out[151]: $f(x_0) + (x - x_0) \dfrac{d}{d\xi_1} f(\xi_1) \Big|_{\xi_1 = x_0} + \mathcal{O}\left((x - x_0)^2 ; x \to x_0\right)$

This also specified n=2, to request a series expansion with only terms up to and including the second order. Note that the order object represents the errors due to the truncated terms $\mathcal{O}(\ldots)$. The order object is useful for keeping track of the order of an expression when computing with series expansions, such as multiplying or adding different expansions. However, for concrete numerical evolution, removing the order term from the expression is necessary, which can be done using the remove0 method.

In [152]: f.series(x, x0, n=2).removeO()

Out[152]: $f(x_0)+(x-x_0)\dfrac{d}{d\xi_1}f(\xi_1)\Big|_{\xi_1=x_0}$

While the expansions shown in the preceding text were computed for an unspecified $f(x)$ function, we can naturally also compute the series expansions of specific functions and expressions. In those cases, we obtain specific evaluated results. For example, we can easily generate the well-known expansions of many standard mathematical functions.

In [153]: sympy.cos(x).series()

Out[153]: $1-\dfrac{x^2}{2}+\dfrac{x^4}{24}+\mathcal{O}(x^6)$

In [154]: sympy.sin(x).series()

Out[154]: $x-\dfrac{x^3}{6}+\dfrac{x^5}{120}+\mathcal{O}(x^6)$

In [155]: sympy.exp(x).series()

Out[155]: $1+x+\dfrac{x^2}{2}+\dfrac{x^3}{6}+\dfrac{x^4}{24}+\dfrac{x^5}{120}+\mathcal{O}(x^6)$

In [156]: (1/(1+x)).series()

Out[156]: $1-x+x^2-x^3+x^4-x^5+\mathcal{O}(x^6)$

Arbitrary expressions of symbols and functions can generally be multivariable functions.

In [157]: expr = sympy.cos(x) / (1 + sympy.sin(x * y))
In [158]: expr.series(x, n=4)

Out[158]: $1-xy+x^2\left(y^2-\dfrac{1}{2}\right)+x^3\left(-\dfrac{5y^3}{6}+\dfrac{y}{2}\right)+\mathcal{O}(x^4)$

In [159]: expr.series(y, n=4)

Out[159]: $\cos(x)-xy\cos(x)+x^2y^2\cos(x)-\dfrac{5x^3y^3\cos(x)}{6}+\mathcal{O}(y^4)$

Limits

Another important tool in calculus is limits, which denotes the value of a function as one of its dependent variables approaches a specific value or as the variable's value approaches negative or positive infinity. An example of a limit is one of the definitions of the derivative.

$$\dfrac{d}{dx}f(x)=\lim_{h\to0}\dfrac{f(x+h)-f(x)}{h}$$

While limits are more of a theoretical tool and do not have as many practical applications as series expansions, it is still useful to compute limits using SymPy. In SymPy, limits can be evaluated using the sympy.limit function, which takes an expression, a symbol it depends on, and the value that the symbol approaches in the limit. For example, to compute the limit of the $\sin(x)/x$ function, as the variable x goes to zero, that is, $\lim_{x \to 0} \sin(x)/x$, we can use the following.

```
In [161]: sympy.limit(sympy.sin(x) / x, x, 0)
Out[161]: 1
```

Here we obtained the well-known answer 1 for this limit. We can also use `sympy.limit` to compute symbolic limits, which can be illustrated by computing derivatives using the previous definition (although it is more efficient to use `sympy.diff`).

```
In [162]: f = sympy.Function('f')
     ...: x, h = sympy.symbols("x, h")
In [163]: diff_limit = (f(x + h) - f(x))/h
In [164]: sympy.limit(diff_limit.subs(f, sympy.cos), h, 0)
Out[164]: -sin(x)
In [165]: sympy.limit(diff_limit.subs(f, sympy.sin), h, 0)
Out[165]: cos(x)
```

A more practical example of using limits is to find the asymptotic behavior as a function, for example, as its dependent variable approaches infinity. As an example, consider the $f(x) = (x^2 - 3x)/(2x - 2)$ function, and suppose we are interested in the large-x dependence of this function. It is in the form $f(x) \to px + q$, and we can compute p and q using `sympy.limit` as in the following.

```
In [166]: expr = (x**2 - 3*x) / (2*x - 2)
In [167]: p = sympy.limit(expr/x, x, sympy.oo)
In [168]: q = sympy.limit(expr - p*x, x, sympy.oo)
In [169]: p, q
```

$$\text{Out[169]: } \left(\frac{1}{2}, -1 \right)$$

Thus, the asymptotic behavior of $f(x)$ as x becomes large is the linear function $f(x) \to x/2 - 1$.

Sums and Products

Sums and products can be symbolically represented using the SymPy classes `sympy.Sum` and `sympy.Product`. They both take an expression as their first argument and as a second argument, they take a tuple of the form (n, n1, n2), where n is a symbol and n1 and n2 are the lower and upper limits for the symbol n, in the sum or product, respectively. After `sympy.Sum` or `sympy.Product` objects have been created, they can be evaluated using the `doit` method.

```
In [171]: n = sympy.symbols("n", integer=True)
In [172]: x = sympy.Sum(1/(n**2), (n, 1, oo))
In [173]: x
```

$$\text{Out[173]: } \sum_{n=1}^{\infty} \frac{1}{n^2}$$

```
In [174]: x.doit()
```

$$\text{Out[174]: } \frac{\pi^2}{6}$$

```
In [175]: x = sympy.Product(n, (n, 1, 7))
In [176]: x
```

$$\text{Out[176]: } \prod_{n=1}^{7} n$$

```
In [177]: x.doit()
Out[177]: 5040
```

Note that the sum in the previous example was specified with an upper limit of infinity. Therefore, this sum was not evaluated by explicit summation but was computed analytically. SymPy can evaluate many summations of this type, including when the summand contains symbolic variables other than the summation index, such as in the following example.

```
In [178]: x = sympy.Symbol("x")
In [179]: sympy.Sum((x)**n/(sympy.factorial(n)), (n, 1, oo)).doit().simplify()
Out[179]: eˣ - 1
```

Equations

Equation solving is a fundamental part of mathematics with applications in nearly every branch of science and technology, and it is immensely important. SymPy can solve a wide variety of equations symbolically, although many equations cannot be solved analytically, even in principle. If an equation, or a system of equations, can be solved analytically, there is a good chance that SymPy can find the solution. If not, numerical methods might be the only option.

In its simplest form, equation solving involves a single equation with a single unknown variable and no additional parameters: for example, finding the value of x that satisfies the second-degree polynomial equation $x^2 + 2x - 3 = 0$. This equation is easy to solve, but in SymPy, we can use the sympy.solve function to find the solutions of x that satisfy this equation using the following.

```
In [170]: x = sympy.Symbol("x")
In [171]: sympy.solve(x**2 + 2*x - 3)
Out[171]: [-3,1]
```

That is, the solutions are x=-3 and x=1. The argument to the sympy.solve function is an expression that is solved under the assumption that it equals zero. When this expression contains more than one symbol, the variable to be solved must be given as a second argument. The following is an example.

```
In [172]: a, b, c = sympy.symbols("a, b, c")
In [173]: sympy.solve(a * x**2 + b * x + c, x)
```

$$\text{Out[173]: } \left[\frac{1}{2a}\left(-b+\sqrt{-4ac+b^2}\right), -\frac{1}{2a}\left(b+\sqrt{-4ac+b^2}\right) \right]$$

In this case, the resulting solutions are expressions that depend on the symbols representing the parameters in the equation.

The sympy.solve function can also solve other types of equations, including trigonometric expressions.

```
In [174]: sympy.solve(sympy.sin(x) - sympy.cos(x), x)
```

$$\text{Out[174]: } \left[-\frac{3\pi}{4}, \right.$$

The sympy.solve function can also solve equations whose solution can be expressed in terms of special functions.

```
In [180]: sympy.solve(sympy.exp(x) + 2 * x, x)
```

$$\text{Out[180]: } \left[-\text{LambertW}\left(\frac{1}{2}\right) \right]$$

However, when dealing with general equations, even for a univariate case, it is not uncommon to encounter equations that are not solvable algebraically or that SymPy cannot solve. In these cases, SymPy returns a formal solution, which can be numerically evaluated if needed or raise an error if no method is available for that particular type of equation.

```
In [176]: sympy.solve(x**5 - x**2 + 1, x)
Out[176]: [RootOf(x⁵ - x² + 1,0), RootOf(x⁵ - x² + 1,1), RootOf(x⁵ - x² + 1,2), RootOf(x⁵ - x²
+ 1,3), RootOf(x⁵ - x² + 1,4)]
In [177]: sympy.solve(sympy.tan(x) + x, x)
---------------------------------------------------------------------------
NotImplementedError                        Traceback (most recent call last)
...
NotImplementedError: multiple generators [x, tan(x)] No algorithms are implemented to solve
equation x + tan(x)
```

Solving a system of equations for more than one unknown variable in SymPy is a straightforward generalization of the procedure used for univariate equations. Instead of passing a single expression as the first argument to sympy.solve, a list of expressions representing the system of equations is used, and in this case, the second argument should be a list of symbols to solve for. For example, the following two examples demonstrate how to solve two systems that are linear and nonlinear equations in x and y, respectively.

```
In [178]: eq1 = x + 2 * y - 1
     ...: eq2 = x - y + 1
In [179]: sympy.solve([eq1, eq2], [x, y], dict=True)
```

$$\text{Out[179]: } \left[\left\{ x:-\frac{1}{3}, y:\frac{2}{3} \right\} \right]$$

```
In [180]: eq1 = x**2 - y
     ...: eq2 = y**2 - x
In [181]: sols = sympy.solve([eq1, eq2], [x, y], dict=True)
In [182]: sols
```

$$\text{Out[182]: } \left[\{x:0, y:0\}, \{x:1, y:1\}, \left\{ x:-\frac{1}{2}+\frac{\sqrt{3}i}{2}, y:-\frac{1}{2}-\frac{\sqrt{3}i}{2} \right\}, \left\{ x:\frac{\left(1-\sqrt{3}i\right)^2}{4}, y:-\frac{1}{2}+\frac{\sqrt{3}i}{2} \right\} \right]$$

Note that in both these examples, the sympy.solve function returns a list where each element represents a solution to the equation system. The optional keyword argument dict=True was also used to request that each solution be returned in dictionary format, which maps the symbols that have been solved to their values. This dictionary can conveniently be used in, for example, calls to subs, as in the following code that verifies that each solution indeed satisfies the two equations.

```
In [183]: [eq1.subs(sol).simplify() == 0 and eq2.subs(sol).simplify() == 0 for sol in sols]
Out[183]: [True, True, True, True]
```

Linear Algebra

Linear algebra is another fundamental branch of mathematics with important applications throughout scientific and technical computing. It concerns vectors, vector spaces, and linear mappings between vector spaces, which can be represented as matrices. In SymPy, we can represent vectors and matrices symbolically using the `sympy.Matrix` class, whose elements can be represented by numbers, symbols, or arbitrary symbolic expressions. To create a matrix with numerical entries—as in the case of NumPy arrays in Chapter 2, pass a Python list to `sympy.Matrix`.

```
In [184]: sympy.Matrix([1, 2])
```
$$\text{Out[184]:} \begin{bmatrix} 1 \\ 2 \end{bmatrix}$$

```
In [185]: sympy.Matrix([[1, 2]])
```
$$\text{Out[185]:} \begin{bmatrix} 1 & 2 \end{bmatrix}$$

```
In [186]: sympy.Matrix([[1, 2], [3, 4]])
```
$$\text{Out[186]:} \begin{bmatrix} 1 & 2 \\ 3 & 4 \end{bmatrix}$$

As this example demonstrates, a single list generates a column vector, while a matrix requires a nested list of values. Note that unlike the multidimensional arrays in NumPy discussed in Chapter 2, the `sympy.Matrix` object in SymPy can only have one or two dimensions: vectors and matrices.

Another way of creating new `sympy.Matrix` objects is to pass as arguments the number of rows, the number of columns, and a function that takes the row and column index as arguments and returns the value of the corresponding element.

```
In [187]: sympy.Matrix(3, 4, lambda m, n: 10 * m + n)
```
$$\text{Out[187]:} \begin{bmatrix} 0 & 1 & 2 & 3 \\ 10 & 11 & 12 & 13 \\ 20 & 21 & 22 & 23 \end{bmatrix}$$

The most powerful features of SymPy's matrix objects, which distinguish it from, for example, NumPy arrays, are that the elements can be symbolic expressions. For example, an arbitrary 2×2 matrix can be represented with a symbolic variable for each element.

```
In [188]: a, b, c, d = sympy.symbols("a, b, c, d")
In [189]: M = sympy.Matrix([[a, b], [c, d]])
In [190]: M
```
$$\text{Out[190]:} \begin{bmatrix} a & b \\ c & d \end{bmatrix}$$

Such matrices can also be used in computations, which then remain parameterized with the symbolic values of the elements. The usual arithmetic operators are implemented for matrix objects, but note that multiplication operator * denotes matrix multiplication.

```
In [191]: M * M
```

$$\text{Out[191]:} \quad \begin{bmatrix} a^2 + bc & ab + bd \\ ac + cd & bc + d^2 \end{bmatrix}$$

```
In [192]: x = sympy.Matrix(sympy.symbols("x_1, x_2"))
In [194]: M * x
```

$$\text{Out[194]:} \quad \begin{bmatrix} ax_1 + bx_2 \\ cx_1 + dx_2 \end{bmatrix}$$

In addition to arithmetic operations, many standard linear algebra operations on vectors and matrices are implemented as SymPy functions and methods of the `sympy.Matrix` class. Table 3-4 summarizes the frequently used linear algebra-related functions (see the docstring for `sympy.Matrix` for a complete list). SymPy matrices can also be used in an element-oriented fashion with indexing and slicing operations that closely resemble those discussed for NumPy arrays in Chapter 2.

Table 3-4. *Selected Functions and Methods for Operating on SymPy Matrices*

Function/Method	Description
transpose/T	Compute the transpose of a matrix.
adjoint/H	Compute the adjoint of a matrix.
Trace	Compute the trace (sum of diagonal elements) of a matrix.
Det	Compute the determinant of a matrix.
Inv	Compute the inverse of a matrix.
LUdecomposition	Compute the LU decomposition of a matrix.
LUsolve	Solve a linear system of equations in the form $Mx = b$, for the unknown vector x, using LU factorization.
QRdecomposition	Compute the QR decomposition of a matrix.
QRsolve	Solve a linear system of equations in the form $Mx = b$, for the unknown vector x, using QR factorization.
diagonalize	Diagonalize matrix M, such that it can be written in the form $D = P^{-1}MP$, where D is diagonal.
Norm	Compute the norm of a matrix.
nullspace	Compute a set of vectors that span the null space of a Matrix.
Rank	Compute the rank of a matrix.
singular_values	Compute the singular values of a matrix.
Solve	Solve a linear system of equations in the form $Mx = b$.

Consider the following parameterized linear equation system as an example of a problem that can be solved with symbolic linear algebra using SymPy, but which is not directly solvable with purely numerical approaches:

$$x + p\ y = b_1,$$

$$q\ x + y = b_2,$$

which we would like to solve for the unknown variables x and y. Here p, q, b_1, and b_2 are unspecified parameters. In matrix form, we can write these two equations as follows.

$$\begin{pmatrix} 1 & p \\ q & 1 \end{pmatrix} \begin{pmatrix} x \\ y \end{pmatrix} = \begin{pmatrix} b_1 \\ b_2 \end{pmatrix}.$$

With purely numerical methods, we would have to choose values of the parameters p and q before we could begin to solve this problem, for example, using an LU factorization (or by computing the inverse) of the matrix on the left-hand side of the equation. With a symbolic computing approach, on the other hand, we can directly proceed with computing the solution as if we carried out the calculation analytically by hand. With SymPy, we can simply define symbols for the unknown variables and parameters and set up the required matrix objects.

```
In [195]: p, q = sympy.symbols("p, q")
In [196]: M = sympy.Matrix([[1, p], [q, 1]])
In [203]: M
```
$$\text{Out[203]:} \quad \begin{bmatrix} 1 & p \\ q & 1 \end{bmatrix}$$

```
In [197]: b = sympy.Matrix(sympy.symbols("b_1, b_2"))
In [198]: b
Out[198]: [b_1 b_2]
```

We can then use, for example, the LUsolve method to solve the linear equation system.

```
In [199]: x = M.LUsolve(b)
In [200]: x
```
$$\text{Out[200]:} \quad \begin{bmatrix} b_1 - \dfrac{p(-b_1 q + b_2)}{-pq + 1} \\[2ex] \dfrac{-b_1 q + b_2}{-pq + 1} \end{bmatrix}$$

Alternatively, we could also directly compute the inverse of the matrix M and multiply it with the vector b.

```
In [201]: x = M.inv() * b
In [202]: x
```
$$\text{Out[202]:} \quad \begin{bmatrix} b_1 \left(\dfrac{pq}{-pq+1} + 1 \right) - \dfrac{b_2 p}{-pq+1} \\[2ex] -\dfrac{b_1 q}{-pq+1} + \dfrac{b_2}{-pq+1} \end{bmatrix}$$

However, computing the inverse of a matrix is more difficult than performing the LU factorization, so if solving the equation $Mx = b$ is the objective, as it was here, then using LU factorization is more efficient. This becomes particularly noticeable for larger equation systems. With both methods considered here, we obtain a symbolic expression for the solution that can be evaluated for any parameter values without recomputing the solution. This is the strength of symbolic computing and an example of how it sometimes can excel over direct numerical computing. The example considered here could also be solved easily by hand. However, as the number of equations and unspecified parameters grows, analytical treatment by hand quickly becomes prohibitively lengthy and tedious. With the help of a computer algebra system such as SymPy, we can push the limits of which problems can be treated analytically.

Summary

This chapter introduced computer-assisted symbolic computing using Python and the SymPy library. Although analytical and numerical techniques are often considered separately, it is a fact that analytical methods underpin everything in computing and are essential in developing algorithms and numerical methods. Whether analytical mathematics is carried out by hand or using a computer algebra system such as SymPy, it is an essential tool for computational work.

I would like to encourage the following approach. Analytical and numerical methods are closely intertwined, and it is often worthwhile to start analyzing a computational problem with analytical and symbolic methods. When such methods are unfeasible, it is time to use numerical methods. Furthermore, by directly applying numerical methods to a problem before analyzing it analytically, one likely ends up solving a more difficult computational problem than is necessary.

Further Reading

Instant SymPy Starter by Ronan Lamy (Packt, 2013) is a short introduction to SymPy. The official SymPy documentation also provides a great tutorial for starting SymPy; it is available at `http://docs.sympy.org/latest/tutorial/index.html`.

CHAPTER 4

■ ■ ■

Plotting and Visualization

Visualization is a universal tool for investigating problems and communicating the results of computational studies. It is hardly an exaggeration to say that the end product of nearly all computations—numeric or symbolic—is a plot or a graph. When visualized in a graphical form, knowledge and insights can be easily gained from computational results. Visualization is a tremendously important part of the workflow in all fields of computational studies.

There are many high-quality visualization libraries in the scientific computing environment for Python. The most popular general-purpose visualization library is Matplotlib, which mainly focuses on generating static publication-quality 2D and 3D graphs. Many other libraries focus on niche areas of visualization. A few prominent examples are Bokeh (http://bokeh.pydata.org) and Plotly (http://plot.ly), which both primarily focus on interactivity and web connectivity, Seaborn (http://seaborn.pydata.org) which is a high-level plotting library that targets statistical data analysis and which is based on the Matplotlib library, and the Mayavi library (http://docs.enthought.com/mayavi/mayavi) for high-quality 3D visualization, which uses the venerable VTK software (http://www.vtk.org) for heavy-duty scientific visualization. It is also worth noting that other VTK-based visualization software, such as ParaView (http://www.paraview.org), is scriptable with Python and can also be controlled from Python applications. In the 3D visualization space, there are also more recent players, such as VisPy (http://vispy.org), an OpenGL-based 2D and 3D visualization library with great interactivity and connectivity with browser-based environments, such as the Jupyter Notebook.

The visualization landscape in the scientific computing environment for Python is vibrant and diverse, providing ample options for various visualization needs. This chapter explores traditional scientific visualization in Python using the Matplotlib library, including plots and figures commonly used to visualize results and data in scientific and technical disciplines, such as line plots, bar plots, contour plots, colormap plots, and 3D surface plots.

■ **Matplotlib** Matplotlib is a Python library for publication-quality 2D and 3D graphics, supporting various output formats. At the time of writing, the latest version is 3.7.2. More information about Matplotlib is available at www.matplotlib.org. This website contains detailed documentation and an extensive gallery showcasing the various graphs that can be generated using the Matplotlib library, together with the code for each example. This gallery is a great source of inspiration for visualization ideas, and I highly recommend exploring Matplotlib by browsing this gallery.

There are two common approaches to creating scientific visualizations: using a graphical user interface to manually build up graphs and using a programmatic approach where the graphs are created with code. Both approaches have their advantages and disadvantages. This chapter takes the programmatic approach and explores how to use the Matplotlib API to create graphs and control every aspect of their appearance. The programmatic approach is a particularly suitable method for creating graphics for scientific

R. Johansson, *Numerical Python*, https://doi.org/10.1007/979-8-8688-0413-7_4

and technical applications, particularly for creating publication-quality figures. An important part of the motivation for this is that programmatically created graphics can guarantee consistency across multiple figures, make reproducible, and easily be revised and adjusted without having to redo potentially lengthy and tedious procedures in a graphical user interface.

Importing Modules

Unlike most Python libraries, Matplotlib provides multiple entry points into the library with different application programming interfaces (APIs). Specifically, it provides a stateful API and an object-oriented API, both provided by the matplotlib.pyplot module. I strongly recommend only using the object-oriented approach, and the remainder of this chapter solely focuses on this part of Matplotlib.[1]

To use the object-oriented Matplotlib API, we first need to import its Python modules. The following assumes that Matplotlib is imported using the following standard convention.

```
In [1]: %matplotlib inline
In [2]: import matplotlib as mpl
In [3]: import matplotlib.pyplot as plt
In [4]: from mpl_toolkits.mplot3d.axes3d import Axes3D
```

The first line assumes that we are working in an IPython environment, specifically in the Jupyter Notebook or the IPython QtConsole. The IPython magic command %matplotlib inline configures the Matplotlib to use the "inline" backend, which results in the created figures being displayed directly in, for example, the Jupyter Notebook rather than in a new window. The statement import matplotlib as mpl imports the main Matplotlib module, and the import statement import matplotlib.pyplot as plt, is for convenient access to the submodule matplotlib.pyplot that provides the functions used to create new Figure instances.

This chapter makes frequent use of the NumPy library and, as in Chapter 2, assumes that NumPy is imported using the following.

```
In [5]: import numpy as np
```

The SymPy library is also used, imported as follows.

```
In [6]: import sympy
```

Getting Started

Before delving deeper into creating graphics with Matplotlib, let's begin with an example of creating a simple but typical graph. Then, let's cover some of the fundamental principles of the Matplotlib library, to build up an understanding of how graphics can be produced with the library.

A graph in Matplotlib is structured in terms of a Figure instance and one or more Axes instances within the figure. The Figure instance provides a canvas area for drawing, and the Axes instances provide coordinate systems assigned to fixed regions of the total figure canvas; see Figure 4-1.

---

[1] Although the stateful API may be convenient and simple for small examples, the readability and maintainability of code written for stateful APIs scale poorly, and the context-dependent nature of such code makes it hard to rearrange or reuse. I therefore recommend to avoid it altogether and to only use the object-oriented API.

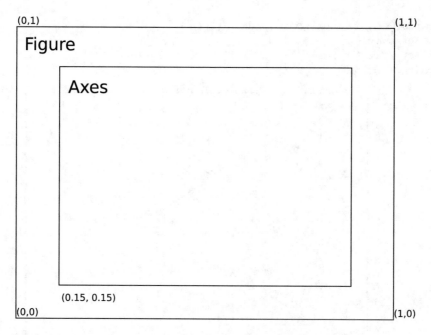

Figure 4-1. *Illustration of the arrangement of a Matplotlib* Figure *instance and an* Axes *instance. The* Axes *instance provides a coordinate system for plotting, and it is assigned to a region within the figure canvas. The figure canvas has a simple coordinate system where (0, 0) is the lower-left corner and (1,1) is the upper-right corner. This coordinate system is only used when placing elements, such as* Axes, *directly on the figure canvas*

A Figure can contain multiple Axes instances, for example, to show multiple panels in a figure or to show insets within another Axes instance. An Axes instance can manually be assigned to an arbitrary region of a figure canvas, or Axes instances can be automatically added to a figure canvas using one of several layout managers provided by Matplotlib. The Axes instance provides a coordinate system that can be used to plot data in various plot styles, including line graphs, scatter plots, bar plots, and many others. In addition, the Axes instance also determines how the coordinate axes are displayed, for example, with respect to the axis labels, ticks and tick labels, and so on. In fact, when working with Matplotlib's object-oriented API, most functions needed to tune a graph's appearance are methods of the Axes class.

As a simple example for getting started with Matplotlib, let's say that we would like to graph the $y(x) = x^3+5x^2+10$ function, together with its first- and second-order derivatives, over the range $x \in [-5, 2]$. To do this, we first create NumPy arrays for the x range and then compute the three functions we want to graph. When the data for the graph is prepared, we need to create Matplotlib Figure and Axes instances and then use the plot method of the Axes instance to plot the data. We can set basic graph properties such as x and y-axis labels using the set_xlabel and set_ylabel methods and generate a legend using the legend method. These steps are carried out in the following code, and the resulting graph is shown in Figure 4-2.

```
In [7]: x = np.linspace(-5, 2, 100)
   ...: y1 = x**3 + 5*x**2 + 10
   ...: y2 = 3*x**2 + 10*x
   ...: y3 = 6*x + 10
   ...:
   ...: fig, ax = plt.subplots()
   ...: ax.plot(x, y1, color="blue", label="y(x)")
   ...: ax.plot(x, y2, color="red", label="y'(x)")
```

```
...: ax.plot(x, y3, color="green", label="y"(x)")
...: ax.set_xlabel("x")
...: ax.set_ylabel("y")
...: ax.legend()
```

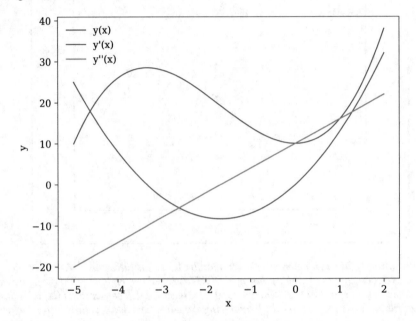

Figure 4-2. *Example of a simple graph created with Matplotlib*

The `plt.subplots` function generated the `Figure` and `Axes` instances. This function can be used to create grids of Axes instances within a newly created `Figure` instance, but here, it was merely used as a convenient way of creating a `Figure` and an `Axes` instance in one function call. Once the Axes instance is available, note that all the remaining steps involve the calling methods of this Axes instance. To graph the data, we use `ax.plot`, which takes as first and second arguments NumPy arrays with numerical data for the *x* and *y* values of the graph, and it draws a line connecting these data points. We also used the optional `color` and `label` keyword arguments to specify the color and assign a text label to each line used in the legend. These few lines of code are enough to generate the graph we set out to produce, but as a bare minimum, we should also set labels on the *x* and *y* axes and, if suitable, add a legend for the curves we have plotted. The axis labels are set with `ax.set_xlabel` and `ax.set_ylabel` methods, which take a text string with the corresponding label as an argument. The legend is added using the `ax.legend` method, which does not require any arguments since the `label` keyword argument was used when plotting the curves.

These are the typical steps required to create a graph using Matplotlib. While the graph in Figure 4-2 is complete and fully functional, there is certainly room for improvements in many aspects of its appearance. For example, to meet publication or production standards, we may need to change the font and the font size of the axis labels, the tick labels, and the legend, and we should probably move the legend to a part of the graph where it does not interfere with the curves we are plotting. We might even want to change the number of axis ticks and labels and add annotations and helplines to emphasize certain aspects of the graph. With a few changes along these lines, the figure may, for example, appear like in Figure 4-3, which is considerably more presentable. The remainder of this chapter examines how to fully control the appearance of the graphics produced using Matplotlib.

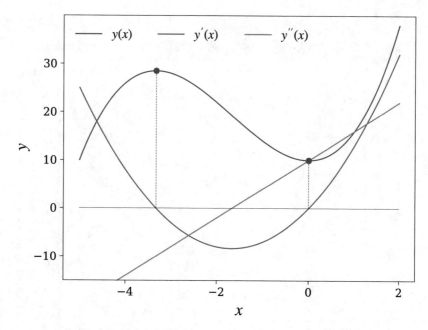

Figure 4-3. *Revised version of Figure 4-2*

Interactive and Noninteractive Modes

The Matplotlib library is designed to work well with many different environments and platforms. As such, the library not only contains routines for generating graphs but also supports displaying graphs in different graphical environments. To this end, Matplotlib provides *backends* for generating graphics in different formats (e.g., PNG, PDF, Postscript, and SVG) and for displaying graphics in a graphical user interface using a variety of different widget toolkits (e.g., Qt, GTK, wxWidgets, and Cocoa for macOS) that are suitable for different platforms.

The backend can be selected in that Matplotlib resource file[2] or using the mpl.use function, which must be called right after importing matplotlib, before importing the matplotlib.pyplot module. For example, to select the Qt5Agg backend, we can use the following.

```
import matplotlib as mpl
mpl.use('Qt5Agg')
import matplotlib.pyplot as plt
```

---

[2] The Matplotlib resource file, matplotlibrc, can be used to set default values of many Matplotlib parameters, including which backend to use. The location of the file is platform dependent. For details, see http://matplotlib.org/stable/tutorials/introductory/customizing.html.

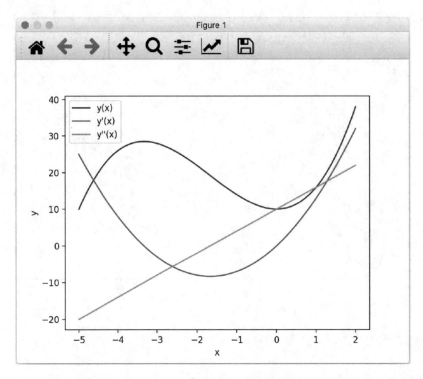

Figure 4-4. *A screenshot of the Matplotlib graphical user interface for displaying figures, using the Qt5 backend on macOS. The detailed appearance varies across platforms and backends, but the basic functionality is the same*

The graphical user interface for displaying Matplotlib figures, as shown in Figure 4-4, is useful for interactive use with Python script files or the IPython console, and it allows us to interactively explore figures, for example, by zooming and panning. When using an interactive backend, which displays the figure in a graphical user interface, it is necessary to call the `plt.show` function to get the window to appear on the screen. By default, the `plt.show` function call hangs until the window is closed. For a more interactive experience, we can activate *interactive mode* by calling the `plt.ion` function. This instructs Matplotlib to take over the GUI event loop and show a window for a figure as soon as it is created, returning the control flow to the Python or IPython interpreter. To have the changes to a figure take effect, we must issue a redraw command using the `plt.draw` function. We can deactivate the interactive mode using the `plt.ioff` function, and we can use the `mpl.is_interactive` function to check if Matplotlib is in interactive or noninteractive mode.While the interactive graphical user interfaces have unique advantages, when working the Jupyter Notebook or the Qtconsole, it is often more convenient to display Matplotlib-produced graphics embedded directly in the notebook. This behavior is the default in the current version of Jupyter, but the "inline backend" can also be activated explicitly using the IPython command `%matplotlib inline`. This configures Matplotlib to use a noninteractive backend to generate graphics images, which are then displayed as static images in, for example, the Jupyter Notebook. The inline backend for Matplotlib can be fine-tuned using the IPython `%config` command. For example, we can select the output format for the generated graphics using the `InlineBackend.figure_format` option,[3] which, for example, we can set to 'svg' to generate SVG graphics rather than PNG files.

---

[3] For macOS users, `%config InlineBackend.figure_format='retina'` is another useful option, which improves the quality of the Matplotlib graphics when viewed on retina displays.

```
In [8]: %matplotlib inline
In [9]: %config InlineBackend.figure_format='svg'
```

With this approach, the interactive aspect of the graphical user interface is lost (e.g., zooming and panning), but embedding the graphics directly in the notebook has many other advantages. For example, keeping the code that was used to generate a figure together with the resulting figure in the same document eliminates the need for rerunning the code to display a figure, and the interactive nature of the Jupyter Notebook itself replaces some of the interactivity of Matplotlib's graphical user interface.

When using the IPython inline backend, it is unnecessary to use `plt.show` and `plt.draw`, since the IPython rich display system is responsible for triggering the rendering and displaying of the figures. This book assumes that code examples are executed in the Jupyter Notebooks, and the calls to the `plt.show` function are not in the code examples. When using an interactive backend, adding this function call at the end of each example is necessary.

Figure

As introduced in the previous section, the `Figure` object is used in Matplotlib to represent a graph. In addition to providing a canvas to place `Axes` instances on, the `Figure` object also provides methods for performing actions on figures, and it has several attributes that can be used to configure the properties of a figure.

A `Figure` object can be created using the `plt.figure` function, which takes several optional keyword arguments for setting figure properties. Notably, it accepts the `figsize` keyword argument, which should be assigned to a tuple on the form `(width, height)`, specifying the width and height of the figure canvas in inches. It can also be useful to specify the color of the figure canvas by setting the `facecolor` keyword argument.

Once a `Figure` is created, we can use the `add_axes` method to create a new `Axes` instance and assign it to a region on the figure canvas. The `add_axes` method takes one mandatory argument: a list containing the coordinates of the lower-left corner and the width and height of the `Axes` in the figure canvas coordinate system in the `(left, bottom, width, height)` format.[4] The coordinates and the width and height of the `Axes` object are expressed as fractions of the total canvas width and height; see Figure 4-1. For example, an `Axes` object that completely fills the canvas corresponds to `(0, 0, 1, 1)`, leaving no space for axis labels and ticks. A more practical size could be `(0.1, 0.1, 0.8, 0.8)`, corresponding to a centered `Axes` instance that covers 80% of the width and height of the canvas. The `add_axes` method takes a large number of keyword arguments for setting properties of the new `Axes` instance. These are described in more detail later in this chapter when I discuss the `Axes` object in depth. However, one keyword argument worth emphasizing here is `facecolor`, with which we can assign a background color for the `Axes` object. Together with the `facecolor` argument of `plt.figure`, this allows selecting colors of both the canvas and the regions covered by `Axes` instances.

With the `Figure` and `Axes` objects obtained from `plt.figure` and `fig.add_axes`, we have the necessary preparations to start plotting data using the methods of the `Axes` objects. See the next section of this chapter for more details on this. However, once the required plots have been created, more methods in the Figure objects are important in the graph-building workflow. For example, to set an overall figure title, we can use `suptitle`, which takes a title as a string as an argument. To save a figure to a file, we can use the `savefig` method. This method takes a string with the output filename as the first argument and several optional keyword arguments. By default, the output file format is determined by the file extension of the filename argument, but we can also specify the format explicitly using the `format` argument. The available output formats depend on which Matplotlib backend is used, but commonly available options are PNG, PDF, EPS,

---

[4] An alternative to passing a coordinate and size tuple to `add_axes` is to pass an existing Axes instance to add it to the figure.

and SVG formats. The resolution of the generated image can be set with the dpi argument. DPI stands for "dots per inch," and since the figure size is specified in inches using the figsize argument, multiplying these numbers gives the output image size in pixels. For example, with figsize=(8, 6) and dpi=100, the size of the generated image is 800x600 pixels. The savefig method also takes some arguments similar to those of the plt.figure function, such as the facecolor argument. Finally, the figure canvas can be made transparent using the transparent=True argument to savefig. The following code listing illustrates these techniques; the result is shown in Figure 4-5.

```
In [10]: fig = plt.figure(figsize=(8, 2.5), facecolor="#f1f1f1")
    ...:
    ...: # axes coordinates as fractions of the canvas width and height
    ...: left, bottom, width, height = 0.1, 0.1, 0.8, 0.8
    ...: ax = fig.add_axes((left, bottom, width, height), facecolor="#e1e1e1")
    ...:
    ...: x = np.linspace(-2, 2, 1000)
    ...: y1 = np.cos(40 * x)
    ...: y2 = np.exp(-x**2)
    ...:
    ...: ax.plot(x, y1 * y2)
    ...: ax.plot(x, y2, 'g')
    ...: ax.plot(x, -y2, 'g')
    ...: ax.set_xlabel("x")
    ...: ax.set_ylabel("y")
    ...:
    ...: fig.savefig("graph.png", dpi=100)
```

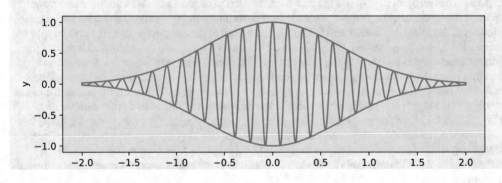

Figure 4-5. *Graph showing the result of setting the size of a figure with figsize, adding a new Axes instance with add_axes, setting the background colors of the Figure and Axes objects using facecolor, and finally saving the figure to file using savefig*

Axes

The Figure object introduced in the previous section provides the backbone of a Matplotlib graph, but all the interesting content is organized within or around Axes instances. We have already encountered Axes objects in this chapter. The Axes object is central to most plotting activities with the Matplotlib library. It provides a coordinate system to plot data and mathematical functions. In addition, it contains the axis objects that determine where the axis labels and the axis ticks are placed. The functions for drawing different types of plots are also methods of this Axes class. This section first explores different types of plots that can be drawn using Axes methods and how to customize the appearance of the *x* and *y* axes and the coordinate systems used with an Axes object.

We have seen how new Axes instances can be added to a figure explicitly using the add_axes method. This is a flexible and powerful method for placing Axes objects at arbitrary positions, which has several important applications, as explained later in the chapter. However, for most common uses, it is tedious to specify explicitly the coordinates of the Axes instances within the figure canvas. This is especially true when using multiple panels of Axes instances within a figure, for example, in a grid layout. Matplotlib provides several Axes layout managers, creating and placing Axes instances within a figure canvas following different strategies. Later in this chapter, we learn about using such layout managers. However, to facilitate the forthcoming examples, we briefly look at one of these layout managers: the plt.subplots function. Earlier in this chapter, this function conveniently generated new Figure and Axes objects in one function call. However, the plt.subplots function is also capable of filling a figure with a grid of Axes instances, which is specified using the first and the second arguments or with the nrows and ncols arguments, which, as the names imply, create a grid of Axes objects, with the given number of rows and columns. For example, to generate a grid of Axes instances in a newly created Figure object with three rows and two columns, we can use the following.

```
fig, axes = plt.subplots(nrows=3, ncols=2)
```

Here, the plt.subplots function returns a tuple (fig, axes), where fig is a Figure instance, and axes is a NumPy array of size (nrows, ncols), in which each element is an Axes instance that has been appropriately placed in the corresponding figure canvas. At this point, we can also specify that columns and/or rows should share the *x* and *y* axes using the sharex and sharey arguments, which can be set to True or False.

The plt.subplots function also takes two special keyword arguments fig_kw and subplot_kw, which are dictionaries with keyword arguments used when creating the Figure and Axes instances, respectively. This allows setting the properties of the Figure and Axes objects with plt.subplots similarly to when directly using plt.figure and the make_axes method.

Plot Types

Effective scientific and technical data visualization requires a wide variety of graphing techniques. Matplotlib implements many types of plotting techniques as methods of the Axes object. For example, the previous examples used the plot method, which draws curves in the coordinate system provided by the Axes object. The following sections explore some of Matplotlib's plotting functions in more depth by using these functions in example graphs. A summary of commonly used 2D plot functions is shown in Figure 4-6. Other types of graphs, such as color maps and 3D graphs, are discussed later in this chapter. All plotting functions in Matplotlib expect data as NumPy arrays as input, typically arrays with *x* and *y* coordinates as the first and second arguments. For details, see the docstrings for each method shown in Figure 4-6, using, for example, help(plt.Axes.bar).

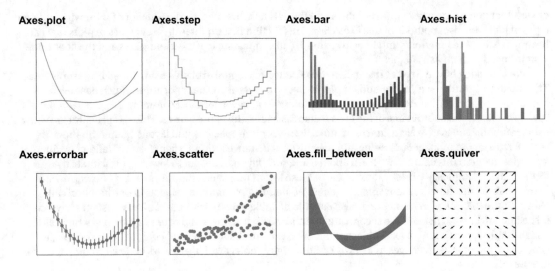

Figure 4-6. *Overview of selected 2D graph types. The name of the* Axes *method for generating each type of graph is shown together with the corresponding graph*

Line Properties

The most basic type of plot is the simple line plot. It may, for example, be used to depict the graph of a univariate function or to plot data as a function of a control variable. In line plots, we frequently need to configure the properties of the lines in the graph, for example, the line width, line color, and line style (solid, dashed, dotted, etc.). In Matplotlib we set these properties with keyword arguments to the plot methods, such as plot, step, and bar. A few of these graph types are shown in Figure 4-6. Many plot methods have specific arguments, but basic properties such as colors and line width are shared among most plotting methods. These properties and the corresponding keyword arguments are summarized in Table 4-1.

Table 4-1. *Basic Line Properties and Their Corresponding Argument Names for Use with the Matplotlib Plotting Methods*

Argument	Example Values	Description
color	A color specification can be a string with a color name, such as "red," "blue," or an RGB color code on the form "#aabbcc"	A color specification
alpha	Float number between 00 (completely transparent) and 1.0 (completely opaque)	The amount of transparency
linewidth, lw	Float number	The width of a line
linestyle, ls	"-" – solid "--" – dashed ":" – dotted ".-" – dash-dotted	The style of the line (i.e., whether the line is to be drawn as a solid line or if it should be, for example, dotted or dashed)
marker	+, o, * = cross, circle, star s = square . = small dot 1, 2, 3, 4, ... = triangle-shaped symbols with different angles	Whether or not connected to adjacent data points, each data point can be represented with a marker symbol as specified with this argument
markersize	Float number	The marker size
markerfacecolor	Color specification	The fill color for the marker
markeredgewidth	Float number	The line width of the marker edge
markeredgecolor	Color specification	The marker edge color

To illustrate these properties and arguments, consider the following code, which draws horizontal lines with various line width values, line style, marker symbol, color, and size. The resulting graph is shown in Figure 4-7.

```
In [11]: x = np.linspace(-5, 5, 5)
    ...: y = np.ones_like(x)
    ...:
    ...: def axes_settings(fig, ax, title, ymax):
    ...:     ax.set_xticks([])
    ...:     ax.set_yticks([])
    ...:     ax.set_ylim(0, ymax+1)
    ...:     ax.set_title(title)
    ...:
    ...: fig, axes = plt.subplots(1, 4, figsize=(16,3))
    ...:
    ...: # Line width
    ...: linewidths = [0.5, 1.0, 2.0, 4.0]
    ...: for n, linewidth in enumerate(linewidths):
    ...:     axes[0].plot(x, y + n, color="blue", linewidth=linewidth)
    ...: axes_settings(fig, axes[0], "linewidth", len(linewidths))
    ...:
```

```
...: # Line style
...: linestyles = ['-', '-.', ':']
...: for n, linestyle in enumerate(linestyles):
...:     axes[1].plot(x, y + n, color="blue", lw=2, linestyle=linestyle)
...:
...: # custom dash style
...: line, = axes[1].plot(x, y + 3, color="blue", lw=2)
...: length1, gap1, length2, gap2 = 10, 7, 20, 7
...: line.set_dashes([length1, gap1, length2, gap2])
...: axes_settings(fig, axes[1], "linetypes", len(linestyles) + 1)
...:
...: # marker types
...: markers = ['+', 'o', '*', 's', '.', '1', '2', '3', '4']
...: for n, marker in enumerate(markers):
...:     # lw = shorthand for linewidth, ls = shorthand for linestyle
...:     axes[2].plot(x, y + n, color="blue", lw=2, ls='None', marker=marker)
...: axes_settings(fig, axes[2], "markers", len(markers))
...:
...: # marker size and color
...: markersizecolors = [(4, "white"), (8, "red"), (12, "yellow"),
...:                     (16, "lightgreen")]
...: for n, (markersize, markerfacecolor) in enumerate (markersizecolors):
...:     axes[3].plot(x, y + n, color="blue", lw=1, ls='-',
...:                  marker='o', markersize=markersize,
...:                  markerfacecolor=markerfacecolor, markeredgewidth=2)
...: axes_settings(fig, axes[3], "marker size/color", len (markersizecolors))
```

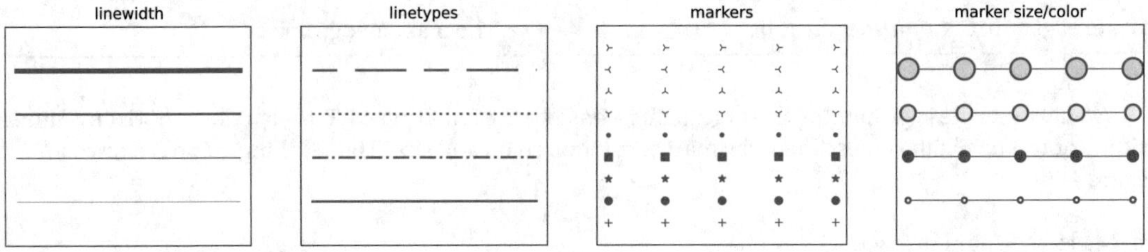

Figure 4-7. *Graphs showing the result of setting the properties line width, line style, marker type, and marker size, and color*

In practice, using different colors, line widths, and line styles are important tools for making a graph easily readable. In a graph with many lines, we can use a combination of colors and style to make each line uniquely identifiable, for example, via a legend. The line width property is best used to emphasize important lines. Consider the following example, where the $\sin(x)$ function is plotted with its first few series expansions around $x = 0$, as shown in Figure 4-8.

```
In [12]: # a symbolic variable for x,
    ...: # and a numerical array with specific values of x
    ...: sym_x = sympy.Symbol("x")
    ...: x = np.linspace(-2 * np.pi, 2 * np.pi, 100)
    ...:
    ...: def sin_expansion(x, n):
    ...:     """
```

```
...:        Evaluate the nth order Taylor series expansion
...:        of sin(x) for the numerical values in the array x.
...:        """
...:        return sympy.lambdify(sym_x,
...:                             sympy.sin(sym_x).series(n=n+1).removeO(),
...:                             'numpy')(x)
...:
...: fig, ax = plt.subplots()
...:
...: ax.plot(x, np.sin(x), linewidth=4, color="red", label='exact')
...:
...: colors = ["blue", "black"]
...: linestyles = [':', '-.', '--']
...: for idx, n in enumerate(range(1, 12, 2)):
...:        ax.plot(x, sin_expansion(x, n), color=colors[idx // 3],
...:               linestyle=linestyles[idx % 3], linewidth=3,
...:               label="order %d approx." % (n+1))
...:
...: ax.set_ylim(-1.1, 1.1)
...: ax.set_xlim(-1.5*np.pi, 1.5*np.pi)
...:
...: # place a legend outsize of the Axes
...: ax.legend(bbox_to_anchor=(1.02, 1), loc=2, borderaxespad=0.0)
...: # make room for the legend to the right of the Axes
...: fig.subplots_adjust(right=.75)
```

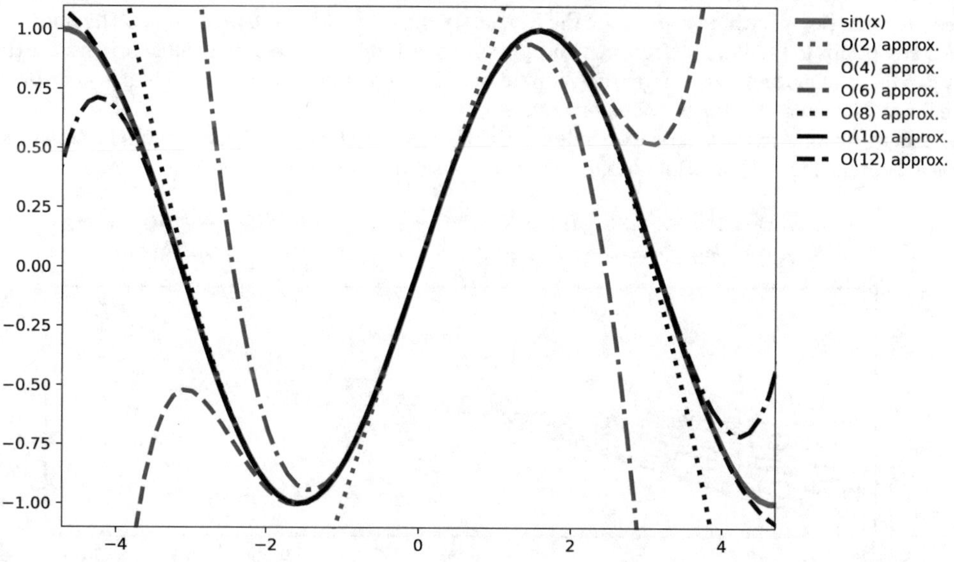

Figure 4-8. *Graph for sin(x) together with its Taylor series approximation of the few lowest orders*

Legends

A graph with multiple lines may often benefit from a legend, which displays a label along each line type somewhere within the figure. As shown in the previous example, using the legend method, a legend may be added to an Axes instance in a Matplotlib figure. Only lines with assigned labels are included in the legend (to assign a label to a line, use the `label` argument of, for example, `Axes.plot`). The legend method accepts a large number of optional arguments. See `help(plt.legend)` for details. Let's emphasize a few of the more useful arguments. The example in the previous section used the `loc` argument, which allows specifying where in the Axes area the legend is to be added: `loc=1` for upper-right corner, `loc=2` for upper-left corner, `loc=3` for the lower-left corner, and `loc=4` for the lower-right corner, as shown in Figure 4-9.

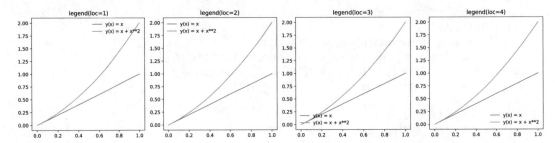

Figure 4-9. *Legend at different positions within an* Axes *instance specified using the* loc *argument of the legend method*

The example in the previous section also used the bbox_to_anchor argument, which helps place the legend at an arbitrary location within the figure canvas. The bbox_to_anchor argument takes the value of a tuple on the form (x, y), where x and y are the *canvas coordinates* within the Axes object. The point (0, 0) corresponds to the lower-left corner, and (1, 1) corresponds to the upper-right corner. Note that x and y can be smaller than 0 and larger than 1 in this case, which indicates that the legend is to be placed outside the Axes area, like in the previous section.

By default, all lines in the legend are shown in a vertical arrangement. Using the ncols argument, it is possible to split the legend labels into multiple columns, as illustrated in Figure 4-10.

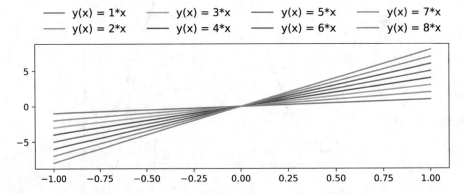

Figure 4-10. *Legend displayed outside the* Axes *object and shown with four columns instead of the single one, using* ax.legend(ncol=4, loc=3, bbox_to_anchor=(0, 1))

Text Formatting and Annotations

Text labels, titles, and annotations are important components in most graphs, and having full control of, for example, the font types and font sizes that are used to render such texts is necessary for producing publication-quality graphs. Matplotlib provides several methods to configure font properties. The default values can, for instance, be set in the Matplotlib resource file, and session-wide configuration can be set in the mpl.rcParams dictionary. This dictionary is a cache of the Matplotlib resource file and changes to parameters within this dictionary are valid until the Python interpreter is restarted and Matplotlib is imported again. Parameters relevant to how text is displayed include, for example, 'font.family' and 'font.size'.

■ **Tip** Use print(mpl.rcParams) to get a list of possible configuration parameters and their current values. Updating a parameter is as simple as assigning a new value to the corresponding item in the dictionary mpl.rcParams, for example, mpl.rcParams['savefig.dpi'] = 100. See also the mpl.rc function, which can be used to update the mpl.rcParams dictionary, and mpl.rcdefaults for restoring the default values.

It is also possible to set text properties on a case-to-case basis by passing a set of standard keyword arguments to functions that create text labels in a graph. Most Matplotlib functions that deal with text labels, in one way or another, accept the keyword arguments summarized in Table 4-2 (this list is an incomplete selection of common arguments; see help(mpl.text.Text) for a complete reference). For example, these arguments can be used with the Axes.text method, which creates a new text label at a given coordinate. They may also be used with set_title, set_xlabel, set_ylabel, and so on. For more information on these methods, see the next section.

Table 4-2. *Summary of Selected Font Properties and the Corresponding Keyword Arguments*

Argument	Description
Fontsize	The size of the font (in points)
family or fontname	The font type
Backgroundcolor	Color specification for the background color of the text label
Color	Color specification for the font color
Alpha	Transparency of the font color
Rotation	Rotation angle of the text label

In scientific and technical visualization, it is important to be able to render mathematical symbols and expressions in text labels. Matplotlib provides excellent support for this through LaTeX markup within its text labels: any text label in Matplotlib can include LaTeX math expressions by enclosing it within $ signs, for example, "Regular text: $f(x)=1-x^2$". By default, Matplotlib uses an internal LaTeX rendering, which supports a subset of the LaTeX language. However, by setting the configuration parameter mpl.rcParams["text.usetex"]=True, it is also possible to use an external full-featured LaTeX engine (if it is available on your system).

When embedding LaTeX code in strings in Python, there is a common stumbling block: Python uses \ as the escape character, whereas in LaTeX, it is used to denote the start of commands. To prevent the Python interpreter from escaping characters in strings containing LaTeX expressions, it is convenient to use raw strings, which are literal string expressions that are prepended with an r, for example, r"$\int f(x) dx$" and r'$x_{\rm A}$'.

The following example demonstrates how to add text labels and annotations to a Matplotlib figure using ax.text and ax.annotate and how to render a text label that includes an equation typeset in LaTeX. The resulting graph is shown in Figure 4-11.

```
In [13]: fig, ax = plt.subplots(figsize=(12, 3))
    ...:
    ...: ax.set_yticks([])
    ...: ax.set_xticks([])
    ...: ax.set_xlim(-0.5, 3.5)
    ...: ax.set_ylim(-0.05, 0.25)
    ...: ax.axhline(0)
    ...:
    ...: # text label
    ...: ax.text(0, 0.1, "Text label", fontsize=14, family="serif")
    ...:
    ...: # annotation
    ...: ax.plot(1, 0, "o")
    ...: ax.annotate("Annotation",
    ...:             fontsize=14, family="serif",
    ...:             xy=(1, 0), xycoords="data",
    ...:             xytext=(+20, +50), textcoords="offset points",
    ...:             arrowprops=dict(arrowstyle="->",
    ...:                             connectionstyle="arc3,rad=.5"))
    ...:
    ...: # equation
    ...: ax.text(2, 0.1, r"Equation: $i\hbar\partial_t \Psi = \hat{H}\Psi$",
    ...:             fontsize=14, family="serif")
```

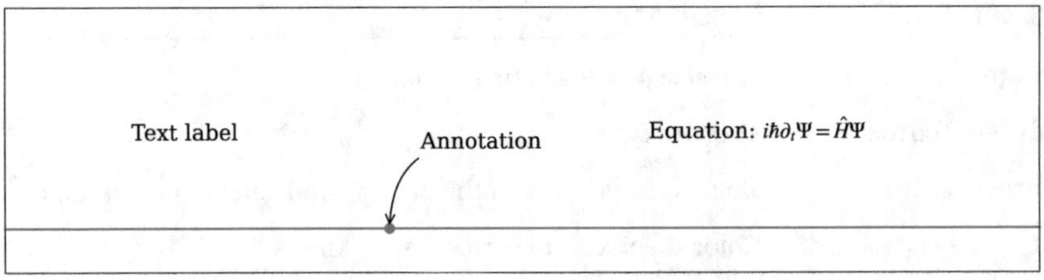

Figure 4-11. *Example demonstrating the result of adding text labels and annotations using* ax.text *and* ax.annotation *and including LaTeX formatted equations in a Matplotlib text label*

Axis Properties

After having created Figure and Axes objects, the data or functions are plotted using some of the many plot functions provided by Matplotlib, and the appearance of lines and markers are customized—the last major aspect of a graph that remains to be configured and fine-tuned is the Axis instances. A two-dimensional graph has two axis objects: the horizontal *x axis* and the vertical *y axis*. Each axis can be individually configured with respect to attributes such as the axis labels, the placement of ticks and the tick labels, and the location and appearance of the axis itself. This section examines how to control these properties of a graph.

Axis Labels and Titles

Arguably, the axis label is the most important property of an axis, which needs to be set in nearly all cases. We can set the axis labels using the set_xlabel and set_ylabel methods: they both take a string with the label as the first argument. In addition, the optional labelpad argument specifies the spacing from the axis to the label in units of points. This padding is occasionally necessary to avoid overlap between the axis label and the axis tick labels. The set_xlabel and set_ylabel methods also take additional arguments for setting text properties, such as color, fontsize, and fontname. The following code, which produces Figure 4-12, demonstrates how to use the set_xlabel and set_ylabel methods and the keyword arguments discussed here.

```
In [14]: x = np.linspace(0, 50, 500)
    ...: y = np.sin(x) * np.exp(-x/10)
    ...:
    ...: fig, ax = plt.subplots(figsize=(8, 2),
    ...:                        subplot_kw={'facecolor': "#ebf5ff"})
    ...:
    ...: ax.plot(x, y, lw=2)
    ...:
    ...: ax.set_xlabel("x", labelpad=5, fontsize=18, fontname='serif',
    ...:               color="blue")
    ...: ax.set_ylabel("f(x)", labelpad=15, fontsize=18, fontname='serif',
    ...:               color="blue")
    ...: ax.set_title("axis labels and title example", fontsize=16,
    ...:              fontname='serif', color="blue")
```

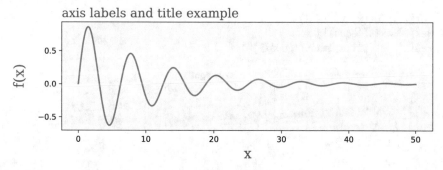

Figure 4-12. *Graph demonstrating the result of using* set_xlabel *and* set_ylabel *for setting the x and y axes labels*

In addition to labels on the x and y axes, we can also set the title of an Axes object using the set_title method. This method takes mostly the same arguments as set_xlabel and set_ylabel, except the loc argument, which can be assigned to 'left', 'centered', to 'right', and which dictates that the title is to be left aligned, centered, or right aligned.

Axis Range

By default, the range of the *x* and *y* axes of a Matplotlib is automatically adjusted to the data plotted in the Axes object. In many cases, these default ranges are sufficient, but in some situations, explicitly setting the axis ranges may be necessary. In such cases, we can use the set_xlim and set_ylim methods of the Axes

111

object. These methods take two arguments that specify the lower and upper limit to be displayed on the axis, respectively. An alternative to set_xlim and set_ylim is the axis method, which, for example, accepts the string argument 'tight', for a coordinate range that tightly fits the lines it contains and 'equal' for a coordinate range where one unit length along each axis corresponds to the same number of pixels (i.e., a ratio preserving coordinate system).

It is also possible to use the autoscale method to selectively turn on and off autoscaling by passing True and False as the first argument for the *x* and/or *y-axis* by setting its axis argument to 'x', 'y', or 'both'. The following example shows how to use these methods to control axis ranges. The resulting graphs are shown in Figure 4-13.

```
In [15]: x = np.linspace(0, 30, 500)
    ...: y = np.sin(x) * np.exp(-x/10)
    ...:
    ...:
    ...: fig, axes = plt.subplots(1, 3, figsize=(9, 3),
    ...:                          subplot_kw={'facecolor': "#ebf5ff"})
    ...:
    ...: axes[0].plot(x, y, lw=2)
    ...: axes[0].set_xlim(-5, 35)
    ...: axes[0].set_ylim(-1, 1)
    ...: axes[0].set_title("set_xlim / set_y_lim")
    ...:
    ...: axes[1].plot(x, y, lw=2)
    ...: axes[1].axis('tight')
    ...: axes[1].set_title("axis('tight')")
    ...:
    ...: axes[2].plot(x, y, lw=2)
    ...: axes[2].axis('equal')
    ...: axes[2].set_title("axis('equal')")
```

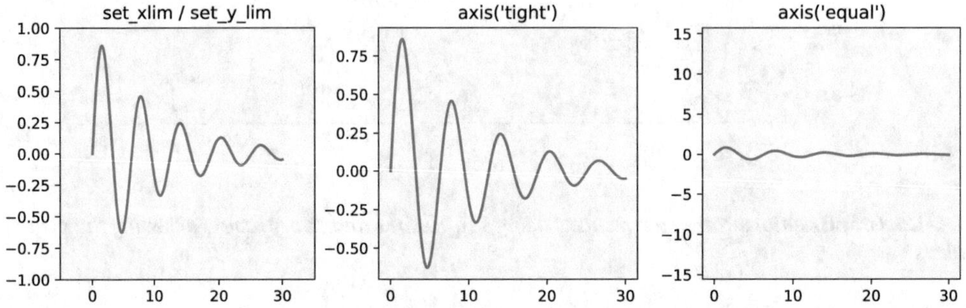

Figure 4-13. *Graphs that show the result of using the* set_xlim, set_ ylim, *and* axis *methods for setting the axis ranges that are shown in a graph*

Axis Ticks, Tick Labels, and Grids

The final basic properties of the axis that remain to be specified are the placement of axis ticks and the placement and formatting of the corresponding tick labels. The axis ticks are an important part of the overall appearance of a graph. When preparing publication and production-quality graphs, detailed control over the axis ticks is often necessary. Matplotlib module mpl.ticker provides a general and extensible tick management system that gives full control of the tick placement. Matplotlib distinguishes between major ticks and minor ticks. By default, every major tick has a corresponding label, and the distances between major ticks may be further marked with minor ticks that do not have labels, although this feature must be explicitly turned on. Figure 4-14 illustrates major and minor ticks.

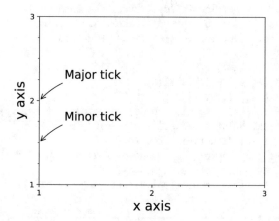

Figure 4-14. *The difference between major and minor ticks*

When configuring graph ticks, a common design requirement is determining where the major ticks with labels should be placed along the coordinate axis. The `mpl.ticker` module provides classes for different tick placement strategies. For example, the `mpl.ticker.MaxNLocator` can set the maximum number of ticks (at unspecified locations), the `mpl.ticker.MultipleLocator` can be used for setting ticks at multiples of a given base, and the `mpl.ticker.FixedLocator` can be used to place ticks at explicitly specified coordinates. To change the ticker strategy, use the `set_major_locator` and the `set_minor_locator` methods in `Axes.xaxis` and `Axes.yaxis`. These methods accept an instance of a ticker class defined in `mpl.ticker` or a custom class that is derived from one of those classes.

When explicitly specifying tick locations, we can also use the `set_xticks` and `set_yticks` methods, which accept a list of coordinates for where to place major ticks. In this case, it is also possible to set custom labels for each tick using the `set_xticklabels` and `set_yticklabels` methods, which expect lists of strings to be used as labels for the corresponding ticks. It is a good idea to use generic tick placement strategies, for example, `mpl.ticker.MaxNLocator`, because they dynamically adjust if the coordinate range is changed, whereas explicit tick placement using `set_xticks` and `set_yticks` then would require manual code updates. However, when the exact placement of ticks must be controlled, then `set_xticks` and `set_yticks` are convenient methods.

The following code demonstrates how to change the default tick placement using combinations of the methods discussed in this section, and the resulting graphs are shown in Figure 4-15.

```
In [16]: x = np.linspace(-2 * np.pi, 2 * np.pi, 500)
    ...: y = np.sin(x) * np.exp(-x**2/20)
    ...:
    ...: fig, axes = plt.subplots(1, 4, figsize=(12, 3))
    ...:
    ...: axes[0].plot(x, y, lw=2)
    ...: axes[0].set_title("default ticks")
    ...: axes[1].plot(x, y, lw=2)
    ...: axes[1].set_title("set_xticks")
    ...: axes[1].set_yticks([-1, 0, 1])
    ...: axes[1].set_xticks([-5, 0, 5])
```

```
...:
...: axes[2].plot(x, y, lw=2)
...: axes[2].set_title("set_major_locator")
...: axes[2].xaxis.set_major_locator(mpl.ticker.MaxNLocator(4))
...: axes[2].xaxis.set_minor_locator(mpl.ticker.MaxNLocator(8))
...: axes[2].yaxis.set_major_locator(mpl.ticker.FixedLocator([-1, 0, 1]))
...: axes[2].yaxis.set_minor_locator(mpl.ticker.MaxNLocator(8))
...:
...: axes[3].plot(x, y, lw=2)
...: axes[3].set_title("set_xticklabels")
...: axes[3].set_yticks([-1, 0, 1])
...: axes[3].set_xticks([-2 * np.pi, -np.pi, 0, np.pi, 2 * np.pi])
...: axes[3].set_xticklabels([r'$-2\pi$', r'$-\pi$', 0, r'$\pi$', r'$2\pi$'])
...: x_minor_ticker = mpl.ticker.FixedLocator(
...:     [-3 * np.pi / 2, -np.pi / 2, 0, np.pi / 2, 3 * np.pi / 2])
...: axes[3].xaxis.set_minor_locator(x_minor_ticker)
...: axes[3].yaxis.set_minor_locator(mpl.ticker.MaxNLocator(4))
```

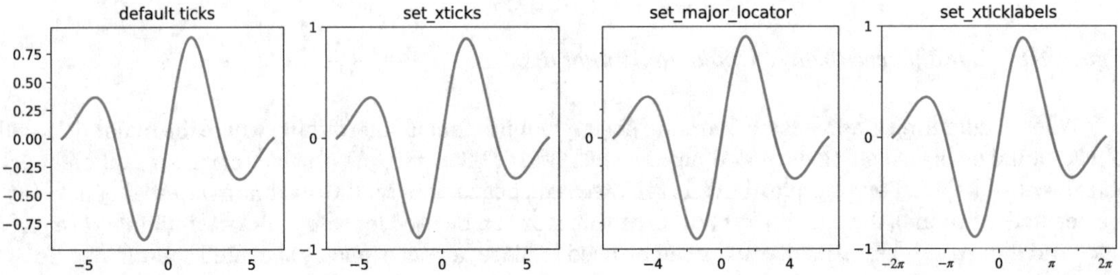

Figure 4-15. *Graphs that demonstrate different ways of configuring the placement and appearance of major and minor ticks along the x axis and the y axis*

A commonly used design element in graphs is grid lines, which are intended as a visual guide when reading values from the graph. Grids and grid lines are closely related to axis ticks since they are drawn at the same coordinate values and are essentially extensions of the ticks that span across the graph. In Matplotlib, we can turn on axis grids using the grid method of an axes object. The grid method takes optional keyword arguments used to control the grid's appearance. For example, like many of the plot functions in Matplotlib, the grid method accepts the color, linestyle, and linewidth arguments for specifying the properties of the grid lines. In addition, it takes the which and axis arguments, which can be assigned values 'major', 'minor', or 'both', and 'x', 'y', or 'both', respectively. These arguments indicate which ticks the given style is to be applied along an axis. If several different styles for the grid lines are required, multiple calls to grid can be used, with different values of which and axis. For an example of how to add grid lines and style them in different ways, see the following code listing, which produces the graphs shown in Figure 4-16.

```
In [17]: fig, axes = plt.subplots(1, 3, figsize=(12, 4))
    ...: x_major_ticker = mpl.ticker.MultipleLocator(4)
    ...: x_minor_ticker = mpl.ticker.MultipleLocator(1)
    ...: y_major_ticker = mpl.ticker.MultipleLocator(0.5)
    ...: y_minor_ticker = mpl.ticker.MultipleLocator(0.25)
    ...:
    ...: for ax in axes:
    ...:     ax.plot(x, y, lw=2)
    ...:     ax.xaxis.set_major_locator(x_major_ticker)
```

```
...:        ax.yaxis.set_major_locator(y_major_ticker)
...:        ax.xaxis.set_minor_locator(x_minor_ticker)
...:        ax.yaxis.set_minor_locator(y_minor_ticker)
...:
...: axes[0].set_title("default grid")
...: axes[0].grid()
...:
...: axes[1].set_title("major/minor grid")
...: axes[1].grid(color="blue", which="both", linestyle=':', linewidth=0.5)
...:
...: axes[2].set_title("individual x/y major/minor grid")
...: axes[2].grid(color="grey", which="major", axis='x',
...:              linestyle='-', linewidth=0.5)
...: axes[2].grid(color="grey", which="minor", axis='x',
...:              linestyle=':', linewidth=0.25)
...: axes[2].grid(color="grey", which="major", axis='y',
...:              linestyle='-', linewidth=0.5)
```

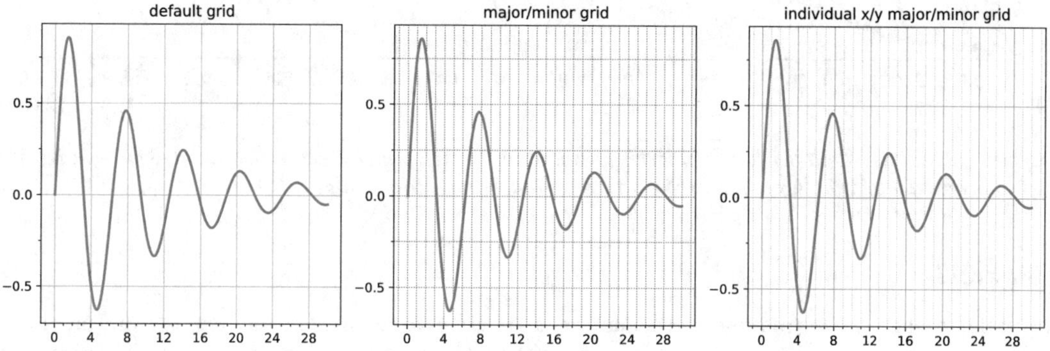

Figure 4-16. *Graphs demonstrating the result of using grid lines*

In addition to controlling the tick placements, the Matplotlib mpl.ticker module also provides classes for customizing the tick labels. For example, the ScalarFormatter from the mpl.ticker module can be used to set several useful properties related to displaying tick labels with scientific notation, for displaying axis labels for large numerical values. If scientific notation is activated using the set_scientific method, we can control the threshold for when scientific notation is used with the set_powerlimits method (by default, tick labels for small numbers are not displayed using the scientific notation). We can use the useMathText=True argument when creating the ScalarFormatter instance to have the exponents shown in math style rather than code style exponents (e.g., 1e10). The formatter object is applied to an Axes object using the set_major_formatter method. See the following code for an example of using scientific notation in tick labels. The resulting graphs are shown in Figure 4-17.

```
In [19]: fig, axes = plt.subplots(1, 2, figsize=(8, 3))
    ...:
    ...: x = np.linspace(0, 1e5, 100)
    ...: y = x ** 2
    ...:
    ...: axes[0].plot(x, y, 'b.')
    ...: axes[0].set_title("default labels", loc='right')
    ...:
```

```
...: axes[1].plot(x, y, 'b')
...: axes[1].set_title("scientific notation labels", loc='right')
...:
...: formatter = mpl.ticker.ScalarFormatter(useMathText=True)
...: formatter.set_scientific(True)
...: formatter.set_powerlimits((-1,1))
...: axes[1].xaxis.set_major_formatter(formatter)
...: axes[1].yaxis.set_major_formatter(formatter)
```

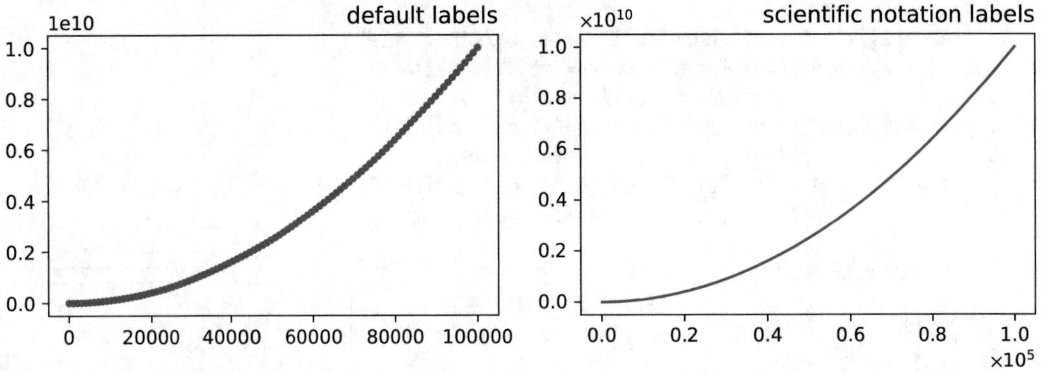

Figure 4-17. *Graphs with tick labels in scientific notation. The left panel uses the default label formatting, while the right panel uses tick labels in scientific notation, rendered as math text*

Log Plots

It is useful to work with logarithmic coordinate systems in visualizations of data that spans several orders of magnitude. In Matplotlib, there are several plot functions for graphing functions in such coordinate systems, for example, loglog, semilogx, and semilogy, which use logarithmic scales for both the x and y axes, for only the x *axis*, and only the y *axis*, respectively. Apart from the logarithmic axis scales, these functions behave similarly to the standard plot method. An alternative approach is using the standard plot method and separately configuring the axis scales to be logarithmic using the set_xscale and/or set_yscale method with 'log' as the first argument. Examples of using methods to produce log-scale plots are shown in the following code listing, and the resulting graphs are shown in Figure 4-18.

```
In [20]: fig, axes = plt.subplots(1, 3, figsize=(12, 3))
    ...:
    ...: x = np.linspace(0, 1e3, 100)
    ...: y1, y2 = x**3, x**4
    ...:
    ...: axes[0].set_title('loglog')
    ...: axes[0].loglog(x, y1, 'b', x, y2, 'r')
    ...:
    ...: axes[1].set_title('semilogy')
    ...: axes[1].semilogy(x, y1, 'b', x, y2, 'r')
    ...:
    ...: axes[2].set_title('plot / set_xscale / set_yscale')
    ...: axes[2].plot(x, y1, 'b', x, y2, 'r')
    ...: axes[2].set_xscale('log')
    ...: axes[2].set_yscale('log')
```

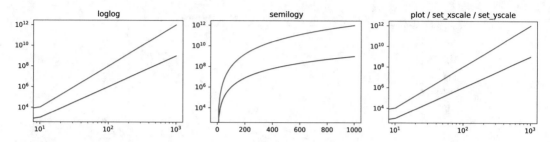

Figure 4-18. *Examples of log-scale plots*

Twin Axes

An interesting trick with axes that Matplotlib provides is the twin axis feature, which allows displaying two independent axes overlaid on each other. This is useful when plotting two different quantities, for example, with different units, within the same graph. A simple example demonstrating this feature is shown as follows, and the resulting graph is shown in Figure 4-19. Let's use the twinx method (there is also a twiny method) to produce a second Axes instance with a shared *x axis* and a new independent *y axis*, which is displayed on the right side of the graph.

```
In [21]: fig, ax1 = plt.subplots(figsize=(8, 4))
    ...:
    ...: r = np.linspace(0, 5, 100)
    ...: a = 4 * np.pi * r ** 2   # area
    ...: v = (4 * np.pi / 3) * r ** 3   # volume
    ...:
    ...: ax1.set_title("surface area and volume of a sphere", fontsize=16)
    ...: ax1.set_xlabel("radius [m]", fontsize=16)
    ...:
    ...: ax1.plot(r, a, lw=2, color="blue")
    ...: ax1.set_ylabel(r"surface area ($m^2$)", fontsize=16, color="blue")
    ...: for label in ax1.get_yticklabels():
    ...:     label.set_color("blue")
    ...:
    ...: ax2 = ax1.twinx()
    ...: ax2.plot(r, v, lw=2, color="red")
    ...: ax2.set_ylabel(r"volume ($m^3$)", fontsize=16, color="red")
    ...: for label in ax2.get_yticklabels():
    ...:     label.set_color("red")
```

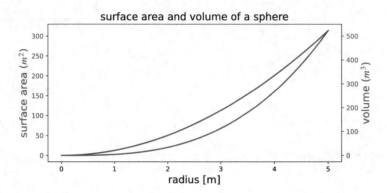

Figure 4-19. Example of a graph with twin axes

Spines

In all graphs generated so far, there was always a box surrounding the Axes region. This is indeed a common style for scientific and technical graphs, but in some cases, for example, moving these coordinate lines may be desired when representing schematic graphs. The lines that make up the surrounding box are called axis spines in Matplotlib, and we can use the Axes.spines attribute to change their properties. For example, we might want to remove the top and the right spines and move the spines to coincide with the origin of the coordinate systems.

The spines attribute of the Axes object is a dictionary with the keys right, left, top, and bottom that can be used to access each spine individually. We can use the set_color method to set the color to 'None' to indicate that a particular spine should not be displayed. In this case, we also need to remove the ticks associated with that spine, using the set_ticks_position method of Axes.xaxis and Axes.yaxis (which accepts the 'both', 'top', or 'bottom' and 'both', 'left', or 'right' arguments, respectively). These methods allow we to transform the surrounding box to *x* and *y* coordinate axes, as demonstrated in the following example. The resulting graph is shown in Figure 4-20.

```
In [22]: x = np.linspace(-10, 10, 500)
    ...: y = np.sin(x) / x
    ...:
    ...: fig, ax = plt.subplots(figsize=(8, 4))
    ...:
    ...: ax.plot(x, y, linewidth=2)
    ...:
    ...: # remove top and right spines
    ...: ax.spines['right'].set_color('none')
    ...: ax.spines['top'].set_color('none')
    ...:
    ...: # remove top and right spine ticks
    ...: ax.xaxis.set_ticks_position('bottom')
    ...: ax.yaxis.set_ticks_position('left')
    ...:
    ...: # move bottom and left spine to x = 0 and y = 0
    ...: ax.spines['bottom'].set_position(('data', 0))
    ...: ax.spines['left'].set_position(('data', 0))
    ...:
```

```
...: ax.set_xticks([-10, -5, 5, 10])
...: ax.set_yticks([0.5, 1])
...:
...: # give each label a solid background of white,
...: # to not overlap with the plot line
...: for label in ax.get_xticklabels() + ax.get_yticklabels():
...:     label.set_bbox({'facecolor': 'white',
...:                     'edgecolor': 'white'})
```

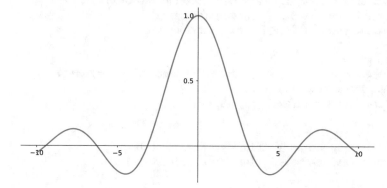

Figure 4-20. *Example of a graph with axis spines*

Advanced Axes Layouts

So far, plt.figure, Figure.make_axes, and plt.subplots have been used to create new Figure and Axes instances, which we then used for producing graphs. In scientific and technical visualization, it is common to pack together multiple figures in different panels, for example, in a grid layout. Matplotlib has functions for automatically creating Axes objects and placing them on a figure canvas using various layout strategies. We have already used the plt.subplots function to generate a uniform grid of Axes objects. This section explores additional features of the plt.subplots function and introduces the subplot2grid and GridSpec layout managers, which are more flexible in how the Axes objects are distributed within a figure canvas.

Insets

Before diving into using more advanced Axes layout managers, it is worth taking a step back and considering an important use for the first approach used to add Axes instances to a figure canvas: the Figure.add_axes method. This approach is well suited for creating an *inset*, a smaller graph displayed within the region of another graph. Insets are, for example, frequently used for displaying a magnified region of special interest in the larger graph or for displaying some related graphs of secondary importance.

In Matplotlib, we can place additional Axes objects at arbitrary locations within a figure canvas, even if they overlap with existing Axes objects. To create an inset, add a new Axes object with Figure.make_axes and the (figure canvas) coordinates for where the inset should be placed. The following code produces a typical example of a graph with an inset, as shown in Figure 4-21. When creating the Axes object for the inset, it may be useful to use the facecolor='none' argument, which indicates that there should be no background color and that the Axes background of the inset should be transparent.

```
In [23]: fig = plt.figure(figsize=(8, 4))
    ...:
    ...: def f(x):
    ...:     return 1/(1 + x**2) + 0.1/(1 + ((3 - x)/0.1)**2)
    ...:
    ...: def plot_and_format_axes(ax, x, f, fontsize):
    ...:     ax.plot(x, f(x), linewidth=2)
    ...:     ax.xaxis.set_major_locator(mpl.ticker.MaxNLocator(5))
    ...:     ax.yaxis.set_major_locator(mpl.ticker.MaxNLocator(4))
    ...:     ax.set_xlabel(r"$x$", fontsize=fontsize)
    ...:     ax.set_ylabel(r"$f(x)$", fontsize=fontsize)
    ...:
    ...: # main graph
    ...: ax = fig.add_axes([0.1, 0.15, 0.8, 0.8], facecolor="#f5f5f5")
    ...: x = np.linspace(-4, 14, 1000)
    ...: plot_and_format_axes(ax, x, f, 18)
    ...:
    ...: x0, x1 = 2.5, 3.5
    ...: ax.axvline(x0, ymax=0.3, color="grey", linestyle=":")
    ...: ax.axvline(x1, ymax=0.3, color="grey", linestyle=":")
    ...:
    ...: # inset
    ...: ax_insert = fig.add_axes([0.5, 0.5, 0.38, 0.42], facecolor='none')
    ...: x = np.linspace(x0, x1, 1000)
    ...: plot_and_format_axes(ax_insert, x, f, 14)
```

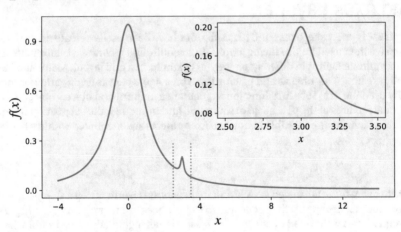

Figure 4-21. *Example of a graph with an inset*

Subplots

We have already used plt.subplots extensively. It returns a tuple with a Figure instance and a NumPy array with the Axes objects for each row and column requested in the function call. It is often the case when plotting grids of subplots that either the *x* or the *y* axis, or both, is shared among the subplots. Using the sharex and sharey arguments to plt.subplots can be useful in such situations since it prevents the same axis labels from being repeated across multiple Axes.

It is also worth noting that the dimension of the NumPy array with Axes instances returned by plt. subplots is "squeezed" by default; the dimensions with length 1 are removed from the array. If both the requested numbers of column and row are greater than one, then a two-dimensional array is returned, but if either (or both) the number of columns or rows is one, then a one-dimensional (or scalar, i.e., the only Axes object itself) is returned. We can turn off the squeezing of the dimensions of the NumPy arrays by passing the squeeze=False argument to the plt.subplots function. In this case, the axes variable in fig, axes = plt.subplots(nrows, ncols) is always a two-dimensional array.

A final touch of configurability can be achieved using the plt.subplots_adjust function, which allows explicitly setting the left, right, bottom, and top coordinates of the overall Axes grid, as well as the width (wspace) and height spacing (hspace) between Axes instances in the grid. The following code and the corresponding Figure 4-22 provide a step-by-step example of setting up an Axes grid with shared x and y axes and adjusted Axes spacing.

```
In [24]: fig, axes = plt.subplots(2, 2, figsize=(6, 6),
    ...:                          sharex=True, sharey=True, squeeze=False)
    ...:
    ...: x1 = np.random.randn(100)
    ...: x2 = np.random.randn(100)
    ...:
    ...: axes[0, 0].set_title("Uncorrelated")
    ...: axes[0, 0].scatter(x1, x2)
    ...:
    ...: axes[0, 1].set_title("Weakly positively correlated")
    ...: axes[0, 1].scatter(x1, x1 + x2)
    ...:
    ...: axes[1, 0].set_title("Weakly negatively correlated")
    ...: axes[1, 0].scatter(x1, -x1 + x2)
    ...:
    ...: axes[1, 1].set_title("Strongly correlated")
    ...: axes[1, 1].scatter(x1, x1 + 0.15 * x2)
    ...:
    ...: axes[1, 1].set_xlabel("x")
    ...: axes[1, 0].set_xlabel("x")
    ...: axes[0, 0].set_ylabel("y")
    ...: axes[1, 0].set_ylabel("y")
    ...:
    ...: plt.subplots_adjust(left=0.1, right=0.95, bottom=0.1, top=0.95,
    ...:                     wspace=0.1, hspace=0.2)
```

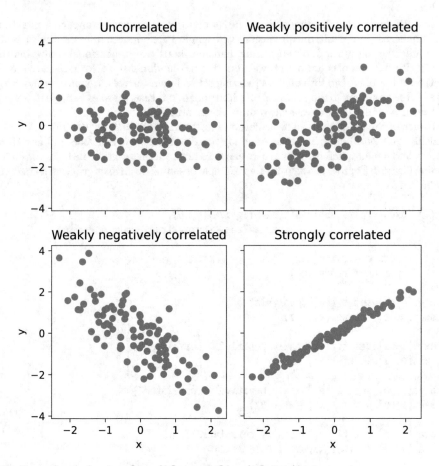

Figure 4-22. *Example graph using* `plt.subplot` *and* `plt.subplot_adjust`

Subplot2grid

The `plt.subplot2grid` function is an intermediary between `plt.subplots` and `gridspec` (see the next section) that provides a more flexible Axes layout management than `plt.subplots` while at the same time being simpler to use than `gridspec`. In particular, `plt.subplot2grid` can create grids with Axes instances that span multiple rows and/or columns. The `plt.subplot2grid` takes two mandatory arguments: the first argument is the shape of the Axes grid in the form on a tuple (`nrows, ncols`), and the second argument is a tuple (`row, col`) that specifies the starting position within the grid. The two optional keyword arguments `colspan` and `rowspan` can indicate how many rows and columns the new Axes instance should span. An example of how to use the `plt.subplot2grid` function is given in Table 4-3. Note that each call to the `plt.subplot2grid` function results in one new Axes instance, unlike `plt.subplots`, which creates all Axes instances in one function call and returns them in a NumPy array.

Table 4-3. *Example of a Grid Layout Created with* `plt.subplot2grid` *and the Corresponding Code*

Axes Grid Layout	Code

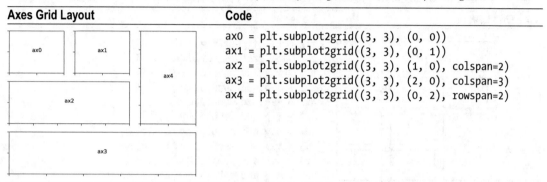	```ax0 = plt.subplot2grid((3, 3), (0, 0))```
	```ax1 = plt.subplot2grid((3, 3), (0, 1))```
	```ax2 = plt.subplot2grid((3, 3), (1, 0), colspan=2)```
	```ax3 = plt.subplot2grid((3, 3), (2, 0), colspan=3)```
	```ax4 = plt.subplot2grid((3, 3), (0, 2), rowspan=2)```

GridSpec

The final grid layout manager covered here is GridSpec from the mpl.gridspec module. This is the most general grid layout manager in Matplotlib, and it allows the creation of grids where not all rows and columns have equal width and height, which is not easily achieved with the grid layout managers used earlier in this chapter.

A GridSpec object is only used to specify the grid layout and does not create any Axes objects. When creating a new instance of the GridSpec class, we must specify the number of rows and columns in the grid. Like for other grid layout managers, we can also set the position of the grid using the keyword arguments left, bottom, right, and top, and we can set the width and height spacing between subplots using wspace and hspace. Additionally, GricSpec allows specifying the relative width and heights of columns and rows using the width_ratios and height_ratios arguments. These should both be lists with relative weights for each column and row size in the grid. For example, to generate a grid with two rows and two columns, where the first row and column is twice as big as the second row and column, we could use mpl.gridspec. GridSpec(2, 2, width_ratios=[2, 1], height_ratios=[2, 1]).

Once a GridSpec instance has been created, we can use the Figure.add_subplot method to create Axes objects and place them on a figure canvas. As an argument to add_subplot, we need to pass an mpl.gridspec.SubplotSpec instance, which we can generate from the GridSpec object using an array-like indexing: for example, given a GridSpec instance gs, we obtain a SubplotSpec instance for the upper-left grid element using gs[0, 0] and for a SubplotSpec instance that covers the first row we use gs[:, 0] and so on. Table 4-4 provides concrete examples of how to use GridSpec and add_subplot to create an Axes instance.

Table 4-4. *Examples of How to Use the Subplot Grid Manager* `mpl.gridspec.GridSpec`

Axes Grid Layout	Code
| 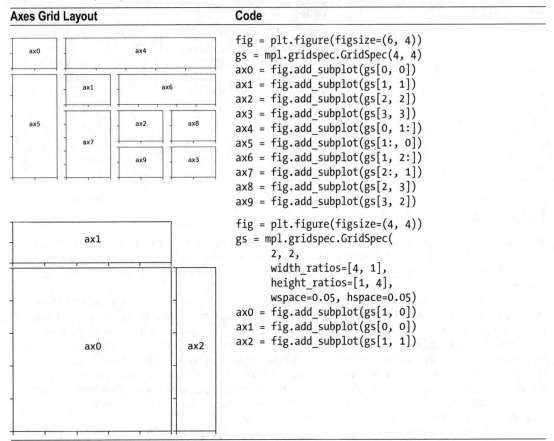 | ```python
fig = plt.figure(figsize=(6, 4))
gs = mpl.gridspec.GridSpec(4, 4)
ax0 = fig.add_subplot(gs[0, 0])
ax1 = fig.add_subplot(gs[1, 1])
ax2 = fig.add_subplot(gs[2, 2])
ax3 = fig.add_subplot(gs[3, 3])
ax4 = fig.add_subplot(gs[0, 1:])
ax5 = fig.add_subplot(gs[1:, 0])
ax6 = fig.add_subplot(gs[1, 2:])
ax7 = fig.add_subplot(gs[2:, 1])
ax8 = fig.add_subplot(gs[2, 3])
ax9 = fig.add_subplot(gs[3, 2])
``` |
| | ```python
fig = plt.figure(figsize=(4, 4))
gs = mpl.gridspec.GridSpec(
        2, 2,
        width_ratios=[4, 1],
        height_ratios=[1, 4],
        wspace=0.05, hspace=0.05)
ax0 = fig.add_subplot(gs[1, 0])
ax1 = fig.add_subplot(gs[0, 0])
ax2 = fig.add_subplot(gs[1, 1])
``` |

Colormap Plots

We have so far only worked with graphs of univariate functions or, equivalently, twodimensional data in *x-y* format. The two-dimensional Axes objects that are used for this purpose can also be used to visualize bivariate functions or three-dimensional data on *x-y-z* format, using *color maps* (or *heat maps*), where each pixel in the Axes area is colored according to the *z* value corresponding to that point in the coordinate system. Matplotlib provides the `pcolor` and `imshow` functions for these types of plots, and the `contour` and `contourf` functions graph data in the same format by drawing contour lines rather than color maps. Examples of graphs generated with these functions are shown in Figure 4-23.

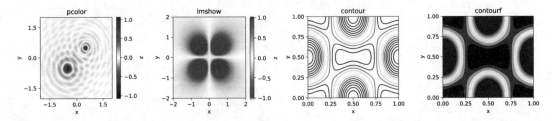

Figure 4-23. *Example graphs generated with* `pcolor`, `imshow`, `contour`, *and* `contourf`

124

To produce a colormap graph, for example, using pcolor, we must prepare the data in the appropriate format. While standard two-dimensional graphs expect one-dimensional coordinate arrays with x and y values, in the present case, we need to use two-dimensional coordinate arrays, as, for example, generated using the NumPy meshgrid function. To plot a bivariate function or data with two dependent variables, we start by defining one-dimensional coordinate arrays, x and y, that span the desired coordinate range or correspond to the values for which data is available. The x and y arrays can then be passed to the np.meshgrid function, which produces the required two-dimensional coordinate arrays X and Y. If necessary, we can use NumPy array computations with X and Y to evaluate bivariate functions to obtain a data array Z, as done in lines 1 to 3 in cell In [25] (see the following).

Once the two-dimensional coordinate and data arrays are prepared, they are easily visualized using, for example, pcolor, contour, or contourf, by passing the X, Y, and Z arrays as the first three arguments. The imshow method works similarly but only expects the data array Z as an argument, and the relevant coordinate ranges must instead be set using the extent argument, which should be set to a list in the format [xmin, xmax, ymin, ymax]. Additional keyword arguments important for controlling the appearance of colormap graphs are vmin, vmax, norm, and cmap: the vmin and vmax can be used to set the range of values mapped to the color axis. This can also be achieved by setting norm=mpl.colors.Normalize(vmin, vmax). The cmap argument specifies a color map for mapping the data values to colors in the graph. This argument can either be a string with a predefined colormap name or a colormap instance. The predefined color maps in Matplotlib are available in mpl.cm. Try help(mpl.cm) or to autocomplete in IPython on the mpl.cm module for a full list of available color maps.[5]

The last piece required for a complete colormap plot is the colorbar element, which allows the viewer to read off the numerical values that different colors correspond to. In Matplotlib we can use the plt.colorbar function to attach a colorbar to a plotted colormap graph. It takes a handle to the plot as first argument, and it takes two optional arguments ax and cax, which can be used to control where in the graph the colorbar is to appear. If ax is given, the space is taken from this Axes object for the new colorbar. If, on the other hand, cax is given, then the colorbar draws on this Axes object. A colorbar instance cb has its own axis object, and the standard methods for setting axis attributes can be used on the cb.ax object, and we can use, for example, the set_label, set_ticks, and set_ticklabels method in the same manner as for x and y axes.

The steps outlined in the previous paragraphs are shown in the following code, and the resulting graph is shown in Figure 4-24. The imshow, contour, and contourf functions can be used in a nearly similar manner, although these functions take additional arguments for controlling their characteristic properties. For example, the contour and contourf functions additionally take an argument N that specifies the number of contour lines to draw.

```
In [25]: x = y = np.linspace(-10, 10, 150)
    ...: X, Y = np.meshgrid(x, y)
    ...: Z = np.cos(X) * np.cos(Y) * np.exp(-(X/5)**2-(Y/5)**2)
    ...:
    ...: fig, ax = plt.subplots(figsize=(6, 5))
    ...:
    ...: norm = mpl.colors.Normalize(-abs(Z).max(), abs(Z).max())
    ...: p = ax.pcolor(X, Y, Z, norm=norm, cmap=mpl.cm.bwr)
    ...:
    ...: ax.axis('tight')
    ...: ax.set_xlabel(r"$x$", fontsize=18)
    ...: ax.set_ylabel(r"$y$", fontsize=18)
    ...: ax.xaxis.set_major_locator(mpl.ticker.MaxNLocator(4))
    ...: ax.yaxis.set_major_locator(mpl.ticker.MaxNLocator(4))
```

---

[5] A nice visualization of all the available color maps is available at http://scipy-cookbook.readthedocs.io/items/Matplotlib_Show_colormaps.html. This page also describes how to create new color maps.

```
   ...:
   ...: cb = fig.colorbar(p, ax=ax)
   ...: cb.set_label(r"$z$", fontsize=18)
   ...: cb.set_ticks([-1, -.5, 0, .5, 1])
```

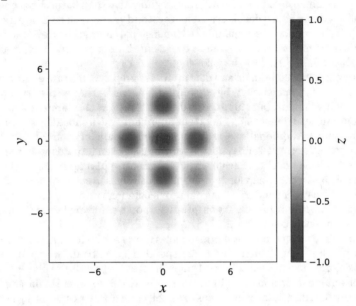

Figure 4-24. *Example using* `pcolor` *to produce a colormap graph*

3D Plots

The colormap graphs discussed in the previous section were used to visualize data with two dependent variables by color-coding data in 2D graphs. Another way of visualizing the same type of data is to use 3D graphs, where a third axis, *z*, is introduced, and the graph is displayed in a perspective on the screen. In Matplotlib, drawing 3D graphs requires using a different axes object, namely, the `Axes3D` object available from the `mpl_toolkits.mplot3d` module. We can create a 3D-aware `Axes` instance explicitly using the constructor of the `Axes3D` class, by passing a `Figure` instance as an argument: `ax = Axes3D(fig)`. Alternatively, we can use the `add_subplot` function with the `projection='3d'` argument.

```
ax = fig.add_subplot(1, 1, 1, projection='3d')
```

Or use `plt.subplots` with the `subplot_kw={'projection': '3d'}` argument.

```
fig, ax = plt.subplots(1, 1, figsize=(8, 6), subplot_kw={'projection': '3d'})
```

In this way, we can use all of the axes layout approaches previously used for 2D graphs, if only we specify the `projection` argument appropriately. Note that using `add_subplot`, it is possible to mix axes objects with 2D and 3D projections within the same figure, but when using `plt.subplots`, the `subplot_kw` argument applies to all the subplots added to a figure.

Having created and added 3D-aware `Axes` instances to a figure, for example, using one of the methods described in the previous paragraph, the `Axes3D` class methods—such as `plot_surface`, `plot_wireframe`, and `contour`— can be used to plot data as surfaces in a 3D perspective. These functions are used in a manner that is nearly the same as how the color map was used in the previous section: these 3D plotting functions all take two-dimensional coordinates and data arrays X, Y, and Z as first arguments. Each function also takes additional parameters for tuning specific properties. For example, the plot_surface function takes the rstride and cstride arguments (row and column stride) for selecting data from the input arrays (to avoid data points that are too dense). The contour and contourf functions take optional arguments zdir and offset, which are used to select a projection direction (the allowed values are "x," "y," and "z") and the plane to display the projection.

In addition to the methods for 3D surface plotting, there are also straightforward generalizations of the line and scatter plot functions that are available for 2D axes, for example, plot, scatter, bar, and bar3d, which in the version that is available in the Axes3D class takes an additional argument for the z coordinates. Like their 2D relatives, these functions expect one-dimensional data arrays rather than the two-dimensional coordinate arrays used for surface plots.

When it comes to axes titles, labels, ticks, and tick labels, all the methods used for 2D graphs, as described earlier, are straightforwardly generalized to 3D graphs. For example, there are new methods set_zlabel, set_zticks, and set_zticklabels for manipulating the attributes of the new z axis. The Axes3D object also provides new class methods for 3D-specific actions and attributes. The view_init method can be used to change the angle from which the graph is viewed, and it takes the elevation and the azimuth, in degrees, as the first and second arguments.

Examples of how to use these 3D plotting functions are given in the following section, and the produced graphs are shown in Figure 4-25.

```
In [26]: fig, axes = plt.subplots(1, 3, figsize=(14, 4),
    ...:                          subplot_kw={'projection': '3d'})
    ...:
    ...: def title_and_labels(ax, title):
    ...:     ax.set_title(title)
    ...:     ax.set_xlabel("$x$", fontsize=16)
    ...:     ax.set_ylabel("$y$", fontsize=16)
    ...:     ax.set_zlabel("$z$", fontsize=16)
    ...:
    ...: x = y = np.linspace(-3, 3, 74)
    ...: X, Y = np.meshgrid(x, y)
    ...:
    ...: R = np.sqrt(X**2 + Y**2)
    ...: Z = np.sin(4 * R) / R
    ...:
    ...: norm = mpl.colors.Normalize(-abs(Z).max(), abs(Z).max())
    ...:
    ...: p = axes[0].plot_surface(
    ...:     X, Y, Z, rstride=1, cstride=1, linewidth=0, antialiased=False,
    ...:     norm=norm, cmap=mpl.cm.Blues)
    ...:
    ...: cb = fig.colorbar(p, ax=axes[0], shrink=0.6)
    ...: title_and_labels(axes[0], "plot_surface")
    ...:
    ...: p = axes[1].plot_wireframe(X, Y, Z, rstride=2, cstride=2,
    ...:                            color="darkgrey")
    ...: title_and_labels(axes[1], "plot_wireframe")
    ...:
    ...: cset = axes[2].contour(X, Y, Z, zdir='z', offset=0, norm=norm,
    ...:                        cmap=mpl.cm.Blues)
    ...: cset = axes[2].contour(X, Y, Z, zdir='y', offset=3, norm=norm,
    ...:                        cmap=mpl.cm.Blues)
    ...: title_and_labels(axes[2], "contour")
```

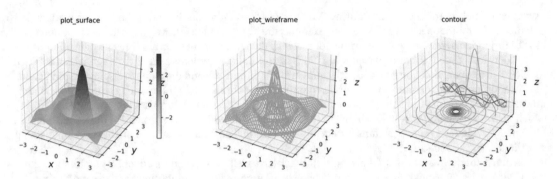

Figure 4-25. *3D surface and contour graphs generated by using* `plot_surface`, `plot_wireframe`, *and* `contour`

Summary

This chapter covered the basics of producing 2D and 3D graphics using Matplotlib. Visualization is one of the most important tools for computational scientists and engineers, both as an analysis tool while working on computational problems and for presenting and communicating computational results. Visualization is an integral part of the computational workflow, and it is equally important to be able to quickly visualize and explore data and to be able to produce picture-perfect publication-quality graphs, with detailed control over every graphical element. Matplotlib is a great general-purpose tool for both exploratory visualization and for producing publication-quality graphics. However, there are limitations to what can be achieved with Matplotlib, especially with respect to interactivity and high-quality 3D graphics. For more specialized cases, I therefore recommend exploring some of the other graphic libraries available in the scientific Python ecosystem, some of which were briefly mentioned at the beginning of this chapter.

Further Reading

Books dedicated to the Matplotlib library include *Matplotlib for Python Developers* by S. Tosi (Packt, 2009) and *Matplotlib Plotting Cookbook* by A. Devert (Packt, 2014). It is also widely covered in *Python Data Visualization Cookbook* by I. Milovanovi (Packt, 2013) and *Python for Data Analysis* by W. McKinney (O'Reilly, 2013). For interesting discussions on data visualization, style guides, and good practices in visualization, see *Visualize This* by N. Yau (Wiley, 2011) and *Beautiful Visualization* by J. Steele (O'Reilly, 2010).

CHAPTER 5

■ ■ ■

Equation Solving

The previous chapters discussed general methodologies and techniques, namely, array-based numerical computing, symbolic computing, and visualization. These methods are the cornerstones of scientific computing that make up a fundamental toolset we have at our disposal when attacking computational problems.

This chapter begins our exploration of solving problems from different domains of applied mathematics and computational sciences using the basic techniques. The topic of this chapter is algebraic equation solving. This is a broad topic that requires the application of theory and techniques from multiple fields of mathematics. When discussing equation solving, we must distinguish between univariate and multivariate equations (i.e., equations containing one unknown variable or many unknown variables). In addition, we need to distinguish between linear and nonlinear equations. This classification is useful because solving equations of these types requires applying different mathematical methods and approaches.

Let's begin with linear equation systems, which are tremendously useful and have important applications in every field of science. Linear algebra theory enables us to solve linear equations straightforwardly, whereas nonlinear equations are generally difficult to solve and typically require more complicated and computationally demanding methods. Important reasons why linear systems are so universal are because they are readily solvable and can be used in local approximations of more complicated nonlinear systems. For example, by considering small variations from an expansion point, a nonlinear system can often be approximated by a linear system in the local vicinity of the expansion point. However, linearization can only describe local properties, and other techniques are required for the global analysis of nonlinear problems. Such methods typically employ iterative approaches for gradually constructing an increasingly accurate estimate of the solution.

This chapter uses SymPy to solve equations symbolically when possible and uses the linear algebra module from the SciPy library to solve linear equation systems numerically. The root-finding functions in the `optimize` module of SciPy are used for tackling nonlinear problems.

■ **SciPy** SciPy is a Python library, the collective name of the scientific computing environment for Python, and the umbrella organization for many of the core libraries for scientific computing with Python. The SciPyi library is rather a collection of libraries for high-level scientific computing, which are more or less independent of each other. The SciPy library is built on top of NumPy, which provides the basic array data structures and fundamental operations on such arrays. The modules in SciPy provide domain-specific high-level computation methods, such as routines for linear algebra, optimization, interpolation, integration, and much more. At the time of writing, the most recent version of SciPy is 1.11.1. See `www.scipy.org` for more information.

© Robert Johansson 2024
R. Johansson, *Numerical Python*, https://doi.org/10.1007/979-8-8688-0413-7_5

Importing Modules

The SciPy module scipy should be considered a collection of selectively imported modules when required. This chapter uses the scipy.linalg module for solving linear systems of equations, and the scipy.optimize module for solving nonlinear equations. Assume that these SciPy modules are imported as follows.

```
In [1]: from scipy import linalg as la
In [2]: from scipy import optimize
```

This chapter also uses the NumPy, SymPy, and Matplotlib libraries introduced in earlier chapters. Assume that those libraries are imported using the following convention.

```
In [3]: import sympy
In [4]: sympy.init_printing()
In [5]: import numpy as np
In [6]: import matplotlib.pyplot as plt
```

Linear Equation Systems

An important application of linear algebra is solving systems of linear equations. We have already encountered linear algebra functionality in the SymPy library in Chapter 3. There are also linear algebra modules in the NumPy and SciPy libraries, numpy.linalg and scipy.linalg, which together provide linear algebra routines for numerical problems that are completely specified in terms of numerical factors and parameters.

A linear equation system can generally be written in the form

$$a_{11}x_1 + a_{12}x_2 + \ldots + a_{1n}x_n = b_1,$$

$$a_{21}x_1 + a_{22}x_2 + \ldots + a_{2n}x_n = b_2,$$

$$\ldots$$

$$a_{m1}x_1 + a_{m2}x_2 + \ldots + a_{mn}x_n = b_m.$$

This is a linear system of m equations in n unknown variables $\{x_1, x_2, \ldots, x_n\}$, where a_{mn} and b_m are known parameters or constant values. When working with linear equation systems, it is convenient to write them in matrix form:

$$\begin{pmatrix} a_{11} & a_{12} & \cdots & a_{1n} \\ a_{21} & a_{22} & \cdots & a_{2n} \\ \vdots & \vdots & \ddots & \vdots \\ a_{m1} & a_{m2} & \cdots & a_{mn} \end{pmatrix} \begin{pmatrix} x_1 \\ x_2 \\ \vdots \\ x_n \end{pmatrix} = \begin{pmatrix} b_1 \\ b_2 \\ \vdots \\ b_m \end{pmatrix},$$

or simply $Ax = b$, where A is a $m \times n$ matrix, b is a $m \times 1$ matrix (or m-vector), and x is the unknown $n \times 1$ solution matrix (or n-vector). Depending on the properties of the matrix A, the solution vector x may or may not exist, and if a solution does exist, it is not necessarily unique. However, if a solution exists, then vector b can be interpreted as a linear combination of the columns of matrix A, where the coefficients are given by the elements in the solution vector x.

A system for which $n < m$ is said to be underdetermined because it has fewer equations than unknown and therefore cannot completely determine a unique solution. If, on the other hand, $m > n$, then the equations are said to be overdetermined. This generally leads to conflicting constraints, and no solution exists to the equation system.

Square Systems

Square systems with $m = n$ is an important special case. It corresponds to the situation where the number of equations equals the number of unknown variables and can potentially have a unique solution. For a unique solution to exist, the matrix A must be *nonsingular*, in which case the inverse of A exists, and the solution can be written as $x = A^{-1}b$. If the matrix A is singular, that is, the rank of the matrix is less than n, rank$(A) < n$, or, equivalently, if its determinant is zero, det$A = 0$, then the equation $Ax = b$ can either have no solution or infinitely many solutions, depending on the right-hand side vector b. For a matrix with rank deficiency, rank$(A) < n$, some columns or rows can be expressed as linear combinations of other columns or vectors. Therefore, they correspond to equations without new constraints, and the system is underdetermined. Computing the rank of the matrix A that defines a linear equation system is a useful method that can tell us whether the matrix is singular or not and whether a solution exists.

When A has full rank, the solution is guaranteed to exist. However, it may be impossible to accurately compute the solution. The *condition number* of the matrix, cond(A), measures how well or poorly conditioned a linear equation system is. If the conditioning number is close to 1, if the system is said to be *well-conditioned* (condition number 1 is ideal), and if the condition number is large, the system is *ill-conditioned*. The solution to an equation system that is ill-conditioned can have large errors. A simple error analysis can provide an intuitive interpretation of the condition number. Assume a linear equation system in the form of $Ax = b$, where x is the solution vector. Now consider a small variation of b, say δb, which gives a corresponding change in the solution, δx, given by $A(x + \delta x) = b + \delta b$. Because of the linearity of the equation, we have $A\delta x = \delta b$. An important question to consider now is: how large is the relative change in x compared to the relative change in b? Mathematically, we can formulate this question in terms of the ratios of the norms of these vectors. Specifically, we are interested in comparing $\|\delta x\|/\|x\|$ and $\|\delta b\|/\|b\|$, where $\|x\|$ denotes the norm of x. Using the matrix norm relation $\|Ax\| \le \|A\| \cdot \|x\|$, we can write the following.

$$\frac{\|\delta x\|}{\|x\|} = \frac{\|A^{-1}\delta b\|}{\|x\|} \le \frac{\|A^{-1}\| \cdot \|\delta b\|}{\|x\|} = \frac{\|A^{-1}\| \cdot \|b\|}{\|x\|} \cdot \frac{\|\delta b\|}{\|b\|} \le \|A^{-1}\| \cdot \|A\| \cdot \frac{\|\delta b\|}{\|b\|}$$

A bound of the relative error in the solution x, given a relative error in the b vector, is therefore given by cond$(A) \equiv \|A^{-1}\| \cdot \|A\|$, which by definition is the condition number of the matrix A. This means that for linear equation systems characterized by an ill-conditioned matrix A, even a small perturbation in the b vector can give large errors in the solution vector x. This is particularly relevant in numerical solutions using floating-point numbers, which are only approximations of real numbers. When numerically solving a system of linear equations, it is therefore important to look at the condition number to estimate the accuracy of the solution.

The rank, condition number, and norm of a symbolic matrix can be computed in SymPy using the `Matrix` methods `rank`, `condition_number`, and `norm`, and for numerical problems, we can use the NumPy functions `np.linalg.matrix_rank`, `np.linalg.cond`, and `np.linalg.norm`. For example, consider the following system of two linear equations.

$$2x_1 + 3x_2 = 4$$
$$5x_1 + 4x_2 = 3$$

These two equations correspond to lines in the (x_1, x_2) plane, and their intersection is the solution to the equation system. As shown in Figure 5-1, which graphs the lines corresponding to the two equations, the lines intersect at $(-1, 2)$.

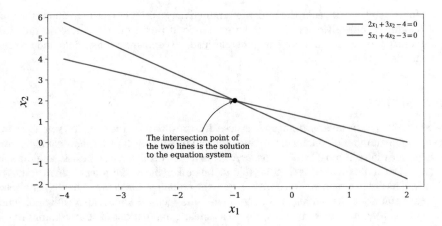

Figure 5-1. *Graphical representation of a system of two linear equations*

We can define this problem in SymPy by creating matrix objects for A and b and computing the rank, condition number, and norm of the matrix A using the following.

```
In [8]: A = sympy.Matrix([[2, 3], [5, 4]])
In [9]: b = sympy.Matrix([4, 3])
In [10]: A.rank()
Out[10]: 2
In [11]: A.condition_number()
```

$$\text{Out[11]:}\quad \frac{\sqrt{27+2\sqrt{170}}}{\sqrt{27-2\sqrt{170}}}$$

```
In [12]: sympy.N(_)
Out[12]: 7.58240137440151
In [13]: A.norm()
```

$\text{Out[13]:}\ 3\sqrt{6}$

We can do the same thing in NumPy/SciPy using NumPy arrays for A and b and functions from the np.linalg and scipy.linalg modules.

```
In [14]: A = np.array([[2, 3], [5, 4]])
In [15]: b = np.array([4, 3])
In [16]: np.linalg.matrix_rank(A)
Out[16]: 2
In [17]: np.linalg.cond(A)
Out[17]: 7.5824013744
In [18]: np.linalg.norm(A)
Out[18]: 7.34846922835
```

A direct approach to solving the linear problem is to compute the inverse of the matrix A and multiply it with the vector b, as used, for example, in the previous analytical discussions. However, this is not the most efficient computational method to find the solution vector x. A better method is LU factorization of the matrix A, such that $A = LU$ where L is a lower triangular matrix and U is an upper triangular matrix.

Given L and U, the solution vector x can be efficiently constructed by first solving $Ly = b$ with forward substitution and then solving $Ux = y$ with backward substitution. Since L and U are triangular matrices, these two procedures are computationally efficient.

In SymPy, we can perform a symbolic LU factorization by using the LUdecomposition method of the sympy.Matrix class. This method returns new Matrix objects for the L and U matrices, as well as a row swap matrix. When we are interested in solving an equation system $Ax = b$, we do not explicitly need to calculate the L and U matrices. But we can use the LUsolve method, which performs the LU factorization internally and solves the equation system using those factors. Returning to the previous example, we can compute the L and U factors and solve the equation system using the following.

```
In [19]: A = sympy.Matrix([[2, 3], [5, 4]])
In [20]: b = sympy.Matrix([4, 3])
In [21]: L, U, _ = A.LUdecomposition()
In [22]: L
```
$$\text{Out}[22]: \begin{bmatrix} 1 & 0 \\ 5/2 & 1 \end{bmatrix}$$

```
In [23]: U
```
$$\text{Out}[23]: \begin{bmatrix} 2 & 3 \\ 0 & -7/2 \end{bmatrix}$$

```
In [24]: L * U
```
$$\text{Out}[24]: \begin{bmatrix} 2 & 3 \\ 5 & 4 \end{bmatrix}$$

```
In [25]: x = A.solve(b); x  # equivalent to A.LUsolve(b)
```
$$\text{Out}[25]: \begin{bmatrix} -1 \\ 2 \end{bmatrix}$$

For numerical problems, we can use the la.lu function from SciPy's linear algebra module. It returns a permutation matrix P and the L and U matrices, such that $A = PLU$. Like with SymPy, we can solve the linear system $Ax = b$ without explicitly calculating the L and U matrices using the la.solve function, which takes the A matrix and the b vector as arguments. This is generally the preferred method for solving numerical linear equation systems using SciPy.

```
In [26]: A = np.array([[2, 3], [5, 4]])
In [27]: b = np.array([4, 3])
In [28]: P, L, U = la.lu(A)
In [29]: L
Out[29]: array([[ 1. ,  0. ],
               [ 0.4,  1. ]])
In [30]: U
Out[30]: array([[ 5. ,  4. ],
               [ 0. ,  1.4]])
In [31]: P.dot(L.dot(U))
Out[31]: array([[ 2., 3.],
               [ 5., 4.]])
In [32]: la.solve(A, b)
Out[32]: array([-1.,  2.])
```

The advantage of using SymPy is that we may obtain exact results, and we can also include symbolic variables in the matrices. However, not all problems are solvable symbolically or some may give exceedingly lengthy results. On the other hand, the advantage of using a numerical approach with NumPy/SciPy is that we are guaranteed to obtain a result. However, it will be an approximate solution due to floating-point errors. See the following code (In [38]) for an example that illustrates the differences between the symbolic and numerical approaches and that numerical approaches can be sensitive for equation systems with large condition numbers. This example solves the equation system

$$\begin{pmatrix} 1 & \sqrt{p} \\ 1 & \dfrac{1}{\sqrt{p}} \end{pmatrix} \begin{pmatrix} x_1 \\ x_2 \end{pmatrix} = \begin{pmatrix} 1 \\ 2 \end{pmatrix}$$

which for $p = 1$ is singular and for p in the vicinity of one is ill-conditioned. Using SymPy, the solution is easily found to be as follows.

```
In [33]: p = sympy.symbols("p", positive=True)
In [34]: A = sympy.Matrix([[1, sympy.sqrt(p)], [1, 1/sympy.sqrt(p)]])
In [35]: b = sympy.Matrix([1, 2])
In [36]: x = A.solve(b)
In [37]: x
```

$$\text{Out[37]:} \quad \begin{pmatrix} \dfrac{2p-1}{p-1} \\ -\dfrac{\sqrt{p}}{p-1} \end{pmatrix}$$

A comparison of this symbolic solution and the numerical solution is shown in Figure 5-2. Here, the errors in the numerical solution are due to numerical floatingpoint errors, and the numerical errors are significantly larger in the vicinity of $p = 1$, where the system has a large condition number. Also, if there are other sources of errors in either A or b, the corresponding errors in x can be even more severe.

```
In [38]: # Symbolic problem specification
    ...: p = sympy.symbols("p", positive=True)
    ...: A = sympy.Matrix([[1, sympy.sqrt(p)], [1, 1/sympy.sqrt(p)]])
    ...: b = sympy.Matrix([1, 2])
    ...:
    ...: # Solve symbolically
    ...: x_sym_sol = A.solve(b)
    ...: Acond = A.condition_number().simplify()
    ...:
    ...: # Numerical problem specification
    ...: AA = lambda p: np.array([[1, np.sqrt(p)], [1, 1/np.sqrt(p)]])
    ...: bb = np.array([1, 2])
    ...: x_num_sol = lambda p: np.linalg.solve(AA(p), bb)
    ...:
    ...: # Graph the difference between the symbolic (exact) and numerical results
    ...: fig, axes = plt.subplots(1, 2, figsize=(12, 4))
    ...:
    ...: p_vec = np.linspace(0.9, 1.1, 200)
    ...: for n in range(2):
    ...:     x_sym = np.array([x_sym_sol[n].subs(p, pp).evalf() for pp in p_vec])
```

```
...:     x_num = np.array([x_num_sol(pp)[n] for pp in p_vec])
...:     axes[0].plot(p_vec, (x_num - x_sym)/x_sym, 'k')
...: axes[0].set_title("Error in solution\n(numerical - symbolic)/symbolic")
...: axes[0].set_xlabel(r'$p$', fontsize=18)
...:
...: axes[1].plot(p_vec, [Acond.subs(p, pp).evalf() for pp in p_vec])
...: axes[1].set_title("Condition number")
...: axes[1].set_xlabel(r'$p$', fontsize=18)
```

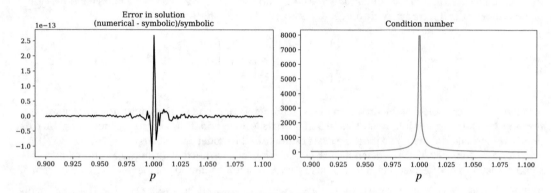

Figure 5-2. *Graph of the relative numerical errors (left) and condition number (right) as a function of the parameter p*

Rectangular Systems

Rectangular systems, with $m \neq n$, can be either underdetermined or overdetermined. Underdetermined systems have more variables than equations, so the solution cannot be fully determined. Therefore, for such a system, the solution must be given in terms of the remaining free variables. This makes it difficult to treat this type of problem numerically, but a symbolic approach can often be used instead.

For example, consider the underdetermined linear equation system

$$\begin{pmatrix} 1 & 2 & 3 \\ 4 & 5 & 6 \end{pmatrix} \begin{pmatrix} x_1 \\ x_2 \\ x_3 \end{pmatrix} = \begin{pmatrix} 7 \\ 8 \end{pmatrix}.$$

There are three unknown variables, but only two equations impose constraints on the relations between these variables. By writing this equation as $Ax - b = 0$, we can use the SymPy sympy.solve function to obtain a solution for x_1 and x_2 parameterized by the remaining free variable x_3.

```
In [39]: x_vars = sympy.symbols("x_1, x_2, x_3")
In [40]: A = sympy.Matrix([[1, 2, 3], [4, 5, 6]])
In [41]: x = sympy.Matrix(x_vars)
In [42]: b = sympy.Matrix([7, 8])
In [43]: sympy.solve(A*x - b, x_vars)
Out[43]: {x_1 = x_3 - 19/3, 0.5x_2 = -2x_3 + 20/3}
```

135

This obtained the symbolic solution $x_1 = x_3 - 19/3$ and $x_2 = -2x_3 + 20/3$, which defines a line in the three-dimensional space spanned by $\{x_1, x_2, x_3\}$. Any point on this line, therefore, satisfies this underdetermined equation system.

On the other hand, if the system is overdetermined and has more equations than unknown variables, $m > n$, then we may have more constraints than degrees of freedom, and in general, there is no exact solution to such a system. However, finding an approximate solution to an overdetermined system is often interesting. An example of when this situation arises is data fitting: let's say we have a model where a variable y is a quadratic polynomial in the variable x so that $y = A + Bx + Cx^2$. We would like to fit this model to experimental data. Here, y is nonlinear in x, but y is linear in the three unknown coefficients A, B, and C, and this fact can be used to write the model as a linear equation system. If we collect data for m pairs $\{(x_i, y_i)\}_{i=1}^{m}$ of the variables x and y, we can write the model as an $m \times 3$ equation system:

$$\begin{pmatrix} 1 & x_1 & x_1^2 \\ \vdots & \vdots & \vdots \\ 1 & x_m & x_m^2 \end{pmatrix} \begin{pmatrix} A \\ B \\ C \end{pmatrix} = \begin{pmatrix} y_1 \\ \vdots \\ y_m \end{pmatrix}.$$

If $m = 3$, we can solve for the unknown model parameters A, B, and C, assuming the system matrix is nonsingular. However, it is intuitively clear that if the data is noisy and if we were to use more than three data points, we should be able to get a more accurate estimate of the model parameters.

However, for $m > 3$, there is generally no exact solution, and we need to introduce an approximate solution that best fits the overdetermined system $Ax \approx b$. A natural definition of the best fit for this system is to minimize the sum of square errors, $\min_x \sum_{i=1}^{m} (r_i)^2$, where $r = b - Ax$ is the residual vector. This leads to the *least square* solution of the problem $Ax \approx b$, which minimizes the distances between the data points and the linear solution. In SymPy, we can solve for the least square solution of an overdetermined system using the solve_least_squares method; for numerical problems, we can use the SciPy function la.lstsq.

The following code demonstrates how the SciPy la.lstsq method can be used to fit the example model considered in the preceding section, and the result is shown in Figure 5-3. First, we define the true parameters of the model, and then we simulate measured data by adding random noise to the true model relation. The least-square problem is then solved using the la.lstsq function, which in addition to the solution vector x also returns the total sum of square errors (the residual r), the rank rank, and the singular values sv of the matrix A. However, the following example only uses the solution vector x.

```
In [44]: # define true model parameters
    ...: x = np.linspace(-1, 1, 100)
    ...: a, b, c = 1, 2, 3
    ...: y_exact = a + b * x + c * x**2
    ...:
    ...: # simulate noisy data
    ...: m = 100
    ...: X = 1 - 2 * np.random.rand(m)
    ...: Y = a + b * X + c * X**2 + np.random.randn(m)
    ...:
    ...: # fit the data to the model using linear least square
    ...: A = np.vstack([X**0, X**1, X**2])  # see np.vander for alternative
    ...: sol, r, rank, sv = la.lstsq(A.T, Y)
    ...:
    ...: y_fit = sol[0] + sol[1] * x + sol[2] * x**2
    ...: fig, ax = plt.subplots(figsize=(12, 4))
    ...:
```

```
...: ax.plot(X, Y, 'go', alpha=0.5, label='Simulated data')
...: ax.plot(x, y_exact, 'k', lw=2, label='True value $y = 1 + 2x + 3x^2$')
...: ax.plot(x, y_fit, 'b', lw=2, label='Least square fit')
...: ax.set_xlabel(r"$x$", fontsize=18)
...: ax.set_ylabel(r"$y$", fontsize=18)
...: ax.legend(loc=2)
```

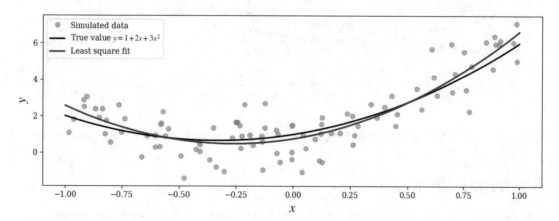

Figure 5-3. *Linear least square fit*

A good data fit to a model requires the model used to describe the data to correspond well to the underlying process that produced the data. The following example (In [45]) and Figure 5-4 fit the same data used in the previous example to a linear model and to a higher-order polynomial model (up to order 15). The former case corresponds to underfitting, where a too-simple model is used for the data, and the latter case corresponds to overfitting, where a too-complex model is used for the data, and thus fits the model not only to the trend and relevant patterns but also to the measurement noise. Using an appropriate model is an important and delicate aspect of data fitting.

```
In [45]: # fit the data to the model using linear least square:
...: # 1st order polynomial
...: A = np.vstack([X**n for n in range(2)])
...: sol, r, rank, sv = la.lstsq(A.T, Y)
...: y_fit1 = sum([s * x**n for n, s in enumerate(sol)])
...:
...: # 15th order polynomial
...: A = np.vstack([X**n for n in range(16)])
...: sol, r, rank, sv = la.lstsq(A.T, Y)
...: y_fit15 = sum([s * x**n for n, s in enumerate(sol)])
...:
...: fig, ax = plt.subplots(figsize=(12, 4))
...: ax.plot(X, Y, 'go', alpha=0.5, label='Simulated data')
...: ax.plot(x, y_exact, 'k', lw=2, label='True value $y = 1 + 2x + 3x^2$')
...: ax.plot(x, y_fit1, 'b', lw=2, label='Least square fit [1st order]')
...: ax.plot(x, y_fit15, 'm', lw=2, label='Least square fit [15th order]')
...: ax.set_xlabel(r"$x$", fontsize=18)
...: ax.set_ylabel(r"$y$", fontsize=18)
...: ax.legend(loc=2)
```

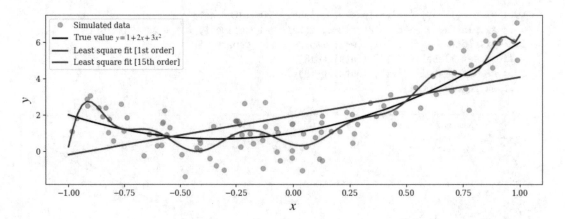

Figure 5-4. *Graph demonstrating underfitting and overfitting of data using the linear least square method*

Eigenvalue Problems

A special system of equations of great theoretical and practical importance is the eigenvalue equation $Ax = \lambda x$, where A is an $N \times N$ square matrix, x is an unknown vector, and λ is an unknown scalar. Here x is an eigenvector and λ an eigenvalue of the matrix A. The eigenvalue equation $Ax = \lambda x$ closely resembles the linear equation system $Ax = b$, but note that both x and λ are unknown here, so we cannot directly apply the same techniques to solve this equation. A standard approach to solve this eigenvalue problem is to rewrite the equation as $(A - I\lambda)x = 0$ and note that for there to exist a nontrivial solution, $x \neq 0$, the matrix $A - I\lambda$ must be singular, and its determinant must be zero, $\det(A - I\lambda) = 0$. This gives a polynomial equation (the characteristic polynomial) of Nth order whose N roots give the N eigenvalues. $\{\lambda_n\}_{n=1}^{N}$ Once the eigenvalues are known, standard forward substitution can solve the equation $(A - I\lambda_n)x_n = 0$ for the Nth eigenvector x_n.

SymPy and the linear algebra package in SciPy contain solvers for eigenvalue problems. In SymPy, we can use the `eigenvals` and `eigenvects` methods of the `Matrix` class, which can compute the eigenvalues and eigenvectors of some matrices with elements that are symbolic expressions. For example, we can use the following to compute the eigenvalues and eigenvectors of a symmetric 2×2 matrix with symbolic elements.

```
In [46]: eps, delta = sympy.symbols("epsilon, Delta")
In [47]: H = sympy.Matrix([[eps, delta], [delta, -eps]])
In [48]: H
```

$$\text{Out[48]:} \begin{pmatrix} \varepsilon & \Delta \\ \Delta & -\varepsilon \end{pmatrix}$$

```
In [49]: H.eigenvals()
```

$$\text{Out[49]:} \left\{ -\sqrt{\varepsilon^2 + \Delta^2} : 1, \sqrt{\varepsilon^2 + \Delta^2} : 1 \right\}$$

```
In [50]: H.eigenvects()
```

$$\text{Out[50]:} \left[\left(-\sqrt{\varepsilon^2 + \Delta^2}, 1, \left[\begin{bmatrix} -\dfrac{\Delta}{\varepsilon + \sqrt{\varepsilon^2 + \Delta^2}} \\ 1 \end{bmatrix} \right] \right), \left(\sqrt{\varepsilon^2 + \Delta^2}, 1, \left[\begin{bmatrix} -\dfrac{\Delta}{\varepsilon - \sqrt{\varepsilon^2 + \Delta^2}} \\ 1 \end{bmatrix} \right] \right) \right]$$

The return value of the `eigenvals` method is a dictionary where each eigenvalue is a key, and the corresponding value is the multiplicity of that particular eigenvalue. Here the eigenvalues are $-\sqrt{\varepsilon^2 + \Delta^2}$ and $\sqrt{\varepsilon^2 + \Delta^2}$, each with multiplicity one. The return value of `eigenvects` is more involved: a list is returned where each element is a tuple containing an eigenvalue, the multiplicity of the eigenvalue, and a list of eigenvectors. The number of eigenvectors for each eigenvalue equals the multiplicity. For the current example, we can unpack the value returned by eigenvects and verify that the two eigenvectors are orthogonal using, for example, as follows.

```
In [51]: (eval1, _, evec1), (eval2, _, evec2) = H.eigenvects()
In [52]: sympy.simplify(evec1[0].T * evec2[0])
Out[52]: [0]
```

Obtaining analytical expressions for eigenvalues and eigenvectors using these methods is often very desirable, but unfortunately, it only works for small matrices. For anything larger than a 3×3, the analytical expression typically becomes extremely lengthy and cumbersome to work with, even using a computer algebra system such as SymPy. Therefore, we must resort to a fully numerical approach for larger systems. We can use the `la.eigvals` and `la.eig` functions in the SciPy linear algebra package. Matrices that are either Hermitian or real and symmetric have real-valued eigenvalues. For such matrices, it is advantageous to use the `la.eigvalsh` and `la.eigh` functions instead, which guarantees that the eigenvalues returned by the function are stored in a NumPy array with real values. For example, to solve a numerical eigenvalue problem with `la.eig`, we can use the following.

```
In [53]: A = np.array([[1, 3, 5], [3, 5, 3], [5, 3, 9]])
In [54]: evals, evecs = la.eig(A)
In [55]: evals
Out[55]: array([ 13.35310908+0.j,  -1.75902942+0.j,   3.40592034+0.j])
In [56]: evecs
Out[56]: array([[ 0.42663918,  0.90353276, -0.04009445],
                [ 0.43751227, -0.24498225, -0.8651975 ],
                [ 0.79155671, -0.35158534,  0.49982569]])
In [57]: la.eigvalsh(A)
Out[57]: array([ -1.75902942,   3.40592034,  13.35310908])
```

Since the matrix in this example is symmetric, we could use `la.eigh` and `la.eigvalsh`, giving real-valued eigenvalue arrays, as shown in the cell `Out[57]` in the preceding code listing.

Nonlinear Equations

This section considers *nonlinear* equations. Systems of linear equations, as considered in the prior sections, are of fundamental importance in scientific computing because they are easily solved and can be used as important building blocks in many computational methods and techniques. However, in natural sciences and engineering disciplines, many, if not most, systems are intrinsically nonlinear.

A linear function $f(x)$, by definition, satisfies additivity $f(x + y) = f(x) + f(y)$ and homogeneity $f(\alpha x) = \alpha f(x)$, which can be written together as the superposition principle $f(\alpha x + \beta y) = \alpha f(x) + \beta f(y)$. This gives a precise definition of linearity. A *nonlinear* function, in contrast, is a function that does not satisfy these conditions. Nonlinearity is a much broader concept; a function can be nonlinear in many ways. However, an expression that contains a variable with a power greater than one is nonlinear. For example, $x^2 + x + 1$ is nonlinear because of the x^2 term.

A nonlinear equation can always be written in the form $f(x) = 0$, where $f(x)$ is a nonlinear function that seeks the value of x (which can be a scalar or a vector) such that $f(x)$ is zero. This x is the root of the $f(x)$ function, and equation-solving is often called *root finding*. In contrast to the previous section of this chapter, here we need to distinguish between univariate equation solving and multivariate equations, in addition to single equations and systems of equations.

Univariate Equations

A univariate function $f(x)$ is a function that depends only on a single scalar variable x, and the corresponding univariate equation is $f(x) = 0$. Typical examples of this type of equation are polynomials, such as $x^2 - x + 1 = 0$, and expressions containing elementary functions, such as $x^3 - 3 \sin(x) = 0$ and $\exp(x) - 2 = 0$. Unlike linear systems, there are no general methods for determining if a nonlinear equation has a solution or multiple solutions or if a given solution is unique. This can be understood intuitively from the fact that graphs of nonlinear functions correspond to curves that can intersect $y = 0$ in an arbitrary number of ways.

Because of the vast number of possible situations, it is difficult to develop a completely automatic approach to solving nonlinear equations. Analytically, only equations on special forms can be solved exactly. For example, polynomials of up to 4th order, and in some special cases also higher orders, can be solved analytically, and some equations containing trigonometric and other elementary functions may be solvable analytically. In SymPy, we can solve many solvable univariate and nonlinear equations using the `sympy.solve` function. For example, to solve the standard quadratic equation $a + bx + cx^2 = 0$, define an expression for the equation, and pass it to the `sympy.solve` function.

```
In [58]: x, a, b, c = sympy.symbols("x, a, b, c")
In [59]: sympy.solve(a + b*x + c*x**2, x)
Out[59]: [(-b + sqrt(-4*a*c + b**2))/(2*c), -(b + sqrt(-4*a*c + b**2))/(2*c)]
```

The solution is the well-known formula for the solution of this equation. The same method can be used to solve some trigonometric equations.

```
In [60]: sympy.solve(a * sympy.cos(x) - b * sympy.sin(x), x)
Out[60]: [-2*atan((b - sqrt(a**2 + b**2))/a), -2*atan((b + sqrt(a**2 + b**2))/a)]
```

However, in general, nonlinear equations are typically not solvable analytically. For example, equations that contain both polynomial expressions and elementary functions, such as $\sin(x) = x$, are often transcendental and do not have an algebraic solution. If we attempt to solve such an equation using SymPy, we obtain an error in the form of an exception.

```
In [61]: sympy.solve(sympy.sin(x)-x, x)
...
NotImplementedError: multiple generators [x, sin(x)]
No algorithms are implemented to solve equation -x + sin(x)
```

In this situation, we must resort to various numerical techniques. As a first step, graphing the function is often very useful. This can give important clues about the number of solutions to the equation and their approximate locations. This information is often necessary when applying numerical techniques to find good approximations to the roots of the equations. For example, consider the following example (In [62]), which plots four examples of nonlinear functions, as shown in Figure 5-5. From these graphs, we can immediately conclude that the plotted functions, from left to right, have two, three, one, and a large number of roots (at least within the interval being graphed).

```
In [62]: x = np.linspace(-2, 2, 1000)
    ...: # four examples of nonlinear functions
    ...: f1 = x**2 - x - 1
    ...: f2 = x**3 - 3 * np.sin(x)
    ...: f3 = np.exp(x) - 2
    ...: f4 = 1 - x**2 + np.sin(50 / (1 + x**2))
    ...:
    ...: # plot each function
    ...: fig, axes = plt.subplots(1, 4, figsize=(12, 3), sharey=True)
    ...:
    ...: for n, f in enumerate([f1, f2, f3, f4]):
    ...:     axes[n].plot(x, f, lw=1.5)
    ...:     axes[n].axhline(0, ls=':', color='k')
    ...:     axes[n].set_ylim(-5, 5)
    ...:     axes[n].set_xticks([-2, -1, 0, 1, 2])
    ...:     axes[n].set_xlabel(r'$x$', fontsize=18)
    ...:
    ...: axes[0].set_ylabel(r'$f(x)$', fontsize=18)
    ...:
    ...: titles = [r'$f(x)=x^2-x-1$', r'$f(x)=x^3-3\sin(x)$',
    ...:           r'$f(x)=\exp(x)-2$',
    ...:           r'$f(x)=\sin\left(50/(1+x^2)\right)+1x^2$']
    ...: for n, title in enumerate(titles):
    ...:     axes[n].set_title(title)
```

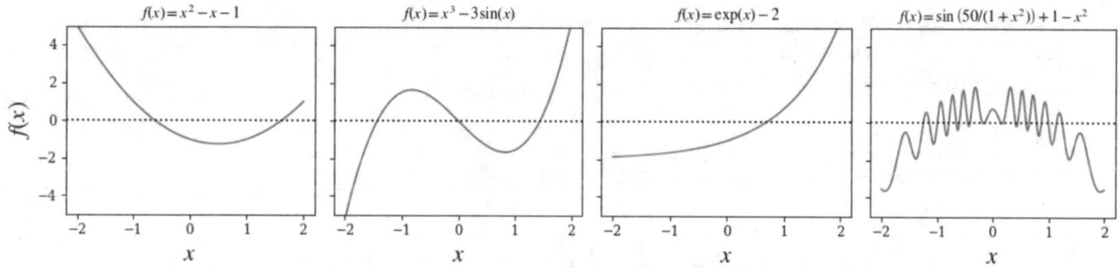

Figure 5-5. *Graphs of four examples of nonlinear functions*

To find the approximate location of a root to an equation, we can apply one of the many techniques for numerical root finding, which typically applies an iterative scheme where the function is evaluated at successive points until the algorithm has narrowed in on the solution to the desired accuracy. Two standard methods that illustrate the basic idea of how many numerical root-finding methods work are the bisection method and the Newton method.

The bisection method requires a starting interval $[a, b]$ such that $f(a)$ and $f(b)$ have different signs. This guarantees that there is at least one root within this interval. In each iteration, the function is evaluated in the middle point m between a and b, and the sign of the function is different at a and m, and then the new interval $[a, b = m]$ is chosen for the next iteration. Otherwise, the interval $[a = m, b]$ is chosen for the next iteration. This guarantees that in each iteration, the function has a different sign at the two endpoints of the interval. In each iteration, the interval is halved and converges toward the root of the equation. The following code example demonstrates a simple implementation of the bisection method with a graphical visualization of each step, as shown in Figure 5-6.

```
In [63]: # define a function, desired tolerance and starting interval [a, b]
   ...: f = lambda x: np.exp(x) - 2
   ...: tol = 0.1
   ...: a, b = -2, 2
   ...: x = np.linspace(-2.1, 2.1, 1000)
   ...:
   ...: # graph the function f
   ...: fig, ax = plt.subplots(1, 1, figsize=(12, 4))
   ...:
   ...: ax.plot(x, f(x), lw=1.5)
   ...: ax.axhline(0, ls=':', color='k')
   ...: ax.set_xticks([-2, -1, 0, 1, 2])
   ...: ax.set_xlabel(r'$x$', fontsize=18)
   ...: ax.set_ylabel(r'$f(x)$', fontsize=18)
   ...:
   ...: # find the root using the bisection method and visualize
   ...: # the steps in the method in the graph
   ...: fa, fb = f(a), f(b)
   ...:
   ...: ax.plot(a, fa, 'ko')
   ...: ax.plot(b, fb, 'ko')
   ...: ax.text(a, fa + 0.5, r"$a$", ha='center', fontsize=18)
   ...: ax.text(b, fb + 0.5, r"$b$", ha='center', fontsize=18)
   ...:
   ...: n = 1
   ...: while b - a > tol:
   ...:     m = a + (b - a)/2
   ...:     fm = f(m)
   ...:
   ...:     ax.plot(m, fm, 'ko')
   ...:     ax.text(m, fm - 0.5, r"$m_%d$" % n, ha='center')
   ...:     n += 1
   ...:
   ...:     if np.sign(fa) == np.sign(fm):
   ...:         a, fa = m, fm
   ...:     else:
   ...:         b, fb = m, fm
   ...:
   ...: ax.plot(m, fm, 'r*', markersize=10)
   ...: ax.annotate(
   ...:     "Root approximately at %.3f" % m,
   ...:     fontsize=14, family="serif", xy=(a, fm), xycoords='data',
   ...:     xytext=(-150, +50), textcoords='offset points',
   ...:     arrowprops=dict(arrowstyle="->", connectionstyle="arc3, rad=-.5"))
   ...:
   ...: ax.set_title("Bisection method")
```

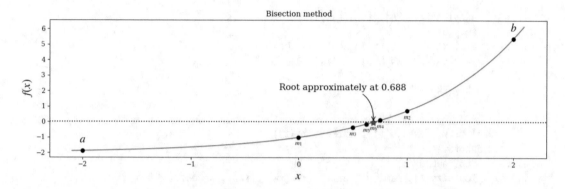

Figure 5-6. *Graphical visualization of how the bisection method works*

Another standard root-finding method is Newton's method, which converges faster than the bisection method discussed in the previous paragraph. While the bisection method only uses the sign of the function at each point, Newton's method uses the actual function values to obtain a more accurate approximation of the nonlinear function. It approximates the $f(x)$ function with its first-order Taylor expansion $f(x + dx) = f(x) + dx f'(x)$, which is a linear function whose root is easily found to be $x - f(x)/f'(x)$. Of course, this does not need to be a root of the $f(x)$ function, but in many cases, it is a good approximation for getting closer to a root of $f(x)$. By iterating this scheme, $x_{k+1} = x_k - f(x_k)/f'(x_k)$, we may approach the root of the function. A potential problem with this method is that it fails if $f'(x_k)$ is zero at some point x_k. This special case would have to be dealt with in a real implementation of this method. The following example (In [64]) demonstrates how this method can be used to solve for the root of the equation $\exp(x) - 2 = 0$, using SymPy to evaluate the derivative of the $f(x)$ function, and Figure 5-7 visualizes the steps in this root-finding process.

```
In [64]: # define a function, desired tolerance and starting point xk
    ...: tol = 0.01
    ...: xk = 2
    ...:
    ...: s_x = sympy.symbols("x")
    ...: s_f = sympy.exp(s_x) - 2
    ...:
    ...: f = lambda x: sympy.lambdify(s_x, s_f, 'numpy')(x)
    ...: fp = lambda x: sympy.lambdify(s_x, sympy.diff(s_f, s_x), 'numpy')(x)
    ...:
    ...: x = np.linspace(-1, 2.1, 1000)
    ...:
    ...: # setup a graph for visualizing the root finding steps
    ...: fig, ax = plt.subplots(1, 1, figsize=(12, 4))
    ...: ax.plot(x, f(x))
    ...: ax.axhline(0, ls=':', color='k')
    ...:
    ...: # iterate Newton's method until convergence to the desired tolerance
    ...: # has been reached
    ...: n = 0
    ...: while f(xk) > tol:
    ...:     xk_new = xk - f(xk) / fp(xk)
    ...:
```

```
...:        ax.plot([xk, xk], [0, f(xk)], color='k', ls=':')
...:        ax.plot(xk, f(xk), 'ko')
...:        ax.text(xk, -.5, r'$x_%d$' % n, ha='center')
...:        ax.plot([xk, xk_new], [f(xk), 0], 'k-')
...:
...:        xk = xk_new
...:        n += 1
...:
...: ax.plot(xk, f(xk), 'r*', markersize=15)
...: ax.annotate(
...:     "Root approximately at %.3f" % xk,
...:     fontsize=14, family="serif", xy=(xk, f(xk)), xycoords='data',
...:     xytext=(-150, +50), textcoords='offset points',
...:     arrowprops=dict(arrowstyle="->", connectionstyle="arc3, rad=-.5"))
...:
...: ax.set_title("Newtown's method")
...: ax.set_xticks([-1, 0, 1, 2])
```

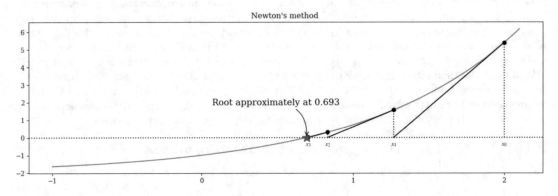

Figure 5-7. *Visualization of the root-finding steps in Newton's method for the equation exp(x) – 2 = 0*

A potential issue with Newton's method is that it requires both the function values and the values of the function's derivative in each iteration. The previous example used SymPy to compute the derivatives symbolically. In an all-numerical implementation, this is not possible, and a numerical approximation of the derivative would be necessary, which would require further function evaluations. A variant of Newton's method that bypasses the requirement to evaluate function derivatives is the secant method, which uses two previous function evaluations to obtain a linear approximation of the function, which can be used to compute a new root estimate. The iteration formula for the secant method is $x_{k+1} = x_k - f(x_k)\dfrac{x_k - x_{k-1}}{f(x_k) - f(x_{k-1})}$. This is only one example of the many variants and possible refinements to the basic idea of Newton's method. State-of-the-art implementations of numerical root-finding functions typically use either the bisection method of Newton's method or a combination of both but also use various additional strategies, such as higher-order interpolations of the function, to achieve faster convergence.

The SciPy optimize module provides multiple functions for numerical root finding. The `optimize.bisect` and `optimize.newton` functions implement variants of bisection and Newton methods. The `optimize.bisect` takes three arguments: first a Python function (e.g., a lambda function) that represents the mathematical function for the equation for which a root is to be calculated, and the second and third arguments are the lower and upper values of the interval for which to perform the bisection method.

Note that the sign of the function must be different at points a and b for the bisection method to work, as discussed earlier. Using the `optimize.bisect` function, we can calculate the root of the equation $\exp(x) - 2 = 0$ from previous examples.

```
In [65]: optimize.bisect(lambda x: np.exp(x) - 2, -2, 2)
Out[65]: 0.6931471805592082
```

As long as $f(a)$ and $f(b)$ have different signs, this is guaranteed to give a root within the interval $[a, b]$. In contrast, the `optimize.newton` function for Newton's method takes a function as the first argument and an initial guess for the root of the function as the second argument. Optionally, it also takes an argument for specifying the function's derivative using the `fprime` keyword argument. If `fprime` is given, Newton's method is used; otherwise, the secant method is used instead. To find the root of the equation $\exp(x) - 2 = 0$, with and without specifying its derivative, we can use the following.

```
In [66]: x_root_guess = 2
In [67]: f = lambda x: np.exp(x) - 2
In [68]: fprime = lambda x: np.exp(x)
In [69]: optimize.newton(f, x_root_guess)
Out[69]: 0.69314718056
In [70]: optimize.newton(f, x_root_guess, fprime=fprime)
Out[70]: 0.69314718056
```

Note that this method gives we less control over which root is being computed if the function has multiple roots. For instance, there is no guarantee that the root the function returns is closest to the initial guess; we cannot know in advance if the root is larger or smaller than the initial guess.

The SciPy `optimize` module provides additional functions for root finding. In particular, the `optimize.brentq` and `optimize.brenth` functions, which are variants of the bisection method, also work on an interval where the function changes sign. The `optimize.brentq` function is generally considered the preferred all-around root-finding function in SciPy. To find a root of the same equation considered previously, use the `optimize.brentq` and `optimize.brenth` functions, as follows.

```
In [71]: optimize.brentq(lambda x: np.exp(x) - 2, -2, 2)
Out[71]: 0.6931471805599453
In [72]: optimize.brenth(lambda x: np.exp(x) - 2, -2, 2)
Out[72]: 0.6931471805599381
```

Note that these two functions take a Python function for the equation as the first argument and the lower and upper values of the sign-changing interval as the second and third arguments.

Systems of Nonlinear Equations

In contrast to a linear system of equations, we cannot generally write a system of nonlinear equations as a matrix-vector multiplication. Instead, we represent a system of multivariate nonlinear equations as a vector-valued function, for example, $f: \mathbb{R}^N \to \mathbb{R}^N$, that takes an N-dimensional vector and maps it to another N-dimensional vector. Multivariate systems of equations are much more complicated to solve than univariate equations, partly because there are so many more possible behaviors. As a consequence, no method strictly guarantees convergence to a solution, such as the bisection method for a univariate nonlinear equation, and the existing methods are much more computationally demanding than the univariate case, especially as the number of variables increases.

Not all methods discussed for univariate equation solving can be generalized to the multivariate case. For example, the bisection method cannot be directly generalized to a multivariate equation system. On the other hand, Newton's method can be used for multivariate problems. In this case, its iteration formula is $x_{k+1} = x_k - J_f(x_k)^{-1}f(x_k)$, where $J_f(x_k)$ is the Jacobian matrix of the $f(x)$ function, with elements $[J_f(x_k)]_{ij} = \partial f_i(x_k)/\partial x_j$. Instead of inverting the Jacobian matrix, it is sufficient to solve the linear equation system $J_f(x_k)\delta x_k = -f(x_k)$ and update x_k using $x_{k+1} = x_k + \delta x_k$. Like the secant variant of the Newton method for univariate equation systems, there are also variants of the multivariate method that avoid computing the Jacobian by estimating it from previous function evaluations. Broyden's method is a popular example of this secant updating method for multivariate equation systems. In the SciPy optimize module, broyden1 and broyden2 provide two implementations of Broyden's method using different approximations of the Jacobian and the optimize.fsolve function implements a Newton-like method, where optionally, the Jacobian can be specified, if available. The functions all have a similar function signature: The first argument is a Python function that represents the equation to be solved, and it should take a NumPy array as the first argument and return an array of the same shape. The second argument is an initial guess for the solution as a NumPy array. The optimize.fsolve function also takes an optional keyword argument fprime, which can be used to provide a function that returns the Jacobian of the $f(x)$ function. In addition, all these functions take numerous optional keyword arguments for tuning their behavior (see the docstrings for details).

For example, consider the following system of two multivariate and nonlinear equations

$$\begin{cases} y - x^3 - 2x^2 + 1 = 0 \\ y + x^2 - 1 = 0 \end{cases},$$

which can be represented by the vector-valued function $f([x_1, x_2]) = [x_2 - x_1^3 - 2x_1^2 + 1, x_2 + x_1^2 - 1]$. To solve this equation system using SciPy, we need to define a Python function for $f([x_1, x_2])$ and call, for example, the optimize.fsolve using the function and an initial guess for the solution vector.

```
In [73]: def f(x):
   ...:     return [x[1] - x[0]**3 - 2 * x[0]**2 + 1, x[1] + x[0]**2 - 1]
In [74]: optimize.fsolve(f, [1, 1])
Out[74]: array([ 0.73205081,  0.46410162])
```

The optimize.broyden1 and optimize.broyden2 can be used similarly. To specify a Jacobian for optimize.fsolve, we need to define a function that evaluates the Jacobian for a given input vector. This requires that we first derive the Jacobian by hand or, for example, using SymPy.

```
In [75]: x, y = sympy.symbols("x, y")
In [76]: f_mat = sympy.Matrix([y - x**3 -2*x**2 + 1, y + x**2 - 1])
In [77]: f_mat.jacobian(sympy.Matrix([x, y]))
```

$$\text{Out[77]: } \begin{pmatrix} -3x^2 - 4x & 1 \\ 2x & 1 \end{pmatrix}$$

which we can then easily implement as a Python function that can be passed to the optimize.fsolve function.

```
In [78]: def f_jacobian(x):
   ...:     return [[-3*x[0]**2-4*x[0], 1], [2*x[0], 1]]
In [79]: optimize.fsolve(f, [1, 1], fprime=f_jacobian)
Out[79]: array([ 0.73205081,  0.46410162])
```

As with Newton's method for a univariate nonlinear equation system, the initial guess for the solution is important, and different initial guesses may result in different solutions being found for the equations. There is no guarantee that any solution is found, although the proximity of the initial guess to the true solution is often correlated with convergence to that particular solution. When possible, it is often a good approach to graph the equations being solved to visually indicate the number of solutions and their locations. For example, the following code demonstrates how three different solutions can be found to the equation systems we are considering here, by using different initial guesses with the optimize.fsolve function. The result is shown in Figure 5-8.

```
In [80]: def f(x):
   ...:         return [x[1] - x[0]**3 - 2 * x[0]**2 + 1,
   ...:                 x[1] + x[0]**2 - 1]
   ...:
   ...: x = np.linspace(-3, 2, 5000)
   ...: y1 = x**3 + 2 * x**2 -1
   ...: y2 = -x**2 + 1
   ...:
   ...: fig, ax = plt.subplots(figsize=(8, 4))
   ...:
   ...: ax.plot(x, y1, 'b', lw=1.5, label=r'$y = x^3 + 2x^2 - 1$')
   ...: ax.plot(x, y2, 'g', lw=1.5, label=r'$y = -x^2 + 1$')
   ...:
   ...: x_guesses = [[-2, 2], [1, -1], [-2, -5]]
   ...: for x_guess in x_guesses:
   ...:         sol = optimize.fsolve(f, x_guess)
   ...:         ax.plot(sol[0], sol[1], 'r*', markersize=15)
   ...:
   ...:         ax.plot(x_guess[0], x_guess[1], 'ko')
   ...:         ax.annotate("", xy=(sol[0], sol[1]), xytext=(x_guess[0], x_guess[1]),
   ...:                     arrowprops=dict(arrowstyle="->", linewidth=2.5))
   ...:
   ...: ax.legend(loc=0)
   ...: ax.set_xlabel(r'$x$', fontsize=18)
```

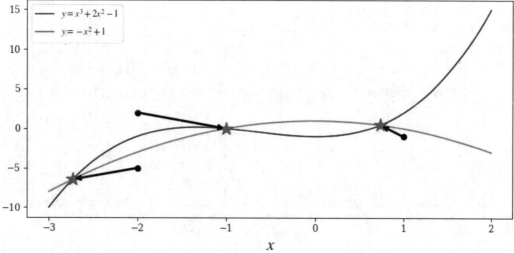

Figure 5-8. *Graph of a system of two nonlinear equations. The solutions are indicated with red stars, and the initial guess with a black dot and an arrow to the solution each initial guess eventually converged to*

By systematically solving the equation systems with different initial guesses, we can build a visualization of how different initial guesses converge to different solutions. This is done in the following code example, and the result is shown in Figure 5-9. This example demonstrates that even for this relatively simple example, the regions of initial guesses that converge to different solutions are highly nontrivial, and there are also missing dots corresponding to initial guesses for which the algorithm fails to converge to any solution. Nonlinear equation solving is a complex task, and visualizations of different types can often be a valuable tool when discovering a particular problem's characteristics.

```
In [81]: fig, ax = plt.subplots(figsize=(8, 4))
    ...:
    ...: ax.plot(x, y1, 'k', lw=1.5)
    ...: ax.plot(x, y2, 'k', lw=1.5)
    ...:
    ...: sol1 = optimize.fsolve(f, [-2,  2])
    ...: sol2 = optimize.fsolve(f, [ 1, -1])
    ...: sol3 = optimize.fsolve(f, [-2, -5])
    ...: sols = [sol1, sol2, sol3]
    ...: colors = ['r', 'b', 'g']
    ...: for idx, s in enumerate(sols):
    ...:     ax.plot(s[0], s[1], colors[idx]+'*', markersize=15)
    ...:
    ...: for m in np.linspace(-4, 3, 80):
    ...:     for n in np.linspace(-15, 15, 40):
    ...:         x_guess = [m, n]
    ...:         sol = optimize.fsolve(f, x_guess)
    ...:         idx = (abs(sols - sol)**2).sum(axis=1).argmin()
    ...:         ax.plot(x_guess[0], x_guess[1], colors[idx]+'.')
    ...:
    ...: ax.set_xlabel(r'$x$', fontsize=18)
```

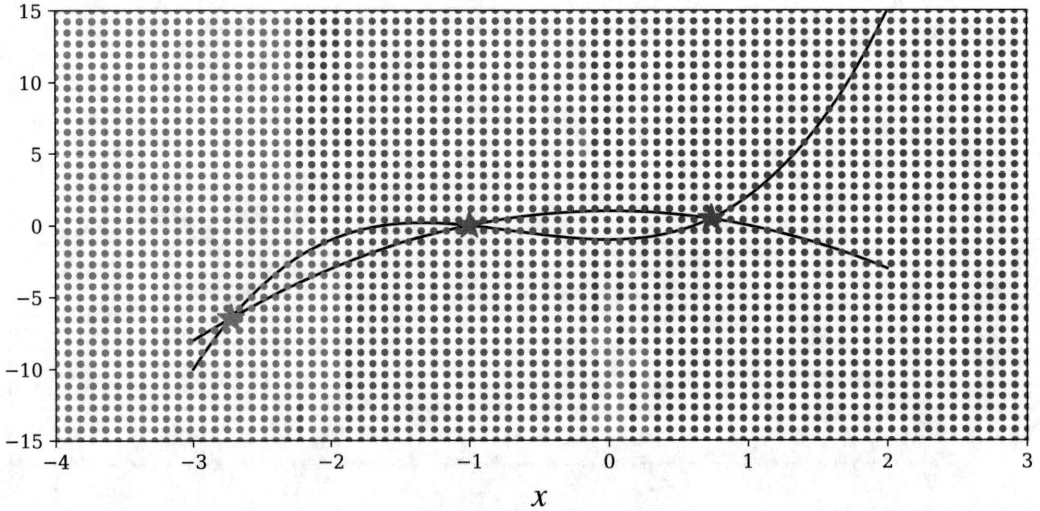

Figure 5-9. *Visualization of the convergence to different solutions for different initial guesses. Each dot represents an initial guess, and its color encodes which solution the solver eventually converges to. The solutions are marked with correspondingly colorcoded stars*

Summary

This chapter explored methods for solving algebraic equations using the SymPy and SciPy libraries. Equation solving is one of the most elementary mathematical tools for computational sciences, and it is an important component in many algorithms and methods and has direct applications in many problem-solving situations. In some cases, analytical algebraic solutions exist, especially for equations that are polynomials or contain certain combinations of elementary functions. Such equations can often be handled symbolically with SymPy. Numerical methods are usually the only feasible approach for equations with no algebraic solution and larger systems of equations. Linear equation systems can always be systematically solved. For this reason, there are many important applications for linear equation systems, be it for originally linear systems or as approximations to originally nonlinear systems. Nonlinear equation solving requires a different set of methods, and it is generally much more complex and computationally demanding compared to linear equation systems. Solving linear equation systems is an important step in the iterative methods employed in many of the methods that exist to solve nonlinear equation systems. For numerical equation solving, we can use the linear algebra and optimization modules in SciPy, which provide efficient and well-tested methods for numerical root finding and equation solving of both linear and nonlinear systems.

Further Reading

Equation solving is a basic numerical technique whose methods are covered in most introductory numerical analysis texts. A good example of a book that covers these topics is *Scientific Computing* by M. Heath (McGraw-Hill, 2001), and *Numerical Recipes: The Art of Scientific Computing* by W. H. Press (Cambridge University Press, 2007) gives a practical introduction with implementation details.

CHAPTER 6

Optimization

This chapter builds on Chapter 5 about equation solving and explores the related topic of solving optimization problems. Optimization is the process of finding and selecting the optimal element from a set of feasible candidates. In mathematical optimization, this problem is usually formulated as determining the extreme value of a function on a given domain. An extreme value, or an optimal value, can refer to either the minimum or maximum of the function, depending on the application and the specific problem. The chapter covers the optimization of real-valued functions of one or several variables, which optionally can be subject to constraints that restrict the function's domain.

The applications of mathematical optimization are many and varied, and so are the methods and algorithms that must be employed to solve optimization problems. Since optimization is a universally important mathematical tool, it has been developed and adapted for use in many fields of science and engineering, and the terminology used to describe optimization problems varies between fields. For example, the optimized mathematical function may be called a cost function, loss function, energy function, or objective function, to mention a few. Here, we use the generic term objective function.

Optimization is closely related to equation solving because at an optimal value of a function, its derivative, or gradient in the multivariate case, is zero. The converse, however, is not necessarily true. But one method to solve optimization problems is to solve for the zeros of the derivative or the gradient and test the resulting candidates for optimality. Although this approach is only sometimes feasible and often requires other numerical techniques, many of these are closely related to the numerical methods for root finding covered in Chapter 5.

This chapter discusses using SciPy's `optimize` optimization module for nonlinear optimization problems, and it briefly explores using `cvxopt`, the convex optimization library for linear optimization problems with linear constraints. This library also has powerful solvers for quadratic programming problems.

■ **cvxopt** This convex optimization library provides solvers for linear and quadratic optimization problems. At the time of writing, the latest version is 1.3.2. For more information, visit the project's website at http:// cvxopt.org. This chapter used the library for constrained linear optimization.

Importing Modules

Let's use the `optimize` module from the SciPy library again, assuming that this module is imported in the following manner.

```
In [1]: from scipy import optimize
```

The latter part of this chapter looks at linear programming using the cvxopt library, which is assumed to be entirely imported without any alias.

```
In [2]: import cvxopt
```

For basic numerics, symbolics, and plotting, let's use the NumPy, SymPy, and Matplotlib libraries, which are imported and initialized using the conventions introduced in earlier chapters.

```
In [3]: import matplotlib.pyplot as plt
In [4]: import numpy as np
In [5]: import sympy
In [6]: sympy.init_printing()
```

Classification of Optimization Problems

Let's restrict our attention to mathematical optimization of real-valued functions with one or more dependent variables. While many mathematical optimization problems can be formulated in this way, a notable exception is the optimization of functions over discrete variables, for example, integers, which is beyond the scope of this book.

A general optimization problem of the type considered here can be formulated as a minimization problem, $\min_x f(x)$, subject to sets of m equality constraints $g(x) = 0$ and p inequality constraints $h(x) \leq 0$. Here $f(x)$ is a real-valued function of x, which can be a scalar or a vector $x = (x_0, x_1, ..., x_n)^T$, while $g(x)$ and $h(x)$ can be vector-valued functions: $f: \mathbb{R}^n \longrightarrow \mathbb{R}$, $g: \mathbb{R}^n \longrightarrow \mathbb{R}^m$ and $h: \mathbb{R}^n \longrightarrow \mathbb{R}^p$. Note that maximizing $f(x)$ is equivalent to minimizing $-f(x)$, so without loss of generality, it is sufficient to consider only minimization problems.

Depending on the properties of the objective function $f(x)$ and the equality and inequality constraints $g(x)$ and $h(x)$, this formulation includes a wide variety of problems. A general mathematical optimization on this form is difficult to solve, and there are no efficient methods for solving completely generic optimization problems. However, there are efficient methods for many important special cases. In optimization, it is therefore important to know as much as possible about the objective functions and constraints to solve a problem.

Optimization problems are classified depending on the properties of the $f(x)$, $g(x)$, and $h(x)$ functions. First and foremost, the problem is *univariate* or *one-dimensional* if x is a scalar, $x \in \mathbb{R}$, and *multivariate* or *multidimensional* if x is a vector, $x \in \mathbb{R}^n$. For high-dimensional objective functions with larger n, the optimization problem is harder and more computationally demanding to solve. If the objective function and the constraints are all linear, the problem is a linear optimization problem or a *linear programming* problem.[1] If either the objective function or the constraints are nonlinear, it is a nonlinear optimization problem or a *nonlinear programming* problem. With respect to constraints, important subclasses of optimization are unconstrained problems and those with linear and nonlinear constraints. Finally, handling equality and inequality constraints requires different approaches.

As usual, nonlinear problems are much harder to solve than linear problems because they have a wider variety of possible behaviors. A general nonlinear problem can have both local and global minima, making it very difficult to find the global minima. Iterative solvers may often converge to local minima rather than the global minima or may even fail to converge altogether if there are both local and global minima. However, an important subclass of nonlinear problems that can be solved efficiently is *convex problems*. Convexity is directly related to the absence of strictly local minima and the existence of a unique global minimum.

---

[1] For historical reasons, optimization problems are often referred to as *programming* problems, which are not directly related to computer programming in the modern sense.

By definition, a function is convex on an interval $[a, b]$ if the values of the function on this interval lie below the line through the endpoints $(a, f(a))$ and $(b, f(b))$. This condition, which can be readily generalized to the multivariate case, implies several important properties, such as a unique minimum on the interval. Because of strong properties like this one, convex problems can be solved efficiently despite being nonlinear. The concepts of local and global minima and convex and nonconvex functions are illustrated in Figure 6-1.

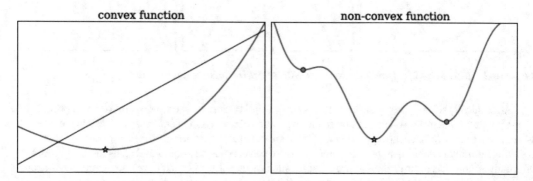

Figure 6-1. *Illustration of a convex function (left) and a nonconvex function (right) with a global minimum and two local minima*

Whether the objective function $f(x)$ and the constraints $g(x)$ and $h(x)$ are continuous and smooth are properties that have significant implications for the methods and techniques that can be used to solve an optimization problem. Discontinuities in these functions, or their derivatives or gradients, cause difficulties for many of the available optimization methods. The following assumes these functions are continuous and smooth. On a related note, if the function itself is not known exactly but contains noise due to measurements or other reasons, many of the methods discussed in the following may not be suitable.

Optimization of continuous and smooth functions is closely related to nonlinear equation solving because extremal values of an $f(x)$ function correspond to points where its derivative, or gradient, is zero. Finding candidates for the optimal value of $f(x)$ is therefore equivalent to solving the (in general nonlinear) equation system $\nabla f(x) = 0$. However, a solution to $\nabla f(x) = 0$, known as a stationary point, does not necessarily correspond to a minimum of $f(x)$; it can also be a maximum or a saddle point; see Figure 6-2. Therefore, candidates obtained by solving $\nabla f(x) = 0$ should be tested for optimality. For unconstrained objective functions, the higher-order derivatives, or the Hessian matrix,

$$\left\{ H_f\left(x\right)\right\}_{ij} = \frac{\partial^2 f\left(x\right)}{\partial x_i \partial x_j},$$

for the multivariate case, can be used to determine if a stationary point is a local minimum. If the second-order derivative is positive, or the Hessian positive definite, when evaluated at stationary point x^*, then x^* is a local minimum. A negative second-order derivative, or negative definite Hessian, corresponds to a local maximum, and a zero second-order derivative, or an indefinite Hessian, corresponds to saddle point.

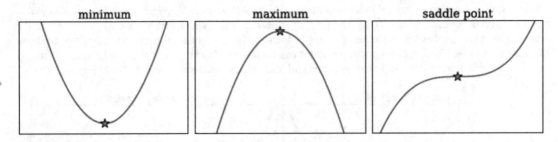

Figure 6-2. *Illustration of different stationary points of a one-dimensional function*

Algebraically solving the equation system $\nabla f(x) = 0$ and testing the candidate solutions for optimality is one possible strategy for solving an optimization problem. However, it is not always a feasible method. We may not have an analytical expression for $f(x)$ from which we can compute the derivatives, and the resulting nonlinear equation system may not be easy to solve, especially not to find all its roots. There are alternative numerical optimization approaches for such cases, some of which have analogs among the root-finding methods discussed in Chapter 5. The remaining part of this chapter explores the various classes of optimization problems and how such problems can be solved using available optimization libraries for Python.

Univariate Optimization

Optimization of a function that only depends on a single variable is relatively easy. In addition to the analytical approach of seeking the roots of the function's derivative, we can employ techniques similar to the root-finding methods for univariate functions, namely, bracketing methods and Newton's method. Like the bisection method for univariate root finding, it is possible to use bracketing and iteratively refine an interval using function evaluations alone. Refining an interval $[a, b]$ that contains a minimum can be achieved by evaluating the function at two interior points x_1 and x_2, $x_1 < x_2$, and selecting $[x_1, b]$ as a new interval if $f(x_1) > f(x_2)$, and $[a, x_2]$ otherwise. This idea is used in the *golden section search* method, which additionally uses the trick of choosing x_1 and x_2 such that their relative positions in the $[a, b]$ interval satisfy the golden ratio. This has the advantage of allowing the reuse of one function evaluation from the previous iteration and thus only requiring one new function evaluation in each iteration and reducing the interval with a constant factor. This approach is guaranteed to converge to an optimal point for functions with a unique minimum on the given interval, but this is not guaranteed for more complicated functions. It is, therefore, important to carefully select the initial interval, ideally relatively close to an optimal point. In the SciPy `optimize` module, the `golden` function implements the golden search method.

Like the bisection method for root finding, the golden search method is a (relatively) reliable but slowly converging method. Methods with better convergence can be constructed if the values of the function evaluations are used rather than only comparing the values to each other (similar to using only the sign of the functions, as in the bisection method). The function values can be used to fit a polynomial, for example, a quadratic polynomial, which can be interpolated to find a new approximation for the minimum, resulting in a candidate for a new function evaluation, after which the process can be iterated. This approach can converge faster but is riskier than bracketing and may not converge at all or converge to local minima outside the given bracket interval.

Newton's method for root finding is an example of a quadratic approximation method that can be applied to find a function minimum by applying the method to the derivative rather than the function itself. This yields the iteration formula $x_{k+1} = x_k - f'(x_k)/f''(x_k)$, which can quickly converge if started close to an optimal point but may not converge at all if started too far from the optimal value. This formula also requires evaluating both the derivative and the second-order derivative in each iteration. This can be a suitable method if analytical expressions for these derivatives are available. If only function evaluations are available, the derivatives may be approximated using an analog of the secant method for root finding.

A combination of the two previous methods is typically used in practical implementations of univariate optimization routines, giving both stability and fast convergence. In SciPy's optimize module, the brent function is such a hybrid method, and it is generally the preferred method for optimizing univariate functions with SciPy. This variant of the golden section search method uses inverse parabolic interpolation to obtain faster convergence.

Instead of calling the optimize.golden and optimize.brent functions directly; it is convenient to use the unified interface function optimize.minimize_scalar, which dispatches to the optimize.golden and optimize.brent functions depending on the value of the method keyword argument, where the currently allowed options are 'Golden', 'Brent', or 'Bounded'. The last option dispatches to optimize.fminbound, which performs optimization on a bounded interval, corresponding to an optimization problem with inequality constraints that limit the domain of objective function $f(x)$. Note that the optimize.golden and optimize.brent functions may converge to a local minimum outside the initial bracket interval, but optimize.fminbound would return the value at the end of the allowed range in such cases.

As an example for illustrating these techniques, consider the following classic optimization problem: Minimize the area of a cylinder with unit volume. Here, suitable variables are the radius r and height h of the cylinder, and the objective function is $f([r, h]) = 2\pi r^2 + 2\pi rh$, subject to the equality constraint $g([r, h]) = \pi r^2 h - 1 = 0$. This problem is formulated here as a two-dimensional optimization problem with an equality constraint. However, we can algebraically solve the constraint equation for one of the dependent variables, for example, $h = 1/\pi r^2$, and substitute this into the objective function to obtain an unconstrained one-dimensional optimization problem: $f(r) = 2\pi r^2 + 2/r$. To begin with, we can solve this problem symbolically using SymPy, using the method of equating the derivative of $f(r)$ to zero.

```
In [7]: r, h = sympy.symbols("r, h")
In [8]: Area = 2 * sympy.pi * r**2 + 2 * sympy.pi * r * h
In [9]: Volume = sympy.pi * r**2 * h
In [10]: h_r = sympy.solve(Volume - 1)[0]
In [11]: Area_r = Area.subs(h_r)
In [12]: rsol = sympy.solve(Area_r.diff(r))[0]
In [13]: rsol
```

$$\text{Out[13]}: \frac{2^{2/3}}{2\sqrt[3]{\pi}}$$

```
In [14]: _.evalf()
Out[14]: 0.541926070139289
```

Now verify that the second derivative is positive and that rsol corresponds to a minimum.

```
In [15]: Area_r.diff(r, 2).subs(r, rsol)
Out[15]: 12π
In [16]: Area_r.subs(r, rsol)
```

$$\text{Out[16]}: 3\sqrt[3]{2\pi}$$

```
In [17]: _.evalf()
Out[17]: 5.53581044593209
```

This approach is often feasible for simple problems, but for more realistic problems, we typically need to resort to numerical techniques. To solve this problem using SciPy's numerical optimization functions, we first define a Python function f that implements the objective function. To solve the optimization problem, we then pass this function to, for example, optimize.brent. Optionally, we can use the brack keyword argument to specify a starting interval for the algorithm.

```
In [18]: def f(r):
    ...:     return 2 * np.pi * r**2 + 2 / r
In [19]: r_min = optimize.brent(f, brack=(0.1, 4))
In [20]: r_min
Out[20]: 0.541926077256
In [21]: f(r_min)
Out[21]: 5.53581044593
```

Instead of calling optimize.brent directly, we could use the generic interface for scalar minimization problems optimize.minimize_scalar. Note that we must use the bracket keyword argument to specify a starting interval in this case.

```
In [22]: optimize.minimize_scalar(f, bracket=(0.1, 4))
Out[22]:  nit: 13
          fun: 5.5358104459320856
            x: 0.54192606489766715
         nfev: 14
```

All these methods give that the radius that minimizes the area of the cylinder is approximately 0.54 (the exact result from the symbolic calculation is $2^{2/3} / 2\sqrt[3]{\pi}$) and a minimum area of approximately 5.54 (the exact result is $3\sqrt[3]{2\pi}$). The objective function we minimized in this example is plotted in Figure 6-3, where the minimum is marked with a red star. It is a good idea to visualize the objective function before attempting a numerical optimization because it can help identify a suitable initial interval or a starting point for the numerical optimization routine.

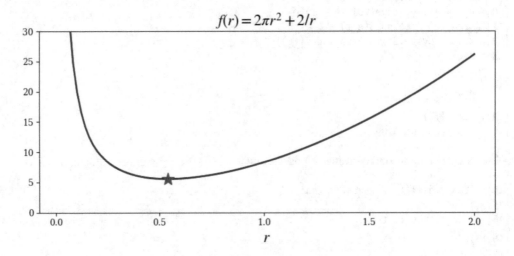

$$f(r) = 2\pi r^2 + 2/r$$

Figure 6-3. *The surface area of a cylinder with unit volume as a function of the radius r. A star denotes the location of the minimum*

Unconstrained Multivariate Optimization

Multivariate optimization is significantly more complex than the univariate optimization discussed in the previous section. In particular, the analytical approach of solving the nonlinear equations for roots of the gradient is rarely feasible in the multivariate case, and the bracketing scheme used in the golden search method is also not directly applicable. Instead, we must resort to techniques that start at some point in the coordinate space and use different strategies to move toward a better approximation of the minimum point. The most basic approach of this type is to consider the gradient $\nabla f(x)$ of the objective function $f(x)$ at a given point x. In general, the negative of the gradient, $-\nabla f(x)$, always points in the direction in which the $f(x)$ function decreases the most. As a minimization strategy, it is therefore sensible to move along this direction for some distance α_k and then iterate this scheme at the new point. This method is known as the *steepest descent method*, and it gives the iteration formula $x_{k+1} = x_k - \alpha_k \nabla f(x_k)$, where α_k is a free parameter known as the *line search parameter* that describes how far along the given direction to move in each iteration. An appropriate α_k can, for example, be selected by solving the one-dimensional optimization problem $\min_{\alpha} f(x_k - \alpha_k \nabla f(x_k))$. This method is guaranteed to progress and eventually converge to a minimum of the function. But the convergence can be quite slow because this method tends to overshoot along the direction of the gradient, giving a zigzag approach to the minimum. Nonetheless, the steepest descent method is the conceptual basis for many multivariate optimization algorithms, and with suitable modifications, the convergence can be sped up.

Newton's method for multivariate optimization is a modification of the steepest descent method that can improve convergence. As in the univariate case, Newton's method can be viewed as a local quadratic approximation of the function, which, when minimized, gives an iteration scheme. In the multivariate case, the iteration formula is $x_{k+1} = x_k - H_f^{-1}(x_k)\nabla f(x_k)$, where compared to the steepest descent method, the gradient has been replaced with the gradient multiplied from the left with the inverse of the Hessian matrix for the function.[2] In general, this alters both the direction and the length of the step, so this method is not strictly the steepest descent method and may not converge if started too far from a minimum. However, when close to a minimum, it converges quickly. As usual, there is a trade-off between convergence rate and stability. As formulated here, Newton's method requires both the gradient and the Hessian of the function.

In SciPy, Newton's method is implemented in the `optimize.fmin_ncg` function. This function takes the following arguments: a Python function for the objective function, a starting point, a Python function for evaluating the gradient, and (optionally) a Python function for evaluating the Hessian. To see how this method can be used to solve an optimization problem, consider the following problem: $\min_x f(x)$ where the objective function is $f(x) = (x_1 - 1)^4 + 5(x_2 - 1)^2 - 2x_1 x_2$. To apply Newton's method, we need to calculate the gradient and the Hessian. For this case, the calculation can easily be done by hand. However, for generality, the following uses SymPy to compute symbolic expressions for the gradient and the Hessian. To this end, let's begin by defining symbols and a symbolic expression for the objective function and then use the `sympy.diff` function for each variable to obtain the gradient and Hessian in symbolic form.

```
In [23]: x1, x2 = sympy.symbols("x_1, x_2")
In [24]: f_sym = (x1-1)**4 + 5 * (x2-1)**2 - 2*x1*x2
In [25]: fprime_sym = [f_sym.diff(x_) for x_ in (x1, x2)]
In [26]: # Gradient
   ...: sympy.Matrix(fprime_sym)
```
$$\text{Out[26]:} \begin{bmatrix} -2x_2 + 4(x_1 - 1)^3 \\ -2x_1 + 10x_2 - 10 \end{bmatrix}$$
```
In [27]: fhess_sym = [[f_sym.diff(x1_, x2_) for x1_ in (x1, x2)]
   ...:                 for x2_ in (x1, x2)]
In [28]: # Hessian
   ...: sympy.Matrix(fhess_sym)
```
$$\text{Out[28]:} \begin{bmatrix} 12(x_1 - 1)^2 & -2 \\ -2 & 10 \end{bmatrix}$$

---

[2] In practice, the inverse of the Hessian does not need to be computed, and instead we can solve the linear equation system $H_f(x_k)y_k = -\nabla f(x_k)$ and use the integration formula $x_{k+1} = x_k + y_k$.

Now that there is a symbolic expression for the gradient and the Hessian, we can create vectorized functions for these expressions using sympy.lambdify.

```
In [29]: f_lmbda = sympy.lambdify((x1, x2), f_sym, 'numpy')
In [30]: fprime_lmbda = sympy.lambdify((x1, x2), fprime_sym, 'numpy')
In [31]: fhess_lmbda = sympy.lambdify((x1, x2), fhess_sym, 'numpy')
```

However, the functions produced by sympy.lambdify take one argument for each variable in the corresponding expression, and the SciPy optimization functions expect a vectorized function where all coordinates are packed into one array. To obtain functions compatible with the SciPy optimization routines, we wrap each of the functions generated by sympy.lambdify with a Python function that rearranges the arguments.

```
In [32]: def func_XY_to_X_Y(f):
    ...:     """
    ...:     Wrapper for f(X) -> f(X[0], X[1])
    ...:     """
    ...:     return lambda X: np.array(f(X[0], X[1]))
In [33]: f = func_XY_to_X_Y(f_lmbda)
In [34]: fprime = func_XY_to_X_Y(fprime_lmbda)
In [35]: fhess = func_XY_to_X_Y(fhess_lmbda)
```

Now the f, fprime, and fhess functions are vectorized Python functions on the form that optimize.fmin_ncg expects, and we can proceed with a numerical optimization of the problem at hand by calling this function. In addition to the functions prepared from SymPy expressions, we must also give a starting point for the Newton method. Let's use $(0, 0)$ as the starting point.

```
In [36]: x_opt = optimize.fmin_ncg(f, (0, 0), fprime=fprime, fhess=fhess)
        Optimization terminated successfully.
                Current function value: -3.867223
                Iterations: 8
                Function evaluations: 10
                Gradient evaluations: 17
                Hessian evaluations: 8
In [37]: x_opt
Out[37]: array([ 1.88292613,  1.37658523])
```

The routine found a minimum point at $(x_1, x_2) = (1.88292613, 1.37658523)$, and diagnostic information about the solution was also printed to standard output, including the number of iterations and the number of function, gradient, and Hessian evaluations that were required to arrive at the solution. As usual, it is illustrative to visualize the objective function $g(\beta) = \sum_{i=0}^{m} r_i(\beta)^2 = \|r(\beta)\|^2$ and the solution (see Figure 6-4).

```
In [38]: fig, ax = plt.subplots(figsize=(6, 4))
    ...: x_ = y_ = np.linspace(-1, 4, 100)
    ...: X, Y = np.meshgrid(x_, y_)
    ...: c = ax.contour(X, Y, f_lmbda(X, Y), 50)
    ...: ax.plot(x_opt[0], x_opt[1], 'r*', markersize=15)
    ...: ax.set_xlabel(r"$x_1$", fontsize=18)
    ...: ax.set_ylabel(r"$x_2$", fontsize=18)
    ...: plt.colorbar(c, ax=ax)
```

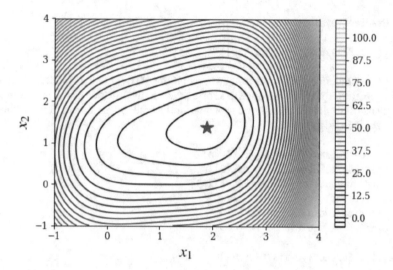

Figure 6-4. *Contour plot of the objective function f(x) = (x₁ – 1)⁴ + 5(x₂ – 1)² – 2x₁x₂. A red star marks the minimum point*

In practice, it may not always be possible to provide functions for evaluating both the gradient and the Hessian of the objective function, and often, it is convenient with a solver that only requires function evaluations. For such cases, several methods exist to estimate the gradient, Hessian, or both numerically. Methods that approximate the Hessian are known as quasi-Newton methods, and there are also alternative iterative methods that avoid entirely using the Hessian. Two popular methods are the Broyden-Fletcher-Goldfarb-Shanno (BFGS) and the conjugate gradient methods, which are implemented in SciPy as the `optimize.fmin_bfgs` and `optimize.fmin_cg` functions. The BFGS method is a quasi-Newton method that can gradually build up numerical estimates of the Hessian and the gradient, if necessary. The conjugate gradient method is a variant of the steepest descent method and does not use the Hessian. It can be used with numerical estimates of the gradient obtained from only function evaluations. With these methods, the number of function evaluations required to solve a problem is much larger than for Newton's method, which, on the other hand, also evaluates the gradient and the Hessian. Both `optimize.fmin_bfgs` and `optimize.fmin_cg` can optionally accept a function for evaluating the gradient, but if not provided, the gradient is estimated from function evaluations.

The preceding problem, which was solved with the Newton method, can also be solved using the `optimize.fmin_bfgs` and `optimize.fmin_cg`, without providing a function for the Hessian.

```
In [39]: x_opt = optimize.fmin_bfgs(f, (0, 0), fprime=fprime)
         Optimization terminated successfully.
             Current function value: -3.867223
             Iterations: 10
             Function evaluations: 14
             Gradient evaluations: 14
In [40]: x_opt
Out[40]: array([ 1.88292605,  1.37658523])
In [41]: x_opt = optimize.fmin_cg(f, (0, 0), fprime=fprime)
         Optimization terminated successfully.
             Current function value: -3.867223
             Iterations: 7
             Function evaluations: 17
```

159

```
               Gradient evaluations: 17
In [42]: x_opt
Out[42]: array([ 1.88292613,  1.37658522])
```

Note that the number of function and gradient evaluations is larger than for Newton's method, as shown in the diagnostic output from the optimization solvers in the previous examples. As mentioned, both methods can also be used without providing a function for the gradient, as shown in the following example using the optimize.fmin_bfgs solver.

```
In [43]: x_opt = optimize.fmin_bfgs(f, (0, 0))
         Optimization terminated successfully.
             Current function value: -3.867223
             Iterations: 10
             Function evaluations: 56
             Gradient evaluations: 14
In [44]: x_opt
Out[44]: array([ 1.88292604,  1.37658522])
```

In this case, the number of function evaluations is even larger, but it is convenient not to implement functions for the gradient and the Hessian.

In general, the BFGS method is often a good first approach to try, particularly if neither the gradient nor the Hessian is known. If only the gradient is known, then the BFGS method is still the generally recommended method, although the conjugate gradient method is often a competitive alternative to the BFGS method. If both the gradient and the Hessian are known, Newton's method is the method with the fastest convergence in general. However, it should be noted that although the BFGS and the conjugate gradient methods theoretically have slower convergence than Newton's method, they can sometimes offer improved stability and be preferable. Each iteration can also be more computationally demanding with Newton's method compared to quasi-Newton methods and the conjugate gradient method, and especially for large problems, these methods can be faster despite requiring more iterations.

The methods for multivariate optimization discussed so far converge to a local minimum. Problems with many local minima can easily lead to a situation when the solver easily gets stuck in a local minimum, even if a global minimum exists. Although there is no complete and general solution to this problem, a practical approach to partially alleviate this problem is using a brute-force search over a coordinate grid to find a suitable starting point for an iterative solver. At least this gives a systematic approach to finding a global minimum within given coordinate ranges. In SciPy, the optimize.brute function can perform a systematic search. To illustrate this method, consider the problem of minimizing the $4 \sin x\pi + 6 \sin y\pi + (x - 1)^2 + (y - 1)^2$ function, which has a large number of local minima. This can make it tricky to pick a suitable initial point for an iterative solver. To solve this optimization problem with SciPy, define a Python function for the objective function.

```
In [45]: def f(X):
    ...:     x, y = X
    ...:     return (4 * np.sin(np.pi * x) +
    ...:             6 * np.sin(np.pi * y)) + (x - 1)**2 + (y - 1)**2
```

To systematically search for the minimum over a coordinate grid, we call optimize.brute with the objective function f as the first parameter and a tuple of slice objects as the second argument, one for each coordinate. The slice objects specify the coordinate grid to search for a minimum value. Let's also set the keyword argument finish=None, which prevents the optimize.brute from automatically refining the best candidate.

```
In [46]: x_start = optimize.brute(
    ...:        f, (slice(-3, 5, 0.5), slice(-3, 5, 0.5)), finish=None)
In [47]: x_start
Out[47]: array([ 1.5,   1.5])
In [48]: f(x_start)
Out[48]: -9.5
```

On the coordinate grid specified by the given tuple of slice objects, the optimal point is $(x_1, x_2) =$ (1.5, 1.5), with a corresponding objective function minimum of –9.5. This is now a good starting point for a more sophisticated iterative solver, such as optimize.fmin_bfgs.

```
In [49]: x_opt = optimize.fmin_bfgs(f, x_start)
         Optimization terminated successfully.
                Current function value: -9.520229
                Iterations: 4
                Function evaluations: 28
                Gradient evaluations: 7
In [50]: x_opt
Out[50]: array([ 1.47586906,   1.48365788])
In [51]: f(x_opt)
Out[51]: -9.52022927306
```

Here, the BFGS method gave the final minimum point $(x_1, x_2) =$ (1.47586906, 1.48365788), with the minimum value of the objective function –9.52022927306. For this type of problem, guessing the initial starting point easily results in the iterative solver converging to a local minimum, and the systematic approach that optimize.brute provides is frequently useful.

As always, it is important to visualize the objective function and the solution when possible. The following code listings plot a contour graph of the current objective function and mark the obtained solution with a red star (see Figure 6-5). As in the previous example, we need a wrapper function for reshuffling the parameters of the objective function because of the different conventions of how the coordinated vectors are passed to the function (separate arrays and packed into one array, respectively).

```
In [52]: def func_X_Y_to_XY(f, X, Y):
    ...:        """
    ...:        Wrapper for f(X, Y) -> f([X, Y])
    ...:        """
    ...:        s = np.shape(X)
    ...:        return f(np.vstack([X.ravel(), Y.ravel()])).reshape(*s)
In [53]: fig, ax = plt.subplots(figsize=(6, 4))
    ...: x_ = y_ = np.linspace(-3, 5, 100)
    ...: X, Y = np.meshgrid(x_, y_)
    ...: c = ax.contour(X, Y, func_X_Y_to_XY(f, X, Y), 25)
    ...: ax.plot(x_opt[0], x_opt[1], 'r*', markersize=15)
    ...: ax.set_xlabel(r"$x_1$", fontsize=18)
    ...: ax.set_ylabel(r"$x_2$", fontsize=18)
    ...: plt.colorbar(c, ax=ax)
```

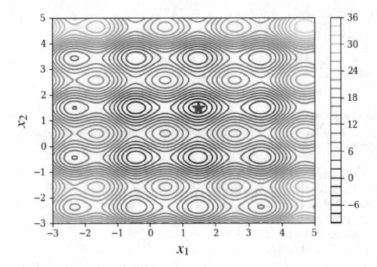

Figure 6-5. *Contour plot of the objective function f(x) = 4 sin xπ + 6 sin yπ + (x – 1)² + (y – 1)². The minimum is marked with a red star*

This section explicitly calls functions for specific solvers, for example, `optimize.fmin_bfgs`. However, like for scalar optimization, SciPy also provides a unified interface for all multivariate optimization solvers with the `optimize.minimize` function, which dispatches out to the solver-specific functions depending on the value of the `method` keyword argument (remember, the univariate minimization function that provides a unified interface is `optimize.scalar_minimize`). For clarity, here I have favored explicitly calling functions for specific solvers, but in general, it is a good idea to use `optimize.minimize`, as this makes it easier to switch between different solvers. For example, the previous example used `optimize.fmin_bfgs` in the following way.

```
In [54]: x_opt = optimize.fmin_bfgs(f, x_start)
```

We could just as well have used the following.

```
In [55]: result = optimize.minimize(f, x_start, method='BFGS')
In [56]: x_opt = result.x
```

The `optimize.minimize` function returns an instance of `optimize.OptimizeResult` that represents the result of the optimization. The solution is available via the x attribute of this class.

Nonlinear Least Square Problems

In Chapter 5, we encountered linear least square problems and explored how they can be solved using linear algebra. In general, a least square problem can be viewed as an optimization problem with the objective function $g(\beta) = \sum_{i=0}^{m} r_i(\beta)^2 = \|r(\beta)\|^2$, where $r(\beta)$ is a vector with the residuals $r_i(\beta) = y_i - f(x_i, \beta)$ for a set of m observations (x_i, y_i). Here, β is a vector with unknown parameters that specifies the $f(x, \beta)$ function. If this problem is nonlinear in the parameters β, it is known as a nonlinear least square problem, and since it is nonlinear, it cannot be solved with the linear algebra techniques discussed in Chapter 5. Instead, we can use the multivariate optimization techniques described in the previous section, such as Newton's or quasi-Newton methods. However, this nonlinear least square optimization problem has a specific structure, and several methods have been developed to solve this particular optimization problem. One example is the Levenberg-Marquardt method, which is based on successive linearizations of the problem in each iteration.

In SciPy, the `optimize.leastsq` function provides a nonlinear least square solver that uses the Levenberg-Marquardt method. To illustrate how this function can be used, consider a nonlinear model on the form $f(x, \beta) = \beta_0 + \beta_1 \exp(-\beta_2 x^2)$ and a set of observations (x_i, y_i). The following example simulates the observations with random noise added to the actual values and solves the minimization problem that gives the best least square estimates of the parameters β. First, we define a tuple with the actual values of the parameter vector β and a Python function for the model function. This function, which should return the y value corresponding to a given x value, takes as the first argument the variable x, and the following arguments are the unknown function parameters.

```
In [57]: beta = (0.25, 0.75, 0.5)
In [58]: def f(x, b0, b1, b2):
    ...:     return b0 + b1 * np.exp(-b2 * x**2)
```

Once the model function is defined, generate randomized data points that simulate the observations.

```
In [59]: xdata = np.linspace(0, 5, 50)
In [60]: y = f(xdata, *beta)
In [61]: ydata = y + 0.05 * np.random.randn(len(xdata))
```

We can start solving the nonlinear least square problem with the model function and observation data prepared. The first step is to define a function for the residuals given the data and the model function specified in terms of the yet-to-be-determined model parameters β.

```
In [62]: def g(beta):
    ...:     return ydata - f(xdata, *beta)
```

Next, define an initial guess for the parameter vector and let the `optimize.leastsq` function solves for the best least square fit for the parameter vector.

```
In [63]: beta_start = (1, 1, 1)
In [64]: beta_opt, beta_cov = optimize.leastsq(g, beta_start)
In [65]: beta_opt
Out[65]: array([ 0.25733353,  0.76867338,  0.54478761])
```

The best fit is close to the actual parameter values (0.25, 0.75, 0.5), as defined earlier. By plotting the observation data and the model function for the actual and fitted function parameters, we can visually confirm that the fitted model describes the data well (see Figure 6-6).

```
In [66]: fig, ax = plt.subplots()
    ...: ax.scatter(xdata, ydata, label='samples')
    ...: ax.plot(xdata, y, 'r', lw=2, label='true model')
    ...: ax.plot(xdata, f(xdata, *beta_opt), 'b', lw=2, label='fitted model')
    ...: ax.set_xlim(0, 5)
    ...: ax.set_xlabel(r"$x$", fontsize=18)
    ...: ax.set_ylabel(r"$f(x, \beta)$", fontsize=18)
    ...: ax.legend()
```

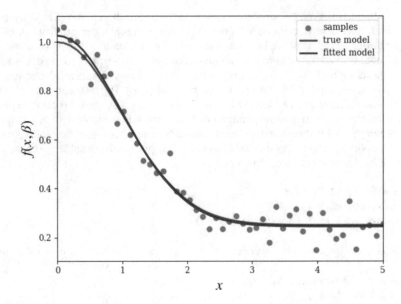

Figure 6-6. *Nonlinear least square fit to the $f(x, \beta) = \beta_0 + \beta_1 \exp(-\beta_2 x^2)$ function with $\beta = (0.25, 0.75, 0.5)$*

The SciPy optimize module provides an alternative interface to nonlinear least square fitting through the optimize.curve_fit function. This convenience wrapper around optimize.leastsq eliminates the need to explicitly define the residual function for the least square problem. The previous problem could, therefore, be solved more concisely using the following.

```
In [67]: beta_opt, beta_cov = optimize.curve_fit(f, xdata, ydata)
In [68]: beta_opt
Out[68]: array([ 0.25733353,  0.76867338,  0.54478761])
```

Constrained Optimization

Constraints add another level of complexity to optimization problems, requiring their own classification. A simple form of constrained optimization is where the coordinate variables are subject to some bounds; for example, $\min f(x)$ subject to $0 \leq x \leq 1$. The constraint $0 \leq x \leq 1$ is simple because it only restricts the range of the coordinate without dependencies on the other variables. This type of problem can be solved using the L-BFGS-B method in SciPy, a variant of the BFGS method used earlier. This solver is available through the optimize.fmin_l_bgfs_b function or via optimize.minimize with the method argument set to 'L-BFGS-B'. To define the coordinate boundaries, the bound keyword argument must be used, and its value should be a list of tuples that contain the minimum and maximum value of each constrained variable. If the minimum or maximum value is set to None, it is interpreted as an unbounded.

As an example of solving a bounded optimization problem with the L-BFGS-B solver, consider minimizing the objective function $f(x) = (x_1 - 1)^2 - (x_2 - 1)^2$ subject to the constraints $2 \leq x_1 \leq 3$ and $0 \leq x_2 \leq 2$. First, to solve this problem, define a Python function for the objective functions and tuples with the boundaries for each of the two variables in this problem, according to the given constraints. For comparison, the following code also solves the unconstrained optimization problem with the same objective function and plots a contour graph of the objective function where the unconstrained and constrained minimum values are marked with blue and red stars, respectively (see Figure 6-7).

```
In [69]: def f(X):
    ...:     x, y = X
    ...:     return (x - 1)**2 + (y - 1)**2
In [70]: x_opt = optimize.minimize(f, [0, 0], method='BFGS').x
In [71]: bnd_x1, bnd_x2 = (2, 3), (0, 2)
In [72]: x_cons_opt = optimize.minimize(f, [0, 0], method='L-BFGS-B',
    ...:                                 bounds=[bnd_x1, bnd_x2]).x
In [73]: fig, ax = plt.subplots(figsize=(6, 4))
    ...: x_ = y_ = np.linspace(-1, 3, 100)
    ...: X, Y = np.meshgrid(x_, y_)
    ...: c = ax.contour(X, Y, func_X_Y_to_XY(f, X, Y), 50)
    ...: ax.plot(x_opt[0], x_opt[1], 'b*', markersize=15)
    ...: ax.plot(x_cons_opt[0], x_cons_opt[1], 'r*', markersize=15)
    ...: bound_rect = plt.Rectangle(
    ...:     (bnd_x1[0], bnd_x2[0]),
    ...:     bnd_x1[1] - bnd_x1[0], bnd_x2[1] - bnd_x2[0], facecolor="grey")
    ...: ax.add_patch(bound_rect)
    ...: ax.set_xlabel(r"$x_1$", fontsize=18)
    ...: ax.set_ylabel(r"$x_2$", fontsize=18)
    ...: plt.colorbar(c, ax=ax)
```

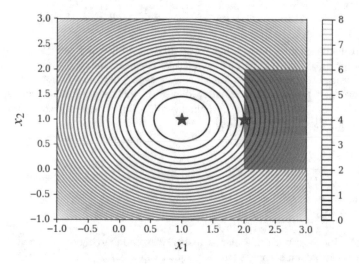

Figure 6-7. *Contours of the objective function f(x), with the unconstrained (blue star) and constrained minima (red star). The feasible region of the constrained problem is shaded in gray*

Constraints defined by equalities or inequalities that include more than one variable are somewhat more complicated. However, there are general techniques also for this type of problem. For example, it is possible to convert a constrained optimization problem to an unconstrained problem using the Lagrange multipliers by introducing additional variables. For example, consider the optimization problem $\min_x f(x)$ methods, linear programming problems with thousands $g(x) = 0$. In an unconstrained optimization problem, the gradient of $f(x)$ vanishes at the optimal points, $\nabla f(x) = 0$. It can be shown that the corresponding condition for constrained problems is that the negative gradient lies in the space supported by the constraint normal, such as $-\nabla f(x) = \lambda J_g^T(x)$. Here, $J_g(x)$ is the Jacobian matrix of the constraint function $g(x)$, and λ is the vector of Lagrange multipliers (new variables). This condition arises from equating to zero the gradient of the $\Lambda(x, \lambda) = f(x) + \lambda^T g(x)$ function, which is known as the Lagrangian function. Therefore, if both $f(x)$ and $g(x)$ are continuous and smooth, a stationary point (x_0, λ_0) of the $\Lambda(x, \lambda)$ function corresponds to an x_0 that is an optimum of the original constrained optimization problem. Note that if $g(x)$ is a scalar function (i.e., there is only one constraint), then the Jacobian $J_g(x)$ reduces to the gradient $\nabla g(x)$.

To illustrate this technique, consider the problem of maximizing the volume of a rectangle with sides of length x_1, x_2, and x_3, subject to the constraint that the total surface area should be unity: $g(x) = 2x_1 x_2 + 2x_0 x_2 + 2x_1 x_0 - 1 = 0$. To solve this optimization problem using Lagrange multipliers, form the Lagrangian $\Lambda(x) = f(x) + \lambda g(x)$ and seek the stationary points for $\nabla \Lambda(x) = 0$. With SymPy, we can carry out this task by first defining the symbols for the variables in the problem, then constructing expressions for $f(x)$, $g(x)$, and $\Lambda(x)$.

```
In [74]: x = x0, x1, x2, l = sympy.symbols("x_0, x_1, x_2, lambda")
In [75]: f = x0 * x1 * x2
In [76]: g = 2 * (x0 * x1 + x1 * x2 + x2 * x0) - 1
In [77]: L = f + l * g
```

Finally, compute $\nabla \Lambda(x)$ using `sympy.diff` and solve the equation $\nabla \Lambda(x) = 0$ using `sympy.solve`.

```
In [78]: grad_L = [sympy.diff(L, x_) for x_ in x]
In [79]: sols = sympy.solve(grad_L)
In [80]: sols
```

$$\text{Out[80]:} \left[\left\{ \lambda : -\frac{\sqrt{6}}{24}, x_0 : \frac{\sqrt{6}}{6}, x_1 : \frac{\sqrt{6}}{6}, x_2 : \frac{\sqrt{6}}{6} \right\}, \left\{ \lambda : \frac{\sqrt{6}}{24}, x_0 : -\frac{\sqrt{6}}{6}, x_1 : -\frac{\sqrt{6}}{6}, x_2 : -\frac{\sqrt{6}}{6} \right\} \right]$$

This procedure gives two stationary points. We could determine which one corresponds to the optimal solution by evaluating the objective function for each case. However, here, only one of the stationary points corresponds to a physically acceptable solution: since x_i is the length of a rectangle side in this problem, it must be positive. We can, therefore, immediately identify the interesting solution corresponding to the intuitive result $x_0 = x_1 = x_2 = \sqrt{6}/6$ (a cube). As a final verification, evaluate the constraint function and the objective function using the obtained solution.

```
In [81]: g.subs(sols[0])
Out[81]: 0
In [82]: f.subs(sols[0])
```

$$\text{Out[82]:} \quad \frac{\sqrt{6}}{36}$$

The method of using Lagrange multipliers can also be extended to handle inequality constraints, and several numerical approaches apply such techniques. One example is the sequential least square programming method, abbreviated as SLSQP, which is available in the SciPy as the `optimize.slsqp` function and via `optimize.minimize` with `method='SLSQP'`. The `optimize.minimize` function takes the keyword argument `constraints`, which should be a list of dictionaries that each specifies a constraint. The allowed keys (values) in this dictionary are `type` (`'eq'` or `'ineq'`), `fun` (constraint function), `jac` (Jacobian of the constraint function), and `args` (additional arguments to constraint function and the function for evaluating its Jacobian). For example, the constraint dictionary describing the constraint in the previous problem would be `dict(type='eq', fun=g)`.

To solve the full problem numerically using SciPy's SLSQP solver, we need to define Python functions for the objective function and the constraint function.

```
In [83]: def f(X):
    ...:     return -X[0] * X[1] * X[2]
```

```
In [84]: def g(X):
   ...:        return 2 * (X[0]*X[1] + X[1] * X[2] + X[2] * X[0]) - 1
```

Note that since the SciPy optimization functions solve minimization problems, and here we are interested in maximization, the f function is the negative of the original objective function. Next, let's define the constraint dictionary for $g(x) = 0$ and finally call the optimize.minimize function.

```
In [85]: constraint = dict(type='eq', fun=g)
In [86]: result = optimize.minimize(f, [0.5, 1, 1.5], method='SLSQP',
   ...:                             constraints=[constraint])
In [87]: result
Out[87]:   status: 0
          success: True
             njev: 18
             nfev: 95
              fun: -0.068041368623352985
                x: array([ 0.40824187,  0.40825127,  0.40825165])
          message: 'Optimization terminated successfully.'
              jac: array([-0.16666925, -0.16666542, -0.16666527,  0.])
              nit: 18
In [88]: result.x
Out[88]: array([ 0.40824187,  0.40825127,  0.40825165])
```

As expected, the solution agrees well with the analytical result of the symbolic calculation using Lagrange multipliers.

To solve problems with inequality constraints, we must set type='ineq' in the constraint dictionary and provide the corresponding inequality function. To demonstrate the minimization of a nonlinear objective function with a nonlinear inequality constraint, return to the quadratic problem considered previously, but in this case, with inequality constraint $g(x) = x_1 - 1.75 - (x_0 - 0.75)^4 \geq 0$. As usual, begin by defining the objective function, the constraint function, and the constraint dictionary.

```
In [89]: def f(X):
   ...:        return (X[0] - 1)**2 + (X[1] - 1)**2
In [90]: def g(X):
   ...:        return X[1] - 1.75 - (X[0] - 0.75)**4
In [91]: constraints = [dict(type='ineq', fun=g)]
```

Next, we can solve the optimization problem by calling the optimize.minimize function. For comparison, let's also solve the corresponding unconstrained problem.

```
In [92]: x_opt = optimize.minimize(f, (0, 0), method='BFGS').x
In [93]: x_cons_opt = optimize.minimize(f, (0, 0), method='SLSQP',
constraints=constraints).x
```

To verify the soundness of the obtained solution, we plot the contours of the objective function together with a shaded area representing the feasible region (where the inequality constraint is satisfied). The constrained and unconstrained solutions are marked with a red and a blue star, respectively (see Figure 6-8).

```
In [94]: fig, ax = plt.subplots(figsize=(6, 4))
In [95]: x_ = y_ = np.linspace(-1, 3, 100)
   ...: X, Y = np.meshgrid(x_, y_)
```

```
...: c = ax.contour(X, Y, func_X_Y_to_XY(f, X, Y), 50)
...: ax.plot(x_opt[0], x_opt[1], 'b*', markersize=15)
...: ax.plot(x_, 1.75 + (x_-0.75)**4, 'k-', markersize=15)
...: ax.fill_between(x_, 1.75 + (x_-0.75)**4, 3, color='grey')
...: ax.plot(x_cons_opt[0], x_cons_opt[1], 'r*', markersize=15)
...:
...: ax.set_ylim(-1, 3)
...: ax.set_xlabel(r"$x_0$", fontsize=18)
...: ax.set_ylabel(r"$x_1$", fontsize=18)
...: plt.colorbar(c, ax=ax)
```

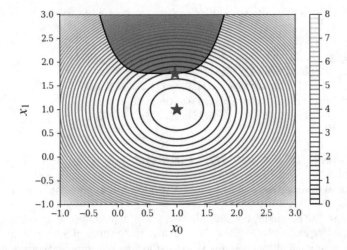

Figure 6-8. *Contour plot of the objective function with the feasible region of the constrained problem shaded gray. The red and blue stars are the optimal points in the constrained and unconstrained problems, respectively*

For optimization problems with *only* inequality constraints, SciPy provides an alternative solver using the constrained optimization by linear approximation (COBYLA) method. This solver is accessible through `optimize.fmin_cobyla` or `optimize.minimize` with `method='COBYLA'`. It could have solved the previous example by replacing `method='SLSQP'` with `method='COBYLA'`.

Linear Programming

The previous section considered methods for general optimization problems where the objective and constraint functions can be nonlinear. However, at this point, it is worth taking a step back and considering a much more restricted type of optimization problem, namely, *linear programming*, where the objective function is linear, and all constraints are linear equality or inequality constraints. The class of problems is much less general. But it turns out that linear programming has many important applications that can be solved vastly more efficiently than general nonlinear problems. This is because linear problems have properties that enable entirely different methods. In particular, the solution to a linear optimization problem must necessarily lie on a constraint boundary, so it is sufficient to search the vertices of the intersections of the linear constraint functions. This can be done efficiently in practice. A popular algorithm for this type of problem is known as *simplex*, which systematically moves from one vertex to another until the optimal vertex has been reached. More recent interior point methods also efficiently solve linear programming problems. These methods make linear programming problems with thousands of variables and constraints readily solvable.

Linear programming problems are typically written in the standard form: $\min_x c^T x$ where $Ax \leq b$ and $x \geq 0$. Here, c and x are vectors of length n, A is a $m \times n$ matrix, and b is a m-vector. For example, consider the problem of minimizing the $f(x) = -x_0 + 2x_1 - 3x_2$ function, subject to the three inequality constraints $x_0 + x_1 \leq 1$, $-x_0 + 3x_1 \leq 2$, and $-x_1 + x_2 \leq 3$. The standard form has $c = (-1, 2, -3)$, $b = (1, 2, 3)$, and

$$A = \begin{pmatrix} 1 & 1 & 0 \\ -1 & 3 & 0 \\ 0 & -1 & 1 \end{pmatrix}.$$

To solve this problem, use the cvxopt library, which provides the linear programming solver with the cvxopt.solvers.lp function. This solver expects the c, A, and b vectors and matrix used in the standard form introduced in the preceding text in the given order as arguments. The cvxopt library uses its own classes for representing matrices and vectors. Fortunately, they are interoperable with NumPy arrays via the array interface[3] and can, therefore, be cast from one form to another using the cvxopt.matrix and np.array functions. Since the NumPy array is the de facto standard array format in the scientific Python environment, it is sensible to use the NumPy array as far as possible and only convert to cvxopt matrices when necessary (i.e., before calling one of the solvers in cvxopt.solvers).

To solve the stated example problem using the cvxopt library, first, create NumPy arrays for the A matrix and the c and b vectors and convert them to cvxopt matrices using the cvxopt.matrix function.

```
In [96]: c = np.array([-1.0, 2.0, -3.0])
In [97]: A = np.array([[ 1.0, 1.0, 0.0],
                       [-1.0, 3.0, 0.0],
                       [ 0.0, -1.0, 1.0]])
In [98]: b = np.array([1.0, 2.0, 3.0])
In [99]: A_ = cvxopt.matrix(A)
In [100]: b_ = cvxopt.matrix(b)
In [101]: c_ = cvxopt.matrix(c)
```

The cvxopt compatible matrices and vectors c_, A_, and b_ can now be passed to the linear programming solver cvxopt.solvers.lp.

```
In [102]: sol = cvxopt.solvers.lp(c_, A_, b_)
In [103]: sol
Out[103]: {'dual infeasibility': 1.4835979218054372e-16,
           'dual objective': -10.0,
           'dual slack': 0.0,
           'gap': 0.0,
           'iterations': 0,
           'primal infeasibility': 0.0,
           'primal objective': -10.0,
           'primal slack': -0.0,
           'relative gap': 0.0,
           'residual as dual infeasibility certificate': None,
           'residual as primal infeasibility certificate': None,
           's': <3x1 matrix, tc='d'>,
           'status': 'optimal',
           'x': <3x1 matrix, tc='d'>,
```

---

[3] For details, see http://docs.scipy.org/doc/numpy/reference/arrays.interface.html.

```
              'y': <0x1 matrix, tc='d'>,
              'z': <3x1 matrix, tc='d'>}
In [104]: x = np.array(sol['x'])
In [105]: x
Out[105]: array([[ 0.25],
                  [ 0.75],
                  [ 3.75]])
In [106]: sol['primal objective']
Out[106]: -10.0
```

The solution to the optimization problem is given in terms of the vector x, which in this example is $x =$ (0.25, 0.75, 3.75), which corresponds to the $f(x)$ value −10. With this method and the `cvxopt.solvers.lp` solver, linear programming problems with hundreds or thousands of variables can readily be solved. All that is needed is to write the optimization problem on the standard form and create the c, A, and b arrays.

Summary

Optimization—selecting the best option from a set of alternatives—is fundamental in many applications in science and engineering. Mathematical optimization provides a rigorous framework for systematically treating optimization problems if they can be formulated as mathematical problems. Computational optimization methods are the tools for solving such optimization problems in practice. In a scientific computing environment, optimization, therefore, plays a very important role. For scientific computing with Python, the SciPy library provides efficient routines for solving many standard optimization problems, which can be used to solve many computational optimization problems. However, optimization is a large field in mathematics, requiring different methods for solving different types of problems, and several optimization libraries for Python provide specialized solvers to specific kinds of optimization problems. The SciPy `optimize` module offers excellent and flexible general-purpose solvers for a wide variety of optimization problems. But, for particular optimization problems, many specialized libraries provide better performance or more features. An example of such a library is `cvxopt`, which complements the general-purpose optimization routines in SciPy with efficient solvers for linear and quadratic problems.

Further Reading

For an accessible introduction to optimization, with more detailed discussions of the numerical properties of several of the methods introduced in this chapter, see, for example, *Scientific Computing: An Introductory Survey* by M. Heath (McGraw-Hill, 2002). For a more rigorous and in-depth introduction to optimization, see, for example, *An Introduction to Optimization* by E. K. P. Chong (Wiley, 2013). The creators of the `cvxopt` library give a thorough treatment of convex optimization in the excellent book *Convex Optimization* by S. Boyd (Cambridge University Press, 2004), which is also available online at `http://stanford.edu/~boyd/cvxbook`.

CHAPTER 7

Interpolation

Interpolation is a mathematical method for constructing a function from discrete data points. The interpolation function, or interpolant, should coincide with the given data points and can be evaluated for other intermediate input values within the sampled range. There are many applications of interpolation. A typical use case that provides an intuitive picture is plotting a smooth curve through a given set of data points. Another use case is to approximate complicated functions, which, for example, could be computationally demanding to evaluate. In that case, evaluating the original function only at a limited number of points and using interpolation to approximate the function when evaluating it for intermediary points can be beneficial.

Interpolation may, at first glance, look a lot like least square fitting, which we saw in both Chapter 5 (linear least square) and Chapter 6 (nonlinear least square). Indeed, there are many similarities between interpolation and curve fitting with least square methods, but important conceptual differences distinguish these two methods. In least-square fitting, we are interested in approximately fitting a function to data points to minimize the sum of square errors, using many data points and an overdetermined system of equations. In interpolation, on the other hand, we require a function that exactly coincides with the given data points and only uses the number of data points that equals the number of free parameters in the interpolation function. Least square fitting is, therefore, more suitable for fitting a large number of data points to a model function, and interpolation is a mathematical tool for creating a functional representation for a given minimum number of data points. Interpolation is an important component in many mathematical methods, including some equation-solving and optimization methods used in Chapters 5 and 6.

Extrapolation is a concept that is related to interpolation. It refers to evaluating the estimated function outside of the sampled range, while interpolation relates only to evaluating the function within the range spanned by the given data points. Extrapolation can often be riskier than interpolation because it involves estimating a function in a region where it has yet to be sampled. Here, we are only concerned with interpolation. To perform interpolation in Python, let's use the `polynomial` module from NumPy and the `interpolate` module from SciPy.

Importing Modules

Let's continue with the convention of importing submodules from the SciPy library explicitly. In this chapter, we need the `interpolate` module from SciPy and the `polynomial` module from NumPy, which provides functions and classes for polynomials. Import these modules as follows.

```
In [1]: from scipy import interpolate
In [2]: from numpy import polynomial as P
```

© Robert Johansson 2024
R. Johansson, *Numerical Python*, https://doi.org/10.1007/979-8-8688-0413-7_7

In addition, we also need the rest of the NumPy library, the linalg linear algebra module from SciPy, and the Matplotlib library for plotting.

```
In [3]: import numpy as np
In [4]: from scipy import linalg
In [5]: import matplotlib.pyplot as plt
```

Interpolation

Before diving into the details of how to perform interpolation with NumPy and SciPy, let's first state the interpolation problem in mathematical form. For notational brevity, let's only consider one-dimensional interpolation, which can be formulated as follows: for a given set of n data points $\left\{(x_i, y_i)\right\}_{i=1}^{n}$, find a $f(x)$ function such that $f(x_i) = y_i$, for $i \in [1, n]$. The $f(x)$ function is known as the interpolant and is not unique. An infinite number of functions satisfy the interpolation criteria. Typically, we can write the interpolant as a linear combination of some $\phi_j(x)$ basis functions, such that $f(x) = \sum_{j=1}^{n} c_j \phi_j(x)$, where c_j are unknown coefficients. Substituting the given data points into this linear combination results in a linear equation system for the unknown coefficients: $\sum_{j=1}^{n} c_j \phi_j(x_i) = y_i$. This equation system can be written in explicit matrix form as follows:

$$\begin{bmatrix} \phi_1(x_1) & \phi_2(x_1) & \cdots & \phi_n(x_1) \\ \phi_1(x_2) & \phi_2(x_2) & \cdots & \phi_n(x_2) \\ \vdots & \vdots & \ddots & \vdots \\ \phi_1(x_n) & \phi_2(x_n) & \cdots & \phi_n(x_n) \end{bmatrix} \begin{bmatrix} c_1 \\ c_2 \\ \vdots \\ c_n \end{bmatrix} = \begin{bmatrix} y_1 \\ y_2 \\ \vdots \\ y_n \end{bmatrix},$$

or in a more compact implicit matrix form as $\Phi(x)c = y$, where the elements of the matrix $\Phi(x)$ are $\{\Phi(x)\}_{ij} = \phi_j(x_i)$. Note that the number of basis functions is the same as the number of data points, and $\Phi(x)$ is a square matrix. Assuming that this matrix has full rank, we can solve for the unique vector c using the standard methods discussed in Chapter 5. If the number of data points is larger than the number of basis functions, then the system is overdetermined, and in general, there is no solution that satisfies the interpolation criteria. In this situation, it is instead more suitable to consider a least square fit than an exact interpolation; see Chapter 5.

The choice of basis functions affects the properties of the resulting equation system, and a suitable choice of basis depends on the properties of the fitted data. Common choices of basis functions for interpolation are various types of polynomials, for example, the power basis $\phi_i(x) = x^{i-1}$, or orthogonal polynomials such as Legendre polynomials $\phi_i(x) = P_{i-1}(x)$, Chebyshev polynomials $\phi_i(x) = T_{i-1}(x)$, or piecewise polynomials. Note that f (x) is not unique in general, but for n data points, there is a unique interpolating polynomial of order $n - 1$, regardless of which polynomial basis we use. For power basis $\phi_i(x) = x^{i-1}$, the matrix $\Phi(x)$ is the Vandermonde matrix, which we have seen applications of in the least square fitting in Chapter 5. For other polynomial bases, $\Phi(x)$ is a generalized Vandermonde matrix, which, for each basis, defines the linear equation system matrix that must be solved in the interpolation problem. The structure of the $\Phi(x)$ matrix differs for different polynomial bases, and its condition number and the computational cost of solving the interpolation problem vary correspondingly. Polynomials thus play an essential role in interpolation, and before we can start to solve interpolation problems, we need a convenient way of working with polynomials in Python. This is the topic of the following section.

Polynomials

The NumPy library contains the submodule polynomial (here imported as P), which provides functions and classes for working with polynomials. In particular, it provides implementations of many standard orthogonal polynomials. These functions and classes are useful when working with interpolation. Let's review how to use this module before looking at polynomial interpolation.

■ **Note**   There are two modules for polynomials in NumPy: numpy.poly1d and numpy.polynomial. There is a significant overlap in functionality in these two modules, but they are incompatible. Specifically, the coordinate arrays have reversed order in the two representations. The numpy.poly1d module is older and has been superseded by numpy.polynomial, which is now recommended for new code. Here, let's only focus on numpy.polynomial, but it is also worth being aware of numpy.poly1d.

The np.polynomial module contains several classes for representing polynomials in different polynomial bases. Standard polynomials, written in the usual power basis $\{x^i\}$, are represented with the Polynomial class. We can pass a coefficient array to its constructor to create an instance of this class. In the coefficient array, the ith element is the coefficient of x^i. For example, we can represent the polynomial $1 + 2x + 3x^2$ by passing the list [1, 2, 3] to the Polynomial class.

```
In [6]: p1 = P.Polynomial([1, 2, 3])
In [7]: p1
Out[7]: x↦1.0+2.0x+3.0x²
```

Alternatively, we can initialize a polynomial by specifying its roots using the P.Polynomial.fromroots class method. For example, the polynomial with roots at x = −1 and x = 1 can be created using the following.

```
In [8]: p2 = P.Polynomial.fromroots([-1, 1])
In [9]: p2.__repr__()
Out[9]: Polynomial([-1.,  0.,  1.], domain=[-1.,  1.], window=[-1.,  1.])
```

The result is the polynomial with the coefficient array [-1, 0, 1], corresponding to $-1 + x^2$. The roots of a polynomial can be computed using the roots method. For example, the roots of the two previously created polynomials are as follows.

```
In [10]: p1.roots()
Out[10]: array([-0.33333333-0.47140452j, -0.33333333+0.47140452j])
In [11]: p2.roots()
Out[11]: array([-1.,  1.])
```

As expected, the roots of the polynomial p2 are $x = -1$ and $x = 1$, as was requested when it was created using the fromroots class method.

In the preceding example, the representation of a polynomial when using __repr__() is on the form Polynomial([-1., 0., 1.], domain=[-1., 1.], window=[-1., 1.]). The first of the lists in this representation is the coefficient array. The second and third lists are the domain and window attributes, which can be used to map the input domain of a polynomial to another interval. Specifically, the input domain interval [domain[0], domain[1]] is mapped to the interval [window[0], window[1]] through a linear transformation (scaling and translation). The default values are domain=[-1, 1] and window=[-1, 1], which correspond to an identity transformation (no change). The domain and window arguments are particularly useful when working with orthogonal polynomials with respect to a scalar product defined on a specific interval. It is then desirable to map the domain of the input data onto this interval. This is important when interpolating with orthogonal polynomials, such as the Chebyshev or Hermite polynomials, because performing this transformation can vastly improve the conditioning number of the Vandermonde matrix for the interpolation problem.

The properties of a Polynomial instance can be directly accessed using the coeff, domain, and window attributes. For example, the p1 polynomial defined in the preceding example is as follows.

```
In [12]: p1.coef
Out[12]: array([ 1.,   2.,   3.])
In [13]: p1.domain
Out[13]: array([-1,   1])
In [14]: p1.window
Out[14]: array([-1,   1])
```

A polynomial represented as a Polynomial instance can easily be evaluated at arbitrary values of x by calling the class instance as a function. The x variable can be specified as a scalar, a list, or an arbitrary NumPy array. For example, to evaluate the polynomial p1 at the points $x = \{1.5, 2.5, 3.5\}$, we simply call the p1 class instance with an array of x values as the argument.

```
In [15]: p1(np.array([1.5, 2.5, 3.5]))
Out[15]: array([ 10.75,   24.75,   44.75])
```

Instances of Polynomial can be operated on using the standard arithmetic operators +, -, *, /, and so on. The // operator is used for polynomial division. To see how this works, consider the division of the polynomial $p_1(x) = (x-3)(x-2)(x-1)$ with the polynomial $p_2(x) = (x-2)$. The answer, which is obvious when written in the factorized form, is $(x-3)(x-1)$. We can compute and verify this using NumPy in the following manner: first, create Polynomial instances for the p1 and p2, and then use the // operator to compute the polynomial division.

```
In [16]: p1 = P.Polynomial.fromroots([1, 2, 3])
In [17]: p1
Out[17]: Polynomial([ -6.,   11.,   -6.,    1.], domain=[-1.,   1.], window=[-1.,   1.])
In [18]: p2 = P.Polynomial.fromroots([2])
In [19]: p2
Out[19]: Polynomial([-2.,   1.], domain=[-1.,   1.], window=[-1.,   1.])
In [20]: p3 = p1 // p2
In [21]: p3
Out[21]: Polynomial([ 3.,   -4.,   1.], domain=[-1.,   1.], window=[-1.,   1.])
```

The result is a new polynomial with coefficient array [3, -4, 1], and if we compute its roots, we find that they are 1 and 3, so this polynomial is indeed $(x-3)(x-1)$.

```
In [22]: p3.roots()
Out[22]: array([ 1.,   3.])
```

In addition to the Polynomial class for polynomials in the standard power basis, the polynomial module also has classes for representing polynomials in Chebyshev, Legendre, Laguerre, and Hermite bases, with the names Chebyshev, Legendre, Laguerre, Hermite (Physicists'), and HermiteE (Probabilists'), respectively. For example, the Chebyshev polynomial with the coefficient list $[1, 2, 3]$, that is, the polynomial $1T_0(x) + 2T_1(x) + 3T_2(x)$, where $T_i(x)$ is the Chebyshev polynomial of order i, can be created using the following.

```
In [23]: c1 = P.Chebyshev([1, 2, 3])
In [24]: c1
Out[24]: Chebyshev([ 1.,   2.,   3.], domain=[-1,  1], window=[-1,  1])
```

Its roots can be computed using the roots attribute.

```
In [25]: c1.roots()
Out[25]: array([-0.76759188,  0.43425855])
```

All the polynomial classes have the same methods, attributes, and operators as the Polynomial class, and they can all be used in the same manner. For example, to create the Chebyshev and Legendre representations of the polynomial with roots $x = -1$ and $x = 1$, we can use the fromroots attribute the same way we did with the Polynomial class.

```
In [26]: c1 = P.Chebyshev.fromroots([-1, 1])
In [27]: c1
Out[27]: Chebyshev([-0.5,  0. ,  0.5], domain=[-1.,  1.], window=[-1.,  1.])
In [28]: l1 = P.Legendre.fromroots([-1, 1])
In [29]: l1
Out[29]: Legendre([-0.66666667,  0.  ,  0.66666667], domain=[-1.,  1.], window=[-1.,  1.])
```

Note that the same polynomial, here with the roots at $x = -1$ and $x = 1$ (a unique polynomial), has different coefficient arrays when represented in different bases. But when evaluated at specific values of x, the result is always the same.

```
In [30]: c1(np.array([0.5, 1.5, 2.5]))
Out[30]: array([-0.75,  1.25,  5.25])
In [31]: l1(np.array([0.5, 1.5, 2.5]))
Out[31]: array([-0.75,  1.25,  5.25])
```

Polynomial Interpolation

The polynomial classes discussed in the previous section provide helpful functions for interpolation. For instance, recall the linear equation for the polynomial interpolation problem: $\Phi(x)c = y$, where x and y are vectors containing the x_i and y_i data points, and c is the unknown coefficient vector. To solve the interpolation problem, we must first evaluate the matrix $\Phi(x)$ for a given basis and then solve the resulting linear equation system. Each of the polynomial classes in polynomial conveniently provides a function for computing the (generalized) Vandermonde matrix for the corresponding basis. For example, for polynomials in the power basis, we can use np.polynomial.polynomial.polyvander; for polynomials in the Chebyshev basis, we can use the corresponding np.polynomial.chebyshev.chebvander function, and so on. See the docstrings for np.polynomial and its submodules for the complete list of generalized Vandermonde matrix functions for the various polynomial bases.

Using the functions for generating the Vandermonde matrices, we can easily perform polynomial interpolations in different bases. For example, consider the data points $(1, 1)$, $(2, 3)$, $(3, 5)$, and $(4, 4)$. Let's begin with creating a NumPy array for the x and y coordinates for the data points.

```
In [32]: x = np.array([1, 2, 3, 4])
In [33]: y = np.array([1, 3, 5, 4])
```

To interpolate a polynomial through these points, we need to use a polynomial of third degree (number of data points minus one). For interpolation in the power basis, seek the coefficient c_i such that $f(x) = \sum_{i=1}^{4} c_i x^{i-1} = c_1 x^0 + c_2 x^1 + c_3 x^2 + c_4 x^3$, and to find this coefficient, evaluate the Vandermonde matrix and solve the interpolation equation system.

```
In [34]: deg = len(x) - 1
In [35]: A = P.polynomial.polyvander(x, deg)
In [36]: c = linalg.solve(A, y)
In [37]: c
Out[37]: array([ 2. , -3.5,  3. , -0.5])
```

The sought coefficient vector is [2, -3.5, 3, -0.5], and the interpolation polynomial is thus $f(x) = 2 - 3.5x + 3x^2 - 0.5x^3$. Given the coefficient array c, we can now create a polynomial representation that can be used for interpolation.

```
In [38]: f1 = P.Polynomial(c)
In [39]: f1(2.5)
Out[39]: 4.1875
```

To perform this polynomial interpolation with another polynomial basis, all we need to change in the previous example is the name of the function used to generate the Vandermonde matrix A. For example, to interpolate using the Chebyshev basis polynomials, we can do the following.

```
In [40]: A = P.chebyshev.chebvander(x, deg)
In [41]: c = linalg.solve(A, y)
In [42]: c
Out[42]: array([ 3.5  , -3.875,  1.5  , -0.125])
```

As expected, the coefficient array has different values in this basis, and the interpolation polynomial in the Chebyshev basis is $f(x) = 3.5T_0(x) - 3.875T_1(x) + 1.5T_2(x) - 0.125T_3(x)$. However, regardless of the polynomial basis, the interpolation polynomial is unique, and evaluating the interpolant always results in the same values.

```
In [43]: f2 = P.Chebyshev(c)
In [44]: f2(2.5)
Out[44]: 4.1875
```

The following demonstrates that the interpolation with the two bases results in the same interpolation function by plotting the f1 and f2 together with the data points (see Figure 7-1).

```
In [45]: xx = np.linspace(x.min(), x.max(), 100)  # supersampled [x[0], x[-1]] interval
In [45]: fig, ax = plt.subplots(1, 1, figsize=(12, 4))
    ...: ax.plot(xx, f1(xx), 'b', lw=2, label='Power basis interp.')
    ...: ax.plot(xx, f2(xx), 'r--', lw=2, label='Chebyshev basis interp.')
    ...: ax.scatter(x, y, label='data points')
    ...: ax.legend(loc=4)
    ...: ax.set_xticks(x)
```

```
...: ax.set_ylabel(r"$y$", fontsize=18)
...: ax.set_xlabel(r"$x$", fontsize=18)
```

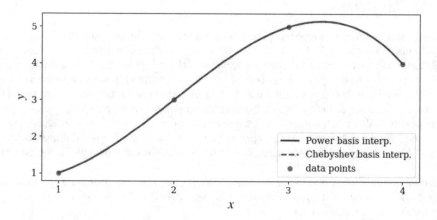

Figure 7-1. *Polynomial interpolation of four data points, using power basis and the Chebyshev basis*

While interpolation with different polynomial bases is convenient due to the functions for the generalized Vandermonde matrices, an even simpler and better method is available. Each polynomial class provides a class method `fit` that can be used to compute an interpolation polynomial.[1] The two interpolation functions computed manually in the previous example could, therefore, be computed in the following manner using the power basis and its `Polynomial` class.

```
In [46]: f1b = P.Polynomial.fit(x, y, deg)
In [47]: f1b
Out[47]: Polynomial([ 4.1875,   3.1875, -1.6875, -1.6875],
                     domain=[ 1.,   4.], window=[-1.,   1.])
```

Using the `fit` class method from the `Chebyshev` class instead obtains the following.

```
In [48]: f2b = P.Chebyshev.fit(x, y, deg)
In [49]: f2b
Out[49]: Chebyshev([ 3.34375 ,   1.921875, -0.84375 , -0.421875],
                    domain=[ 1.,   4.], window=[-1.,   1.])
```

Note that with this method, the `domain` attribute of the resulting instances is automatically set to the appropriate x values of the data points (in this example, the input range is [1, 4]), and the coefficients are adjusted accordingly. As mentioned, mapping the interpolation data into the range most suitable for a specific basis can significantly improve the numerical stability of the interpolation. For example, using the Chebyshev basis with x values that are scaled such that $x \in [-1, 1]$, rather than the original x values in the previous example, reduces the condition number from almost 4660 to about 1.85.

---

[1] If the requested polynomial degree of the interpolant is smaller than the number of data points minus one, then a least square fit is computed rather than an exact interpolation.

```
In [50]: np.linalg.cond(P.chebyshev.chebvander(x, deg))
Out[50]: 4659.7384241399586
In [51]: np.linalg.cond(P.chebyshev.chebvander((2*x-5)/3.0, deg))
Out[51]: 1.8542033440472896
```

Polynomial interpolation of a few data points is a powerful and valuable mathematical tool integral to many mathematical methods. However, when the number of data points increases, we need to use increasingly high-order polynomials for exact interpolation, which is problematic in several ways. To begin with, determining and evaluating the interpolant for increasing polynomial order becomes increasingly demanding. However, a more serious issue is that high-order polynomial interpolation can have undesirable behavior between the interpolation points. Although the interpolation is exact at the given data points, a high-order polynomial can vary wildly between the specified points. This is famously illustrated by polynomial interpolation of Runge's function $f(x) = 1/(1 + 25x^2)$ using evenly spaced sample points in the interval $[-1, 1]$. The result is an interpolant that nearly diverges between the data points near the end of the interval.

To illustrate this behavior, create a Python function runge that implements Runge's function and a runge_interpolate function that interpolates an nth-order polynomial in the power basis to Runge's function at evenly spaced sample points.

```
In [52]: def runge(x):
    ...:     return 1/(1 + 25 * x**2)
In [53]: def runge_interpolate(n):
    ...:     x = np.linspace(-1, 1, n + 1)
    ...:     p = P.Polynomial.fit(x, runge(x), deg=n)
    ...:     return x, p
```

Next, plot Runge's function with the 13th and 14th order polynomial interpolations at supersampled x values in the $[-1, 1]$ interval. The resulting plot is shown in Figure 7-2.

```
In [54]: xx = np.linspace(-1, 1, 250)
In [55]: fig, ax = plt.subplots(1, 1, figsize=(8, 4))
    ...: ax.plot(xx, runge(xx), 'k', lw=2, label="Runge's function")
    ...: # 13th order interpolation of the Runge function
    ...: n = 13
    ...: x, p = runge_interpolate(n)
    ...: ax.plot(x, runge(x), 'ro')
    ...: ax.plot(xx, p(xx), 'r', label='interp. order %d' % n)
    ...: # 14th order interpolation of the Runge function
    ...: n = 14
    ...: x, p = runge_interpolate(n)
    ...: ax.plot(x, runge(x), 'go')
    ...: ax.plot(xx, p(xx), 'g', label='interp. order %d' % n)
    ...:
    ...: ax.legend(loc=8)
    ...: ax.set_xlim(-1.1, 1.1)
    ...: ax.set_ylim(-1, 2)
    ...: ax.set_xticks([-1, -0.5, 0, 0.5, 1])
    ...: ax.set_ylabel(r"$y$", fontsize=18)
    ...: ax.set_xlabel(r"$x$", fontsize=18)
```

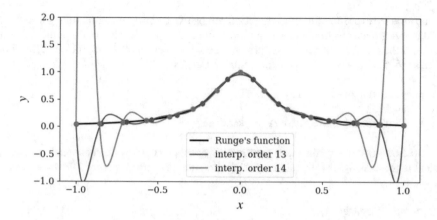

Figure 7-2. *The Runge's function, together with two high-order polynomial interpolations*

In Figure 7-2, the interpolants exactly agree with Runge's function at the sample points, but between these points, they oscillate wildly near the ends of the interval. This is an undesirable property of an interpolant, and it defeats the purpose of the interpolation. A solution to this problem is to use piecewise low-order polynomials when interpolating with a large number of data points. In other words, instead of fitting all the data points to a single high-order polynomial, a different low-order polynomial is used to describe each subinterval bracketed by two consecutive data points. This is the topic of the following section.

Spline Interpolation

For a set of n data points $\{x_i, y_i\}$, there are $n - 1$ subintervals $[x_i, x_{i+1}]$ in the full range of the data $[x_0, x_{n-1}]$. An interior data point that connects two such subintervals is known as a *knot* in the terminology of piecewise polynomial interpolation. To interpolate the n data points using piecewise polynomials of degree k on each of the subintervals, we must determine $(k + 1)(n - 1)$ unknown parameters. The values at the knots give $2(n - 1)$ equations. These equations are only sufficient to determine a piecewise polynomial of order one (i.e., a piecewise linear function). However, additional equations can be obtained by requiring that derivatives and higher-order derivatives are continuous at the knots. This condition ensures that the resulting piecewise polynomial has a smooth appearance.

A spline is a particular type of piecewise polynomial interpolant: a piecewise polynomial of degree k is a spline if it is continuously differentiable $k - 1$ times. The most popular choice is the third-order spline, $k = 3$, which requires $4(n - 1)$ parameters. For this case, the continuity of two derivatives at the $n - 2$ knots gives $2(n - 2)$ additional equations, bringing the total number of equations to $2(n - 1) + 2(n - 2) = 4(n - 1)$ $- 2$. Therefore, two remaining undetermined parameters must be determined by other means. A common approach is to require that the second-order derivatives at the endpoints are zero (resulting in the *natural* spline). This gives two more equations, which closes the equation system.

The SciPy `interpolate` module provides several functions and classes for performing spline interpolation. For example, we can use the `interpolate.interp1d` function, which takes x and y arrays for the data points as first and second arguments. The optional keyword argument `kind` can be used to specify the type and order of the interpolation. We can set `kind=3` (or, equivalently, `kind='cubic'`) to compute the cubic spline. This function returns a class instance that can be called like a function and evaluated for different values of x using function calls. An alternative spline function is `interpolate.InterpolatedUnivariateSpline`, which also takes x and y arrays as the first and second arguments but uses the keyword argument `k` (instead of `kind`) to specify the order of the spline interpolation.

To see how the `interpolate.interp1d` function can be used, consider again Runge's function, and we now want to interpolate this function with a third-order spline polynomial. To this end, we first create NumPy arrays for the *x* and *y* coordinates of the sample points. Next, call the `interpolate.interp1d` function with `kind=3` to obtain the third-order spline for the given data.

```
In [56]: x = np.linspace(-1, 1, 11)
In [57]: y = runge(x)
In [58]: f_i = interpolate.interp1d(x, y, kind=3)
```

To evaluate how good this spline interpolation is (here represented by the instance f_i) class, plot the interpolant together with the original Runge's function and the sample points. The result is shown in Figure 7-3.

```
In [59]: xx = np.linspace(-1, 1, 100)
In [60]: fig, ax = plt.subplots(figsize=(8, 4))
    ...: ax.plot(xx, runge(xx), 'k', lw=1, label="Runge's function")
    ...: ax.plot(x, y, 'ro', label='sample points')
    ...: ax.plot(xx, f_i(xx), 'r--', lw=2, label='spline order 3')
    ...: ax.legend()
    ...: ax.set_xticks([-1, -0.5, 0, 0.5, 1])
    ...: ax.set_ylabel(r"$y$", fontsize=18)
    ...: ax.set_xlabel(r"$x$", fontsize=18)
```

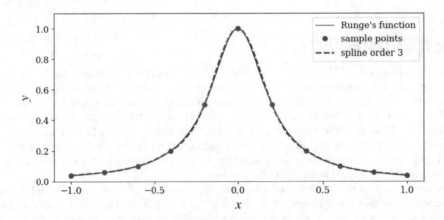

Figure 7-3. *Runge's function with a third-order spline interpolation using 11 data points*

Figure 7-3 used 11 data points and a spline of the third order. The interpolant agrees very well with the original function in Figure 7-3. Typically, spline interpolation of order three or less does not suffer from the same type of oscillations that we observed with high-order polynomial interpolation, and it is usually sufficient to use splines of order three if we have a sufficient number of data points.

To illustrate the effect of the order of a spline interpolation, consider the problem of interpolating the data (0, 3), (1, 4), (2, 3.5), (3, 2), (4, 1), (5, 1.5), (6, 1.25), and (7, 0.9) with splines of increasing order. First, define the x and y arrays and then loop over the required spline orders, computing the interpolation and plotting it for each order (see Figure 7-4).

```
In [61]: x = np.array([0, 1, 2, 3, 4, 5, 6, 7])
In [62]: y = np.array([3, 4, 3.5, 2, 1, 1.5, 1.25, 0.9])
In [63]: xx = np.linspace(x.min(), x.max(), 100)
In [64]: fig, ax = plt.subplots(figsize=(8, 4))
    ...: ax.scatter(x, y)
    ...:
    ...: for n in [1, 2, 3, 5]:
    ...:     f = interpolate.interp1d(x, y, kind=n)
    ...:     ax.plot(xx, f(xx), label='order %d' % n)
    ...:
    ...: ax.legend()
    ...: ax.set_ylabel(r"$y$", fontsize=18)
    ...: ax.set_xlabel(r"$x$", fontsize=18)
```

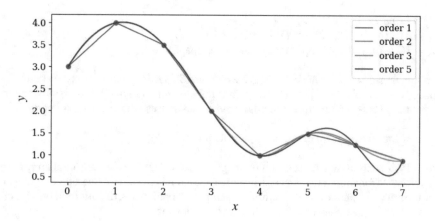

Figure 7-4. *Spline interpolations of different orders*

The spline interpolation shown in Figure 7-4 shows that spline order 2 or 3 provides a rather good interpolation, with relatively small errors between the original and interpolant functions. For higher-order splines, the same problem as we saw for high-order polynomial interpolation resurfaces. In practice, it is, therefore, often suitable to use third-order spline interpolation.

Multivariate Interpolation

Polynomial and spline interpolations can be straightforwardly generalized to multivariate situations. In analogy with the univariate case, let's seek a function whose values are given at a set of specified points and that can be evaluated for intermediary points within the sampled range. SciPy provides several functions and classes for multivariate interpolation. The following two examples explore two of the most useful functions for bivariate interpolation: the `interpolate.interp2d` and `interpolate.griddata` functions. See the docstring for the `interpolate` module and its reference manual for further information on other interpolation options.

Let's begin by looking at `interpolate.interp2d`, a straightforward generalization of the interp1d function previously used. This function takes the x and y coordinates of the available data points as separate one-dimensional arrays, followed by a two-dimensional array of values for each combination of x and y coordinates. This presumes the data points are given on a regular and uniform grid of x and y coordinates.

To illustrate how the `interp2d` function can be used, let's simulate noisy measurements by adding random noise to a known function, which in the following example is taken to be $f(x, y) = \exp\left(-(x + 1/2)^2 - 2(y + 1/2)^2\right) - \exp\left(-(x - 1/2)^2 - 2(y - 1/2)^2\right)$. To form an interpolation problem, sample this function at 10 points in the interval $[-2, 2]$, along the x and y coordinates, and then add a small normal-distributed noise to the exact values. First, create NumPy arrays for the x and y coordinates of the sample points and define a Python function for $f(x, y)$.

```
In [65]: x = y = np.linspace(-2, 2, 10)
In [66]: def f(x, y):
    ...:        return (np.exp(-(x + .5)**2 - 2*(y + .5)**2) -
    ...:                np.exp(-(x - .5)**2 - 2*(y - .5)**2))
```

Next, evaluate the function at the sample points and add the random noise to simulate uncertain measurements.

```
In [67]: X, Y = np.meshgrid(x, y)
In [68]: # simulate noisy data at fixed grid points X, Y
    ...: Z = f(X, Y) + 0.05 * np.random.randn(*X.shape)
```

At this point, there is a matrix of data points Z with noisy data associated with exactly known and regularly spaced coordinates x and y. To obtain an interpolation function that can be evaluated for intermediary x and y values within the sampled range, we can now use the `interp2d` function.

```
In [69]: f_i = interpolate.interp2d(x, y, Z, kind='cubic')
```

Here, x and y are one-dimensional arrays (of length 10), and Z is a two-dimensional array of shape (10, 10). The `interp2d` function returns a class instance, `f_i`, that behaves as a function that we can evaluate at arbitrary x and y coordinates (within the sampled range). Using the interpolation function, a supersampling of the original data can be obtained in the following way.

```
In [70]: xx = yy = np.linspace(x.min(), x.max(), 100)
In [71]: ZZi = f_i(xx, yy)
In [72]: XX, YY = np.meshgrid(xx, yy)
```

Here, XX and YY are coordinate matrices for the supersampled points, and the corresponding interpolated values are ZZi. These can, for example, be used to plot a smoothed function describing the sparse and noisy data. The following code plots the contours of the original function and the interpolated data. Figure 7-5 shows the resulting contour plot.

```
In [73]: fig, axes = plt.subplots(1, 2, figsize=(12, 5))
    ...: # for reference, first plot the contours of the exact function
    ...: c = axes[0].contourf(XX, YY, f(XX, YY), 15, cmap=plt.cm.RdBu)
    ...: axes[0].set_xlabel(r"$x$", fontsize=20)
    ...: axes[0].set_ylabel(r"$y$", fontsize=20)
    ...: axes[0].set_title("exact / high sampling")
    ...: cb = fig.colorbar(c, ax=axes[0])
    ...: cb.set_label(r"$z$", fontsize=20)
    ...: # next, plot the contours of the supersampled interpolation of the
    ...: # noisy data
    ...: c = axes[1].contourf(XX, YY, ZZi, 15, cmap=plt.cm.RdBu)
    ...: axes[1].set_ylim(-2.1, 2.1)
```

```
...: axes[1].set_xlim(-2.1, 2.1)
...: axes[1].set_xlabel(r"$x$", fontsize=20)
...: axes[1].set_ylabel(r"$y$", fontsize=20)
...: axes[1].scatter(X, Y, marker='x', color='k')
...: axes[1].set_title("interpolation of noisy data / low sampling")
...: cb = fig.colorbar(c, ax=axes[1])
...: cb.set_label(r"$z$", fontsize=20)
```

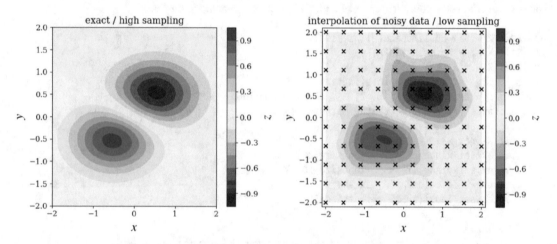

Figure 7-5. *Contours of the exact function (left) and a bivariate cubic spline interpolation (right) of noisy samples from the function on a regular grid (marked with crosses)*

With relatively sparsely spaced data points, we can thus approximate the underlying function using `interpolate.interp2d` to compute the bivariate cubic spline interpolation. This gives a smoothed approximation for the underplaying function, which is frequently useful when dealing with data obtained from measurements or computations that are costly in terms of time or other resources. For higher-dimensional problems, there is the `interpolate.interpnd` function, which is a generalization to N-dimensional problems.

Another common situation that requires multivariate interpolation occurs when sampled data is given on an irregular coordinate grid. This situation frequently arises (e.g., in experiments or other data collection processes) when the exact values at which the observations are collected cannot be directly controlled. To easily plot and analyze such data with existing tools, it may be desirable to interpolate it onto a regular coordinate grid. In SciPy, we can use the `interpolate.griddata` for this task. This function takes as the first argument a tuple of one-dimensional coordinate vectors (`xdata, ydata`) for the data values `zdata`, which are passed to the function in matrix form as the second argument. The third argument is a tuple (`X, Y`) of coordinate vectors or coordinate matrices for the new points at which the interpolant is evaluated. Optionally, we can also set the interpolation method using the `method` keyword argument (`'nearest'`, `'linear'`, or `'cubic'`).

```
In [74]: Zi = interpolate.griddata((xdata, ydata), zdata, (X, Y), method='cubic')
```

To demonstrate how to use the `interpolate.griddata` function for interpolating data at unstructured coordinate points, we take the $f(x, y) = \exp(-x^2 - y^2) \cos 4x \sin 6y$ function and randomly select sampling points in the interval $[-1, 1]$ along the x and y coordinates. The resulting $\{x_i, y_i, z_i\}$ data is then interpolated and evaluated on a supersampled regular grid spanning the $x, y \in [-1, 1]$ region. To this end, we first define a Python function for $f(x, y)$ and then generate the randomly sampled data.

```
In [75]: def f(x, y):
    ...:     return np.exp(-x**2 - y**2) * np.cos(4*x) * np.sin(6*y)
In [76]: N = 500
In [77]: xdata = np.random.uniform(-1, 1, N)
In [78]: ydata = np.random.uniform(-1, 1, N)
In [79]: zdata = f(xdata, ydata)
```

To visualize the function and the density of the sampling points, plot a scatter plot for the sampling locations overlaid on a contour graph of $f(x, y)$. The result is shown in Figure 7-6.

```
In [80]: x = y = np.linspace(-1, 1, 100)
In [81]: X, Y = np.meshgrid(x, y)
In [82]: Z = f(X, Y)
In [83]: fig, ax = plt.subplots(figsize=(8, 6))
    ...: c = ax.contourf(X, Y, Z, 15, cmap=plt.cm.RdBu)
    ...: ax.scatter(xdata, ydata, marker='.')
    ...: ax.set_ylim(-1,1)
    ...: ax.set_xlim(-1,1)
    ...: ax.set_xlabel(r"$x$", fontsize=20)
    ...: ax.set_ylabel(r"$y$", fontsize=20)
    ...: cb = fig.colorbar(c, ax=ax)
    ...: cb.set_label(r"$z$", fontsize=20)
```

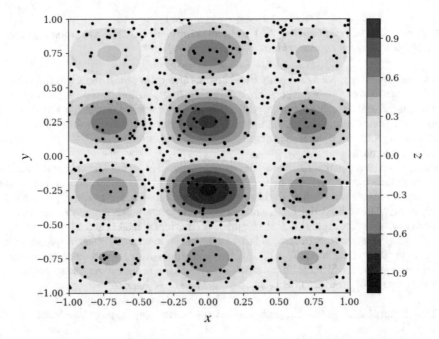

Figure 7-6. *Exact contour plot of a randomly sampled function. The 500 sample points are marked with black dots*

From the contour graph and scatter plots in Figure 7-6, it appears that the randomly chosen sample points cover the coordinate region of interest reasonably well, and it is plausible that we should be able to reconstruct the $f(x, y)$ function relatively accurately by interpolating the data. Let's interpolate the data on the finely spaced (supersampled) regular grid described by the X and Y coordinate arrays. To compare different interpolation methods and the effect of increasing the number of sample points, define the z_interpolate function that interpolates the given data points with the nearest data point, a linear interpolation, and a cubic spline interpolation.

```
In [84]: def z_interpolate(xdata, ydata, zdata):
    ...:        Zi_0 = interpolate.griddata(
    ...:            (xdata, ydata), zdata, (X, Y), method='nearest')
    ...:        Zi_1 = interpolate.griddata(
    ...:            (xdata, ydata), zdata, (X, Y), method='linear')
    ...:        Zi_3 = interpolate.griddata(
    ...:            (xdata, ydata), zdata, (X, Y), method='cubic')
    ...:        return Zi_0, Zi_1, Zi_3
```

Finally, plot a contour graph of the interpolated data for the three different interpolation methods applied to three subsets of the total sample points that use 50, 150, and all 500 points, respectively. The result is shown in Figure 7-7.

```
In [85]: fig, axes = plt.subplots(3, 3, figsize=(12, 12),
    ...:                               sharex=True, sharey=True)
    ...: np.random.seed(115925231)  # use a random seed for reproducibility
    ...: n_vec = [50, 150, 500]
    ...: for idx, n in enumerate(n_vec):
    ...:     Zi_0, Zi_1, Zi_3 = z_interpolate(xdata[:n], ydata[:n], zdata[:n])
    ...:     axes[idx, 0].contourf(X, Y, Zi_0, 15, cmap=plt.cm.RdBu)
    ...:     axes[idx, 0].set_ylabel("%d data points\ny" % n, fontsize=16)
    ...:     axes[idx, 0].set_title("nearest", fontsize=16)
    ...:     axes[idx, 1].contourf(X, Y, Zi_1, 15, cmap=plt.cm.RdBu)
    ...:     axes[idx, 1].set_title("linear", fontsize=16)
    ...:     axes[idx, 2].contourf(X, Y, Zi_3, 15, cmap=plt.cm.RdBu)
    ...:     axes[idx, 2].set_title("cubic", fontsize=16)
    ...: for m in range(len(n_vec)):
    ...:     axes[2, m].set_xlabel("x", fontsize=16)
```

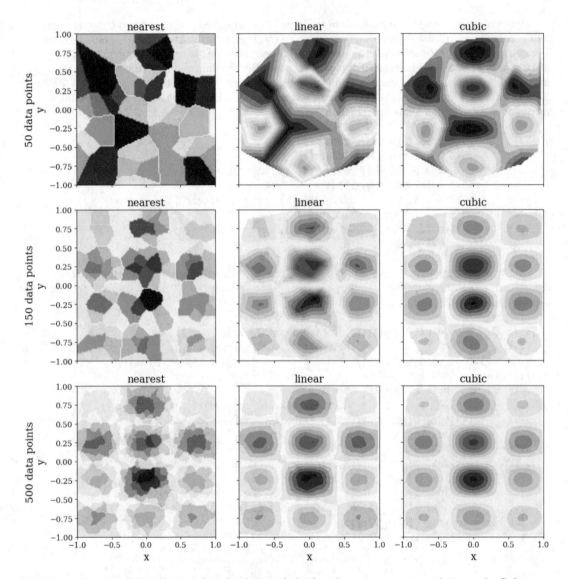

Figure 7-7. *Bivariate interpolation of randomly sampled values for increasing interpolation order (left to right) and increasing the number of sample points (top to bottom)*

Figure 7-7 shows that it is possible to reconstruct a function fairly well from interpolating unstructured samples, as long as the region of interest is well covered. In this example, and quite generally for other situations, it is clear that the cubic spline interpolation is vastly superior to nearest-point and linear interpolation. Although it is more computationally demanding to compute the spline interpolation, it is often worthwhile.

Summary

Interpolation is a fundamental mathematical tool with significant applications throughout scientific and technical computing. It is crucial to many mathematical methods and algorithms. It is also a practical tool, useful when plotting or analyzing data obtained from experiments, observations, or resource-demanding computations. The combination of the NumPy and SciPy libraries provides good coverage of numerical interpolation methods in one or more independent variables. For most practical interpolation problems involving many data points, cubic spline interpolation is the most useful technique. However, polynomial interpolation of low degree is commonly used as a tool in other numerical methods (such as root finding, optimization, and numerical integration). This chapter explored using NumPy's `polynomial` and SciPy's `interpolate` modules to perform interpolation on given datasets with one and two independent variables. Mastering these techniques is an essential skill of a computational scientist, and I encourage further exploring the content in `scipy.interpolate` that was not covered here by studying the docstrings for this module and its many functions and classes.

Further Reading

Interpolation is covered in most texts on numerical methods. For a more thorough theoretical introduction, I recommend *Introduction to Numerical Analysis* by J. Stoer et al. (Springer, 1992) or *Numerical Methods for Scientists and Engineers* by R. Hamming (Dover Publications, 1987).

CHAPTER 8

■ ■ ■

Integration

This chapter covers different aspects of integration, focusing on numerical integration. For historical reasons, numerical integration is also known as *quadrature*. Integration is significantly more complex than its inverse operation—differentiation—and while many examples of integrals can be calculated analytically, in general, we must resort to numerical methods. Depending on the properties of the integrand (the function being integrated) and the integration limits, it can be easy or difficult to compute an integral numerically. In most cases, integrals of continuous functions and finite integration limits can be computed efficiently in one dimension. But integrable functions with singularities or integrals with infinite integration limits are cases that can be difficult to handle numerically, even in a single dimension. Two-dimensional integrals (double integrals) and higher-order integrals can be numerically computed with repeated single-dimension integration or using multidimensional generalizations of the techniques used to solve single-dimensional integrals. However, the computational complexity grows quickly with the number of dimensions to integrate. Such methods are only feasible for low-dimensional integrals like double or triple integrals. Integrals of higher dimensions than that often require entirely different techniques, such as Monte Carlo sampling algorithms.

In addition to the numerical evaluation of integrals with definite integration limits, which results in a single number, integration also has other important applications. For example, equations where the integrand of an integral is the unknown quantity are called integral equations, and such equations frequently appear in science and engineering applications. Integral equations are usually difficult to solve, but they can often be recast into linear equation systems by discretizing the integral. However, this topic is not covered here, but examples of this type of problem are shown in Chapter 10. Another important application of integration is integral transforms, which are techniques for transforming functions and equations between different domains. The end of this chapter briefly discusses how SymPy can compute some integral transforms, such as Laplace transforms and Fourier transforms.

To carry out symbolic integration, we can use SymPy, as briefly discussed in Chapter 3, and to compute numerical integration, the `integrate` module is mainly used in SciPy. However, SymPy (through the `mpmath` multiple-precision library) also has routines for numerical integration, which complement those in SciPy, for example, by offering arbitrary-precision integration. This chapter examines both options, discusses their pros and cons, and briefly looks at Monte Carlo integrations using the `scikit-monaco` library.

■ **scikit-monaco** `scikit-monaco` is a small library that makes Monte Carlo integration convenient and easily accessible. At the time of writing, the most recent version of scikit-monaco is 0.2.1. See `http://scikit-monaco.readthedocs.org` for more information.

R. Johansson, *Numerical Python*, https://doi.org/10.1007/979-8-8688-0413-7_8

Importing Modules

This chapter requires, as usual, the NumPy and the Matplotlib libraries for basic numerical and plotting support, and on top of that, use the `integrate` module from SciPy, the SymPy library, and the `mpmath` arbitrary-precision math library. Here, let's assume that these modules are imported as follows.

```
In [1]: import numpy as np
In [2]: import matplotlib.pyplot as plt
   ...: import matplotlib as mpl
In [3]: from scipy import integrate
In [4]: import sympy
In [5]: import mpmath
```

In addition, for nicely formatted output from SymPy, we must set up its printing system.

```
In [6]: sympy.init_printing()
```

Numerical Integration Methods

Here, we are concerned with evaluating definite integrals on the form $I(f) = \int_a^b f(x)\mathrm{d}x$, with given integration limits a and b. The interval $[a, b]$ can be finite, semi-infinite (where either $a = -\infty$ or $b = \infty$), or infinite (where both $a = -\infty$ and $b = \infty$). The integral $I(f)$ can be interpreted as the area between the curve of the integrand $f(x)$ and the x axis, as illustrated in Figure 8-1.

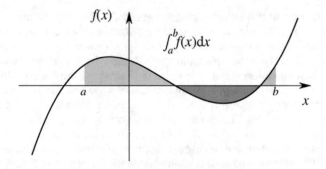

Figure 8-1. *Interpretation of an integral as the area between the curve of the integrand and the x axis, where the area is counted as positive where f(x) > 0 (green/light) and negative otherwise (red/dark)*

A general strategy for numerically evaluating an integral $I(f)$, on the preceding form, is to write the integral as a discrete sum that approximates the value of the integral, as follows:

$$I(f) = \sum_{i=1}^{n} w_i f(x_i) + r_n.$$

Here, w_i are the weights of n evaluations of $f(x)$ at the points $x_i \in [a, b]$, and r_n is the residual due to the approximation. In practice, assume that r_n is small and can be neglected, but it is important to estimate r_n to know how accurately the integral is approximated. This summation formula for $I(f)$ is known as an n-point *quadrature rule*, and the choice of the number of points n, their locations in $[a, b]$, and the weight factors w_i influence the accuracy and the computational complexity of its evaluation. Quadrature rules can be derived from interpolations of $f(x)$ on the interval $[a, b]$. If the points x_i are evenly spaced in the interval $[a, b]$, and a polynomial interpolation is used, then the resulting quadrature rule is known as a *Newton-Cotes quadrature rule*. For instance, approximating $f(x)$ with a zeroth-order polynomial (constant value) using the midpoint value $x_0 = (a+b)/2$, we obtain:

$$\int_a^b f(x)\,dx \approx f\left(\frac{a+b}{2}\right)\int_a^b dx = (b-a)f\left(\frac{a+b}{2}\right).$$

This is known as the *midpoint rule*, and it integrates polynomials of up to order one (linear functions) exactly, and it is therefore said to be of polynomial degree one. Approximating $f(x)$ by a polynomial of degree one, evaluated at the endpoints of the interval, results in the following:

$$\int_a^b f(x)\,dx \approx \frac{b-a}{2}\left(f(a)+f(b)\right).$$

This is known as the *trapezoid rule and* is also of polynomial degree one. Using an interpolation polynomial of second order results in *Simpson's rule*,

$$\int_a^b f(x)\,dx \approx \frac{b-a}{6}\left(f(a)+4f\left(\frac{a+b}{2}\right)+f(b)\right),$$

which uses function evaluations at the endpoints and the midpoint. This method is of polynomial degree three, meaning it integrates exactly polynomials up to order three. The method of arriving at this formula can easily be demonstrated using SymPy. First, define symbols for the a, b, and x variables and the f function.

```
In [7]: a, b, X = sympy.symbols("a, b, x")
In [8]: f = sympy.Function("f")
```

Next, define a tuple x that contains the sample points (the endpoints and the middle point of the interval $[a, b]$) and a list w of weight factors to be used in the quadrature rule, corresponding to each sample point.

```
In [9]: x = a, (a+b)/2, b  # for Simpson's rule
In [10]: w = [sympy.symbols("w_%d" % i) for i in range(len(x))]
```

Given x and w, we can now construct a symbolic expression for the quadrature rule.

```
In [11]: q_rule = sum([w[i] * f(x[i]) for i in range(len(x))])
In [12]: q_rule
```

Out[12]: $w_0 f(a)+w_1 f\left(\dfrac{a}{2}+\dfrac{b}{2}\right)+w_2 f(b)$

To compute the appropriate values of the weight factors w_i, choose the polynomial basis functions $\{\phi_n(x)=x^n\}_{n=0}^2$ for the interpolation of $f(x)$, and here, let's use the sympy.Lambda function to create symbolic representations for each of these basis functions.

```
In [13]: phi = [sympy.Lambda(X, X**n) for n in range(len(x))]
In [14]: phi
Out[14]: [(x ↦ 1), (x ↦ x), (x ↦ x²)]
```

The key to finding the weight factors in the quadrature expression (Out[12]) is that the integral $\int_a^b \phi_n(x)\,dx$ can be computed analytically for each of the basis functions $\phi_n(x)$. By substituting the $f(x)$ function with each of the $\phi_n(x)$ basis functions in the quadrature rule, obtain an equation system for the unknown weight factors.

$$\sum_{i=0}^{2} w_i \phi_n(x_i) = \int_a^b \phi_n(x)\,dx$$

These equations are equivalent to requiring that the quadrature rule exactly integrates all the basis functions and, therefore, (at least) all functions spanned by the basis. This equation system can be constructed with SymPy using the following.

```
In [15]: eqs = [q_rule.subs(f, phi[n]) - sympy.integrate(phi[n](X), (X, a, b))
    ...:         for n in range(len(phi))]
In [16]: eqs
```

Out[16]: $\left[a - b + w_0 + w_1 + w_2, \dfrac{a^2}{2} + aw_0 - \dfrac{b^2}{2} + bw_2 + w_1\left(\dfrac{a}{2} + \dfrac{b}{2}\right), \dfrac{a^3}{3} + a^2w_0 - \dfrac{b^3}{3} + \right.$
$\left. b^2w_2 + w_1\left(\dfrac{a}{2} + \dfrac{b}{2}\right)^2 \right]$

Solving this linear equation system gives analytical expressions for the weight factors.

```
In [17]: w_sol = sympy.solve(eqs, w)
In [18]: w_sol
```

Out[18]: $\left\{w_0 : -\dfrac{a}{6} + \dfrac{b}{6}, w_1 : -\dfrac{2a}{3} + \dfrac{2b}{3}, w_2 : -\dfrac{a}{6} + \dfrac{b}{6}\right\}$

Substituting the solution into the symbolic expression for the quadrature rule obtains the following.

```
In [19]: q_rule.subs(w_sol).simplify()
```

Out[19]: $-\dfrac{1}{6}(a - b)\left(f(a) + f(b) + 4f\left(\dfrac{a}{2} + \dfrac{b}{2}\right)\right)$

We recognize this result as Simpson's quadrature rule given in the preceding section. Choosing different sample points (the x tuple in this code) results in different quadrature rules.

Higher-order quadrature rules can similarly be derived using higher-order polynomial interpolation (more sample points in the [a, b] interval). However, high-order polynomial interpolation can have undesirable behavior between the sample points, as discussed in Chapter 7. Rather than using higher-order quadrature rules, it is therefore often better to divide the integration interval [a, b] into subintervals $[a = x_0, x_1], [x_1, x_2], ..., [x_{N-1}, x_N = b]$ and use a low-order quadrature rule in each of these subintervals. Such methods are known as *composite quadrature rules*. Figure 8-2 shows the three lowest-order Newton-Cotes quadrature rules for the $f(x) = 3 + x + x^2 + x^3 + x^4$ function on the interval [−1, 1] and the corresponding composite quadrature rules with four subdivisions of the original interval.

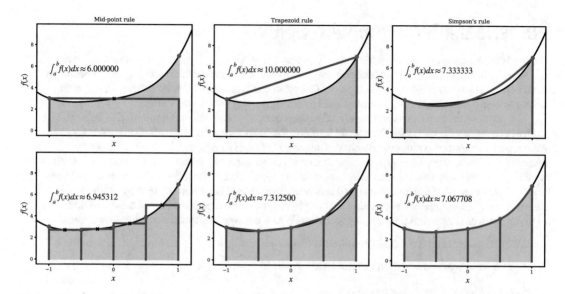

Figure 8-2. *Visualization of quadrature rules (top panel) and composite quadrature rules (bottom panel) of orders zero (the midpoint rule), one (the trapezoid rule), and two (Simpson's rule)*

The subinterval length $h = (b - a)/N$ is an important parameter that characterizes composite quadrature rules. Estimates for the errors in an approximate quadrature rule and the scaling of the error with respect to h can be obtained from Taylor series expansions of the integrand and the analytical integration of the term in the resulting series. An alternative technique simultaneously considers quadrature rules of different orders or subinterval lengths h. The difference between two such results can often be shown to give estimates of the error, and this is the basis for how many quadrature routines produce an estimate of the error in addition to the estimation of the integral, as shown in the examples in the following section.

We have seen that the Newton-Cotes quadrature rules use evenly spaced sample points of the integrand $f(x)$. This is often convenient, especially if the integrand is obtained from measurements or observations at prescribed points and cannot be evaluated at arbitrary points in the interval $[a, b]$. However, this is not necessarily the most efficient choice of quadrature nodes. If the integrand is given as a function that easily can be evaluated at arbitrary values of $x \in [a, b]$, then it can be advantageous to use quadrature rules that do not use evenly spaced sample points. An example of such a method is *Gaussian quadrature*, which also uses polynomial interpolation to determine the values of the weight factors in the quadrature rule but where the quadrature nodes x_i are chosen to maximize the order of polynomials that can be integrated exactly (the polynomial degree) given a fixed number of quadrature points. It turns out that choices x_i that satisfy these criteria are the roots of different orthogonal polynomials, and the sample points x_i are typically located at irrational locations in the integration interval $[a, b]$. This generally is not a problem for numerical implementations. But practically, it requires that the $f(x)$ function is available to be evaluated at arbitrary points decided by the integration routine rather than given as tabulated or precomputed data at regularly spaced x values. Gaussian quadrature rules are typically superior if $f(x)$ can be evaluated at arbitrary values. Yet, for the reason just mentioned, the Newton-Cotes quadrature rules also have important use cases when the integrand is given as tabulated data.

Numerical Integration with SciPy

The numerical quadrature routines in the SciPy `integrate` module can be categorized into two types: those that take the integrand as a Python function and those that take arrays with samples of the integrand at given points. The functions of the first type use Gaussian quadrature (quad, quadrature, fixed_quad), while functions of the second type use Newton-Cotes methods (`trapz`, `simps`, and `romb`).

The quadrature function is an adaptive Gaussian quadrature routine implemented in Python. The quadrature repeatedly calls the `fixed_quad` function for Gaussian quadrature of a fixed order, increasing order until the required accuracy is reached. The quad function is a wrapper for routines from the *FORTRAN* library QUADPACK, which has superior speed performance and more features (such as support for infinite integration limits). It is, therefore, usually preferable to use quad, and the following uses this quadrature function. However, all these functions take similar arguments and can often be replaced with each other. They take as a first argument the function that implements the integrand, and the second and third arguments are the lower and upper integration limits. As a concrete example, consider the numerical evaluation of the integral $\int_{-1}^{1} e^{-x^2} dx$. To evaluate this integral using SciPy's quad function, first define a function for the integrand and then call the quad function.

```
In [20]: def f(x):
    ...:        return np.exp(-x**2)
In [21]: val, err = integrate.quad(f, -1, 1)
In [22]: val
Out[22]: 1.493648265624854
In [23]: err
Out[23]: 1.6582826951881447e-14
```

The quad function returns a tuple that contains the numerical estimate of the integral, `val`, and an estimate of the absolute error, `err`. The tolerances for the absolute and the relative errors can be set using the optional epsabs and epsrel keyword arguments, respectively. If the f function takes more than one variable, the quad routine integrates the function over its first argument. We can optionally specify the values of additional arguments by passing those values to the integrand function via the keyword argument args to the quad function. For example, suppose we wish to evaluate $\int_{-1}^{1} ae^{-(x-b)^2/c^2} dx$ for the specific values of the parameters $a = 1$, $b = 2$, and $c = 3$. We can define a function for the integrand that takes all these additional arguments and specifies the values of a, b, and c by passing args=(1, 2, 3) to the quad function.

```
In [24]: def f(x, a, b, c):
    ...:        return a * np.exp(-((x - b)/c)**2)
In [25]: val, err = integrate.quad(f, -1, 1, args=(1, 2, 3))
In [26]: val
Out[26]: 1.2763068351022229
In [27]: err
Out[27]: 1.4169852348169507e-14
```

When working with functions where the variable we want to integrate over is not the first argument, we can reshuffle the arguments by using a `lambda` function. For example, if we wish to compute the integral $\int_{0}^{5} J_0(x) dx$, where the integrand $J_0(x)$ is the zeroth-order Bessel function of the first kind, it would be convenient to use the jv function from the scipy.special module as an integrand. The jv function takes the v and x arguments and is the Bessel function of the first kind for the real-valued order v and is evaluated at x. To use the jv function as an integrand for quad, we need to reshuffle the arguments of jv. With a lambda function, we can do this in the following manner.

```
In [28]: from scipy.special import jv
In [29]: f = lambda x: jv(0, x)
In [30]: val, err = integrate.quad(f, 0, 5)
In [31]: val
Out[31]: 0.7153119177847678
In [32]: err
Out[32]: 2.47260738289741e-14
```

With this technique, we can arbitrarily reshuffle arguments of any function and always obtain a function where the integration variable is the first argument so that it can be used as an integrand for quad.

The quad routine supports infinite integration limits. To represent infinite integration limits, we use the floating-point representation of infinity, float('inf'), which is conveniently available in NumPy as np.inf.

For example, consider the integral $\int_{-\infty}^{\infty} e^{-x^2} dx$. To evaluate it using quad, we can do

```
In [33]: f = lambda x: np.exp(-x**2)
In [34]: val, err = integrate.quad(f, -np.inf, np.inf)
In [35]: val
Out[35]: 1.7724538509055159
In [36]: err
Out[36]: 1.4202636780944923e-08
```

However, note that the quadrature and fixed_quad functions only support finite integration limits. With extra guidance, the quad function can also handle many integrals with integrable singularities.

For example, consider the integral $\int_{-1}^{1} \frac{1}{\sqrt{|x|}} dx$. The integrand diverges at $x = 0$, but the value of the integral does not diverge, and its value is 4. Naively trying to compute this integral using quad may fail because of the diverging integrand.

```
In [37]: f = lambda x: 1/np.sqrt(abs(x))
In [38]: a, b = -1, 1
In [39]: integrate.quad(f, a, b)
Out[39]: (NaN, NaN)
```

In situations like these, it can be helpful to graph the integrand to gain insights into its behavior, as shown in Figure 8-3.

```
In [40]: fig, ax = plt.subplots(figsize=(8, 3))
    ...: x = np.linspace(a, b, 10000)
    ...: ax.plot(x, f(x), lw=2)
    ...: ax.fill_between(x, f(x), color='green', alpha=0.5)
    ...: ax.set_xlabel("$x$", fontsize=18)
    ...: ax.set_ylabel("$f(x)$", fontsize=18)
    ...: ax.set_ylim(0, 25)
    ...: ax.set_xlim(-1, 1)
```

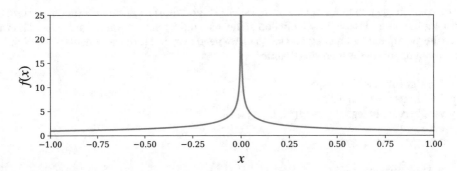

Figure 8-3. *Example of a diverging integrand with finite integral (green/shaded area) that can be computed using the quad function*

In this case, the evaluation of the integral fails because the integrand diverges exactly at one of the sample points in the Gaussian quadrature rule (the midpoint). We can guide the quad routine by specifying a list of points that should be avoided using the points keyword arguments, and using points=[0] in the current example allows quad to evaluate the integral correctly.

```
In [41]: integrate.quad(f, a, b, points=[0])
Out[41]: (4.0,5.684341886080802e-14)
```

Tabulated Integrand

We have seen that the quad routine is suitable for evaluating integrals when the integrand is specified using a Python function that the routine can evaluate at arbitrary points (determined by the specific quadrature rule). However, we may have an integrand that is only specified at predetermined points, such as evenly spaced points in the integration interval [a, b]. For example, this type of situation can occur when the integrand is obtained from experiments or observations that the particular integration routine cannot control. In this case, we can use a Newton-Cotes quadrature, such as the midpoint rule, trapezoid rule, or Simpson's rule described earlier in this chapter.

In the SciPy integrate module, the composite trapezoid rule and Simpson's rule are implemented in the trapz and simps functions. These functions take as first argument an array y with values of the integrand at a set of points in the integration interval, and they optionally take as second argument an array x that specifies the x values of the sample points, or alternatively, the spacing dx between each sample (if uniform). Note that the sample points do not necessarily need to be evenly spaced, but they must be known in advance.

Let's evaluate an integral of a function that is given by sampled values using the integral $\int_0^2 \sqrt{x}\,dx$ by taking 25 samples of the integrand in the integration interval [0, 2], as shown in Figure 8-4.

```
In [42]: f = lambda x: np.sqrt(x)
In [43]: a, b = 0, 2
In [44]: x = np.linspace(a, b, 25)
In [45]: y = f(x)
In [46]: fig, ax = plt.subplots(figsize=(8, 3))
    ...: ax.plot(x, y, 'bo')
    ...: xx = np.linspace(a, b, 500)
    ...: ax.plot(xx, f(xx), 'b-')
```

```
...: ax.fill_between(xx, f(xx), color='green', alpha=0.5)
...: ax.set_xlabel(r"$x$", fontsize=18)
...: ax.set_ylabel(r"$f(x)$", fontsize=18)
```

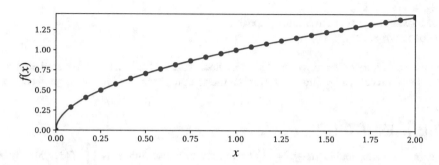

Figure 8-4. *Integrand given as tabulated values marked with dots. The integral corresponds to the shaded area*

We can pass the x and y arrays to the trapz or methods to evaluate the integral. Note that the y array must be passed as the first argument.

```
In [47]: val_trapz = integrate.trapz(y, x)
In [48]: val_trapz
Out[48]: 1.88082171605
In [49]: val_simps = integrate.simps(y, x)
In [50]: val_simps
Out[50]: 1.88366510245
```

The trapz and simps functions do not provide any error estimates. For this example, we can compute the integral analytically and compare it to the numerical values computed with the two methods.

```
In [51]: val_exact = 2.0/3.0 * (b-a)**(3.0/2.0)
In [52]: val_exact
Out[52]: 1.8856180831641267
In [53]: val_exact - val_trapz
Out[53]: 0.00479636711328
In [54]: val_exact - val_simps
Out[54]: 0.00195298071541
```

Since all the information about the integrand is the given sample points, we cannot ask either trapz or simps to compute more accurate solutions. The only options for increasing the accuracy are expanding the number of sample points (which might be difficult if the underlying function is unknown) or possibly using a higherorder method.

The integrate module also implements the Romberg method with the romb function. The Romberg method is a Newton-Cotes method but one that uses Richardson extrapolation to accelerate the convergence of the trapezoid method; however, this method does require that the sample points are evenly spaced and also that there are 2^n+1 sample points, where n is an integer. Like the trapz and simps methods, romb takes an array with integrand samples as the first argument, but the second argument must (if given) be the sample-point spacing dx.

```
In [55]: x = np.linspace(a, b, 1 + 2**6)
In [56]: len(x)
Out[56]: 65
In [57]: y = f(x)
In [58]: dx = x[1] - x[0]
In [59]: val_exact - integrate.romb(y, dx=dx)
Out[59]: 0.000378798422913
```

Among the SciPy integration functions discussed here, simps is the most useful overall. It provides a good balance between ease of use (no constraints on the sample points) and relatively good accuracy.

Multiple Integration

Multiple integrals, such as double integrals $\int_a^b \int_c^d f(x,y)\mathrm{d}x\mathrm{d}y$ and triple integrals $\int_a^b \int_c^d \int_e^f f(x,y,z)\mathrm{d}x\mathrm{d}y\mathrm{d}z$, can be evaluated using the dblquad and tplquad functions from the SciPy integrate module. Also, integration over n variables $\int...\int_D f(\boldsymbol{x})\mathrm{d}\boldsymbol{x}$, over some domain D, can be evaluated using the nquad function. These functions are wrappers around the single-variable quadrature function quad, which is called repeatedly along each integral dimension.

Specifically, the double integral routine dblquad can evaluate integrals on the form

$$\int_a^b \int_{g(x)}^{h(x)} f(x,y)\mathrm{d}x\mathrm{d}y,$$

and it has the dblquad(f, a, b, g, h) function signature, where f is a Python function for the integrand, a and b are constant integration limits along the x dimension, and g and f are Python functions (taking x as argument) that specify the integration limits along the y dimension. For example, consider the integral $\int_0^1 \int_0^1 e^{-x^2-y^2}\mathrm{d}x\mathrm{d}y$. To evaluate this, first define the f function for the integrand and graph the function and the integration region, as shown in Figure 8-5.

```
In [60]: def f(x, y):
    ...:     return np.exp(-x**2 - y**2)
In [61]: fig, ax = plt.subplots(figsize=(6, 5))
    ...: x = y = np.linspace(-1.25, 1.25, 75)
    ...: X, Y = np.meshgrid(x, y)
    ...: c = ax.contour(X, Y, f(X, Y), 15, cmap=mpl.cm.RdBu, vmin=-1, vmax=1)
    ...: bound_rect = plt.Rectangle((0, 0), 1, 1, facecolor="grey")
    ...: ax.add_patch(bound_rect)
    ...: ax.axis('tight')
    ...: ax.set_xlabel('$x$', fontsize=18)
    ...: ax.set_ylabel('$y$', fontsize=18)
```

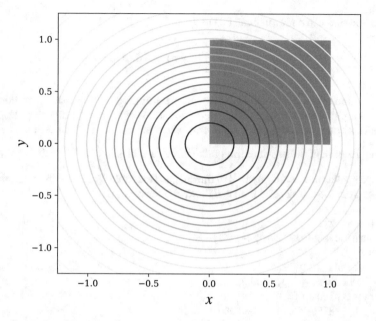

Figure 8-5. *Two-dimensional integrand as a contour plot with integration region shown as a shaded area*

In this example, the integration limits for both the *x* and *y* variables are constants. But since dblquad expects functions for the integration limits for the *y* variable, we must also define the h and g functions, even though, in this case, they only evaluate to constants regardless of the value of x.

```
In [62]: a, b = 0, 1
In [63]: g = lambda x: 0
In [64]: h = lambda x: 1
```

Now, with all the arguments prepared, we can call dblquad to evaluate the integral.

```
In [65]: integrate.dblquad(f, a, b, g, h)
Out[65]: (0.5577462853510337, 6.1922276789587025e-15)
```

We could also do the same thing more concisely, although slightly less readable, by using inline lambda function definitions.

```
In [66]: integrate.dblquad(lambda x, y: np.exp(-x**2-y**2), 0, 1, lambda x: 0, lambda x: 1)
Out[66]: (0.5577462853510337, 6.1922276789587025e-15)
```

Because g and h are functions, we can compute integrals with *x*-dependent integration limits along the *y* dimension. For example, the following is obtained with g(*x*) = x – 1 and h(*x*) = 1 – x.

```
In [67]: integrate.dblquad(f, 0, 1, lambda x: -1 + x, lambda x: 1 - x)
Out[67]: (0.7320931000008094, 8.127866157901059e-15)
```

The `tplquad` function can compute integrals in the form

$$\int_a^b \int_{g(x)}^{h(x)} \int_{q(x,y)}^{r(x,y)} f(x,y,z)\,dxdydz,$$

which is a generalization of the double integral expression computed with `dblquad`. It additionally takes two Python functions as arguments that specify the integration limits along the z dimension. These functions take two arguments, x and y, but note that g and h only take one argument (x). To see how `tplquad` can be used, generalization of the previous integral to three variables: $\int_0^1 \int_0^1 \int_0^1 e^{-x^2-y^2-z^2}\,dxdydz$. We compute this integral using a similar method compared to the `dblquad` example. We first define functions for the integrand and the integration limits and then call the `tplquad` function.

```
In [68]: def f(x, y, z):
    ...:     return np.exp(-x**2-y**2-z**2)
In [69]: a, b = 0, 1
In [70]: g, h = lambda x: 0, lambda x: 1
In [71]: q, r = lambda x, y: 0, lambda x, y: 1
In [72]: integrate.tplquad(f, 0, 1, g, h, q, r)
Out[72]: (0.4165383858866382, 4.624505066515441e-15)
```

For an arbitrary number of integrations, we can use the `nquad` function. It also takes the integrand as a Python function as the first argument. The integrand function should have the function signature `f(x1, x2, ..., xn)`. In contrast to `dplquad` and `tplquad`, the `nquad` function expects a list of integration limit specifications as the second argument. The list should contain a tuple with integration limits for each integration variable or a callable function that returns such a limit. For example, we could use the following to compute the integral previously calculated with `tplquad`.

```
In [73]: integrate.nquad(f, [(0, 1), (0, 1), (0, 1)])
Out[73]: (0.4165383858866382, 8.291335287314424e-15)
```

For an increasing number of integration variables, the computational complexity of a multiple integral grows quickly, for example, when using `nquad`. To see this scaling trend, consider the following generalized version of the integrand studied with `dplquad` and `tplquad`.

```
In [74]: def f(*args):
    ...:     """
    ...:     f(x1, x2, ... , xn) = exp(-x1^2 - x2^2 - ... - xn^2)
    ...:     """
    ...:     return np.exp(-np.sum(np.array(args)**2))
```

Next, let's evaluate the integral for varying dimensions (ranging from 1 up to 5). In the following examples, the length of the list of integration limits determines the number of integrals. To see a rough estimate of the computation time, use the IPython `%time` command.

```
In [75]: %time integrate.nquad(f, [(0,1)] * 1)
CPU times: user 398 µs, sys: 63 µs, total: 461 µs
Wall time: 466 µs
Out[75]: (0.7468241328124271,8.291413475940725e-15)
In [76]: %time integrate.nquad(f, [(0,1)] * 2)
CPU times: user 6.31 ms, sys: 298 µs, total: 6.61 ms
```

```
Wall time: 6.57 ms
Out[76]: (0.5577462853510337,8.291374381535408e-15)
In [77]: %time integrate.nquad(f, [(0,1)] * 3)
CPU times: user 123 ms, sys: 2.46 ms, total: 126 ms
Wall time: 125 ms
Out[77]: (0.4165383858866382,8.291335287314424e-15)
In [78]: %time integrate.nquad(f, [(0,1)] * 4)
CPU times: user 2.41 s, sys: 11.1 ms, total: 2.42 s
Wall time: 2.42 s
Out[78]: (0.31108091882287664,8.291296193277774e-15)
In [79]: %time integrate.nquad(f, [(0,1)] * 5)
CPU times: user 49.5 s, sys: 169 ms, total: 49.7 s
Wall time: 49.7 s
Out[79]: (0.23232273743438786,8.29125709942545e-15)
```

Here, we see that increasing the number of integrations from one to five increases the computation time from hundreds of microseconds to nearly a minute. For an even larger number of integrals, it may become impractical to use direct quadrature routines, and other methods, such as Monte Carlo sampling techniques, can often be superior, especially if the required precision is low. Monte Carlo integration is a simple but powerful technique based on sampling the integrand at randomly selected points in the integral domain and gradually forming an estimate of the integral. Due to the stochastic nature of the algorithm, the conversion rate is typically relatively slow, and it is difficult to achieve very high accuracy. However, Monte Carlo integration scales very well with dimensionality and is often a competitive method for high-dimensional integrals.

To compute an integral using Monte Carlo sampling, we can use the mcquad function from the skmonaco library (known as scikit-monaco). As the first argument, it takes a Python function for the integrand; as the second argument, it takes a list of lower integration limits; and as the third argument, it takes a list of upper integration limits. Note that the way the integration limits are specified is different from for the quad function in SciPy's integrate module. Let's begin by importing the skmonaco (scikit-monaco) module.

```
In [80]: import skmonaco
```

Once the module is imported, we can use the skmonaco.mcquad function for performing a Monte Carlo integration. The following example computes the same integral as in the previous example using nquad.

```
In [81]: %time val, err = skmonaco.mcquad(f, xl=np.zeros(5), xu=np.ones(5), npoints=100000)
CPU times: user 1.43 s, sys: 100 ms, total: 1.53 s
Wall time: 1.5 s
In [82]: val, err
Out[82]: (0.231322502809, 0.000475071311272)
```

While the error is not comparable to the result given by nquad, the computation time is much shorter. By increasing the number of sample points, which we can specify using the npoints argument, we can increase the accuracy of the result. However, the convergence of Monte Carlo integration is very slow, and it is most suitable when high accuracy is not required. However, the beauty of Monte Carlo integration is that its computational complexity is independent of the number of integrals. This is illustrated in the following example, which computes a ten-variable integration at the same time and with a comparable error level as the previous example with a five-variable integration.

```
In [83]: %time val, err = skmonaco.mcquad(f, xl=np.zeros(10), xu=np.ones(10),
npoints=100000)
CPU times: user 1.41 s, sys: 64.9 ms, total: 1.47 s
```

```
Wall time: 1.46 s
In [84]: val, err
Out[84]: (0.0540635928549, 0.000171155166006)
```

Symbolic and Arbitrary-Precision Integration

Chapter 3 presented examples of how SymPy can compute definite and indefinite integrals of symbolic functions using the `sympy.integrate` function. For example, to compute the integral, $\int_{-1}^{1} 2\sqrt{1-x^2}\,dx$ first create a symbol for x and define expressions for the integrand, and the integration limits $a = -1$ and $b = 1$.

```
In [85]: x = sympy.symbols("x")
In [86]: f = 2 * sympy.sqrt(1-x**2)
In [87]: a, b = -1, 1
```

Next, we can compute the closed-form expression for the integral using the following.

```
In [88]: val_sym = sympy.integrate(f, (x, a, b))
In [89]: val_sym
Out[89]: π
```

For this example, SymPy can find the analytic expression for the integral, π. As pointed out earlier, this situation is the exception, and in general, we cannot find an analytical closed-form expression. We then need to resort to numerical quadrature, for example, using SciPy's `integrate.quad`, as discussed earlier in this chapter. However, the mpmath library,[1] which is closely integrated with SymPy, provides an alternative implementation of numerical quadrature using arbitrary-precision computations. With this library, we can evaluate an integral to arbitrary precision without being restricted by the limitations of floating-point numbers. However, the downside is, of course, that arbitrary-precision computations are significantly slower than floating-point computations. But when we require precision beyond what the SciPy quadrature functions can provide, this multiple-precision quadrature offers a solution.

For example, to evaluate the integral $\int_{-1}^{1} 2\sqrt{1-x^2}\,dx$ to a given precision,[2] we can use the `mpmath.quad` function, which takes a Python function for the integrand as the first argument and the integration limits as a tuple (a, b) as the second argument. To specify the precision, set the variable `mpmath.mp.dps` to the required number of accurate decimal places. For example, if we need 75 accurate decimal places, set the following.

```
In [90]: mpmath.mp.dps = 75
```

The integrand must be given as a Python function that uses math functions from the mpmath library to compute the integrand. From a SymPy expression, we can create such a function using `sympy.lambdify` with `'mpmath'` as the third argument, which indicates that we want an mpmath compatible function. Alternatively,

---

[1] For more information about the multiprecision (arbitrary precision) math library mpmath, see the project's web page at http://mpmath.org.

[2] Here, I deliberately choose to work with an integral with a known analytical value to compare the multiprecision quadrature result with the known exact value.

we can directly implement a Python function using the math functions from the mpmath module in SymPy, which would be f_mpmath = lambda x: 2 * mpmath.sqrt(1 - x**2). However, let's use sympy.lambdify to automate this step.

```
In [91]: f_mpmath = sympy.lambdify(x, f, 'mpmath')
```

Next, compute the integral using mpmath.quad and display the resulting value.

```
In [92]: val = mpmath.quad(f_mpmath, (a, b))
In [93]: sympy.sympify(val)
Out[93]: 3.14159265358979323846264338327950288419716939937510582097494459230781640629
```

To verify that the numerically computed value is accurate to the required number of decimal places (75), compare the result with the known analytical value (π). The error is indeed very small.

```
In [94]: sympy.N(val_sym, mpmath.mp.dps+1) - val
Out[94]: 6.90893484407555570030908149024031965689280029154902510801896277613487344253e-77
```

This level of precision cannot be achieved with the quad function in SciPy's integrate module since the precision of floating-point numbers limits it.

The mpmath library's quad function can also evaluate double and triple integrals. To do so, we only need to pass to it an integrand function that takes multiple variables as arguments and pass tuples with integration limits for each integration variable. For example, to compute the double integral

$$\int_0^1\int_0^1 \cos(x)\cos(y)e^{-x^2-y^2}\,\mathrm{d}x\mathrm{d}y$$

and the triple integral

$$\int_0^1\int_0^1\int_0^1 \cos(x)\cos(y)\cos(z)e^{-x^2-y^2-z^2}\,\mathrm{d}x\mathrm{d}y\mathrm{d}z$$

to 30 significant decimals (this example cannot be solved symbolically with SymPy), we could first create SymPy expressions for the integrands, and then use sympy.lambdify to create the corresponding mpmath expressions.

```
In [95]: x, y, z = sympy.symbols("x, y, z")
In [96]: f2 = sympy.cos(x) * sympy.cos(y) * sympy.exp(-x**2 - y**2)
In [97]: f3 = sympy.cos(x) * sympy.cos(y) * sympy.cos(z) * sympy.exp(-x**2 - y**2 - z**2)
In [98]: f2_mpmath = sympy.lambdify((x, y), f2, 'mpmath')
In [99]: f3_mpmath = sympy.lambdify((x, y, z), f3, 'mpmath')
```

The integrals can then be evaluated to the desired accuracy by setting mpmath.mp.dps and calling mpmath.quad.

```
In [100]: mpmath.mp.dps = 30
In [101]: mpmath.quad(f2_mpmath, (0, 1), (0, 1))
Out[101]: mpf('0.430564794306099099242308990195783')
In [102]: res = mpmath.quad(f3_mpmath, (0, 1), (0, 1), (0, 1))
In [103]: sympy.sympify(res)
Out[103]: 0.282525579518426896867622772405
```

203

Again, this gives access to accuracy levels beyond what `scipy.integrate.quad` can achieve, but this additional accuracy comes with a hefty increase in computational cost. Note that the type of the object returned by `mpmath.quad` is a multiprecision float (`mpf`). It can be cast into a SymPy type using `sympy.sympify`.

Line Integrals

SymPy can also compute line integrals on the form $\int_C f(x, y)\,ds$, where C is a curve in the x–y plane, using the `line_integral` function. This function takes the integrand, as a SymPy expression, as the first argument, a `sympy.Curve` instance as the second argument, and a list of integration variables as the third argument. The path of the line integral is specified by the `Curve` instance, which describes a parameterized curve for which the x and y coordinates are given as a function of an independent parameter, t. We can use the following to create a Curve instance that describes a path along the unit circle.

```
In [104]: t, x, y = sympy.symbols("t, x, y")
In [105]: C = sympy.Curve([sympy.cos(t), sympy.sin(t)], (t, 0, 2 * sympy.pi))
```

Once the integration path is specified, we can easily compute the corresponding line integral for a given integrand using `line_integral`. For example, with the integrand $f(x, y) = 1$, the result is the circumference of the unit circle.

```
In [106]: sympy.line_integrate(1, C, [x, y])
Out[106]: 2π
```

The result is less obvious for a nontrivial integrand, such as in the following example, which computes the line integral with the integrand $f(x, y) = x^2 y^2$.

```
In [107]: sympy.line_integrate(x**2 * y**2, C, [x, y])
Out[107]: π/4
```

Integral Transforms

The last application of integrals discussed in this chapter is integral transforms. An integral transform is a procedure that takes a function as input and outputs another function. Integral transforms are the most useful when they can be computed symbolically, and here we explore two examples of integral transforms that can be performed using SymPy: the Laplace transform and the Fourier transform. These two transformations have numerous applications, but the fundamental motivation is to transform problems into a more easily handled form. It can, for example, be a transformation of a differential equation into an algebraic equation, using Laplace transforms, or a transformation of a problem from the time domain to the frequency domain using Fourier transforms.

In general, an integral transform of the $f(t)$ function can be written as

$$T_f(u) = \int_{t_1}^{t_2} K(t,u) f(t)\,dt,$$

where $T_f(u)$ is the transformed function. The choice of the kernel $K(t, u)$ and the integration limits determine the type of integral transform.

The inverse of the integral transform is given by

$$f(u) = \int_{u_1}^{u_2} K^{-1}(u,t) T_f(u) du,$$

where $K^{-1}(u, t)$ is the kernel of the inverse transform. SymPy provides functions for several types of integral transform.

Let's focus on the Laplace transform

$$L_f(s) = \int_0^\infty e^{-st} f(t) dt,$$

with the inverse transform

$$f(t) = \frac{1}{2\pi i} \int_{c-i\infty}^{c+i\infty} e^{st} L_f(s) ds.$$

Let's also focus on the Fourier transform

$$F_f(\omega) = \frac{1}{\sqrt{2\pi}} \int_{-\infty}^{\infty} e^{-i\omega t} f(t) dt,$$

with the inverse transform

$$f(t) = \frac{1}{\sqrt{2\pi}} \int_{-\infty}^{\infty} e^{i\omega t} F_f(\omega) d\omega.$$

With SymPy, we can perform these transforms with the `sympy.laplace_transform` and `sympy.fourier_transform`, respectively, and the corresponding inverse transforms can be computed with the `sympy.inverse_laplace_transform` and `sympy.inverse_fourier_transform`. These functions take a SymPy expression for the function to transform as the first argument and the symbol for the independent variable of the expression to transform as the second argument (e.g., t). As the third argument they take the symbol for the transformation variable (e.g., s). For example, to compute the Laplace transform of the $f(t) = sin(at)$ function, we begin by defining SymPy symbols for the variables a, t, and s and a SymPy expression for the $f(t)$ function.

```
In [108]: s = sympy.symbols("s")
In [109]: a, t = sympy.symbols("a, t", positive=True)
In [110]: f = sympy.sin(a*t)
```

Once we have SymPy objects for the variables and the function, we can call the `laplace_transform` function to compute the Laplace transform.

```
In [111]: sympy.laplace_transform(f, t, s)
```

Out[111]: $(\dfrac{a}{a^2 + s^2}, -\infty, 0 < \Re s)$

By default, the `laplace_transform` function returns a tuple containing the resulting transform; the value A from the convergence condition of the transform, which takes the form $A < \Re s$; and lastly, additional conditions required for the transform to be well defined. These conditions typically depend on the constraints specified when symbols are created. For example, positive=True was used when creating the symbols *a* and *t* to indicate that they represent real and positive numbers. Often, we are only interested in the transform itself, and we can then use the noconds=True keyword argument to suppress the conditions in the return result.

```
In [112]: F = sympy.laplace_transform(f, t, s, noconds=True)
In [113]: F
```

Out[113]: $\dfrac{a}{a^2 + s^2}$

The inverse transformation can be used similarly, except we need to reverse the roles of the symbols s and t. The Laplace transform is a unique one-to-one mapping, so if we compute the inverse Laplace transform of the previously computed Laplace transform, we expect to recover the original function.

```
In [114]: sympy.inverse_laplace_transform(F, s, t, noconds=True)
Out[114]: sin(at)
```

SymPy can compute the transforms for many elementary mathematical functions and combinations of such functions. When solving problems using Laplace transformations by hand, one typically searches for matching functions in reference tables with known Laplace transformations. Using SymPy, this process can conveniently be automated in many, but not all, cases. The following examples show a few additional examples of well-known functions that one finds in Laplace transformation tables. Polynomials have simple Laplace transformation.

```
In [115]: [sympy.laplace_transform(f, t, s, noconds=True) for f in [t, t**2, t**3, t**4]]
```

Out[115]: $[\dfrac{1}{s^2}, \dfrac{2}{s^3}, \dfrac{6}{s^4}, \dfrac{24}{s^5}]$

We can also compute the general result with an arbitrary integer exponent.

```
In [116]: n = sympy.symbols("n", integer=True, positive=True)
In [117]: sympy.laplace_transform(t**n, t, s, noconds=True)
```

Out[117]: $\dfrac{\Gamma(n+1)}{s^{n+1}}$

The Laplace transform of composite expressions can also be computed, as in the following example, which computes the transform of the $f(t) = (1 - at)e^{-at}$ function.

```
In [118]: sympy.laplace_transform((1 - a*t) * sympy.exp(-a*t), t, s, noconds=True)
```

Out[118]: $\dfrac{s}{(a+s)^2}$

The main application of Laplace transforms is to solve differential equations, where the transformation can be used to bring the differential equation into a purely algebraic form, which can then be solved and transformed back to the original domain by applying the inverse Laplace transform. Chapter 9 provides concrete examples of this method. Fourier transforms can also be used for the same purpose.

The Fourier transform function, fourier_tranform, and its inverse, inverse_fourier_transform, are used in much the same way as the Laplace transformation functions. For example, to compute the Fourier transform $f(t) = e^{-at^2}$, we would first define SymPy symbols for the variables *a*, *t*, and *ω*, and the $f(t)$ function, then compute the Fourier transform by calling the sympy.fourier_tranform function.

```
In [119]: a, t, w = sympy.symbols("a, t, omega")
In [120]: f = sympy.exp(-a*t**2)
In [121]: F = sympy.fourier_transform(f, t, w)
In [122]: F
```
Out[122]: $\sqrt{\pi / a}\, e^{-\pi^2 \omega^2 / a}$

As expected, computing the inverse transformation for F recovers the original function.

```
In [123]: sympy.inverse_fourier_transform(F, w, t)
```

Out[123]: e^{-at^2}

SymPy can be used to compute a wide range of Fourier transforms symbolically, and it can be a valuable tool for solving time-dependent equations in the frequency domain. Unfortunately, SymPy does not yet handle transformations involving Dirac delta functions well, either in the original or the resulting transformation. This currently limits its usability in many applications where the resulting equations contain single-frequency components.

Summary

Integration is one of the fundamental tools in mathematical analysis. Numerical quadrature, or numerical evaluation of integrals, has important applications in many fields of science because integrals that occur in practice often cannot be computed analytically and expressed as a closed-form expression. Their computation then requires numerical techniques. This chapter reviewed basic techniques and methods for numerical quadrature and introduced the corresponding functions in the SciPy `integrate` module that can be used to evaluate integrals in practice. When the integrand is given as a function that can be evaluated at arbitrary points, we typically prefer Gaussian quadrature rules.

On the other hand, when the integrand is defined as tabulated data, the more straightforward Newton-Cotes quadrature rules can be used. We also studied symbolic integration and arbitrary-precision quadrature, which can complement floating-point quadrature for specific integrals that can be computed symbolically or when additional precision is required. As usual, a good starting point is to analyze a problem symbolically, and if a particular integral can be solved symbolically by finding its antiderivative, that is generally the most desirable outcome. When symbolic integration fails, we must resort to numerical quadrature, which should first be explored with floating-point-based implementations, like the ones provided by the SciPy `integrate` module. If additional accuracy is required, we can fall back on arbitrary-precision quadrature. Another application of symbolic integration is integral transformation, which can be used to transform problems, such as differential equations, between different domains. This chapter briefly examined how to perform Laplace and Fourier transforms symbolically using SymPy. The next chapter continues to explore this for solving certain types of differential equations.

Further Reading

Numerical quadrature is discussed in many introductory textbooks on numerical computing, such as *Scientific Computing: An Introductory Survey* by M. T. Heath (McGraw-Hill, 2002) and *Introduction to Numerical Analysis* by J. Stoer et al. (Springer, 1992). Quadrature methods and implementations are discussed in *Numerical Recipes in C* by W. H. Press (Cambridge University Press, 2002). For the theory of integral transforms, such as the Fourier and the Laplace transforms, see *Fourier Analysis and Its Applications* by G. B. Folland (American Mathematical Society, 1992).

CHAPTER 9

■ ■ ■

Ordinary Differential Equations

Equations, wherein the unknown quantity is a function rather than a variable and involves derivatives of the unknown function, are known as differential equations. An *ordinary* differential equation is a special case where the unknown function has only one independent variable with respect to which derivatives occur in the equation. If, on the other hand, derivatives of more than one variable occur in the equation, then it is known as a *partial* differential equation, and that is the topic of Chapter 11. This chapter focuses on ordinary differential equations (ODEs) and explores symbolic and numerical methods for solving this type of equation. Analytical closed-form solutions to ODEs often do not exist. But for certain special kinds of ODEs, there are analytical solutions, and in those cases, there is a chance that we can find solutions using symbolic methods. If that fails, we must, as usual, resort to numerical techniques.

Ordinary differential equations are ubiquitous in science and engineering, as well as in many other fields, and they arise, for example, in studies of dynamical systems. A typical example of an ODE is an equation that describes the time evolution of a process where the rate of change (the derivative) can be related to other properties of the process. To learn how the process evolves in time, given some initial state, we must solve or integrate the ODE that describes the process. Examples of applications of ODEs are the laws of mechanical motion in physics, molecular reactions in chemistry and biology, and population modeling in ecology, to mention a few.

This chapter explores both symbolic and numerical approaches to solving ODE problems. The SymPy module is used for symbolic methods; the numerical integration of ODEs uses functions from the `integrate` module in SciPy.

Importing Modules

Here, the NumPy and Matplotlib libraries are required for basic numerical and plotting purposes, and for solving ODEs, we need the SymPy library and SciPy's `integrate` module. As usual, let's assume that these modules are imported in the following manner.

```
In [1]: import numpy as np
In [2]: import matplotlib.pyplot as plt
In [3]: from scipy import integrate
In [4]: import sympy
```

For nicely displayed output from SymPy, we need to initialize its printing system.

```
In [5]: sympy.init_printing()
```

© Robert Johansson 2024
R. Johansson, *Numerical Python*, https://doi.org/10.1007/979-8-8688-0413-7_9

Ordinary Differential Equations

The simplest form of an ordinary differential equation is $\dfrac{dy(x)}{dx} = f(x,y(x))$, where $y(x)$ is the unknown

function and $f(x,y(x))$ is known. It is a differential equation because the derivative of the unknown $y(x)$ function occurs. Only the first derivative occurs in the equation, which is an example of a first-order

ODE. More generally, we can write an ODE of nth order in *explicit* form as $\dfrac{d^n y}{dx^n} = f\left(x,y,\dfrac{dy}{dx},...,\dfrac{d^{n-1}y}{dx^{n-1}}\right)$ or in

implicit form as $F\left(x,y,\dfrac{dy}{dx},...,\dfrac{d^n y}{dx^n}\right) = 0$, where f and F are known functions.

An example of a first-order ODE is Newton's law of cooling $\dfrac{dT(t)}{dt} = -k\left(T(t)-T_a\right)$, which describes

the temperature $T(t)$ of a body in a surrounding with temperature Ta. The solution to this ODE is
$T(t) = T_0 + (T_0 - Ta)e^{-kt}$, where T_0 is the initial temperature of the body. An example of a second-order ODE is

Newton's second law of motion $F = ma$, or more explicitly $F(x(t)) = m\dfrac{d^2 x(t)}{dt^2}$. This equation describes the

position $x(t)$ of an object with mass m when subjected to a position-dependent force $F(x(t))$. To completely specify a solution to this ODE, we would, in addition to finding its general solution, also have to give the initial position and velocity of the object. Similarly, the general solution of an nth order ODE has n free parameters that we need to specify, for example, as initial conditions for the unknown function and $n-1$ of its derivatives.

An ODE can always be rewritten as a system of first-order ODEs. Specifically, the nth order ODE on the

explicit form $\dfrac{d^n y}{dx^n} = g\left(x,y,\dfrac{dy}{dx},...,\dfrac{d^{n-1}y}{dx^{n-1}}\right)$ can be written in the *standard form* by introducing n new functions

$y_1 = y,\ y_2 = \dfrac{dy}{dx},\ ...,\ y_n = \dfrac{d^{n-1}y}{dx^{n-1}}$. This gives the following system of first-order ODEs:

$$\frac{d}{dx}\begin{bmatrix} y_1 \\ y_2 \\ \vdots \\ y_{n-1} \\ y_n \end{bmatrix} = \begin{bmatrix} y_2 \\ y_3 \\ \vdots \\ y_n \\ g(x,y_1,...,y_n) \end{bmatrix},$$

which also can be written in a more compact vector form: $\dfrac{d}{dx}\mathbf{y}(x) = f(x,\mathbf{y}(x))$. This canonical form is

particularly useful for numerical solutions of ODEs, and it is common that numerical methods for solving ODEs take the $f = (f_1, f_2, ..., f_n)$ function, which in the current case is $f = (y_2, y_3, ..., g)$, as the input that specifies

the ODE. For example, the second-order ODE for Newton's second law of motion, $F(x) = m\dfrac{d^2 x}{dt^2}$, can be

written in the standard form using $\mathbf{y} = \left[y_1 = x, y_2 = \dfrac{dx}{dt}\right]^T$ giving $\dfrac{d}{dt}\begin{bmatrix} y_1 \\ y_2 \end{bmatrix} = \begin{bmatrix} y_2 \\ F(y_1)/m \end{bmatrix}$.

If the $f_1, f_2, ..., f_n$ functions are all linear, then the corresponding system of ODEs can be written on the

simple form $\dfrac{d\mathbf{y}(\mathbf{x})}{dx} = A(x)\mathbf{y}(x) + \mathbf{r}(x)$, where $A(x)$ is an $n \times n$ matrix and $\mathbf{r}(x)$ is an n-vector that only

depends on x. In this form $\dfrac{dT(t)}{dt} = -k\left(T(t)-T_a\right)$, the $\mathbf{r}(x)$ is known as the *source term*, and the linear system is known as *homogeneous* if $r(x) = 0$ and *nonhomogeneous* otherwise. Linear ODEs are an important special case that can be solved, for example, using eigenvalue decomposition of $A(x)$. Likewise, for certain properties and forms of the $f(x, y(x))$ function, there may be known solutions and special methods for solving the corresponding ODE problem. But there is no general method for an arbitrary $f(x, y(x))$, other than approximate numerical methods.

In addition to the properties of the $f(x, y(x))$ function, the boundary conditions for an ODE also influence the solvability of the ODE problem, as well as which numerical approaches are available. Boundary conditions are needed to determine the values of the integration constants that appear in a solution. There are two main types of boundary conditions for ODE problems: *initial value conditions* and *boundary value conditions*. For initial value problems, the value of the function and its derivatives are given at a starting point, and the problem is to evolve the function forward in the independent variable (e.g., representing time or position) from this starting point. For boundary value problems, the value of the unknown function, or its derivatives, is given at fixed points. These fixed points are frequently the endpoints of the domain of interest. This chapter mostly focuses on initial value problem, and methods that apply to boundary value problems are discussed in Chapter 10 on partial differential equations.

Symbolic Solution to ODEs

SymPy provides a generic ODE solver `sympy.dsolve`, which can find analytical solutions to many elementary ODEs. The `sympy.dsolve` function attempts to automatically classify a given ODE, and it may attempt a variety of techniques to find its solution. It is also possible to give hints to the `dsolve` function, which can guide it to the most appropriate solution method. While `dsolve` can be used to solve many simple ODEs symbolically, as presented in the following, it is worth keeping in mind that most ODEs cannot be solved analytically. Typical examples of ODEs where one can hope to find a symbolic solution are ODEs of first or second order or linear systems of first-order ODEs with only a few unknown functions. It also helps greatly if the ODE has special symmetries or properties, such as being separable, having constant coefficients, or is in a special form for which known analytical solutions exist. While these types of ODEs are exceptions and special cases, there are many important applications of such ODEs. For these cases, SymPy's `dsolve` can be a very useful complement to traditional analytical methods. This section explores using SymPy and its `dsolve` function to solve simple but commonly occurring ODEs.

To illustrate the method for solving ODEs with SymPy, let's begin with a simple problem and gradually look at more complicated situations. The first example is the simple first-order ODE for Newton's cooling law, $\dfrac{dT(t)}{dt} = -k\left(T(t) - T_a\right)$, with the initial value $T(0) = T_0$. To approach this problem using SymPy, we first need to define symbols for the variables t, k, T_0, and T_a, and to represent the unknown function $T(t)$, we can use a `sympy.Function` object.

```
In [6]: t, k, To, Ta = sympy.symbols("t, k, T_0, T_a")
In [7]: T = sympy.Function("T")
```

Next, we can define the ODE very naturally by simply creating a SymPy expression for the left-hand side of the ODE when written on the form $\dfrac{dT(t)}{dt} + k\left(T(t) - T_a\right) = 0$. Here, to represent the $T(t)$ function, we can now use the SymPy `function` object T. Applying the symbol t to it, using the function call syntax `T(t)`, results in an applied function object that we can take derivatives of using either `sympy.diff` or the `diff` method on the `T(t)` expression.

```
In [8]: ode = T(t).diff(t) + k*(T(t) - Ta)
In [9]: sympy.Eq(ode)
```

$$\text{Out[9]:}\quad k\left(-T_a + T(t)\right) + \frac{dT(t)}{dt} = 0$$

Here sympy.Eq is used to display the equation, including the equality sign and a right-hand side that is zero. Given this representation of the ODE, we can directly pass it to sympy.dsolve, which attempts to automatically find the general solution of the ODE.

```
In [10]: ode_sol = sympy.dsolve(ode)
In [11]: ode_sol
Out[11]: T(t) = C₁e⁻ᵏᵗ + Tₐ
```

For this ODE problem, the sympy.dsolve function indeed finds the general solution, which includes an unknown integration constant C_1 that we have to determine from the initial conditions of the problem. The return value from the sympy.dsolve is an instance of sympy.Eq, which is a symbolic representation of equality. It has the lhs and rhs attributes for accessing the left-hand side and the right-hand side of the equality object.

```
In [12]: ode_sol.lhs
Out[12]: T(t)
In [13]: ode_sol.rhs
Out[13]: C₁e⁻ᵏᵗ + Ta
```

Once the general solution has been found, we need to use the initial conditions to find the values of the yet-to-be-determined integration constants. Here, the initial condition is $T(0) = T_0$. To this end, we first create a dictionary that describes the initial condition, ics = {T(0): T0}, that we can use with SymPy's subs method to apply the initial condition to the solution of the ODE. This results in an equation for the unknown integration constant C_1.

```
In [14]: ics = {T(0): T0}
In [15]: ics
Out[15]: {T(0): T₀}
In [16]: C_eq = ode_sol.subs(t, 0).subs(ics)
In [17]: C_eq
Out[17]: T₀ = C₁ + Ta
```

In the present example, the equation for C_1 is trivial, but for generality, let's solve it using sympy.solve. The result is a list of solutions (in this case a list of only one solution). We can substitute the solution for C_1 into the general solution of the ODE problem to obtain the solution that corresponds to the given initial conditions.

```
In [18]: C_sol = sympy.solve(C_eq)
In [19]: C_sol
Out[19]: [{C₁: T₀ − Tₐ}]
In [20]: ode_sol.subs(C_sol[0])
Out[20]: T(t) = Tₐ + (T₀ − Tₐ)e⁻ᵏᵗ
```

These steps solved the ODE problem symbolically and obtained the solution $T(t) = T_a + (T_0 - T_a)e^{-kt}$. The steps involved in this process are straightforward, but applying the initial conditions and solving for the undetermined integration constants can be slightly tedious, and it is worthwhile to collect these steps in a reusable function. The following function apply_ics is a basic implementation that generalizes these steps to a differential equation of arbitrary order.

```
In [21]: def apply_ics(sol, ics, x, known_params):
    ...:     """
    ...:     Apply the initial conditions (ics), given as a dictionary on
```

```
...:     the form ics = {y(0): y0, y(x).diff(x).subs(x, 0): yp0, ...},
...:     to the solution of the ODE with independent variable x.
...:     The undetermined integration constants C1, C2, ... are extracted
...:     from the free symbols of the ODE solution, excluding symbols in
...:     the known_params list.
...:     """
...:     free_params = sol.free_symbols - set(known_params)
...:     eqs = [(sol.lhs - sol.rhs).diff(x, n).subs(x, 0).subs(ics)
...:            for n in range(len(ics))]
...:     sol_params = sympy.solve(eqs, free_params)
...:     return sol.subs(sol_params)
```

With this function, we can more conveniently single out a particular solution to an ODE that satisfies a set of initial conditions, given the general solution to the same ODE. For the previous example, we get the following.

```
In [22]: ode_sol
Out[22]: T(t) = C₁e⁻ᵏᵗ + Tₐ
```
Out[22]: $T(t) = C_1 e^{-kt} + T_a$
```
In [23]: apply_ics(ode_sol, ics, t, [k, Ta])
```
Out[23]: $T(t) = T_a + (T_0 - T_a)e^{-kt}$

The example we have looked at is almost trivial, but the same method can be used to approach any ODE problem, although there is no guarantee that a solution will be found. As an example of a slightly more complicated problem, consider the ODE for a damped harmonic oscillator, which is a second-order ODE on the form $\dfrac{d^2x(t)}{dt^2} + 2\gamma\omega_0 \dfrac{dx(t)}{dt} + \omega_0^2 x(t) = 0$ where $x(t)$ is the position of the oscillator at time t, ω_0 is the frequency for the undamped case, and γ is the damping ratio. We first define the required symbols, construct the ODE, and then ask SymPy to find the general solution by calling sympy.dsolve.

```
In [24]: t, omega0, gamma= sympy.symbols("t, omega_0, gamma", positive=True)
In [25]: x = sympy.Function("x")
In [26]: ode = x(t).diff(t, 2) + 2*gamma*omega0 * x(t).diff(t) + omega0**2 * x(t)
In [27]: sympy.Eq(ode)
```

Out[27]: $\dfrac{d^2x(t)}{dt^2} + 2\gamma\omega_0 \dfrac{dx(t)}{dt} + \omega_0^2 x(t) = 0$

```
In [28]: ode_sol = sympy.dsolve(ode)
In [29]: ode_sol
```

Out[29]: $x(t) = C_1 e^{\omega_0 t\left(-\gamma - \sqrt{\gamma^2 - 1}\right)} + C_2 e^{\omega_0 t\left(-\gamma + \sqrt{\gamma^2 - 1}\right)}$

Since this is a second-order ODE, the general solution has two undetermined integration constants.

We need to specify initial conditions for the position $x(0)$ and the velocity $\left.\dfrac{dx(t)}{dt}\right|_{t=0}$ to single out a particular solution to the ODE. To do this, create a dictionary with these initial conditions and apply it to the general ODE solution using apply_ics.

```
In [30]: ics = {x(0): 1, x(t).diff(t).subs(t, 0): 0}
In [31]: ics
```

Out[31]: $\left\{x(0):1, \dfrac{dx(t)}{dt}\bigg|_{t=0} :0\right\}$

In [32]: x_t_sol = apply_ics(ode_sol, ics, t, [omega0, gamma])
In [33]: x_t_sol

Out[33]: $x(t) = \left(-\dfrac{\gamma}{2\sqrt{\gamma^2-1}} + \dfrac{1}{2}\right)e^{\omega_0 t\left(-\gamma-\sqrt{\gamma^2-1}\right)} + \left(\dfrac{\gamma}{2\sqrt{\gamma^2-1}} + \dfrac{1}{2}\right)e^{\omega_0 t\left(-\gamma+\sqrt{\gamma^2-1}\right)}$

This is the solution for the oscillator dynamics for arbitrary values of t, ω_0, and γ, where I used the initial condition $x(0) = 1$ and $\dfrac{dx(t)}{dt}\bigg|_{t=0} = 0$. However, substituting $\gamma = 1$, which corresponds to critical damping, directly into this expression results in a division by zero error. For this particular choice of γ, we must be careful and compute the limit where $\gamma \rightarrow 1$.

In [34]: x_t_critical = sympy.limit(x_t_sol.rhs, gamma, 1)
In [35]: x_t_critical

Out[35]: $\dfrac{\omega_0 t + 1}{e^{\omega_0 t}}$

Finally, plot the solutions for $\omega_0 = 2\pi$ and a sequence of different values of the damping ratio γ.

```
In [36]: fig, ax = plt.subplots(figsize=(8, 4))
    ...: tt = np.linspace(0, 3, 250)
    ...: w0 = 2 * sympy.pi
    ...: for g in [0.1, 0.5, 1, 2.0, 5.0]:
    ...:     if g == 1:
    ...:         x_t_expr = x_t_critical.subs({omega0: w0})
    ...:     else:
    ...:         x_t_expr = x_t_sol.rhs.subs({omega0: w0, gamma: g})
    ...:     x_t = sympy.lambdify(t, x_t_expr, 'numpy')
    ...:     ax.plot(tt, x_t(tt).real, label=r"$\gamma = %.1f$" % g)
    ...: ax.set_xlabel(r"$t$", fontsize=18)
    ...: ax.set_ylabel(r"$x(t)$", fontsize=18)
    ...: ax.legend()
```

The solution to the ODE for the damped harmonic oscillator is graphed in Figure 9-1. For $\gamma < 1$, the oscillator is underdamped, and we see oscillatory solutions. For $\gamma > 1$, the oscillator is overdamped and decays monotonically. The crossover between these two behaviors occurs at the critical damping ratio $\gamma = 1$.

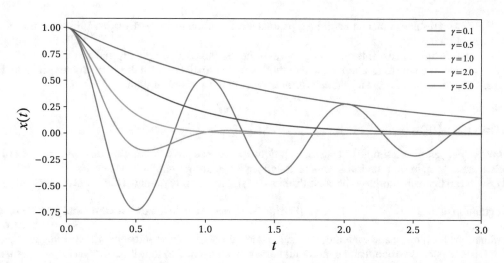

Figure 9-1. *Solutions to the ODE for a damped harmonic oscillator for a sequence of damping ratios*

The two examples of ODEs we have looked at so far could both be solved exactly by analytical means, but this is far from always the case. Even many first-order ODEs cannot be solved exactly in terms of elementary functions. For example, consider $\dfrac{dy(x)}{dx} = x + y(x)^2$, which is an example of an ODE that does not have any closed-form solution. If we try to solve this equation using sympy.dsolve, we obtain an approximate solution in the form of a power series (using the hint keyword argument with value '1st_power_series').

```
In [37]: x = sympy.symbols("x")
In [38]: y = sympy.Function("y")
In [39]: f = y(x)**2 + x
In [40]: sympy.Eq(y(x).diff(x), f)
```

Out[40]: $\dfrac{dy(x)}{dx} = x + y(x)^2$

```
In [41]: sympy.dsolve(y(x).diff(x) - f, hint='1st_power_series')
```

Out[41]: $y(x) = C_1 + C_1 x + \dfrac{1}{2}\left(2C_1 + 1\right)x^2 + \dfrac{7C_1}{6}x^3 + \dfrac{C_1}{12}\left(C_1 + 5\right)x^4$
$+ \dfrac{1}{60}\left(C_1^2\left(C_1 + 45\right) + 20C_1 + 3\right)x^5 + \mathcal{O}\left(x^6\right)$

For many other types of equations, SymPy outright fails to produce any solution at all. For example, if we attempt to solve the second-order ODE $\dfrac{d^2 y(x)}{dx^2} = x + y(x)^2$, we obtain the following error message.

```
In [42]: sympy.Eq(y(x).diff(x, x), f)
```

Out[42]: $\dfrac{d^2 y(x)}{dx^2} = x + y(x)^2$

```
In [43]: sympy.dsolve(y(x).diff(x, x) - f)
---------------------------------------------------------------------
...
NotImplementedError: solve: Cannot solve -x - y(x)**2 + Derivative(y(x), x, x)
```

215

This type of result can mean that there is no analytical solution to the ODE or, just as likely, simply that SymPy cannot handle it.

The dsolve function accepts many optional arguments, and it can frequently make a difference if the solver is guided by giving hints about which methods should be used to solve the ODE problem at hand. See the docstring for sympy.dsolve for more information about the available options.

Direction Fields

A *direction field graph* is a simple but useful technique to visualize possible solutions to arbitrary first-order ODEs. It is made up of short lines that show the slope of the unknown function on a grid in the *x–y* plane. This graph can be easily produced because the slope of $y(x)$ at arbitrary points of the *x–y* plane is given by

the definition of the ODE: $\dfrac{dy(x)}{dx} = f\big(x, y(x)\big)$. That is, we only need to iterate over the *x* and *y* values on the

coordinate grid of interest and evaluate $f(x, y(x))$ to know the slope of $y(x)$ at that point. The direction field graph is useful because smooth and continuous curves that tangent the slope lines (at every point) in the direction field graph are possible solutions to the ODE.

The following function plot_direction_field produces a direction field graph for a first-order ODE, given the independent variable *x*, the unknown function $y(x)$, and the right-hand side $f(x, y(x))$ function. It also takes optional ranges for the *x* and *y* axes (x_lim and y_lim, respectively) and an optional Matplotlib axis instance to draw the graph on.

```
In [44]: def plot_direction_field(
    ...:         x, y_x, f_xy, x_lim=(-5, 5), y_lim=(-5, 5), ax=None):
    ...:     f_np = sympy.lambdify((x, y_x), f_xy, 'numpy')
    ...:     x_vec = np.linspace(x_lim[0], x_lim[1], 20)
    ...:     y_vec = np.linspace(y_lim[0], y_lim[1], 20)
    ...:
    ...:     if ax is None:
    ...:         _, ax = plt.subplots(figsize=(4, 4))
    ...:
    ...:     dx = x_vec[1] - x_vec[0]
    ...:     dy = y_vec[1] - y_vec[0]
    ...:
    ...:     for m, xx in enumerate(x_vec):
    ...:         for n, yy in enumerate(y_vec):
    ...:             Dy = f_np(xx, yy) * dx
    ...:             Dx = 0.8 * dx**2 / np.sqrt(dx**2 + Dy**2)
    ...:             Dy = 0.8 * Dy*dy / np.sqrt(dx**2 + Dy**2)
    ...:             ax.plot([xx - Dx/2, xx + Dx/2],
    ...:                     [yy - Dy/2, yy + Dy/2], 'b', lw=0.5)
    ...:     ax.axis('tight')
    ...:     ax.set_title(r"$%s$" %
    ...:                  (sympy.latex(sympy.Eq(y(x).diff(x), f_xy))),
    ...:                  fontsize=18)
    ...:     return ax
```

With this function, we can produce the direction field graphs for the ODEs on the form $\frac{dy(x)}{dx} = f\big(x, y(x)\big)$. For example, the following code generates the direction field graphs for $f(x, y(x)) = y(x)^2 + x$, $f(x, y(x)) = -x/y(x)$, and $f(x, y(x)) = y(x)^2/x$. The result is shown in Figure 9-2.

```
In [45]: x = sympy.symbols("x")
In [46]: y = sympy.Function("y")
In [47]: fig, axes = plt.subplots(1, 3, figsize=(12, 4))
    ...: plot_direction_field(x, y(x), y(x)**2 + x, ax=axes[0])
    ...: plot_direction_field(x, y(x), -x / y(x), ax=axes[1])
    ...: plot_direction_field(x, y(x), y(x)**2 / x, ax=axes[2])
```

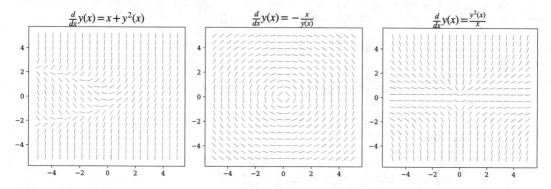

Figure 9-2. *Direction fields for three first-order differential equations*

The direction lines in the graphs in Figure 9-2 suggest how the curves that are solutions to the corresponding ODE behave, and direction field graphs are a valuable tool for visualizing solutions to ODEs that cannot be solved analytically. To illustrate this point, consider again the ODE $\frac{dy(x)}{dx} = x + y(x)^2$ with the initial condition $y(0) = 0$, which we previously saw can be solved inexactly as an approximate power series. Like before, we solve this problem again by defining the symbol x and the $y(x)$ function, which we, in turn, use to construct and display the ODE.

```
In [48]: x = sympy.symbols("x")
In [49]: y = sympy.Function("y")
In [50]: f = y(x)**2 + x
In [51]: sympy.Eq(y(x).diff(x), f)
```

Out[51]: $\dfrac{dy(x)}{dx} = x + y(x)^2$

We want to find the specific power-series solution that satisfies the initial condition. For this problem, we can specify the initial condition directly using the ics keyword argument to the dsolve function[1].

```
In [52]: ics = {y(0): 0}
In [53]: ode_sol = sympy.dsolve(y(x).diff(x)-f, ics=ics, hint='1st_power_series')
In [54]: ode_sol
```

Out[54]: $y(x) = \dfrac{x^2}{2} + \dfrac{x^5}{20} + \mathcal{O}(x^6)$

Plotting the solution together with the direction field for the ODE is a quick and straightforward way to get an idea of the validity range of the power-series approximation. The following code plots the approximate solution and the direction field (see Figure 9-3, left panel). A solution with an extended validity range is also obtained by repeatedly solving the ODE with initial conditions at increasing values of x, taken from a previous power-series solution (see Figure 9-3, right panel).

```
In [55]: fig, axes = plt.subplots(1, 2, figsize=(8, 4))
    ...: # left panel
    ...: plot_direction_field(x, y(x), f, ax=axes[0])
    ...: x_vec = np.linspace(-3, 3, 100)
    ...: axes[0].plot(x_vec,
    ...:                 sympy.lambdify(x, ode_sol.rhs.removeO()) (x_vec), 'b', lw=2)
    ...: axes[0].set_ylim(-5, 5)
    ...:
    ...: # right panel
    ...: plot_direction_field(x, y(x), f, ax=axes[1])
    ...: x_vec = np.linspace(-1, 1, 100)
    ...: axes[1].plot(x_vec,
    ...:                 sympy.lambdify(x, ode_sol.rhs.removeO()) (x_vec), 'b', lw=2)
    ...: # iteratively resolve the ODE with updated initial conditions
    ...: ode_sol_m = ode_sol_p = ode_sol
    ...: dx = 0.125
    ...: # positive x
    ...: for x0 in np.arange(1, 2., dx):
    ...:     x_vec = np.linspace(x0, x0 + dx, 100)
    ...:     ics = {y(x0): ode_sol_p.rhs.removeO().subs(x, x0)}
    ...:     ode_sol_p = sympy.dsolve(
    ...:         y(x).diff(x) - f, ics=ics, n=6, hint='1st_power_series')
    ...:     axes[1].plot(
    ...:         x_vec, sympy.lambdify(x, ode_sol_p.rhs.removeO())(x_vec),
    ...:         'r', lw=2)
    ...: # negative x
    ...: for x0 in np.arange(-1, -5, -dx):
    ...:     x_vec = np.linspace(x0, x0 - dx, 100)
```

---

[1] In the current version of SymPy, the ics keyword argument is only recognized by the power-series solver in dsolve. Solvers for other types of ODEs ignore the ics argument; hence the need for the apply_ics function defined and used earlier in this chapter. To guide dsolve to use a power series method, we can use the hint='1st_power_series' argument.

```
...:    ics = {y(x0): ode_sol_m.rhs.removeO().subs(x, x0)}
...:    ode_sol_m = sympy.dsolve(
...:        y(x).diff(x) - f, ics=ics, n=6, hint='1st_power_series')
...:    axes[1].plot(
...:        x_vec, sympy.lambdify(x, ode_sol_m.rhs.removeO())(x_vec),
...:        'r', lw=2)
```

In the left panel of Figure 9-3, the approximate solution curve aligns well with the direction field lines near $x = 0$ but starts to deviate for $|x| \gtrsim 1$, suggesting that the approximate solution is no longer valid. The solution curve in the right panel aligns better with the direction field throughout the plotted range. The blue (dark gray) curve segment is the original approximate solution, and the red (light gray) curves are continuations obtained from resolving the ODE with an initial condition sequence that starts where the blue (dark gray) curves end.

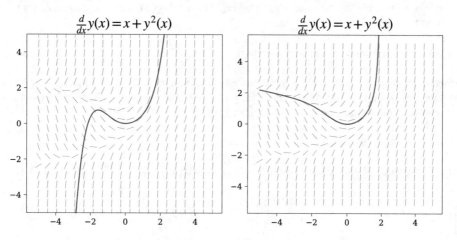

Figure 9-3. *Direction field graph of the ODE with the $\dfrac{dy(x)}{dx} = y(x)^2 + x$ fifth-order power-series solutions around $x = 0$ (left), and consecutive power-series expansions around x between –5 and 2, with a 0.125 spacing (right)*

Solving ODEs Using Laplace Transformations

An alternative to solving ODEs symbolically with SymPy's "black-box" solver[2] dsolve is to use the symbolic capabilities of SymPy to assist in a more manual approach to solving ODEs. A technique that can be used to solve certain ODE problems is to Laplace transform the ODE, which for many problems results in an algebraic equation that is easier to solve. The solution to the algebraic equation can then be transformed back to the original domain with an inverse Laplace transform to obtain the solution to the original problem. The key to this method is that the Laplace transform of the derivative of a function is an algebraic expression in the Laplace transform of the function itself: $L[y'(t)] = sL[y(t)] – y(0)$. However, while SymPy is

---

[2] Or "white-box" solver, since SymPy is open source and the inner workings of dsolve are readily available for inspection.

good at Laplace transforming many types of elementary functions, it does not recognize how to transform derivatives of an unknown function. But defining a function that performs this task easily amends this shortcoming.

For example, consider the following differential equation for a driven harmonic oscillator.

$$\frac{d^2}{dt^2}y(t)+2\frac{d}{dt}y(t)+10y(t)=2\sin 3t$$

To work with this ODE, first create SymPy symbols for the independent variable t and the $y(t)$ function and then use them to construct the symbolic expression for the ODE.

```
In [56]: t = sympy.symbols("t", positive=True)
In [57]: y = sympy.Function("y")
In [58]: ode = y(t).diff(t, 2) + 2 * y(t).diff(t) + 10 * y(t) - 2 * sympy.sin(3*t)
In [59]: sympy.Eq(ode)
```

Out[59]: $10y(t)-2\sin(3t)+2\dfrac{d}{dt}y(t)+\dfrac{d^2}{dt^2}y(t)=0$

Laplace transforming this ODE should yield an algebraic equation. To pursue this approach using SymPy and its function sympy.laplace_transform, we first need to create a symbol s, to be used in the Laplace transformation. At this point, we also create a symbol Y for later use.

```
In [60]: s, Y = sympy.symbols("s, Y", real=True)
```

Next, let's proceed to Laplace transform the unknown function $y(t)$, as well as the entire ODE equation.

```
In [61]: L_y = sympy.laplace_transform(y(t), t, s)
In [62]: L_y
Out[62]: ℒₜ[y(t)](s)
In [63]: L_ode = sympy.laplace_transform(ode, t, s, noconds=True)
In [64]: sympy.Eq(L_ode)
```

Out[64]: $10\mathcal{L}_t\big[y(t)\big](s)+2\mathcal{L}_t\Big[\dfrac{d}{dt}y(t)\Big](s)+\mathcal{L}_t\Big[\dfrac{d^2}{dt^2}y(t)\Big](s)-\dfrac{6}{s^2+9}=0$

When Laplace-transforming the unknown function $y(t)$, we get the undetermined result L$_t$[y(t)](s), which is to be expected. However, applying sympy.laplace_transform on a derivative of $y(t)$, such as $\frac{d}{dt}y(t)$, results in the unevaluated expression, $\mathcal{L}_t\Big[\frac{d}{dt}y(t)\Big](s)$. This form is not the desired result, and we need to work around this issue to obtain the sought-after algebraic equation. The Laplace transformation of the derivative of an unknown function has a well-known form that involves the Laplace transform of the function itself rather than its derivatives. The following formula is the nth derivative of the $y(t)$ function:

$$\mathcal{L}_t\Big[\frac{d^n}{dt^n}y(t)\Big](s)=s^n\mathcal{L}_t\big[y(t)\big](s)-\sum_{m=0}^{n-1}s^{n-m-1}\frac{d^m}{dt^m}y(t)\Big|_{t=0}.$$

By iterating through the SymPy expression tree for L_ode and replacing the occurrences of $\mathcal{L}_t\Big[\frac{d^n}{dt^n}y(t)\Big](s)$ with expressions of the form given by this formula, we can obtain the algebraic form of the ODE that we seek. The following function takes a Laplace-transformed ODE and performs the substitution of the unevaluated Laplace transforms of the derivatives of $y(t)$.

```
In [65]: def laplace_transform_derivatives(e):
    ...:     """
    ...:     Evaluate laplace transforms of derivatives of functions
    ...:     """
    ...:     if isinstance(e, sympy.LaplaceTransform):
    ...:         if isinstance(e.args[0], sympy.Derivative):
    ...:             d, t, s = e.args
    ...:             n = d.args[1][1]
    ...:             return (
    ...:                 (s**n) * sympy.LaplaceTransform(d.args[0], t, s) -
    ...:                 sum([s**(n-i) * sympy.diff(d.args[0], t, i-1).subs(t, 0)
    ...:                     for i in range(1, n+1)]))
    ...:
    ...:     if isinstance(e, (sympy.Add, sympy.Mul)):
    ...:         t = type(e)
    ...:         return t(*[laplace_transform_derivatives(arg)
    ...:                     for arg in e.args])
    ...:
    ...:     return e
```

Applying this function on the Laplace-transformed ODE equation, L_ode, yields the following.

```
In [66]: L_ode_2 = laplace_transform_derivatives(L_ode)
In [67]: sympy.Eq(L_ode_2)
```

Out[67]:
$$s^2 \mathcal{L}_t\big[y(t)\big](s) + 2s\mathcal{L}_t\big[y(t)\big](s) - sy(0)$$
$$+10\mathcal{L}_t\big[y(t)\big](s) - 2y(0) - \frac{d}{dt}y(t)\bigg|_{t=0} - \frac{6}{s^2+9} = 0$$

To simplify the notation, now substitute the expression $L_t[y(t)](s)$ for the symbol Y.

```
In [68]: L_ode_3 = L_ode_2.subs(L_y, Y)
In [69]: sympy.Eq(L_ode_3)
```

Out[69]: $s^2 Y + 2sY - sy(0) + 10Y - 2y(0) - \dfrac{d}{dt}y(t)\bigg|_{t=0} - \dfrac{6}{s^2+9} = 0$

At this point, we need to specify the boundary conditions for the ODE problem. Here, use $y(0) = 1$ and $y'(t) = 0$, and after creating a dictionary that contains these boundary conditions, use it to substitute the values into the Laplace-transformed ODE equation.

```
In [70]: ics = {y(0): 1, y(t).diff(t).subs(t, 0): 0}
In [71]: ics
```

Out[71]: $\left\{ y(0):1, \dfrac{d}{dt}y(t)\bigg|_{t=0} :0 \right\}$

```
In [72]: L_ode_4 = L_ode_3.subs(ics)
In [73]: sympy.Eq(L_ode_4)
```

Out[74]: $Ys^2 + 2Ys + 10Y - s - 2 - \dfrac{6}{s^2+9} = 0$

This is an algebraic equation that can be solved for Y.

```
In [75]: Y_sol = sympy.solve(L_ode_4, Y)
In [76]: Y_sol
```

Out[76]: $\left[\dfrac{s^3 + 2s^2 + 9s + 24}{s^4 + 2s^3 + 19s^2 + 18s + 90} \right]$

The result is a list of solutions, which, in this case, contains only one element. Performing the inverse Laplace transformation on this expression gives the solution to the original problem in the time domain.

```
In [77]: y_sol = sympy.inverse_laplace_transform(Y_sol[0], s, t)
In [78]: sympy.simplify(y_sol)
```

Out[78]: $\dfrac{1}{111e^t}\left(6\left(\sin 3t - 6\cos 3t\right)e^t + 43\sin 3t + 147\cos 3t\right)$

This technique of Laplace transforming an ODE, solving the corresponding algebraic equation, and inverse Laplace transforming the result to obtain the solution to the original problem can be applied to solve many significant ODE problems that arise in, for example, electrical engineering and process control applications. Although these problems can be solved by hand with the help of Laplace transformation tables, using SymPy has the potential of significantly simplifying the process.

Numerical Methods for Solving ODEs

While some ODE problems can be solved with analytical methods, as shown in previous examples, it is much more common that ODE problems cannot be solved analytically. In practice, ODE problems are mainly solved with numerical methods. There are many approaches to solving ODEs numerically, and most of them are designed for problems formulated as a system of first-order ODEs on the standard form[3]

$\dfrac{d\mathbf{y}(x)}{dx} = f\left(x,\mathbf{y}(x)\right)$, where $\mathbf{y}(x)$ is a vector of unknown functions of x. SciPy provides functions for solving

this kind of problem, but before exploring how to use those functions, let's briefly review the fundamental concepts and introduce the terminology used for the numerical integration of ODE problems.

The basic idea of many numerical methods for ODEs is captured in the Euler method. This method can, for example, be derived from a Taylor series expansion of $y(x)$ around the point x,

$$y(x+h) = y(x) + \frac{dy(x)}{dx}h + \frac{1}{2}\frac{d^2y(x)}{dx^2}h^2 + \ldots,$$

where we consider the case when $y(x)$ is a scalar function for notational simplicity. By dropping terms of second order or higher, we get the approximate equation $y(x+h) \approx y(x) + f(x, y(x))h$, which is accurate to first order in the stepsize h. This equation can be turned into an iteration formula by discretizing the x variables, $x_0, x_1, \ldots, x_k$, choosing the stepsize $h_k = x_{k+1} - x_k$, and denoting $y_k = y(x_k)$. The resulting iteration formula $y_{k+1} \approx y_k + f(x_k, y_k)h_k$ is known as the *forward Euler method*, and it is said to be an *explicit* form because, given the value of the y_k, we can directly compute y_{k+1} using the formula. The goal of the numerical solution of an initial value problem is to compute $y(x)$ at some points x_n, given the initial condition $y(x_0) = y_0$. Therefore,

---

[3] Recall that any ODE problem can be written as a system of first-order ODEs on this standard form.

an iteration formula like the forward Euler method can be used to compute successive values of y_k, starting from y_0. There are two types of errors involved in this approach. First, the truncation of the Taylor series gives an error that limits the method's *accuracy*. Second, the approximation of y_k provided by the previous iteration when computing y_{k+1} gives an additional error that may accumulate over successive iterations and affect the method's *stability*.

An alternative form, which can be derived in a similar manner, is the *backward Euler method*, given by the iteration formula $y_{k+1} \approx y_k + f(x_{k+1}, y_{k+1})h_k$. This is an example of a *backward differentiation formula* *(BDF)*, which is *implicit* because y_{k+1} occurs on both sides of the equation. To compute y_{k+1}, we must solve an algebraic equation (e.g., using Newton's method, see Chapter 5). Implicit methods are more complicated to implement than explicit methods, and each iteration requires more computational work. However, the advantage is that implicit methods generally have larger stability regions and better accuracy, which means larger stepsize h_k can be used while still obtaining an accurate and stable solution. Whether explicit or implicit methods are more efficient depends on the problem being solved. Implicit methods are often particularly useful for *stiff* problems, which, loosely speaking, are ODE problems that describe dynamics with multiple disparate timescales (e.g., dynamics that include both fast and slow oscillations).

There are several methods to improve upon the first-order Euler forward and backward methods. One strategy is to keep higher-order terms in the Taylor series expansion of $y(x+h)$, which gives higher-order iteration formulas that can have better accuracy, such as the second-order method

$y_{k+1} \approx y(x_k) + f(x_{k+1}, y_{k+1})h_k + \frac{1}{2}y_k''(x)h_k^2$. However, such methods require evaluating higher-order

derivatives of $y(x)$, which may be a problem if $f(x, y(x))$ is not known in advance (and not given in symbolic form). Ways around this problem include approximating the higher-order derivatives using finite-difference approximations of the derivatives or sampling the $f(x, y(x))$ function at intermediary points in the interval $[x_k, x_{k+1}]$. An example of this type of method is the well-known Runge-Kutta method, a single-step method that uses additional evaluations of $f(x, y(x))$. The most well-known Runge-Kutta method is the fourth-order scheme

$$y_{k+1} = y_k + \frac{1}{6}\left(k_1 + 2k_2 + 2k_3 + k_4\right),$$

where

$$k_1 = f\left(t_k, y_k\right)h_k,$$

$$k_2 = f\left(t_k + \frac{h_k}{2}, y_k + \frac{k_1}{2}\right)h_k,$$

$$k_3 = f\left(t_k + \frac{h_k}{2}, y_k + \frac{k_2}{2}\right)h_k,$$

$$k_4 = f\left(t_k + h_k, y_k + k_3\right)h_k.$$

Here, k_1 to k_4 are four different evaluations of the ODE $f(x, y(x))$ function used in the explicit formula for y_{k+1}. The resulting estimate of y_{k+1} is accurate to the fourth order, with an error of the fifth order. Higher-order schemes that use more function evaluations can also be constructed. By combining two methods of different orders, it can be possible also to estimate the error in the approximation. A popular combination is the Runge-Kutta fourth- and fifth-order schemes, which results in a fourth-order accurate method with error estimates. It is known as RK45 or the Runge-Kutta-Fehlberg method. The Dormand-Prince method is another example of a higher-order method that uses adaptive step-size control. For example, the 8-5-3

223

method combines third- and fifth-order schemes to produce an eighth-order method. An implementation of this method is available in SciPy, which we will see in the next section.

An alternative method is to use more than one previous value of y_k to compute y_{k+1}. Such methods are known as multistep methods and can, in general, be written in the following form.

$$y_{k+s} = \sum_{n=0}^{s-1} a_n y_{k+n} + h \sum_{n=0}^{s} b_n f\left(x_{k+n}, y_{k+n}\right)$$

With this formula, to compute y_{k+s}, the previous s values of y_k and $f(x_k, y_k)$ are used (known as an s-step method). The choices of the coefficients a_n and b_n give rise to different multistep methods. Note that if $b_s = 0$, the method is explicit; if $b_s \neq 0$, it is implicit.

For example, $b_0 = b_1 = \dots = b_{s-1} = 0$ gives the general formula for an s-step BDF formula, where a_n and b_n are chosen to maximize the order of the accuracy by requiring that the method is exact for polynomials up to as high order as possible. This gives an equation system that can be solved for the unknown coefficients a_n and b_n. For example, the one-step BDF method with $b_1 = a_0 = 1$ reduces to the backward Euler method, $y_{k+1} = y_k + hf(x_{k+1}, y_{k+1})$, and the two-step BDF method, $y_{k+2} = a_0 y_k + a_1 y_{k+1} + hb_2 f(x_{k+2}, y_{k+2})$, when solved for the coefficients (a_0, a_1, and b_2), becomes $y_{k+2} = -\dfrac{1}{3} y_k + \dfrac{4}{3} y_{k+1} + \dfrac{2}{3} hf\left(x_{k+2}, y_{k+2}\right)$. Higher-order BDF methods can also be constructed. SciPy provides a BDF solver recommended for stiff problems because of its good stability properties.

Another family of multistep methods is the Adams methods, which result from the choice $a_0 = a_1 = \dots = a_{s-2} = 0$ and $a_{s-1} = 1$, where again the remaining unknown coefficients are chosen to maximize the order of the method. Specifically, the explicit methods with $b_s = 0$ are known as Adams-Bashforth methods, and the implicit methods with $b_s \neq 0$ are known as Adams-Moulton methods. For example, the one-step Adams-Bashforth and Adams-Moulton methods reduce to the forward and backward Euler methods, respectively, and the two-step methods are $y_{k+2} = y_{k+1} + h\left(-\dfrac{1}{2} f\left(x_k, y_k\right) + \dfrac{3}{2} f\left(x_{k+1}, y_{k+1}\right)\right)$ and $y_{k+1} = y_k + \dfrac{1}{2} h\left(f\left(x_k, y_k\right) + f\left(x_{k+1}, y_{k+1}\right)\right)$, respectively. Higher-order explicit and implicit methods can also be constructed in this way. Solvers using these Adams methods are also available in SciPy.

In general, explicit methods are more convenient to implement and less computationally demanding to iterate than implicit methods, which in principle requires solving (a potentially nonlinear) equation in each iteration with an initial guess for the unknown y_{k+1}. However, as mentioned, implicit methods often are more accurate and have superior stability properties. A compromise that retains some of the advantages of both methods is to combine explicit and implicit methods in the following way: first compute y_{k+1} using an explicit method, and then use this y_{k+1} as an initial guess for solving the equation for y_{k+1} given by an implicit method. This equation does not need to be solved exactly, and since the initial guess from the explicit method should be quite good, it could be sufficient with a small number of iterations, using, for example, Newton's method. Methods like these, where the result from an explicit method is used to predict y_{k+1} and an implicit method is used to *correct* the prediction, are called *predictorcorrector* methods.

Finally, an important technique that many advanced ODE solvers employ is *adaptive stepsize* or *stepsize control*: the accuracy and stability of an ODE are strongly related to the stepsize h_k used in the iteration formula for an ODE method, and so is the computational cost of the solution. If the error in y_{k+1} can be estimated together with the computation of y_{k+1} itself, then it is possible to automatically adjust the stepsize h_k so that the solver uses large economical stepsizes when possible and smaller stepsizes when required. A related technique, which is possible with some methods, automatically adjusts the order of the method so that a lower-order method is used when possible and a higher-order method is used when necessary. The Adams methods are examples of methods where the order can be changed easily.

A vast variety of high-quality implementations of ODE solvers exist, and rarely should it be necessary to reimplement any of the methods discussed here. Doing so would probably be a mistake unless it is for educational purposes or if one's primary interest is researching methods for numerical ODE solving.

For practical purposes, it is advisable to use one of the many highly tuned and thoroughly tested ODE suites that already exist, most of which are free and open source and packaged into libraries such as SciPy. However, there are a large number of solvers to choose from, and to be able to make an informed decision on which one to use for a particular problem and to understand many of their options, it is important to be familiar with the basic ideas and methods and the terminology that is used to discuss them.

Numerical Integration of ODEs Using SciPy

After reviewing numerical methods for solving ODEs in the previous section, we are ready to explore the ODE solvers available in SciPy and how to use them. The integrate module of SciPy provides two ODE solver interfaces: integrate.odeint and integrate.ode. The odeint function is an interface to the LSODA solver from ODEPACK,[4] which automatically switches between an Adams predictor-corrector method for nonstiff problems and a BDF method for stiff problems. In contrast, the integrate.ode class provides an object-oriented interface to several different solvers: the VODE and ZVODE solvers[5] (ZVODE is a variant of VODE for complex-valued functions), the LSODA solver, and dopri5 and dop853, which are fourth- and eighth-order Dormand-Prince methods (i.e., types of Runge-Kutta methods) with adaptive stepsize. While the object-oriented interface provided by integrate.ode is more flexible, the odeint function is, in many cases, simpler and more convenient. Let's look at both interfaces, starting with the odeint function.

The odeint function takes three mandatory arguments: a function for evaluating the right-hand side of the ODE on the standard form, an array (or scalar) that specifies the initial condition for the unknown functions, and an array with values of the independent variable where an unknown function is to be computed. The function for the right-hand side of the ODE takes two mandatory arguments and an arbitrary number of optional arguments. The required arguments are the array (or scalar) for the vector $y(x)$ as the first argument and the value of x as the second argument. For example, consider again the scalar ODE $y'(x) = f(x, y(x)) = x + y(x)^2$. To be able to plot the direction field for this ODE again, this time together with a specific solution obtained by numerical integration using odeint, we first define the SymPy symbols required to construct a symbolic expression for $f(x, y(x))$.

```
In [79]: x = sympy.symbols("x")
In [80]: y = sympy.Function("y")
In [81]: f = y(x)**2 + x
```

To be able to solve this ODE with SciPy's odeint, we first and foremost need to define a Python function for $f(x, y(x))$ that takes Python scalars or NumPy arrays as input. From the SymPy expression f, we can generate such a function using sympy.lambdify with the 'numpy' argument.[6]

```
In [82]: f_np = sympy.lambdify((y(x), x), f, 'numpy')
```

Next, we need to define the initial value y0 and a NumPy array with the discrete values of x for which to compute the $y(x)$ function. Let's solve the ODE starting at $x = 0$ in both the positive and negative directions, using the NumPy arrays xp and xm, respectively. We only need to create a NumPy array with negative increments to solve the ODE in the negative direction. Once we have set up the ODE f_np function, initial value y0, and array of x coordinates, for example, xp, we can integrate the ODE problem by calling integrate.odeint(f_np, y0, xp).

---

[4] More information about ODEPACK is available at http://computing.llnl.gov/projects/odepack.
[5] The VODE and ZVODE solvers are available at www.netlib.org/ode.
[6] In this particular case, with a scalar ODE, we could also use the 'math' argument, which produces a scalar function using functions from the standard math library, but more often, array-aware functions are needed. They are obtained by using the 'numpy' argument to sympy.lambdify.

```
In [83]: y0 = 0
In [84]: xp = np.linspace(0, 1.9, 100)
In [85]: yp = integrate.odeint(f_np, y0, xp)
In [86]: xm = np.linspace(0, -5, 100)
In [87]: ym = integrate.odeint(f_np, y0, xm)
```

The results are two one-dimensional NumPy arrays ym and yp, of the same length as the corresponding coordinate arrays xm and xp (i.e., 100), which contain the solution to the ODE problem at the specified points. To visualize the solution, plot the ym and yp arrays together with the direction field for the ODE. The result is shown in Figure 9-4. As expected, the solution aligns with (tangents) the lines in the direction field at every point in the graph.

```
In [88]: fig, ax = plt.subplots(1, 1, figsize=(4, 4))
    ...: plot_direction_field(x, y(x), f, ax=ax)
    ...: ax.plot(xm, ym, 'b', lw=2)
    ...: ax.plot(xp, yp, 'r', lw=2)
```

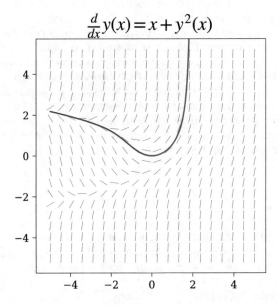

Figure 9-4. *The direction field of the ODE $y'(x) = x + y(x)^2$ and the specific solution that satisfies $y(0) = 0$*

The previous example solved a scalar ODE problem. More often, we are interested in vector-valued ODE problems (systems of ODEs). To see how we can solve that kind of problem using odeint, consider the Lotka-Volterra equations for the dynamics of a population of predator and prey animals (a classic example of coupled ODEs). The equations are $x'(t) = ax - bxy$ and $y'(t) = cxy - dy$, where $x(t)$ is the number of prey animals and $y(t)$ is the number of predator animals, and the coefficients a, b, c, and d describe the rates of the processes in the model. For example, a is the rate at which prey animals are born, and d is the rate at which predators die. The b and c coefficients are the rates at which predators consume prey and the rate at which the predator population grows at the expense of the prey population, respectively. Note that this is a nonlinear system of ODEs because of the xy terms.

To solve this problem with odeint, we first need to write a function for the right-hand side of the ODE in vector form. For this case, we have $f(t, [x, y]^T) = [ax - bxy, cxy - dy]^T$, which we can implement as a Python function in the following way.

```
In [89]: a, b, c, d = 0.4, 0.002, 0.001, 0.7
In [90]: def f(xy, t):
    ...:     x, y = xy
    ...:     return [a * x - b * x * y, c * x * y - d * y]
```

This also defined variables and values for the coefficients a, b, c, and d. Note that here, the first argument of the ODE f function is an array containing the current values of $x(t)$ and $y(t)$. For convenience, we first unpack these variables into separate variables, x and y, which makes the rest of the function easier to read. The function's return value should be an array, or list, containing the values of the derivatives of $x(t)$ and $y(t)$. The f function must also take the t argument, with the current $x(t)$ value of the independent coordinate. However, t is not used in this example. Once the f function is defined, we must define an array xy0 with the initial values $x(0)$ and $y(0)$ and an array t for the points at which we wish to compute the solution to the ODE. Here, let's use the initial conditions $x(0) = 600$ and $y(0) = 400$, corresponding to 600 prey animals and 400 predators at the beginning of the simulation.

```
In [91]: xy0 = [600, 400]
In [92]: t = np.linspace(0, 50, 250)
In [93]: xy_t = integrate.odeint(f, xy0, t)
In [94]: xy_t.shape
Out[94]: (250,2)
```

Calling integrate.odeint(f, xy0, t) integrates the ODE problem and returns an array of shape (250, 2), which contains $x(t)$ and $y(t)$ for each of the 250 values in t. The following code plots the solution as a function of time and in phase space. The result is shown in Figure 9-5.

```
In [95]: fig, axes = plt.subplots(1, 2, figsize=(8, 4))
    ...: axes[0].plot(t, xy_t[:,0], 'r', label="Prey")
    ...: axes[0].plot(t, xy_t[:,1], 'b', label="Predator")
    ...: axes[0].set_xlabel("Time")
    ...: axes[0].set_ylabel("Number of animals")
    ...: axes[0].legend()
    ...: axes[1].plot(xy_t[:,0], xy_t[:,1], 'k')
    ...: axes[1].set_xlabel("Number of prey")
    ...: axes[1].set_ylabel("Number of predators")
```

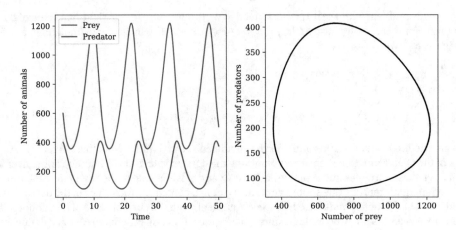

Figure 9-5. *A solution to the Lotka-Volterra ODE for predator-prey populations, as a function of time (left) and in phase space (right)*

In the previous two examples, the function for the right-hand side of the ODE was implemented without additional arguments. In the example with the Lotka-Volterra equation, the f function used globally defined coefficient variables. Rather than using global variables, it is often convenient and elegant to implement the f function so that it takes arguments for all its coefficients or parameters. To illustrate this point, let's consider another famous ODE problem: the Lorenz equations, which are the following system of three coupled nonlinear ODEs, $x'(t) = \sigma(y - x)$, $y'(t) = x(\rho - z) - y$, and $z'(t) = xy - \beta z$. These equations are known for their chaotic solutions, which sensitively depend on the values of the parameters σ, ρ, and β. Suppose we wish to solve these equations for different values of these parameters. In that case, it is useful to write the ODE function so that it additionally takes the values of these variables as arguments. In the following implementation of f, the three arguments—sigma, rho, and beta—for the corresponding parameters, have been added after the mandatory $y(t)$ and t arguments.

```
In [96]: def f(xyz, t, sigma, rho, beta):
    ...:     x, y, z = xyz
    ...:     return [sigma * (y - x),
    ...:             x * (rho - z) - y,
    ...:             x * y - beta * z]
```

Next, define variables with specific values of the parameters, the array with t values to compute the solution for, and the initial conditions for the $x(t)$, $y(t)$, and $z(t)$ functions.

```
In [97]: sigma, rho, beta = 8, 28, 8/3.0
In [98]: t = np.linspace(0, 25, 10000)
In [99]: xyz0 = [1.0, 1.0, 1.0]
```

This time when we call integrate.odeint, we also need to specify the args argument, which needs to be a list, tuple, or array with the same number of elements as the number of additional arguments in the f function defined in the preceding section. In this case, there are three parameters, and we pass a tuple with the values of these parameters via the args argument when calling integrate.odeint. The following solves the ODE for three different sets of parameters (but the same initial conditions).

```
In [100]: xyz1 = integrate.odeint(f, xyz0, t, args=(sigma, rho, beta))
In [101]: xyz2 = integrate.odeint(f, xyz0, t, args=(sigma, rho, 0.6*beta))
In [102]: xyz3 = integrate.odeint(f, xyz0, t, args=(2*sigma, rho, 0.6*beta))
```

The solutions are stored in the NumPy arrays xyz1, xyz2, and xyz3. In this case, these arrays have the shape (10000, 3) because the t array has 10,000 elements, and there are three unknown functions in the ODE problem. The three solutions are plotted in 3D graphs in the following code, and the result is shown in Figure 9-6. The resulting solutions can vary greatly with small changes in the system parameters.

```
In [103]: from mpl_toolkits.mplot3d.axes3d import Axes3D
In [104]: fig, (ax1, ax2, ax3) = plt.subplots(1, 3, figsize=(12, 4),
     ...:                       subplot_kw={'projection':'3d'})
     ...: for ax, xyz, c in [(ax1, xyz1, 'r'), (ax2, xyz2, 'b'),
     ...:                     (ax3, xyz3, 'g')]:
     ...:     ax.plot(xyz[:,0], xyz[:,1], xyz[:,2], c, alpha=0.5)
     ...:     ax.set_xlabel('$x$', fontsize=16)
     ...:     ax.set_ylabel('$y$', fontsize=16)
     ...:     ax.set_zlabel('$z$', fontsize=16)
     ...:     ax.set_xticks([-15, 0, 15])
     ...:     ax.set_yticks([-20, 0, 20])
     ...:     ax.set_zticks([0, 20, 40])
```

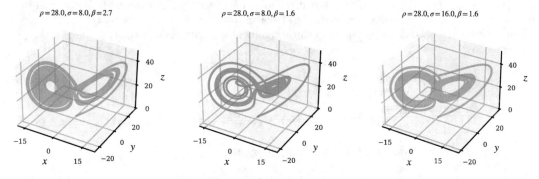

Figure 9-6. *The dynamics for the Lorenz ODE for three different sets of parameters*

The three examples we have looked at use the odeint solver. This function takes a large number of optional arguments that can be used to fine-tune the solver, including options for the maximum number of allowed steps (hmax) and the maximum order for the Adams (mxordn) and BDF (mxords) methods, to mention a few. See the docstring of odeint for further information.

The alternative to odeint in SciPy is the object-oriented interface provided by the integrate.ode class. Like with the odeint function, to use the integrate.ode class, we first need to define the right-hand side function for the ODE and define the initial state array and an array for the values of the independent variable at which we want to compute the solution. However, one small but important difference is that while the for $f(x, y(x))$ function to be used with odeint had to have the f(y, x, ...) function signature, the corresponding function to be used with integrate.ode must have the f(x, y, ...) function signature (i.e., the order of x and y is reversed).

The `integrate.ode` class can work with a collection of different solvers and has specific options for each solver. The docstring of `integrate.ode` describes the available solvers and their options. To illustrate how to use the `integrate.ode` interface, let's first look at the following sets of coupled second-order ODEs.

$$m_1 x_1''(t) + \gamma_1 \; x_1'(t) + k_1 x_1 - k_2 \; (x_2 - x_1) = 0$$

$$m_2 x_2''(t) + \gamma_2 x_2'(t) + k_2 \; (x_2 - x_1) = 0$$

These equations describe the dynamics of two coupled springs, where $x_1(t)$ and $x_2(t)$ are the displacements of two objects, with masses m_1 and m_2, from their equilibrium positions. The object at x_1 is connected to a fixed wall via a spring with spring constant k_1 and connected to the object at x_2 via a spring with spring constant k_2. Both objects are subject to damping forces characterized by γ_1 and γ_2, respectively. To solve this kind of problem with SciPy, we must first write it in standard form, which we can do by introducing $y_0(t) = x_1(t)$, $y_1(t) = x_1'(t)$, $y_2(t) = x_2(t)$, and $y_3(t) = x_2'(t)$, which results in four coupled first-order equations.

$$\frac{d}{dt} \begin{bmatrix} y_0(t) \\ y_1(t) \\ y_2(t) \\ y_3(t) \end{bmatrix} = f(t, y(t)) = \begin{bmatrix} y_1(t) \\ \left(-\gamma_1 y_1(t) - k_1 y_0(t) - k_2 y_0(t) + k_2 y_2(t)\right)/m_1 \\ y_3(t) \\ \left(-\gamma_2 y_3(t) - k_2 y_2(t) + k_2 y_0(t)\right)/m_2 \end{bmatrix}$$

The first task is to write a Python function that implements the $f(t, \mathbf{y}(t))$ function, which also takes the problem parameters as additional arguments. The following implementation bunches all the parameters into a tuple that is passed to the function as a single argument and unpack on the first line of the function body.

```
In [105]: def f(t, y, args):
    ...:     m1, k1, g1, m2, k2, g2 = args
    ...:     return [y[1],
    ...:             - k1/m1 * y[0] + k2/m1 * (y[2] - y[0]) - g1/m1 * y[1],
    ...:             y[3],
    ...:             - k2/m2 * (y[2] - y[0]) - g2/m2 * y[3]]
```

The return value of the `f` function is a list of length 4, whose elements are the derivatives of the ODE $y_0(t)$ to $y_3(t)$ functions. Next, let's create variables with specific values for the parameters and pack them into a tuple `args` that can be passed to the `f` function. Like before, we also need to create arrays for the initial condition `y0` and for the t values we want to compute the solution to the ODE, `t`.

```
In [106]: m1, k1, g1 = 1.0, 10.0, 0.5
In [107]: m2, k2, g2 = 2.0, 40.0, 0.25
In [108]: args = (m1, k1, g1, m2, k2, g2)
In [109]: y0 = [1.0, 0, 0.5, 0]
In [110]: t = np.linspace(0, 20, 1000)
```

The main difference between using `integrate.odeint` and `integrate.ode` starts at this point. Instead of calling the `odeint` function, we need to create an instance of the `integrate.ode` class, passing the ODE `f` function as an argument.

```
In [111]: r = integrate.ode(f)
```

Here, we store the resulting solver instance in the variable r. Before we can use it, we must configure some of its properties. At a minimum, we need to set the initial state using the set_initial_value method, and if the f function takes additional arguments, we need to configure those using the set_f_params method. We can also select a solver using the set_integrator method, which accepts the following solver names as the first argument: vode, zvode, lsoda, dopri5, and dop853. Each solver takes additional optional arguments. See the docstring for integrate.ode for details. Here, let's use the LSODA solver and set the initial state and the parameters to the f function.

```
In [112]: r.set_integrator('lsoda')
In [113]: r.set_initial_value(y0, t[0])
In [114]: r.set_f_params(args)
```

Once the solver is created and configured, we can start solving the ODE step by step by calling the r.integrate method, and the status of the integration can be queried using the r.successful method (which returns True as long as the integration is proceeding fine). We need to keep track of which point to integrate to, and we need to store results by ourselves.

```
In [115]: dt = t[1] - t[0]
     ...: y = np.zeros((len(t), len(y0)))
     ...: idx = 0
     ...: while r.successful() and r.t < t[-1]:
     ...:     y[idx, :] = r.y
     ...:     r.integrate(r.t + dt)
     ...:     idx += 1
```

This is not as convenient as simply calling the odeint, but it offers extra flexibility that sometimes is needed. This example stored the solution in the array y for each corresponding element in t, similar to what odeint would have returned. The following code plots the solution; the result is shown in Figure 9-7.

```
In [116]: fig = plt.figure(figsize=(10, 4))
     ...: ax1 = plt.subplot2grid((2, 5), (0, 0), colspan=3)
     ...: ax2 = plt.subplot2grid((2, 5), (1, 0), colspan=3)
     ...: ax3 = plt.subplot2grid((2, 5), (0, 3), colspan=2, rowspan=2)
     ...: # x_1 vs time plot
     ...: ax1.plot(t, y[:, 0], 'r')
     ...: ax1.set_ylabel('$x_1$', fontsize=18)
     ...: ax1.set_yticks([-1, -.5, 0, .5, 1])
     ...: # x2 vs time plot
     ...: ax2.plot(t, y[:, 2], 'b')
     ...: ax2.set_xlabel('$t$', fontsize=18)
     ...: ax2.set_ylabel('$x_2$', fontsize=18)
     ...: ax2.set_yticks([-1, -.5, 0, .5, 1])
     ...: # x1 and x2 phase space plot
     ...: ax3.plot(y[:, 0], y[:, 2], 'k')
     ...: ax3.set_xlabel('$x_1$', fontsize=18)
     ...: ax3.set_ylabel('$x_2$', fontsize=18)
     ...: ax3.set_xticks([-1, -.5, 0, .5, 1])
     ...: ax3.set_yticks([-1, -.5, 0, .5, 1])
     ...: fig.tight_layout()
```

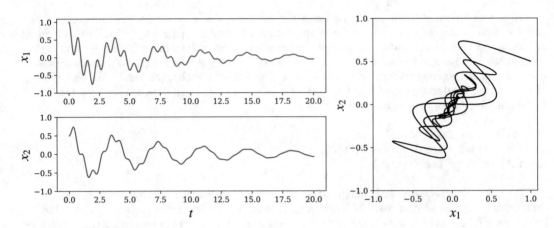

Figure 9-7. *The solution of the ODE for two coupled damped oscillators*

In addition to providing a Python function for the ODE $f(t, y(t))$ function, we can also provide a Python function that computes the Jacobian matrix for a given t and $y(t)$. The solver can, for example, use the Jacobian to solve the system of equations that arise in implicit methods more efficiently. To use a Jacobian function jac, like the one defined for the current problem, we must pass it to the integrate.ode class when it is created, together with the f function. If the Jacobian function jac takes additional arguments, those also must be configured using the set_jac_params method in the resulting integrate.ode instance.

```
In [117]: def jac(t, y, args):
     ...:     m1, k1, g1, m2, k2, g2 = args
     ...:     return [[0, 1, 0, 0],
     ...:             [- k1/m2 - k2/m1, - g1/m1, k2/m1, 0],
     ...:             [0, 0, 0, 1],
     ...:             [k2/m2, 0, - k2/m2, - g2/m2]]
In [118]: r = integrate.ode(f, jac)
In [119]: r.set_jac_params(args)
```

Python functions for $f(t, y(t))$ and its Jacobian can conveniently be generated using SymPy's lambdify, provided that the ODE problem first can be defined as a SymPy expression. This symbolic-numeric hybrid approach is a powerful method for solving ODE problems. To illustrate this approach, consider the rather complicated system of two coupled second-order and nonlinear ODEs for a double pendulum. The equations of motion for the angular deflection, $\theta_1(t)$ and $\theta_2(t)$, for the first and the second pendulum, respectively, are[7]

$$(m_1 + m_2)\, l_1\theta_1''(t) + m_2 l_2\theta_2''(t)\cos(\theta_1 - \theta_2) + m_2 l_2\left(\theta_2'(t)\right)^2\sin(\theta_1 - \theta_2) + g(m_1 + m_2)\sin\theta_1 = 0,$$

$$m_2 l_2\theta_2''(t) + m_2 l_1\theta_1''(t)\cos(\theta_1 - \theta_2) - m_2 l_1\left(\theta_1'(t)\right)^2\sin(\theta_1 - \theta_2) + m_2 g\sin\theta_2 = 0.$$

---

[7] See http://scienceworld.wolfram.com/physics/DoublePendulum.html for details.

The first pendulum is attached to a fixed support, and the second pendulum is attached to the first. Here m_1 and m_2 are the masses and l_1 and l_2 are the lengths of the first and second pendulums, respectively. Let's begin by defining SymPy symbols for the variables and the functions in the problem and then constructing the ode expressions.

```
In [120]: t, g, m1, l1, m2, l2 = sympy.symbols("t, g, m_1, l_1, m_2, l_2")
In [121]: theta1, theta2 = sympy.symbols("theta_1, theta_2", cls=sympy.Function)
In [122]: ode1 = sympy.Eq(
     ...:      (m1+m2)*l1 * theta1(t).diff(t,t) +
     ...:      m2*l2 * theta2(t).diff(t,t) * sympy.cos(theta1(t)-theta2(t)) +
     ...:      m2*l2 * theta2(t).diff(t)**2 * sympy.sin(theta1(t)-theta2(t)) +
     ...:      g*(m1+m2) * sympy.sin(theta1(t)), 0)
     ...: ode1
```

$$\text{Out[122]:} \quad g\left(m_1+m_2\right)\sin\theta_1\left(t\right)+l_1\left(m_1+m_2\right)\frac{d^2}{dt^2}\theta_1\left(t\right)+l_2 m_2 \sin\left(\theta_1\left(t\right)-\theta_2\left(t\right)\right)\left(\frac{d}{dt}\theta_2\left(t\right)\right)^2$$

$$+l_2 m_2 \frac{d^2}{dt^2}\theta_2\left(t\right)\cos\left(\theta_1\left(t\right)-\theta_2\left(t\right)\right)=0$$

```
In [123]: ode2 = sympy.Eq(
     ...:      m2*l2 * theta2(t).diff(t,t) +
     ...:      m2*l1 * theta1(t).diff(t,t) * sympy.cos(theta1(t)-theta2(t))-
     ...:      m2*l1 * theta1(t).diff(t)**2 * sympy.sin(
     ...:          theta1(t) - theta2(t)) + m2*g * sympy.sin(theta2(t)), 0)
     ...: ode2
```

$$\text{Out[123]:} \quad g m_2 \sin\theta_2\left(t\right)-l_1 m_2 \sin\left(\theta_1\left(t\right)-\theta_2\left(t\right)\right)\left(\frac{d}{dt}\theta_1\left(t\right)\right)^2$$

$$+l_1 m_2 \cos\left(\theta_1\left(t\right)-\theta_2\left(t\right)\right)\frac{d^2}{dt^2}\theta_1\left(t\right)+l_2 m_2 \frac{d^2}{dt^2}\theta_2\left(t\right)=0$$

Now, ode1 and ode2 are SymPy expressions for the two second-order ODE equations. Trying to solve these equations with sympy.dsolve is fruitless, and we need to use a numerical method to proceed. However, the equations as they stand here are not in a form that is suitable for numerical solution with the ODE solvers that are available in SciPy. First, we must write the system of two second-order ODEs as a system of four first-order ODEs on the standard form. Rewriting the equations in the standard form is not difficult but can be tedious to do by hand. Fortunately, we can leverage the symbolic capabilities of SymPy to automate this task. To this end, we need to introduce new functions—$y_1(t) = \theta_1(t)$, $y_2(t) = \theta_1'(t)$, $y_3(t) = \theta_2(t)$, and $y_4(t) = \theta_2'(t)$—and rewrite the ODEs in terms of these functions. By creating a dictionary for the variable change and using the SymPy function subs to perform the substitution using this dictionary, we can easily obtain the equations for $y_2'(t)$ and $y_4'(t)$.

```
In [124]: y1, y2, y3, y4 = sympy.symbols("y_1, y_2, y_3, y_4", cls=sympy.Function)
In [125]: varchange = {theta1(t).diff(t, t): y2(t).diff(t),
     ...:              theta1(t): y1(t),
     ...:              theta2(t).diff(t, t): y4(t).diff(t),
     ...:              theta2(t): y3(t)}
In [126]: ode1_vc = ode1.subs(varchange)
In [127]: ode2_vc = ode2.subs(varchange)
```

We also need to introduce two more ODEs for $y_1'(t)$ and $y_3'(t)$.

```
In [128]: ode3 = sympy.Eq(y1(t).diff(t), y2(t))
In [129]: ode4 = sympy.Eq(y3(t).diff(t), y4(t))
```

At this point, we have four coupled first-order ODEs for the y_1 to y_4 functions. It only remains to solve for the derivatives of these functions to obtain the ODEs in standard form. We can do this using sympy.solve.

```
In [130]: y = sympy.Matrix([y1(t), y2(t), y3(t), y4(t)])
In [131]: vcsol = sympy.solve((ode1_vc, ode2_vc, ode3, ode4), y.diff(t), dict=True)
In [132]: f = y.diff(t).subs(vcsol[0])
```

Now, f is the SymPy expression for the ODE $f(t, y(t))$ function. We can display the ODEs using sympy. Eq(y.diff(t), f), but the result is rather lengthy, and in the interest of space, we do not show the output here. The main purpose of constructing f here is to convert it to a NumPy-aware function that can be used with integrate.odeint or integrate.ode. The ODEs are now on a form that we can create such a function using sympy.lambdify. Also, since there is a symbolic representation of the problem so far, it is easy to compute the Jacobian and create a NumPy-aware function. When using sympy.lambdify to create functions for odeint and ode, we must be careful to put t and y in the correct order in the tuple passed to sympy. lambdify. Here, let's use integrate.ode, so we need a function with the signature f(t, y, *args), and thus we pass the tuple (t, y) as the first argument to sympy.lambdify. We wrap the resulting function with a lambda function to receive the additional argument args, which is not used in the SymPy expression.

```
In [133]: params = {m1: 5.0, l1: 2.0, m2: 1.0, l2: 1.0, g: 10.0}
In [134]: _f_np = sympy.lambdify((t, y), f.subs(params), 'numpy')
In [135]: f_np = lambda _t, _y, *args: _f_np(_t, _y)
In [136]: jac = sympy.Matrix([[fj.diff(yi) for yi in y] for fj in f])
In [137]: _jac_np = sympy.lambdify((t, y), jac.subs(params), 'numpy')
In [138]: jac_np = lambda _t, _y, *args: _jac_np(_t, _y)
```

Here, I have also substituted specific values of the system parameters before calling sympy.lambdify. The first pendulum is twice as long and five times as heavy as the second pendulum. With the f_np and jac_np functions, we are now ready to solve the ODE using integrate.ode in the same manner as in the previous examples. Here, let's take the initial state to be $\theta_1(0) = 2$ and $\theta_2(0) = 0$, and with the derivatives set to zero, solve for the time interval [0, 20] with 1000 steps.

```
In [139]: y0 = [2.0, 0, 0, 0]
In [140]: tt = np.linspace(0, 20, 1000)
In [141]: r = integrate.ode(f_np, jac_np).set_initial_value(y0, tt[0])
In [142]: dt = tt[1] - tt[0]
     ...: yy = np.zeros((len(tt), len(y0)))
     ...: idx = 0
     ...: while r.successful() and r.t < tt[-1]:
     ...:     yy[idx, :] = r.y
     ...:     r.integrate(r.t + dt)
     ...:     idx += 1
```

The solution to the ODEs is now stored in the array yy, which has the shape (1000, 4). When visualizing this solution, it is more intuitive to plot the positions of the pendulums in the $x - y$ plane rather than their angular deflections. The transformations between the angular variables θ_1 and θ_2 and x and y coordinates are $x_1 = l_1 \sin \theta_1$, $y_1 = l_1 \cos \theta_1$, $x_2 = x_1 + l_2 \sin \theta_2$, and $y_2 = y_1 + l_2 \cos \theta_2$.

```
In [143]: theta1_np, theta2_np = yy[:, 0], yy[:, 2]
In [144]: x1 = params[l1] * np.sin(theta1_np)
     ...: y1 = -params[l1] * np.cos(theta1_np)
     ...: x2 = x1 + params[l2] * np.sin(theta2_np)
     ...: y2 = y1 - params[l2] * np.cos(theta2_np)
```

Finally, let's plot the dynamics of the double pendulum as a function of time and in the x–y plane. The result is shown in Figure 9-8. As expected, pendulum 1 is confined to moving on a circle (because of its fixed anchor point), while pendulum 2 has a much more complicated trajectory.

```
In [145]: fig = plt.figure(figsize=(10, 4))
     ...: ax1 = plt.subplot2grid((2, 5), (0, 0), colspan=3)
     ...: ax2 = plt.subplot2grid((2, 5), (1, 0), colspan=3)
     ...: ax3 = plt.subplot2grid((2, 5), (0, 3), colspan=2, rowspan=2)
     ...:
     ...: ax1.plot(tt, x1, 'r')
     ...: ax1.plot(tt, y1, 'b')
     ...: ax1.set_ylabel('$x_1, y_1$', fontsize=18)
     ...: ax1.set_yticks([-3, 0, 3])
     ...:
     ...: ax2.plot(tt, x2, 'r')
     ...: ax2.plot(tt, y2, 'b')
     ...: ax2.set_xlabel('$t$', fontsize=18)
     ...: ax2.set_ylabel('$x_2, y_2$', fontsize=18)
     ...: ax2.set_yticks([-3, 0, 3])
     ...:
     ...: ax3.plot(x1, y1, 'r')
     ...: ax3.plot(x2, y2, 'b', lw=0.5)
     ...: ax3.set_xlabel('$x$', fontsize=18)
     ...: ax3.set_ylabel('$y$', fontsize=18)
     ...: ax3.set_xticks([-3, 0, 3])
     ...: ax3.set_yticks([-3, 0, 3])
```

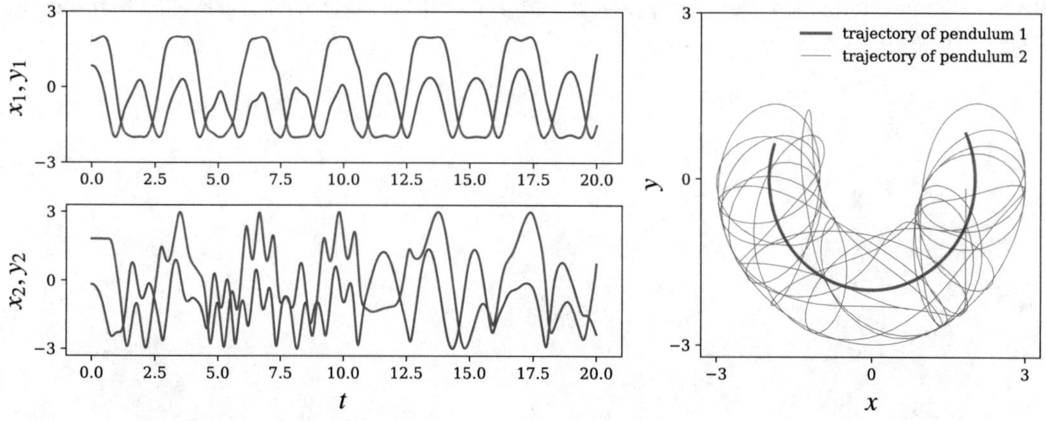

Figure 9-8. *The dynamics of the double pendulum*

Summary

This chapter explored various methods and tools for solving ordinary differential equations (ODEs) using the scientific computing packages for Python. ODEs show up in many areas of science and engineering—particularly modeling and the description of dynamical systems—and mastering the techniques to solve ODE problems is a crucial part of the skill set of a computational scientist. The chapter first looked at solving ODEs symbolically using SymPy, either with the `sympy.dsolve` function or using a Laplace transformation method. The symbolic approach is often a good starting point, and with the symbolic capabilities of SymPy, many fundamental ODE problems can be solved analytically. However, for most practical problems, there is no analytical solution, and the symbolic methods are then doomed to fail. Our remaining option is then to fall back on numerical techniques. Numerical integration of ODEs is a vast field in mathematics, and numerous reliable methods exist for solving ODE problems. The chapter briefly reviewed methods for integrating ODEs, intending to introduce the concepts and ideas behind the Adams and BDF multistep methods used in the solvers provided by SciPy. Finally, we looked at how the `odeint` and ode solvers, available through the SciPy `integrate` module, can solve a few example problems. Although most ODE problems eventually require numerical integration, there can be significant advantages in using a hybrid symbolic-numerical approach, which uses features from both SymPy and SciPy. The last example of this chapter is devoted to demonstrating this approach.

Further Reading

An accessible introduction to many methods for numerically solving ODE problems is given in *Scientific Computing* by M. T. Heath (McGraw-Hill, 2002). For a review of ordinary differential equations with code examples, see Chapter 11 in *Numerical Recipes* by W. H. Press et al. (Cambridge University Press, 2007). For a more detailed survey of numerical methods for ODEs, see, for example, (Kendall Atkinson (2009). The main implementations of ODE solvers used in SciPy are the VODE and LSODA solvers. The source code for these methods is available from Netlib at `www.netlib.org/ode/vode.f` and `www.netlib.org/odepack`, respectively. In addition to these solvers, there is also a well-known suite of solvers called sundials, which the Lawrence Livermore National Laboratory provides and is available at `http://computing.llnl.gov/projects/sundials`. This suite also includes solvers of differential algebraic equations (DAEs). A Python interface for the sundials solvers is provided by the `scikit.odes` library, which can be obtained from `http://github.com/bmcage/odes`. The odespy library also offers a unified interface for many different ODE solvers. For more information about odespy, see the project's website at `http://hplgit.github.io/odespy/doc/web/index.html`.

CHAPTER 10

■ ■ ■

Sparse Matrices and Graphs

We have gone through numerous examples of arrays and matrices that are essential in many aspects of numerical computing. And, we have represented arrays with the NumPy ndarray data structure, a heterogeneous representation that stores all the array elements it represents. This is often the most efficient way to represent an object, such as a vector, matrix, or higher-dimensional array. However, notable exceptions are matrices where most of the elements are zeros. Such matrices are known as *sparse matrices*, and they occur in many applications, for example, in connection networks (such as circuits) and large algebraic equation systems that arise, for example, when solving partial differential equations (see Chapter 11 for examples).

For matrices dominated by zero elements, storing all the zeros in the computer's memory is inefficient, and it is more suitable to store only the nonzero values with additional information about their locations. For non-sparse matrices, known as *dense matrices*, such a representation is less efficient than storing all values consecutively in the memory, but for large sparse matrices, it can be vastly superior.

There are several options for working with sparse matrices in Python. This chapter focuses on the sparse matrix module in SciPy, `scipy.sparse`, which provides a feature-rich and easy-to-use interface for representing sparse matrices and carrying out linear algebra operations on such objects. Another option is PySparse,[1] which offers similar functionality. For very large-scale problems, the PyTrilinos[2] and PETSc[3] packages have powerful parallel implementations of many sparse matrix operations. However, using these packages requires more programming, and they have a steeper learning curve and are more difficult to install and set up. For most basic use cases, SciPy's `sparse` module is the most suitable option or at least an appropriate starting point.

The end of the chapter briefly explores representing and processing graphs using the SciPy `sparse.csgraph` module and the NetworkX library. Graphs can be represented as adjacency matrices, which in many applications are very sparse. Graphs and sparse matrices are, therefore, closely connected topics.

Importing Modules

The main module we work with in this chapter is the `sparse` module in the SciPy library. Let's assume this module is included under the name `sp`, and we also need to explicitly import its `linalg` submodule to make it accessible through `sp.linalg`.

```
In [1]: import scipy.sparse as sp
In [2]: import scipy.sparse.linalg
```

---

[1] http://pysparse.sourceforge.net
[2] http://trilinos.org/packages/pytrilinos
[3] See http://www.mcs.anl.gov/petsc and https://bitbucket.org/petsc/petsc4py for its Python bindings.

R. Johansson, *Numerical Python*, https://doi.org/10.1007/979-8-8688-0413-7_10

We also need the NumPy library, which we, as usual, import under the name np, and the Matplotlib library for plotting:

```
In [3]: import numpy as np
In [4]: import matplotlib.pyplot as plt
```

The last part of this chapter uses the networkx module, which we import under the name nx.

```
In [5]: import networkx as nx
```

Sparse Matrices in SciPy

The basic idea of sparse matrix representation is to avoid storing the excessive zeros in a sparse matrix. In dense matrix representation, where all elements of an array are stored consecutively, it is sufficient to store the values since each element's row and column indices are implicitly known from the position in the array. However, if we store only the nonzero elements, we must also store each element's row and column indices. There are numerous approaches to organizing the storage of the nonzero elements and their corresponding row and column indices. These approaches have different advantages and disadvantages, for example, how easy it is to create the matrices and, perhaps more importantly, how efficiently they can be used to implement mathematical operations on the sparse matrices. A summary and comparison of sparse matrix formats available in the SciPy sparse module is given in Table 10-1.

Table 10-1. *Summary and Comparison of Methods to Represent Sparse Matrices*

Type	Description	Pros	Cons
Coordinate list (COO, sp.coo_matrix)	Nonzero values are stored in a list together with their row and column.	Simple to construct and efficient to add new elements.	Inefficient element access. Not suitable for mathematical operations, such as matrix multiplication.
List of lists (LIL, sp.lil_matrix)	Stores a list of column indices for nonzero elements for each row and a list of the corresponding values.	Supports slicing operations.	Not ideal for mathematical operations.
Dictionary of keys (DOK, sp.dok_matrix)	Nonzero values are stored in a dictionary with a tuple of (row, column) as key.	Simple to construct and fast to add, remove and access elements.	Not ideal for mathematical operations.
Diagonal matrix (DIA, sp.dia_matrix)	Stores lists of diagonals of the matrix.	Efficient for diagonal matrices.	Not suitable for nondiagonal matrices.
Compressed sparse column (CSC, sp.csc_matrix) and compressed sparse row (CSR, sp.csr_matrix)	Stores the values together with arrays with column or row indices.	Relatively complicated to construct.	Efficient matrix-vector multiplication.
Block-sparse matrix (BSR, cp.bsr_matrix)	Similar to CSR, but for sparse matrices with dense submatrices.	Efficient for their specific intended purpose.	Not suitable for general-purpose use.

A simple and intuitive approach for storing sparse matrices is storing lists with column indices and row indices together with the list of nonzero values. This format is called *coordinate list format*, abbreviated as COO in SciPy. The sp.coo_matrix class represents sparse matrices in this format. This format is straightforward to initialize. For instance, with the matrix

$$A = \begin{bmatrix} 0 & 1 & 0 & 0 \\ 0 & 0 & 0 & 2 \\ 0 & 0 & 3 & 0 \\ 4 & 0 & 0 & 0 \end{bmatrix},$$

we can easily identify the nonzero values $[A_{01} = 1, A_{13} = 2, A_{22} = 3, A_{30} = 4]$ and their corresponding rows $[0, 1, 2, 3]$ and columns $[1, 3, 2, 0]$ (note that here we have used Python's zero-based indexing). To create a sp.coo_matrix object, we can create lists (or arrays) for the values, row indices, and column indices and pass them to sp.coo_matrix. Optionally, we can also specify the shape of the array using the shape argument, which is useful when the nonzero elements do not span the entire matrix (for example, if there are columns or rows containing only zeros so that the shape cannot be correctly inferred from the row and column arrays).

```
In [6]: values = [1, 2, 3, 4]
In [7]: rows = [0, 1, 2, 3]
In [8]: cols = [1, 3, 2, 0]
In [9]: A = sp.coo_matrix((values, (rows, cols)), shape=[4, 4])
In [10]: A
Out[10]: <4x4 sparse matrix of type '<type 'numpy.int64'>'
            with 4 stored elements in Coordinate format>
```

The result is a data structure that represents the sparse matrix. All sparse matrix representations in SciPy's sparse module share several common attributes, many of which are derived from NumPy's ndarray object. Examples of such attributes are size, shape, dtype, and ndim, and common to all sparse matrix representations are the nnz (number of nonzero elements) and data (the nonzero values) attributes.

```
In [11]: A.shape, A.size, A.dtype, A.ndim
Out[11]: ((4, 4), 4, dtype('int64'), 2)
In [12]: A.nnz, A.data
Out[12]: (4, array([1, 2, 3, 4]))
```

In addition to the shared attributes, each type of sparse matrix representation also has attributes specific to storing the positions for each nonzero value. For sp.coo_matrix objects, there are row and col attributes for accessing the underlying row and column arrays.

```
In [13]: A.row
Out[13]: array([0, 1, 2, 3], dtype=int32)
In [14]: A.col
Out[14]: array([1, 3, 2, 0], dtype=int32)
```

Many methods are also available for operating on sparse matrix objects. Many of these methods are for applying mathematical functions on the matrix, for example, elementwise math methods like sin, cos, arcsin, and so on; aggregation methods like min, max, sum, and so on; mathematical array operators such as conjugate (conj) and transpose (transpose), and so on; and dot for computing the dot product between sparse matrices or a sparse matrix and a dense vector (the * operator also denotes matrix multiplication for sparse matrices). For further details, see the docstring for the sparse matrix classes (summarized in Table 10-1). Another important family of methods is used to convert sparse matrices between different formats: for example, tocoo, tocsr, tolil, and so on. There are also methods for converting a sparse matrix to NumPy ndarray and NumPy matrix objects (i.e., dense matrix representations): toarray and todense.

For example, to convert the sparse matrix A from COO format to CSR format and a NumPy array, respectively, we can use the following.

```
In [15]: A.tocsr()
Out[15]: <4x4 sparse matrix of type '<type 'numpy.int64'>'
         with 4 stored elements in Compressed Sparse Row format>
In [16]: A.toarray()
Out[16]: array([[0, 1, 0, 0],
                [0, 0, 0, 2],
                [0, 0, 3, 0],
                [4, 0, 0, 0]])
```

The obvious way to access elements in a matrix, which has been used in numerous different contexts so far, is using the indexing syntax (e.g., A[1,2]), the slicing syntax (e.g., A[1:3, 2]), and so on. We can often use this syntax with sparse matrices, too, but not all representations support indexing and slicing, and if it is supported, it may not be an efficient operation. In particular, assigning values to zero-valued elements can be costly, as it may require rearranging the underlying data structures, depending on the format used. To incrementally add new elements to a sparse matrix, the LIL (sp.lil_matrix) format is a suitable choice, but this format is, on the other hand, not ideal for arithmetic operations.

When working with sparse matrices, it is common to face the situation that different tasks—such as construction, updating, and arithmetic operations—are most efficiently handled in other formats. Converting between different sparse formats is relatively efficient, so switching between different formats in different parts of an application is useful. Therefore, efficient use of sparse matrices requires understanding how different formats are implemented and what they are suitable for. Table 10-1 briefly summarizes the pros and cons of the sparse matrix formats available in SciPy's sparse module, and using the conversion methods, it is easy to switch between different formats. For a more in-depth discussion of the merits of the various formats, see the "Sparse Matrices"[4] section in the SciPy reference manual.

For computations, the most important sparse matrix representations in SciPy's sparse module are the CSR (Compressed Sparse Row) and CSC (Compressed Sparse Column) formats because they are well-suited for efficient matrix arithmetic and linear algebra applications. Other formats, like COO, LIL, and DOK, are mainly used for constructing and updating sparse matrices. Once a sparse matrix is ready for computations, it is best to convert it to CSR or CSC format, utilizing the tocsr or tocsc methods, respectively.

In the CSR format, the nonzero values (data) are stored along with an array that contains the column indices of each value (indices) and another array that stores the offsets of the column index array of each row (indptr). For instance, consider the following matrix:

$$A = \begin{bmatrix} 1 & 2 & 0 & 0 \\ 0 & 3 & 4 & 0 \\ 0 & 0 & 5 & 6 \\ 7 & 0 & 8 & 9 \end{bmatrix}.$$

Here the nonzero values are [1, 2, 3, 4, 5, 6, 7, 8, 9] (data), and the *column indices* corresponding to the nonzero values in the first row are [0, 1], the second row [1, 2], the third row [2, 3], and the fourth row [0, 2, 3]. Concatenating these column index lists gives the indices array [0, 1, 1, 2, 2, 3, 0, 2, 3]. To keep track of which

---

[4]http://docs.scipy.org/doc/scipy/reference/sparse.html

row entries in this column index array correspond to, we can store the starting position for each row in a second array. The column indices of the first row are elements 0 to 1, the second-row elements 2 to 3, the third-row elements 4 to 5, and finally the fourth-row elements 6 to 9. Collecting the starting indices in an array gives [0, 2, 4, 6]. For convenience in the implementation, we also add the total number of nonzero elements at the end of this array, which results in the indptr array [0, 2, 4, 6, 9]. The following code creates a dense NumPy array corresponding to matrix A, converts it to a CSR matrix using sp.csr_matrix, and then displays the data, indices, and indptr attributes.

```
In [17]: A = np.array([[1, 2, 0, 0], [0, 3, 4, 0], [0, 0, 5, 6], [7, 0, 8, 9]])
    ...: A
Out[17]: array([[1, 2, 0, 0],
                [0, 3, 4, 0],
                [0, 0, 5, 6],
                [7, 0, 8, 9]])
In [18]: A = sp.csr_matrix(A)
In [19]: A.data
Out[19]: array([1, 2, 3, 4, 5, 6, 7, 8, 9], dtype=int64)
In [20]: A.indices
Out[20]: array([0, 1, 1, 2, 2, 3, 0, 2, 3], dtype=int32)
In [21]: A.indptr
Out[21]: array([0, 2, 4, 6, 9], dtype=int32)
```

With this storage scheme, the nonzero elements in the row with index i are stored in the data array between index indptr[i] and indptr[i+1]-1, and the column indices for these elements are stored at the same indices in the indices array. For example, the elements in the third row, with index i=2, start at indptr[2]=4 and end at indptr[3]-1=5, which gives the element values, data[4]=5 and data[5]=6, and the column indices, indices[4]=2 and indices[5]=3. Thus, $A[2, 2] = 5$ and $A[2, 3] = 6$ (in zero indexbased notation).

```
In [22]: i = 2
In [23]: A.indptr[i], A.indptr[i+1]-1
Out[23]: (4, 5)
In [24]: A.indices[A.indptr[i]:A.indptr[i+1]]
Out[24]: array([2, 3], dtype=int32)
In [25]: A.data[A.indptr[i]:A.indptr[i+1]]
Out[25]: array([5, 6], dtype=int64)
In [26]: A[2, 2], A[2,3]  # check
Out[26]: (5, 6)
```

While the CSR storage method is less intuitive than COO, LIL, or DOK, it is well suited for implementing matrix arithmetic and linear algebra operations. Together with the CSC format, it is, therefore, the main format for use in sparse matrix computations. The CSC format is almost identical to CSR, except that instead of column indices and row pointers, row indices and column pointers are used (i.e., the role of columns and rows is reversed).

Functions for Creating Sparse Matrices

As shown in the examples earlier in this chapter, one way of constructing sparse matrices is to prepare the data structures for a specific sparse matrix format and pass these to the constructor of the corresponding sparse matrix class. While this method is sometimes suitable, composing sparse matrices from predefined template matrices is often more convenient. The sp.sparse module provides a variety of functions for generating such matrices, for example, sp.eye for creating diagonal sparse matrices with ones on the diagonal (optionally offset from the main diagonal), sp.diags for creating diagonal matrices with a specified pattern along the diagonal, sp.kron for calculating the Kronecker (tensor) product of two sparse matrices, and bmat, vstack, and hstack, for building sparse matrices from sparse block matrices, and by stacking sparse matrices vertically and horizontally, respectively.

For example, in many applications, sparse matrices have a diagonal form. To create a sparse matrix of size 10×10 with a main diagonal and an upper and lower diagonal, we can use three calls to sp.eye, using the k argument to specify the offset from the main diagonal.

```
In [27]: N = 10
In [28]: A = sp.eye(N, k=1) - 2 * sp.eye(N) + sp.eye(N, k=-1)
In [29]: A
Out[29]: <10x10 sparse matrix of type '<class 'numpy.float64'>'
                with 28 stored elements in Compressed Sparse Row format>
```

By default, the resulting object is a sparse matrix in the CSR format, but using the format argument, we can specify any other sparse matrix format. The value of the format argument should be a string such as 'csr', 'csc', 'lil', and so on. All functions for creating sparse matrices in sp.sparse accept this argument. For example, in the previous example, we could have produced the same matrix using sp.diags, by specifying the pattern [1, -2, 1] (the coefficients to the sp.eye functions in the previous expression) and the corresponding offsets from the main diagonal [1, 0, -1]. If we want the resulting sparse matrix in CSC format, we can set format='csc'.

```
In [30]: A = sp.diags([1, -2, 1], [1, 0, -1], shape=[N, N], format='csc')
In [31]: A
Out[31]: <10x10 sparse matrix of type '<class 'numpy.float64'>'
                with 28 stored elements in Compressed Sparse Column format>
```

The advantages of using sparse matrix formats rather than dense matrices manifest when working with large matrices. Sparse matrices are, by their nature, therefore, large, and hence, it can be challenging to visualize a matrix by, for example, printing its elements in the terminal. Matplotlib provides the spy function, a valuable tool for visualizing the structure of a sparse matrix. It is available as a function in the pyplot module or as a method for Axes instances. Using it on the previously defined A matrix, obtains the results shown in Figure 10-1.

```
In [32]: fig, ax = plt.subplots()
    ...: ax.spy(A)
```

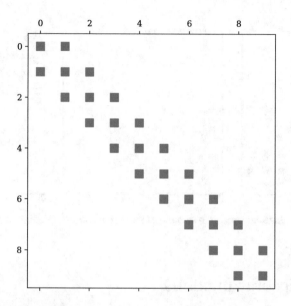

Figure 10-1. *Structure of the sparse matrix with nonzero elements only on the two diagonals closest to the main diagonal and the main diagonal itself*

Sparse matrices are also often associated with tensor product spaces. We can use the sp.kron function to compose a sparse matrix from its smaller components for such cases. For example, to create a sparse matrix for the tensor product between A and the matrix

$$B = \begin{bmatrix} 0 & 1 & 0 \\ 1 & 0 & 1 \\ 0 & 1 & 0 \end{bmatrix},$$

we can use sp.kron(A, B):

```
In [33]: B = sp.diags([1, 1], [-1, 1], shape=[3,3])
In [34]: C = sp.kron(A, B)
In [35]: fig, (ax_A, ax_B, ax_C) = plt.subplots(1, 3, figsize=(12, 4))
    ...: ax_A.spy(A)
    ...: ax_B.spy(B)
    ...: ax_C.spy(C)
```

For comparison, let's plot the sparse matrix structure of *A*, *B*, and *C*; the result is shown in Figure 10-2. For more detailed information on ways to build sparse matrices with the sp.sparse module, see its docstring and the "Sparse Matrices" section in the SciPy reference manual.

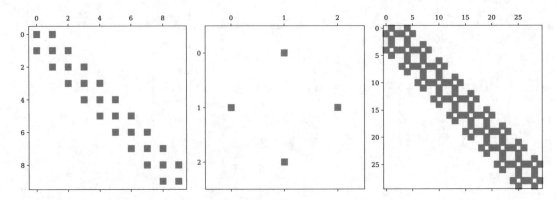

Figure 10-2. *The sparse matrix structures of two matrices, A (left) and B (middle), and their tensor product (right)*

Sparse Linear Algebra Functions

The main application of sparse matrices is to perform linear algebra operations on large matrices that are intractable or inefficient to treat using dense matrix representations. The SciPy sparse module contains a linalg submodule that implements many linear algebra routines. Not all linear algebra operations are suitable for sparse matrices. In some cases, the behavior of the sparse matrix version of operations needs to be modified compared to the dense counterparts. Consequently, there are several differences between the sparse linear algebra module scipy.sparse.linalg and the dense linear algebra module scipy.linalg. For example, the eigenvalue solvers for dense problems typically compute and return all eigenvalues and eigenvectors. This is not manageable for sparse matrices because storing all eigenvectors of a sparse matrix A of size $N \times N$ usually amounts to storing a dense matrix of size $N \times N$. Instead, sparse eigenvalue solvers typically give *a few* eigenvalues and eigenvectors, for example, those with the smallest or largest eigenvalues. In general, sparse matrix methods must retain the sparsity of matrices involved in the computation to be efficient. An example of an operation where the sparsity is usually not retained is the matrix inverse, which should, therefore, be avoided when possible.

Linear Equation Systems

The most important application of sparse matrices is arguably to solve linear equation systems on the form $Ax = b$, where A is a sparse matrix and x and b are dense vectors. The SciPy sparse.linalg module has both direct and iterative solver for this type of problem (sp.linalg.spsolve) and methods to factor a matrix A, using, for example, *LU* factorization (sp.linalg.splu) and incomplete *LU* factorization (sp.linalg.spilu). For example, consider the problem $Ax = b$ where A is the tridiagonal matrix considered in the preceding text and b is a dense vector filled with negative ones (see Chapter 11 for a physical interpretation of this equation). To solve this problem for the system size 10×10, first create the sparse matrix A and the dense vector b.

```
In [36]: N = 10
In [37]: A = sp.diags([1, -2, 1], [1, 0, -1], shape=[N, N], format='csc')
In [38]: b = -np.ones(N)
```

Now, to solve the equation system using the direct solver provided by SciPy, we can use the following.

```
In [39]: x = sp.linalg.spsolve(A, b)
In [40]: x
Out[40]: array([  5.,    9.,   12.,   14.,   15.,   15.,   14.,   12.,    9.,    5.])
```

The solution vector is a dense NumPy array. For comparison, we can also solve this problem using the dense direct solver in NumPy np.linalg.solve (or, similarly, using scipy.linalg.solve). To use the dense solver, we need to convert the sparse matrix A to a dense array using A.todense().

```
In [41]: np.linalg.solve(A.todense(), b)
Out[41]: array([  5.,    9.,   12.,   14.,   15.,   15.,   14.,   12.,    9.,    5.])
```

As expected, the result agrees with what was obtained from the sparse solver. For small problems like this one, there is little to gain from using sparse matrices, but for increasing system size, the merits of using sparse matrices and sparse solvers become apparent. For this particular problem, the threshold system size beyond which sparse methods outperform dense methods is approximately $N = 100$, as shown in Figure 10-3. While the exact threshold varies from problem to problem and hardware and software versions, this behavior is typical for problems where the matrix A is sufficiently sparse.[5]

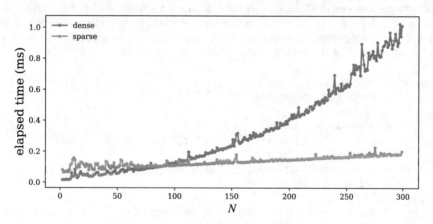

Figure 10-3. *Performance comparison between sparse and dense methods to solve the one-dimensional Poisson problem as a function of problem size*

An alternative to the spsolve interface is explicitly computing the LU factorization using sp.sparse.splu or sp.sparse.spilu (incomplete LU factorization). These functions return an object that contains the L and U factors, and that has a method that solves $LUx = b$ for a given vector b. This is particularly useful when the $Ax = b$ must be solved for multiple vectors b. For example, the LU factorization of the matrix A used previously is computed using the following.

```
In [42]: lu = sp.linalg.splu(A)
In [43]: lu.L
Out[43]: <10x10 sparse matrix of type '<class 'numpy.float64'>'
                with 20 stored elements in Compressed Sparse Column format>
```

---

[5] For a discussion of techniques and methods to optimize Python code, see Chapter 21.

```
In [44]: lu.U
Out[44]: <10x10 sparse matrix of type '<class 'numpy.float64'>'
                    with 20 stored elements in Compressed Sparse Column format>
```

Once the LU factorization is available, we can efficiently solve the equation $LUx = b$ using the solve method for the lu object.

```
In [45]: x = lu.solve(b)
In [46]: x
Out[46]: array([  5.,    9.,   12.,   14.,   15.,   15.,   14.,   12.,    9.,    5.])
```

An important consideration that arises with sparse matrices is that the LU factorization of A may introduce new nonzero elements in L and U compared to the matrix A and, therefore, make L and U less sparse. Elements that exist in L or U but not in A are called fill-ins. The advantage of using sparse matrices may be lost if the number of fill-ins is large. While there is no complete solution to eliminate fill-ins, it is often possible to reduce fill-ins by permuting the rows and columns in A so that the LU factorization takes the form $P_r A P_c = LU$, where P_r and P_c are row and column permutation matrices, respectively. Several such methods for permutations methods are available. The spsolve, splu, and spilu functions all take the permc_spec argument, which can take the values NATURAL, MMD_ATA, MMD_AT_PLUS_A, or COLAMD, which indicates different permutation methods that are built into these methods. The object returned by splu and spilu accounts for such permutations, and the permutation vectors are available via the perm_c and perm_r attributes. Because of these permutations, the product of lu.L and lu.U is not directly equal to A, and to reconstruct A from lu.L and lu.U, we also need to undo the row and column permutations.

```
In [47]: def sp_permute(A, perm_r, perm_c):
    ...:     """ permute rows and columns of A """
    ...:     M, N = A.shape
    ...:     # row permutation matrix
    ...:     Pr = sp.coo_matrix((np.ones(M), (perm_r, np.arange(N)))).tocsr()
    ...:     # column permutation matrix
    ...:     Pc = sp.coo_matrix((np.ones(M), (np.arange(M), perm_c))).tocsr()
    ...:     return Pr.T * A * Pc.T
In [48]: lu.L * lu.U - A  # != 0
Out[48]: <10x10 sparse matrix of type '<class 'numpy.float64'>'
                    with 8 stored elements in Compressed Sparse Column format>
In [49]: sp_permute(lu.L * lu.U, lu.perm_r, lu.perm_c) - A  # == 0
Out[49]: <10x10 sparse matrix of type '<class 'numpy.float64'>'
                    with 0 stored elements in Compressed Sparse Column format>
```

By default, the direct sparse linear solver in SciPy uses the SuperLU[6] package. An alternative sparse matrix solver that can also be used in SciPy is the UMFPACK 7 package. However, this package is not bundled with SciPy and requires the installation of the scikit-umfpack Python library. If scikit-umfpack is available, and if the use_umfpack argument to the sp.linalg.spsolve function is True, the UMFPACK is used instead of SuperLU. Whether SuperLU or UMFPACK gives better performance varies from problem to problem, so it is worth having both installed and tested for any given problem.

The sp.spsolve function is an interface to direct solvers, which internally performs matrix factorization. An alternative approach is to use iterative methods that originate in optimization. The SciPy sparse.linalg module contains several functions for the iterative solution of sparse linear problems: for example, bicg

---

[6]http://crd-legacy.lbl.gov/~xiaoye/SuperLU/

(biconjugate gradient method), bicgstab (biconjugate gradient stabilized method), cg (conjugate gradient), gmres (generalized minimum residual), and lgmres (loose generalized minimum residual method). These functions (and a few others) can solve the problem $Ax = b$ by calling the function with A and b as arguments. They all return a tuple (x, info) where x is the solution, and info contains additional information about the solution process (info=0 indicates success, and it is positive for convergence error and negative for input error). The following is an example.

```
In [50]: x, info = sp.linalg.bicgstab(A, b)
In [51]: x
Out[51]: array([  5.,   9.,  12.,  14.,  15.,  15.,  14.,  12.,   9.,   5.])
In [52]: x, info = sp.linalg.lgmres(A, b)
In [53]: x
Out[53]: array([  5.,   9.,  12.,  14.,  15.,  15.,  14.,  12.,   9.,   5.])
```

In addition, each iterative solver takes its solver-dependent arguments. See the docstring for each function for details. Iterative solvers may have an advantage over direct solvers for large problems, where direct solvers may require excessive memory usage due to undesirable fill-ins. In contrast, iterative solvers only require the evaluation of sparse matrix-vector multiplications and, therefore, do not suffer from fill-in problems. On the other hand, they might have slow convergence for many problems, especially if not adequately preconditioned.

Eigenvalue Problems

Sparse eigenvalue and singular-value problems can be solved using the sp.linalg.eigs and sp.linalg. svds functions, respectively. For real symmetric or complex Hermitian matrices, the eigenvalues (which in this case are real) and eigenvectors can also be computed using sp.linalg.eigsh. These functions do not compute all eigenvalues or singular values but rather compute a given number of eigenvalues and vectors (the default is six). Using the keyword argument k with these functions, we can define how many eigenvalues and vectors should be computed. We can specify the k values to be computed using the which keyword argument. The options for eigs are the largest magnitude LM, smallest magnitude SM, largest real part LR, smallest real part SR, largest imaginary part LI, and smallest imaginary part SI. For svds, only LM and SM are available.

For example, to compute the lowest four eigenvalues for the sparse matrix of the one-dimensional Poisson problem (of system size 10x10), we can use sp.linalg.eigs(A, k=4, which='LM').

```
In [54]: N = 10
In [55]: A = sp.diags([1, -2, 1], [1, 0, -1], shape=[N, N], format='csc')
In [56]: evals, evecs = sp.linalg.eigs(A, k=4, which='LM')
In [57]: evals
Out[57]: array([-3.91898595+0.j, -3.68250707+0.j, -3.30972147+0.j, -2.83083003+0.j])
```

The return value of sp.linalg.eigs (and sp.linalg.eigsh) is a tuple (evals, evecs) whose first element is an array of eigenvalues (evals), and the second element is an array (evecs) of shape $N \times k$, whose columns are the eigenvectors corresponding to the k eigenvalues in evals. Thus, expect the dot product between A and a column in evecs to be equal to the same column in evecs scaled by the corresponding eigenvalue in evals. We can directly confirm that this is indeed the case.

```
In [58]: np.allclose(A.dot(evecs[:,0]), evals[0] * evecs[:,0])
Out[58]: True
```

For this particular example, the sparse matrix A is symmetric, so instead of sp.linalg.eigs, we could use sp.linalg.eigsh, and in doing so, obtain an eigenvalue array with real-valued elements.

```
In [59]: evals, evecs = sp.linalg.eigsh(A, k=4, which='LM')
In [60]: evals
Out[60]: array([-3.91898595, -3.68250707, -3.30972147, -2.83083003])
```

Changing the which='LM' argument (largest magnitude) to which='SM' (smallest magnitude), obtains a different set of eigenvalues and vectors (those with the smallest magnitude).

```
In [61]: evals, evecs = sp.linalg.eigs(A, k=4, which='SM')
In [62]: evals
Out[62]: array([-0.08101405+0.j, -0.31749293+0.j, -0.69027853+0.j, 1.1691699+0.j])
In [63]: np.real(evals).argsort()
Out[63]: array([3, 2, 1, 0])
```

Note that although we requested and obtained the four eigenvalues with the smallest magnitude in the previous example, those eigenvalues and vectors are not necessarily sorted within each other (although they are in this case). Obtaining sorted eigenvalues is often desirable, and this is easily achieved with a small but convenient wrapper function that sorts the eigenvalues using NumPy's argsort method. Let's give such a function, sp_eigs_sorted, which returns the eigenvalues and eigenvectors sorted by the real part of the eigenvalue.

```
In [64]: def sp_eigs_sorted(A, k=6, which='SR'):
    ...:     """ compute and return eigenvalues sorted by the real part """
    ...:     evals, evecs = sp.linalg.eigs(A, k=k, which=which)
    ...:     idx = np.real(evals).argsort()
    ...:     return evals[idx], evecs[idx]
In [65]: evals, evecs = sp_eigs_sorted(A, k=4, which='SM')
In [66]: evals
Out[66]: array([-1.16916997+0.j, -0.69027853+0.j, -0.3174929+0.j, -0.0810140+0.j])
```

As a less trivial example using sp.linalg.eigs and the wrapper function sp_eigs_sorted, consider the spectrum of lowest eigenvalues of the linear combination $(1 - x)M_1 + xM_2$ of random sparse matrices M_1 and M_2. We can use the sp.rand function to generate two random sparse matrices, and by repeatedly using sp_eigs_sorted to find the smallest 25 eigenvalues of the $(1 - x)M_1 + xM_2$ matrix for different values of x, we can build a matrix (evals_mat) that contains the eigenvalues as a function of x.

The following uses 50 values of x in the interval [0, 1].

```
In [67]: N = 100
In [68]: x_vec = np.linspace(0, 1, 50)
In [69]: M1 = sp.rand(N, N, density=0.2)
In [70]: M2 = sp.rand(N, N, density=0.2)
In [71]: evals_mat = np.array(
    ...:     [sp_eigs_sorted((1-x)*M1 + x*M2, k=25)[0] for x in x_vec])
```

Once the matrix evals_mat of eigenvalues as a function of x is computed, we can plot the eigenvalue spectrum. The result is shown in Figure 10-4, which is a complicated eigenvalue spectrum due to the randomness of the matrices M_1 and M_2.

```
In [72]: fig, ax = plt.subplots(figsize=(8, 4))
    ...: for idx in range(evals_mat.shape[1]):
    ...:     ax.plot(x_vec, np.real(evals_mat[:,idx]), lw=0.5)
    ...: ax.set_xlabel(r"$x$", fontsize=16)
    ...: ax.set_ylabel(r"eig.vals. of $(1-x)M_1+xM_2$", fontsize=16)
```

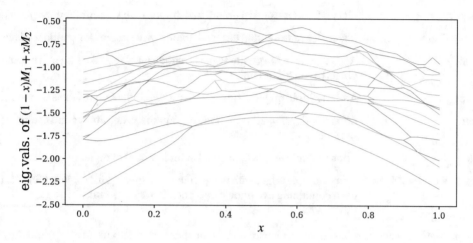

Figure 10-4. *The spectrum of the lowest 25 eigenvalues of the sparse matrix $(1-x)M_1+xM_2$, as a function of x, where M_1 and M_2 are two random matrices*

Graphs and Networks

Representing graphs as adjacency matrices is another important application of sparse matrices. In an adjacency matrix, an element describes which nodes in a graph are connected to each other. Consequently, the adjacency matrix is sparse if each node is only connected to a small set of other nodes. The csgraph module in the SciPy sparse module provides functions for processing such graphs, including methods for traversing a graph using different methods (e.g., breadth-first and depth-first traversals) and for computing shortest paths between nodes in a graph, and so on. For more information about this module, refer to its docstring: help(sp.csgraph).

For a more comprehensive framework for working with graphs, there is the NetworkX Python library. It provides utilities for creating and manipulating undirected and directed graphs and implements many graph algorithms, such as finding minimum paths between nodes in a graph. Here, assume that the NetworkX library is imported under the name nx. Using this library, we can, for example, create an undirected graph by initiating an object of the nx.Graph class. Any hashable Python object can be stored as nodes in a graph object, which makes it a very flexible data structure. However, in the following examples, only use graph objects with integers and strings as node labels. Table 10-2 provides a summary of functions for creating graphs and adding nodes and edges to graph objects.

Table 10-2. *Summary of Objects and Methods for Basic Graph Construction Using NetworkX*

Object/Method	Description
nx.Graph	A class for representing undirected graphs.
nx.DiGraph	A class for representing directed graphs.
nx.MultiGraph	A class for representing undirected graphs with support for multiple edges.
nx.MultiDiGraph	A class for representing directed graphs with support for multiple edges.
add_node	Adds a node to the graph. Expects a node label (e.g., a string or a hashable object) as an argument.
add_nodes_from	Adds multiple nodes. Expects a list (or iterable) of node labels as arguments.
add_edge	Adds an edge. Expects two node arguments as arguments and creates an edge between those nodes.
add_edges_from	Adds multiple edges. Expects a list (or iterable) of tuples of node labels.
add_weighted_edges_from	Adds multiple edges with weight factors. Expects a list (or iterable) of tuples each containing two node labels and the weight factor.

For example, we can create a simple graph with node data of integers using nx.Graph(), and the add_node method, or add_nodes_from, to add multiple nodes in one go. The nodes method returns an iterator object for the nodes, called a NodeView.

```
In [73]: g = nx.Graph()
In [74]: g.add_node(1)
In [75]: g.nodes()
Out[75]: NodeView((1,))
In [76]: g.add_nodes_from([3, 4, 5])
In [77]: g.nodes()
Out[77]: NodeView((1, 3, 4, 5))
```

To connect nodes, we can add edges using add_edge. Pass the labels of the two nodes we want to connect as arguments. To add multiple edges, we can use add_edges_from and pass to it a list of tuples of nodes to connect. The edges method returns an iterator object for the edges called EdgeView.

```
In [78]: g.add_edge(1, 2)
In [79]: g.edges()
Out[79]: EdgeView([(1, 2)])
In [80]: g.add_edges_from([(3, 4), (5, 6)])
In [81]: g.edges()
Out[81]: EdgeView([(1, 2), (3, 4), (5, 6)])
```

To represent edges between nodes with weights associated with them (e.g., a distance), we can use add_weighted_edges_from, to which we pass a list of tuples containing the weight factor for each edge in addition to the two nodes. When calling the edges method, we can additionally give argument data=True to indicate that the edge data should also be included in the resulting view.

```
In [82]: g.add_weighted_edges_from([(1, 3, 1.5), (3, 5, 2.5)])
In [83]: g.edges(data=True)
Out[83]: EdgeDataView([(1, 2, {}),
                        (1, 3, {'weight': 1.5}),
                        (3, 4, {}),
                        (3, 5, {'weight': 2.5}),
                        (5, 6, {})])
```

Note that if we add edges between nodes that do not yet exist in the graph, they are seamlessly added. For example, add a weighted edge between nodes 6 and 7 in the following code. Node 7 does not previously exist in the graph, but when adding an edge, it is automatically created and added.

```
In [84]: g.add_weighted_edges_from([(6, 7, 1.5)])
In [85]: g.nodes()
Out[85]: NodeView((1, 3, 4, 5, 2, 6, 7))
In [86]: g.edges()
Out[86]: EdgeView([(1, 2), (1, 3), (3, 4), (3, 5), (5, 6), (6, 7)])
```

With these fundamentals in place, we can look at a more complicated graph example. The following builds a graph from a dataset stored in a JSON file called tokyo-metro.json (available with the code listings), which we load using the Python standard library module json.[7]

```
In [87]: import json
In [88]: with open("tokyo-metro.json") as f:
    ...:        data = json.load(f)
```

The result of loading the JSON file is a dictionary data that contains metro line descriptions. Each line has a list of travel times between stations (travel_times), possible transfer points to other lines (transfer), and the line color.

```
In [89]: data.keys()
Out[89]: dict_keys(['C', 'T', 'N', 'F', 'Z', 'M', 'G', 'Y', 'H'])
In [90]: data["C"]
Out[90]: {'color': '#149848',
          'transfers': [['C3', 'F15'], ['C4', 'Z2'], ...],
          'travel_times': [['C1', 'C2', 2], ['C2', 'C3', 2], ...]}
```

Here the format of the travel_times list is [['C1', 'C2', 2], ['C2', 'C3', 2], ...], indicating that it takes 2 minutes to travel between the stations C1 and C2, and 2 minutes to travel between C2 and C3, and so on. The format of the transfers list is [['C3', 'F15'], ...], indicating that it is possible to transfer from the C line to the F line at station C3 to station F15. The travel_times and transfers are directly suitable for feeding to add_weighed_edges_from and add_edges_from, and we can easily create a graph for representing the metro network by iterating over each metro line dictionary and calling these methods.

```
In [91]: g = nx.Graph()
    ...: for line in data.values():
    ...:        g.add_weighted_edges_from(line["travel_times"])
    ...:        g.add_edges_from(line["transfers"])
```

---

[7] For more information about the JSON format and the json module, see Chapter 18.

The line transfer edges do not have edge weights, so let's first mark all transfer edges by adding a new Boolean attribute transfer to each edge.

```
In [92]: for n1, n2 in g.edges():
    ...:     g[n1][n2]["transfer"] = "weight" not in g[n1][n2]
```

Next, for plotting purposes, create two lists of edges containing transfer edges and on-train edges, and also create a list with colors corresponding to each node in the network.

```
In [93]: on_foot = [e for e in g.edges() if g.get_edge_data(*e)["transfer"]]
In [94]: on_train = [e for e in g.edges () if not g.get_edge_data(*e)["transfer"]]
In [95]: colors = [data[n[0].upper()]["color"] for n in g.nodes()]
```

To visualize the graph, we can use the Matplotlib-based drawing routines in the Networkx library: use nx.draw to draw each node, nx.draw_networkx_labels to draw the labels to the nodes, and nx.draw_network_edges to draw the edges. Call nx.draw_network_edges twice, with the edge lists for transfers (on_foot) and on-train (on_train) connections, and color the links blue and black, respectively, using the edge_color argument. The pos argument to the drawing functions determines the layout of the graph. Use the graphviz_layout function from networkx.drawing.nx_agraph to lay out the nodes. All drawing functions also accept a Matplotlib axes instance via the ax argument. The resulting graph is shown in Figure 10-5.

```
In [96]: fig, ax = plt.subplots(1, 1, figsize=(14, 10))
    ...: pos = nx.drawing.nx_agraph.graphviz_layout(g, prog="neato")
    ...: nx.draw(g, pos, ax=ax, node_size=200, node_color=colors)
    ...: nx.draw_networkx_labels(g, pos=pos, ax=ax, font_size=6)
    ...: nx.draw_networkx_edges(g, pos=pos, ax=ax, edgelist=on_train, width=2)
    ...: nx.draw_networkx_edges(
    ...:     g, pos=pos, ax=ax, edgelist=on_foot, edge_color="blue")
```

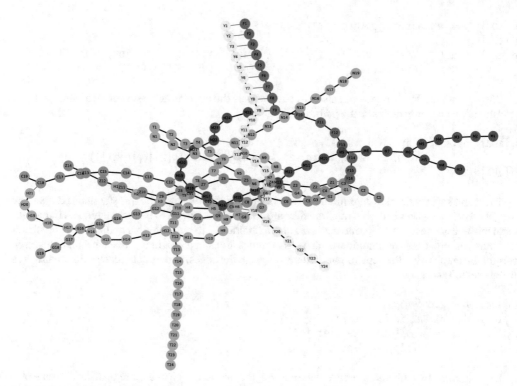

Figure 10-5. *Network graph for the Tokyo Metro stations*

Once the network has been constructed, we can use the many graph algorithms provided by the NetworkX library to analyze the network. For example, to compute the degree (i.e., the number of connections to a node) of each node, we can use the degree method (here, the output is truncated at ... to save space).

```
In [97]: g.degree()
Out[97]: DegreeView({'Y8': 3,  'N18': 2,  'M24': 2,  'G15': 3,  'C18': 3, ... })
```

For this graph, the degree of a node can be interpreted as the number of connections to a station: the more metro lines that connect at a station, the higher the degree of the corresponding node. We can easily search for the most highly connected station in the network using the degree method and the max function to find the highest degree. Next, iterate over the result of the degree method and select the nodes with the maximal degree (which is 6 in this network).

```
In [98]: d_max = max(d for (n, d) in g.degree())
In [99]: [(n, d) for (n, d) in g.degree() if d == d_max]
Out[99]: [('N7', 6), ('G5', 6), ('Y16', 6), ('M13', 6), ('Z4', 6)]
```

The result shows that the most highly connected stations are station numbers 7 on the N line, 5 on the G line, and so on. All these lines intercept at the same station (the Nagatachou station). We can also compute the closest path between two points in the network using nx.shortest_path. For example, the following finds the optimal traveling route (assuming no waiting time and instantaneous transfer) for traveling between Y24 and C19.

```
In [100]: p = nx.shortest_path(g, "Y24", "C19")
In [101]: p
Out[101]: ['Y24', 'Y23', 'Y22', 'Y21', 'Y20', 'Y19', 'Y18', 'C9', 'C10', 'C11',
          'C12', 'C13', 'C14', 'C15', 'C16', 'C17', 'C18', 'C19']
```

Given a path on this form, we can also directly evaluate the travel time by summing up the weight attributes of neighboring nodes in the path.

```
In [102]: np.sum([g[p[n]][p[n+1]]["weight"]
     ...:         for n in range(len(p)-1) if "weight" in g[p[n]][p[n+1]]])
Out[102]: 35
```

The result suggests it takes 35 minutes to travel from Y24 to C19. Since the transfer nodes do not have a weight associated with them, the train transfers are effectively assumed to be instantaneous. It may be reasonable to assume that a train transfer takes about 5 minutes. To consider this in the shortest path and travel time computation, we can update the transfer nodes and add a weight of 5 to each. To do this, create a copy of the graph using the copy method, iterate through the edges, and update those with the transfer attribute set to True.

```
In [103]: h = g.copy()
In [104]: for n1, n2 in h.edges():
     ...:     if h[n1][n2]["transfer"]:
     ...:         h[n1][n2]["weight"] = 5
```

Recomputing the path and the traveling time with the new graph gives a more realistic estimate of the traveling time.

```
In [105]: p = nx.shortest_path(h, "Y24", "C19")
In [106]: p
Out[106]: ['Y24', 'Y23', 'Y22', 'Y21', 'Y20', 'Y19', 'Y18', 'C9', 'C10', 'C11', 'C12',
'C13', 'C14', 'C15', 'C16', 'C17', 'C18', 'C19']
In [107]: np.sum([h[p[n]][p[n+1]]["weight"] for n in range(len(p)-1)])
Out[107]: 40
```

With this method, we can compute the optimal path and travel time between arbitrary nodes in the network. For example, we can compute the shortest path and traveling time between Z1 and H16 (32 minutes).

```
In [108]: p = nx.shortest_path(h, "Z1", "H16")
In [109]: np.sum([h[p[n]][p[n+1]]["weight"] for n in range(len(p)-1)])
Out[109]: 32
```

The NetworkX representation of a graph can be converted to an adjacency matrix in the form of a SciPy sparse matrix using the nx.to_scipy_sparse_array, after which we can also analyze the graph with the routines in the sp.csgraph module. As an example of this, let's convert the Tokyo Metro graph to an adjacency matrix and compute its reverse Cuthill-McKee ordering (using sp.csgraph.reverse_cuthill_mckee, which is a reordering that reduces the maximum distance of the matrix elements from the diagonal) and permute the matrix with this ordering. Plot the result of both matrices using Matplotlib's spy function, as shown in Figure 10-6.

```
In [110]: A = nx.to_scipy_sparse_array(g)
In [111]: A
Out[111]: <184x184 sparse matrix of type '<class 'numpy.int64'>'
                     with 486 stored elements in Compressed Sparse Row format>
In [112]: perm = sp.csgraph.reverse_cuthill_mckee(A)
In [113]: fig, (ax1, ax2) = plt.subplots(1, 2, figsize=(8, 4))
     ...: ax1.spy(A, markersize=2)
     ...: ax2.spy(sp_permute(A, perm, perm), markersize=2)
```

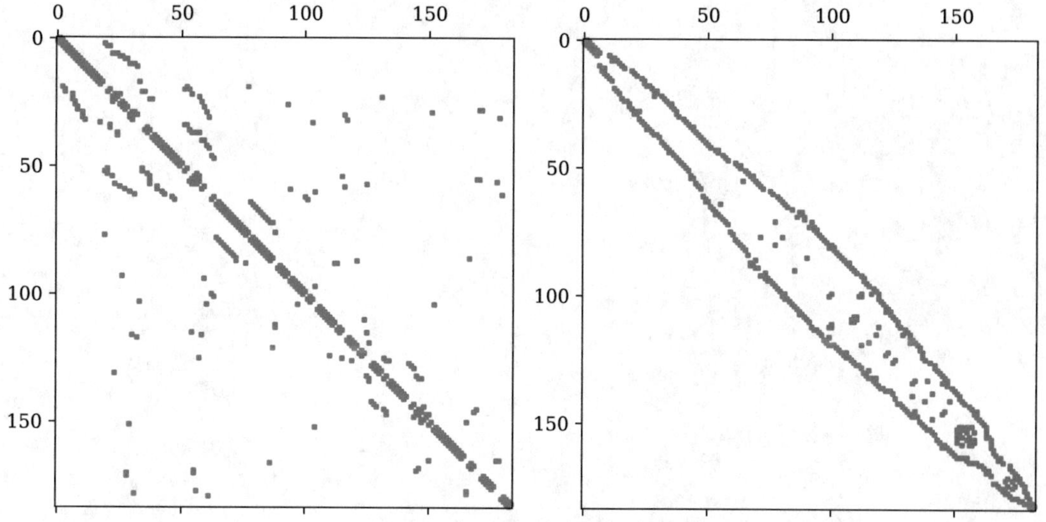

Figure 10-6. *The adjacency matrix of the Tokyo metro graph (left) and the same after RCM ordering (right)*

Summary

This chapter introduced common methods of storing sparse matrices and reviewed how these can be represented using the sparse matrix classes in the SciPy sparse module. It also reviewed the sparse matrix construction functions in the SciPy sparse module and the sparse linear algebra routines in sparse.linalg. To complement the linear algebra routines built into SciPy, the chapter briefly discussed the scikit. umfpack extension package, which makes the UMFPACK solver available to SciPy. The sparse matrix library in SciPy is versatile and very convenient to work with. It also offers good performance because it uses efficient low-level libraries for linear algebra routines (SuperLU or UMFPACK). For large-scale problems that require parallelization to distribute the workload to multiple cores or even multiple computers, the PETSc and Trilinos frameworks, which both have Python interfaces, provide routes for using sparse matrices and sparse linear algebra with Python in high-performance applications. Graph representations and processing using the SciPy sparse.csgraph and NetworkX libraries were also introduced.

Further Reading

A good and accessible introduction to sparse matrices and direct solvers for sparse linear equation systems is given in *Direct Methods for Sparse Linear Systems* by T. Davis (SIAM, 2006). A detailed discussion of sparse matrices and methods is provided in *Numerical Recipes in C: The Art of Scientific Computing* by W. H. Press (Cambridge University Press, 2007). For a thorough introduction to network and graph theory, see *Networks: An Introduction* by M. Newman (Oxford, 2010).

CHAPTER 11

■ ■ ■

Partial Differential Equations

Partial differential equations (PDEs) are multivariate differential equations where derivatives of more than one dependent variable occur. That is, the derivatives in the equation are *partial* derivatives. They are generalizations of ordinary differential equations covered in Chapter 9. Conceptually, the difference between ordinary and partial differential equations is small, but the computational techniques required to deal with ODEs and PDEs are very different, and solving PDEs is typically much more computationally demanding. Most techniques for solving PDEs numerically are based on discretizing the problem in each independent variable that occurs in the PDE, thereby recasting the problem into an algebraic form. This usually results in very large-scale linear algebra problems. Two common techniques for recasting PDEs into algebraic form are the finite-difference methods (FDMs), where the derivatives in the problem are approximated with their finite-difference formula, and the finite-element methods (FEMs), where the unknown function is written as a linear combination of simple basis functions that can be differentiated and integrated easily. The unknown function is described by a set of coefficients for the basis functions in this representation, and by a suitable rewriting of the PDEs, we can obtain algebraic equations for these coefficients.

With both FDMs and FEMs, the resulting algebraic equation system is usually very large, and in the matrix form, such equation systems are usually very sparse. Both FDM and FEM, therefore, heavily rely on sparse matrix representation for the algebraic linear equations, as discussed in Chapter 10. Most general-purpose frameworks for PDEs are based on FEM, or some variant thereof, as this method solves very general problems on complicated problem domains.

Solving PDE problems can be far more resource-demanding compared to other types of computational problems covered so far (e.g., compared to ODEs). It can be resource-demanding partly because of the number of points required to discretize a region of space scale exponentially with the number of dimensions. If a one-dimensional problem requires 100 points to describe, a two-dimensional problem with similar resolution requires $100^2 = 10^4$ points, and a three-dimensional problem requires $100^3 = 10^6$ points. Since each point in the discretized space corresponds to an unknown variable, it is easy to imagine that PDE problems can result in very large equation systems. Defining PDE problems programmatically can also be complicated. One reason is that the possible forms of a PDE vastly outnumber the more limited possible forms of ODEs. Another reason is geometry: while an interval in one-dimensional space is uniquely defined by two points, an area in two-dimensional problems and a volume in three-dimensional problems can have arbitrarily complicated geometries enclosed by curves and surfaces. Defining the problem domain of a PDE and its discretization in a mesh of coordinate points can, therefore, require advanced tools, and there is a large amount of freedom in how boundary conditions can be defined as well. In contrast to ODE problems, there is no standard form to define any PDE problem.

For these reasons, the PDE solvers for Python are only available through libraries and frameworks that are specifically dedicated to PDE problems. For Python, there are at least three significant libraries for solving PDE problems using the FEM method: the FiPy library, the SfePy library, and the FEniCS library. These libraries are extensive and feature-rich, and going into the details of using either is beyond this book's scope. Here, we can only briefly introduce PDE problems and survey prominent examples of PDE

© Robert Johansson 2024
R. Johansson, *Numerical Python*, https://doi.org/10.1007/979-8-8688-0413-7_11

libraries that can be used from Python. The chapter goes through a few examples that illustrate some of the features of one of these libraries (FEniCS). The hope is that this can give the reader interested in solving PDE problems with Python a bird's-eye overview of the available options and some useful pointers on where to look for further information.

Importing Modules

Basic numerical and plotting usage requires the NumPy and Matplotlib libraries. For 3D plotting, we must explicitly import the `mplot3d` module from the `mpl_toolkits` Matplotlib toolkit library. As usual, assume that these libraries are imported in the following manner.

```
In [1]: import numpy as np
In [2]: import matplotlib.pyplot as plt
In [3]: import matplotlib as mpl
In [4]: import mpl_toolkits.mplot3d
```

Let's also use the `linalg` and the `sparse` modules from SciPy, and to use the `linalg` submodule of the `sparse` module, we need to import it explicitly.

```
In [5]: import scipy.sparse as sp
In [6]: import scipy.sparse.linalg
In [7]: import scipy.linalg as la
```

With these imports, we can access the dense linear algebra module as `la`, while the sparse linear algebra module is accessed as `sp.linalg`. Furthermore, later in this chapter, we use the FEniCS FEM framework, which requires that its `dolfin` and `mshr` libraries be imported in the following manner.

```
In [8]: import dolfin
In [9]: import mshr
```

Partial Differential Equations

The unknown quantity in a PDE is a multivariate function, here denoted as u. In an N-dimensional problem, the u function depends on n-independent variables: $u(x_1, x_2, ..., x_n)$. A general PDE can formally be written as

$$F\left(x_1, x_2, ..., x_n, u, \left\{\frac{\partial u}{\partial x_{i_1}}\right\}_{1 \le i_1 \le n}, \left\{\frac{\partial^2 u}{\partial x_{i_1} \partial x_{i_2}}\right\}_{1 \le i_1, i_2 \le n}, ...\right) = 0, x \in \Omega$$

where $\left\{\dfrac{\partial u}{\partial x_{i_1}}\right\}_{1 \le i_1 \le n}$ denotes all first-order derivatives with respect to the independent variables

$x_1, ..., x_n, \left\{\dfrac{\partial^2 u}{\partial x_{i_1} \partial x_{i_2}}\right\}_{1 \le i_1, i_2 \le n}$ denotes all second-order derivatives, and so on. Here F is a known function that

describes the form of the PDE, and Ω is the domain of the PDE problem. Many PDEs that occur in practice only contain up to second-order derivatives, and we typically deal with problems in two or three spatial dimensions and possibly time. When working with PDEs, it is common to simplify the notation by denoting the partial derivatives with respect to an independent variable x using the subscript notation: $u_x = \dfrac{\partial u}{\partial x}$.

Higher-order derivatives are denoted with multiple indices: $u_{xx} = \dfrac{\partial^2 u}{\partial x^2}$, $u_{xy} = \dfrac{\partial^2 u}{\partial x \partial y}$, and so on. An example of a typical PDE is the heat equation, which in a two-dimensional Cartesian coordinate system takes the form $u_t = \alpha(u_{xx} + u_{yy})$. Here, the $u = u(t, x, y)$ function describes the temperature at the spatial point (x, y) at time t, and α is the thermal diffusivity coefficient.

To fully specify a particular solution to a PDE, we need to define its boundary conditions, which are known values of the function or a combination of its derivatives along the boundary of the problem domain Ω, as well as the initial values if the problem is time-dependent. The boundary is often denoted as Γ or $\partial \Omega$, and different boundary conditions can generally be given for different parts of the boundary. Two important types of boundary conditions are Dirichlet boundary conditions, which specify the value of the function at the boundary, $u(x) = h(x)$ for $x \in \Gamma_D$, and Neumann boundary conditions, which specify the normal derivative on the boundary, $\dfrac{\partial u(x)}{\partial n} = g(x)$ for $x \in \Gamma_N$, where n is the outward normal from the boundary. Here $h(x)$ and $g(x)$ are arbitrary functions.

Finite-Difference Methods

The basic idea of the finite-difference method is to approximate the derivatives that occur in a PDE with their finite-difference formulas on a discretized space. For example, the finite-difference formula for the ordinary derivative $\dfrac{du(x)}{dx}$ on a discretization of the continuous variable x into discrete points $\{x_n\}$ can be approximated with the forward difference formula $\dfrac{du(x_n)}{dx} \approx \dfrac{u(x_{n+1}) - u(x_n)}{x_{n+1} - x_n}$, the backward difference formula $\dfrac{du(x_n)}{dx} \approx \dfrac{u(x_n) - u(x_{n-1})}{x_n - x_{n-1}}$, or the centered difference formula $\dfrac{du(x_n)}{dx} \approx \dfrac{u(x_{n+1}) - u(x_{n-1})}{x_{n+1} - x_{n-1}}$. Similarly, we can construct finite-difference formulas for higher-order derivatives, such as the second-order derivative $\dfrac{d^2 u(x_n)}{dx^2} \approx \dfrac{u(x_{n+1}) - 2u(x_n) + u(x_{n-1})}{(x_n - x_{n-1})^2}$. Assuming that the discretization of the continuous variable x into discrete points is fine enough, these finite-difference formulas can give good approximations of the derivatives. Replacing derivatives in an ODE or PDE with their finite-difference formulas recasts the equations from differential equations to algebraic equations. If the original ODE or PDE is linear, the algebraic equations are also linear and can be solved with standard linear algebra methods.

To make this method more concrete, consider the ODE problem $u_{xx} = -5$ in the interval $x \in [0, 1]$ and with boundary conditions $u(x = 0) = 1$ and $u(x = 1) = 2$, which, for example, arises from the steady-state heat equation in one dimension. In contrast to the ODE initial value problem considered in Chapter 9, this is a boundary value problem because the value of u is specified at $x = 0$ and $x = 1$. The methods for initial value problems are, therefore, not applicable here. Instead, we can treat this problem by dividing the interval $[0, 1]$ into discrete points x_n, and the problem is then to find the $u(x_n) = u_n$ function at these points. Writing the ODE problem in finite-difference form gives an equation $(u_{n-1} - 2u_n + u_{n+1})/\Delta x^2 = -5$ for every interior point n and the boundary conditions $u_0 = 1$ and $u_{N+1} = 2$. Here, the interval $[0, 1]$ is discretized into $N + 2$ evenly spaced points, including the boundary points, with separation $\Delta x = 1/(N + 1)$. Since the function is known at the two boundary points, there are N unknown variables u_n corresponding to the function values at the interior points. The set of equations for the interior points can be written in a matrix form as $Au = b$, where $u = [u_1, ..., u_N]^T$, $b = \left[-5 - \dfrac{u_0}{\Delta x^2}, -5, ..., -5, -5 - \dfrac{u_{N+1}}{\Delta x^2} \right]^T$, and

259

$$A = \frac{1}{\Delta x^2} \begin{bmatrix} -2 & 1 & 0 & 0 & \cdots \\ 1 & -2 & 1 & 0 & \cdots \\ 0 & 1 & -2 & 1 & 0 \\ 0 & 0 & 1 & -2 & \ddots \\ \vdots & \vdots & 0 & \ddots & \ddots \end{bmatrix}.$$

Here, the matrix A describes the coupling of the equations for u_n to values at neighboring points due to the finite-difference formula used to approximate the second-order derivative in the ODE. The boundary values are included in the b vector, which also contains the constant right-hand side of the original ODE (the source term). At this point, we can straightforwardly solve the linear equation system $Au = b$ for the unknown vector of u and thereby obtain the approximate values of the $u(x)$ function at the discrete points $\{x_n\}$.

In Python code, we can set up and solve this problem. First, define variables for the number of interior points N, the values of the function at the boundaries u0 and u1, and the spacing between neighboring points dx.

```
In [10]: N = 5
In [11]: u0, u1 = 1, 2
In [12]: dx = 1.0 / (N + 1)
```

Next, construct the matrix A as described in the preceding section. For this, we can use the eye function from NumPy, which creates a two-dimensional array with ones on the diagonal or the upper or lower diagonal that is shifted from the main diagonal by the number given by the k argument.

```
In [13]: A = (np.eye(N, k=-1) - 2 * np.eye(N) + np.eye(N, k=1)) / dx**2
In [14]: A
Out[14]: array([[-72.,  36.,   0.,   0.,   0.],
               [ 36., -72.,  36.,   0.,   0.],
               [  0.,  36., -72.,  36.,   0.],
               [  0.,   0.,  36., -72.,  36.],
               [  0.,   0.,   0.,  36., -72.]])
```

Next, we need to define an array for the vector b, which corresponds to the source term –5 in the differential equation and the boundary condition. The boundary conditions enter the equations via the finite-difference expressions for the derivatives of the first and the last equation (for u_1 and u_N). But these terms are missing from the expression represented by the matrix A and must, therefore, be added to the vector b.

```
In [15]: b = -5 * np.ones(N)
    ...: b[0] -= u0 / dx**2
    ...: b[N-1] -= u1 / dx**2
```

Once the matrix A and the vector b are defined, we can solve the equation system using the linear equation solver from SciPy (we could also use the one provided by NumPy, np.linalg.solve).

```
In [16]: u = la.solve(A, b)
```

This completes the solution of this ODE problem. To visualize the solution, here we first create an array x that contains the discrete coordinate points for which we have solved the problem, including the boundary points, and we also create an array U that combines the boundary values and the interior points in one array. The result is then plotted and shown in Figure 11-1.

```
In [17]: x = np.linspace(0, 1, N+2)
In [18]: U = np.hstack([[u0], u, [u1]])
In [19]: fig, ax = plt.subplots(figsize=(8, 4))
    ...: ax.plot(x, U)
    ...: ax.plot(x[1:-1], u, 'ks')
    ...: ax.set_xlim(0, 1)
    ...: ax.set_xlabel(r"$x$", fontsize=18)
    ...: ax.set_ylabel(r"$u(x)$", fontsize=18)
```

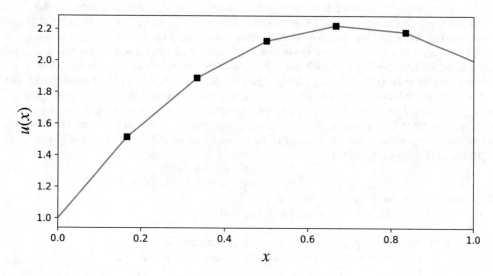

Figure 11-1. *The solution to the second-order ODE boundary value problem introduced in the text*

The finite-difference method can easily be extended to higher dimensions using the finite-difference formula along each discretized coordinate. For a two-dimensional problem, there is a two-dimensional array u for the unknown interior function values, and when using the finite differential formula, obtain a system of coupled equations for the elements in u. We can rearrange the u array into a vector and assemble the corresponding matrix A from the finite-difference equations to write these equations in the standard matrix-vector form.

As an example, consider the following two-dimensional generalization of the previous problem: $u_{xx} + u_{yy} = 0$, with the boundary conditions $u(x = 0) = 3$, $u(x = 1) = -1$, $u(y = 0) = -5$, and $u(y = 1) = 5$. There is no source term here, but the boundary conditions in a two-dimensional problem are more complicated than in the onedimensional problem solved earlier. In finite-difference form, we can write the PDE as $(u_{m-1,n} - 2u_{m,n} + u_{m+1,n})/x^2 + (u_{m,n-1} - 2u_{m,n} + u_{m,n+1})/y^2 = 0$. If we divide the x and y intervals into N interior points $(N + 2$ points including the boundary points), then $\Delta x = \Delta y = \dfrac{1}{N+1}$, and u is an $N \times N$ matrix. To write the equation in the standard form $Av = b$, we can rearrange the matrix u by stacking its rows or columns into a vector of size $N^2 \times 1$. The matrix A is then of size $N^2 \times N^2$, which can be very big if we need to use a fine discretization of the x and y coordinates. For example, using 100 points along both x and y gives an equation system that has 10^4 unknown values u_{mn}, and the matrix A has $100^4 = 10^8$ elements. Fortunately, since the finite-difference formula only couples neighboring points, the matrix A turns out to be very sparse, and here, we can benefit significantly from working with sparse matrices, as shown in the following.

To solve this PDE problem with Python and the finite-element method, start by defining variables for the number of interior points and the values along the four boundaries of the unit square.

```
In [20]: N = 100
In [21]: u0_t, u0_b = 5, -5
In [22]: u0_l, u0_r = 3, -1
In [23]: dx = 1. / (N+1)
```

This also computed the separation dx between the uniformly spaced coordinate points in the discretization of x and y (assumed equal). Because the finite-difference formula couples both neighboring rows and columns, it is slightly more involved to construct the matrix A for this example. However, a relatively direct approach is first to define the matrix A_1d corresponding to the one-dimensional formula along one of the coordinates (say x or the index m in $u_{m,n}$). To distribute this formula along each row, we can take the tensor product of the identity matrix of size $N \times N$ with the A_1d matrix. The result describes all derivatives along the m-index for all indices n. To cover the terms that couple the equation for $u_{m,n}$ to $u_{m,n+1}$ and $u_{m,n-1}$, that is, the derivatives along the index n, we need to add diagonals separated from the main diagonal by N positions. The following steps construct A using the eye and kron functions from the scipy.sparse module. The result is a sparse matrix A that describes the finite-difference equation system for the two-dimensional PDE considered here.

```
In [24]: A_1d = (sp.eye(N, k=-1) + sp.eye(N, k=1) - 4 * sp.eye(N))/dx**2
In [25]: A = sp.kron(sp.eye(N), A_1d) + \
    ...:     (sp.eye(N**2, k=-N) + sp.eye(N**2, k=N))/dx**2
In [26]: A
Out[26]: <10000x10000 sparse matrix of type '<type 'numpy.float64'>'
            with 49600 stored elements in Compressed Sparse Row format>
```

The printout of A shows that it is a sparse matrix with 10^8 elements with 49,600 nonzero elements, so only 1 out of about 2000 elements is nonzero, and A is very sparse. To construct the vector b from the boundary conditions, it is convenient to create an $N \times N$ array of zeros and assign the boundary condition to edge elements of this array (which are the corresponding elements in u that are coupled to the boundaries—i.e., the interior points that are neighbors to the boundary). Once this $N \times N$ array is created and assigned, we can use the reshape method to rearrange it into an $N^2 \times 1$ vector that can be used in the $Av = b$ equation.

```
In [27]: b = np.zeros((N, N))
    ...: b[0, :] += u0_b    # bottom
    ...: b[-1, :] += u0_t   # top
    ...: b[:, 0] += u0_l    # left
    ...: b[:, -1] += u0_r   # right
    ...: b = - b.reshape(N**2) / dx**2
```

When the A and b arrays are created, we can proceed to solve the equation system for the vector v and use the reshape method to arrange it back into the $N \times N$ matrix u.

```
In [28]: v = sp.linalg.spsolve(A, b)
In [29]: u = v.reshape(N, N)
```

For plotting purposes, create a matrix U that combines the u matrix with the boundary conditions. With the coordinate matrices X and Y, plot a colormap graph and a 3D surface view of the solution. The result is shown in Figure 11-2.

```
In [30]: U = np.vstack([np.ones((1, N+2)) * u0_b,
   ...:                  np.hstack([np.ones((N, 1)) * u0_l, u,
   ...:                             np.ones((N, 1)) * u0_r]),
   ...:                  np.ones((1, N+2)) * u0_t])
In [31]: x = np.linspace(0, 1, N+2)
In [32]: X, Y = np.meshgrid(x, x)
In [33]: fig = plt.figure(figsize=(12, 5.5))
   ...: cmap = mpl.cm.get_cmap('RdBu_r')
   ...:
   ...: ax = fig.add_subplot(1, 2, 1)
   ...: c = ax.pcolor(X, Y, U, vmin=-5, vmax=5, cmap=cmap)
   ...: ax.set_xlabel(r"$x_1$", fontsize=18)
   ...: ax.set_ylabel(r"$x_2$", fontsize=18)
   ...:
   ...: ax = fig.add_subplot(1, 2, 2, projection='3d')
   ...: p = ax.plot_surface(X, Y, U, vmin=-5, vmax=5, rstride=3, cstride=3,
   ...:                     linewidth=0, cmap=cmap)
   ...: ax.set_xlabel(r"$x_1$", fontsize=18)
   ...: ax.set_ylabel(r"$x_2$", fontsize=18)
   ...: cb = plt.colorbar(p, ax=ax, shrink=0.75)
   ...: cb.set_label(r"$u(x_1, x_2)$", fontsize=18)
```

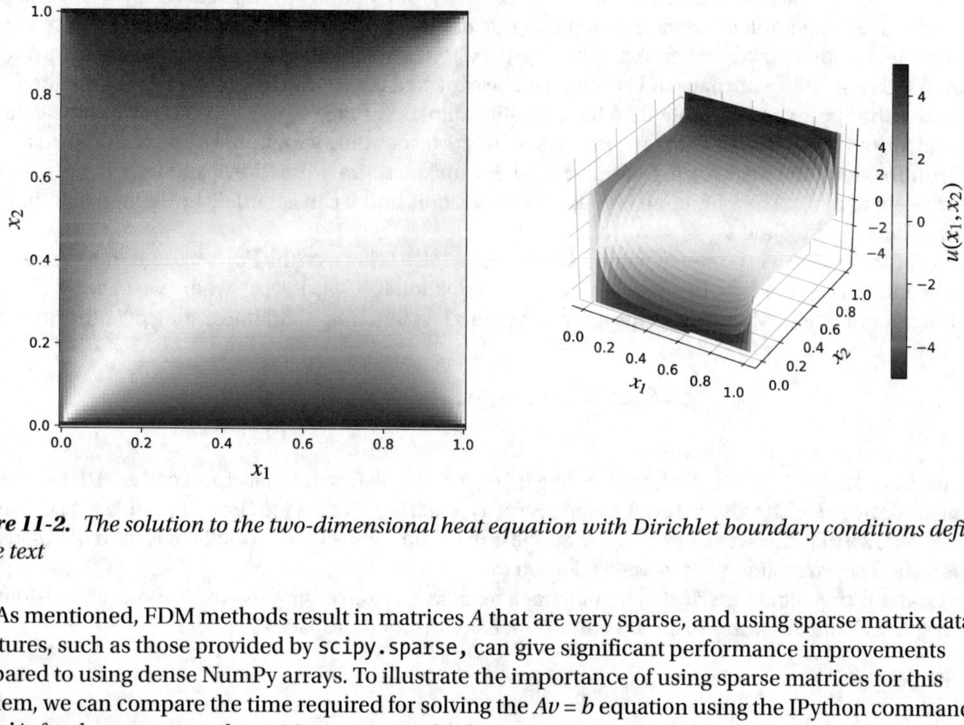

Figure 11-2. *The solution to the two-dimensional heat equation with Dirichlet boundary conditions defined in the text*

As mentioned, FDM methods result in matrices A that are very sparse, and using sparse matrix data structures, such as those provided by `scipy.sparse`, can give significant performance improvements compared to using dense NumPy arrays. To illustrate the importance of using sparse matrices for this problem, we can compare the time required for solving the $Av = b$ equation using the IPython command `%timeit`, for the two cases where A is a sparse and a dense matrix.

```
In [34]: A_dense = A.todense()
In [35]: %timeit la.solve(A_dense, b)
1 loops, best of 3: 10.8 s per loop
In [36]: %timeit sp.linalg.spsolve(A, b)
10 loops, best of 3: 31.9 ms per loop
```

These results show that using sparse matrices for the present problem results in a speedup of several orders of magnitude (this case has a speedup of a factor $10.8/0.0319 \approx 340$).

The finite-difference method used in the last two examples is a powerful and relatively simple method for solving ODE boundary value problems and PDE problems with simple geometries. However, it is not easily adapted to problems on more complicated domains or problems on nonuniform coordinate grids. For such problems, finite-element methods are typically more flexible and convenient. Although FEMs are conceptually more complicated than FDMs, they can be computationally efficient and adapt well to complex problem domains and more involved boundary conditions.

Finite-Element Methods

The finite-element method is a powerful and universal way to convert PDEs into algebraic equations. The basic idea of this method is to represent the domain on which the PDE is defined with a finite set of discrete regions, or *elements,* and to approximate the unknown function as a linear combination of basis functions with local support on each of these elements (or on a small group of neighboring elements). Mathematically, this approximation solution, u_h, represents a projection of the exact solution u in the function space V (e.g., continuous real-valued functions) onto a finite subspace $V_h \subset V$ related to the discretization of the problem domain. If V_h is a suitable subspace of V, then it can be expected that u_h can be a good approximation to u.

To solve the approximate problem on the simplified function space V_h, we can first rewrite the PDE from its original formulation, known as the *strong form,* to its corresponding variational form, also known as the *weak form.* To obtain the weak form, multiply the PDE with an arbitrary function v and integrate it over the entire problem domain. The v function is called a *test function,* and it can generally be defined on function space that differs from V and V_h.

For example, consider the steady-state heat equation (also known as the Poisson equation) solved using the FDM earlier in this chapter: the strong form of this equation is $-\Delta u(\boldsymbol{x}) = f(\boldsymbol{x})$, where we have used the vector operator notation. Multiplying this equation with a test function v and integrating over the domain $\boldsymbol{x} \in \Omega$ obtains the weak form:

$$-\int_\Omega \Delta u \ v \ \mathrm{d}x = \int_\Omega fv \ \mathrm{d}x.$$

Since the exact solution u satisfies the strong form, it also satisfies the weak form of the PDE for any reasonable choice of v. The reverse does not necessarily hold true. But if a function, u_h, called a *trial function* in this context, satisfies the weak form for a large class of suitably chosen test functions v, then it is plausible that it is a good approximation to the exact solution u.

To treat this problem numerically, we first need to make the transition from the infinite-dimensional function spaces V and $\hat{V}$ to approximate finite-dimensional function spaces V_h and $\hat{V}_h$:

$$-\int_\Omega \Delta u_h v_h \ \mathrm{d}x = \int_\Omega fv_h \ \mathrm{d}x,$$

where $u_h \in V_h$ and $v_h \in \hat{V}_h$. The key point here is that V_h and $\hat{V}_h$ are finite-dimensional, so we can use a finite set of basis functions $\{\phi_i\}$ and $\{\hat{\phi}_i\}$ that spans the function spaces V_h and $\hat{V}_h$, respectively, to describe the functions u_h and v_h. We can express u_h as a linear combination of the basis functions that span its function space, $u_h = \sum U_i \phi_i$. Inserting this linear combination in the weak form of the PDE and carrying out the integrals and differential operators on the basis functions instead of directly over terms in the PDE yields a set of algebraic equations.

To obtain an equation system on the simple form $AU = b$, we also must write the weak form of the PDE in bilinear form with respect to the u_h and v_h functions $a(u_h, v_h) = L(v_h)$, for some functions a and L. This is not always possible, but for the current example of the Poisson equation, we can obtain this form by integrating by parts:

$$-\int_\Omega \Delta u_h v_h \ dx = \int_\Omega \nabla u_h . \nabla v_h \ dx - \int_\Omega \nabla . (\nabla u_h v_h) dx = \int_\Omega \nabla u_h . \nabla v_h \ dx - \int_{\partial\Omega} (\nabla u_h . n) v_h \ d\Gamma.$$

where in the second equality, we have also applied Gauss' theorem to convert the second term to an integral over the boundary $\partial\Omega$ of the domain Ω. Here $\boldsymbol{n}$ is the outward normal vector of the boundary $\partial\Omega$. There is no general method for rewriting a PDE from the strong form to the weak form, and each problem must be approached on a case-by-case basis. However, the technique used here, to integrate by part and rewrite the resulting integrals using integral identities, can be used for many frequently occurring PDEs.

We must also deal with the boundary term in the preceding weak-form equation to reach the bilinear form that can be approached with standard linear algebra methods. To this end, assume that the problem satisfies the Dirichlet boundary condition on the part of $\partial\Omega$ denoted Γ_D and Neumann boundary conditions on the remaining part of $\partial\Omega$, denoted Γ_N: $\{u = h, x \in \Gamma_D\}$ and $\{\nabla u \cdot \boldsymbol{n} = g, x \in \Gamma_N\}$. Not all boundary conditions are of Dirichlet or Neumann type, but these cover many physically motivated situations.

Since we can choose the test functions v_h, we can let v_h vanish on the part of the boundary that satisfies Dirichlet boundary conditions. In this case, obtain the following weak form of the PDE problem.

$$\int_\Omega \nabla u_h . \nabla v_h \ dx = \int_\Omega f v_h \ dx + \int_{\Gamma_N} g \ v_h \ d\Gamma$$

If we substitute the u_k function for its expression as a linear combination of basis functions and substitute the test function with one of its basis functions, we obtain an algebraic equation.

$$\sum_j U_j \int_\Omega \nabla \phi_j . \nabla \hat{\phi}_i \ dx = \int_\Omega f \hat{\phi}_i \ dx + \int_{\Gamma_N} g \hat{\phi}_i \ d\Gamma$$

If there are N basis functions in V_k, then there are N unknown coefficients U_j, and we need N-independent test functions $\hat{\phi}_i$ to obtain a closed equation system. This equation system is on the form $AU = b$ with $A_{ij} = \int_\Omega \nabla \phi_j . \nabla \hat{\phi}_i \ dx$ and $b_i = \int_\Omega f \hat{\phi}_i \ dx + \int_{\Gamma_N} g \hat{\phi}_i \ d\Gamma$. Following this procedure we have converted the PDE problem into a system of linear equations that can be readily solved using techniques discussed in previous chapters.

In practice, a very large number of basis functions can be required to obtain a good approximation to the exact solution, and the linear equation system generated by FEMs is, therefore, often very large. However, the fact that each basis function has support only at one or a few nearby elements in the discretization of the problem domain ensures that the matrix A is sparse, which makes it tractable to solve rather large-scale FEM problems. Note that an important property of the basis functions ϕ_i and $\hat{\phi}_i$ is that it should be easy to compute the derivatives and integrals of the expression that occurs in the final weak form of the problem so that the matrix A and vector b can be assembled efficiently. Typical examples of basis functions are low-order polynomial functions that are nonzero only within a single element. Figure 11-3 is a one-dimensional illustration of this type of basis function, where the interval $[0, 6]$ is discretized using five interior points. A continuous function (black solid curve) is approximated as a piecewise linear function (dashed red line) by suitably weighted triangular basic functions (blue solid lines).

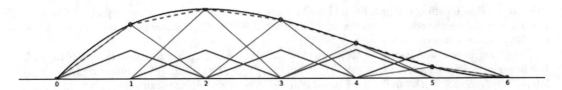

Figure 11-3. *An example of possible basis functions (blue), with local support, for the one-dimensional domain [0, 6]*

When using FEM software for solving PDE problems, it is typically required to convert the PDE to weak form by hand and, if possible, rewrite it on the bilinear form $a(u, v) = L(v)$. It is also necessary to provide a suitable discretization of the problem domain. This discretization is called a mesh, and it is usually made up of triangular partitioning (or their higher-order generalizations) of the total domain. Meshing an intricate problem domain can be a complicated process, and it may require using sophisticated software, especially dedicated to mesh generation. There are tools for programmatically generating meshes for simple geometries, which we see examples of in the following section.

Once a mesh is generated and the PDE problem is written in a suitable weak form, we can feed the problem into a FEM framework, which then automatically assembles the algebraic equation system and applies suitable sparse equation solvers to find the solution. In this process, we often have a choice of what type of basis functions to use and which type of solver to use. Once the algebraic equation is solved, we can construct the approximation solution to the PDE with the help of the basis functions, and we can, for example, visualize the solution or post-process it in some other fashion.

Solving a PDE using FEM typically involves the following steps.

1. Generate a mesh for the problem domain.

2. Write the PDE in weak form.

3. Program the problem in the FEM framework.

4. Solve the resulting algebraic equations.

5. Post-process and/or visualize the solution.

The following section reviews available FEM frameworks that can be used with Python and then look at several examples that illustrate some of the key steps in the PDE solution process using FEM.

Survey of FEM Libraries

For Python, there are at least three significant FEM packages: FiPy, SfePy, and FEniCS. These are all rather full-featured frameworks capable of solving a wide range of PDE problems. Technically, the FiPy library is not FEM software but rather a finite-volume method (FVM) software, but the gist of this method is quite similar to FEM. The FiPy framework can be obtained from www.ctcms.nist.gov/fipy. The SfePy library is a FEM software that takes a slightly different approach to defining PDE problems in that it uses Python files as configuration files for its FEM solver, rather than programmatically setting up a FEM problem (although this mode of operation is technically also supported in SfePy). The SfePy library is available from http://sfepy.org. The third major framework for FEM with Python is FEniCS, which is written for C++ and Python. The FEniCS framework is my personal favorite when it comes to FEM software for Python, as it provides an elegant Python interface to a powerful FEM engine. Like FDM problems, FEM problems typically result in very large-scale equation systems that require using sparse matrix techniques to solve efficiently. A crucial part of a FEM framework is, therefore, to efficiently solve large-scale linear and nonlinear systems, using sparse matrix representation and direct or iterative solvers that work on sparse systems, possibly using parallelization. Each framework mentioned in the preceding section supports multiple backends for such low-level computations. For example, many FEM frameworks can use the PETSc and Trilinos frameworks.

Although I'm not able to comprehensively explain how to use these FEM frameworks here, the following section looks at solving example problems with FEniCS and thereby introduce some of its basic features and usage. The hope is that the examples can give a flavor of how it is to work with FEM problems in Python and provide a starting point for readers interested in learning more about FEM with Python.

Solving PDEs Using FEniCS

This section solves a series of increasingly complicated PDEs using the FEniCS framework and introduces the workflow and a few of the main features of this FEM software. For a thorough introduction to the FEniCS framework, see the documentation on the project websites and the official FEniCS book (Anders Logg, 2012).

■ **FEniCS** FEniCS is a highly capable FEM framework that is made up of a collection of libraries and tools for solving PDE problems. Much of FEniCS is programmed in C++ but provides an official Python interface. Because of the complexity of the many dependencies of the FEniCS libraries to external low-level numerical libraries, FEniCS is usually packaged and installed as an independent environment. However, it can also be installed using conda on some platforms. For more information about the FEniCS, see `http://fenicsproject.org`. At the time of writing, the most recent version is 2019.1.0. A new version called FEniCSx is also under development, but let's use the stable original FEniCS here.

The Python interface to FEniCS is provided by a library named `dolfin`. For mesh generation, let's use the `mshr` library. The following code assumes that these libraries are entirely imported, as shown at the beginning of this chapter. For a summary of these libraries' most important functions and classes, see Table 11-1 and Table 11-2.

Table 11-1. *Summary of Selected Functions and Classes in the dolfin Library*

Function/Class	Description	Example
parameters	Dictionary holding configuration parameters for the FEniCS framework	`dolfin.parameters["reorder_dofs_serial"]`
RectangleMesh	Object for generating a rectangular 2D mesh	`mesh = dolfin.RectangularMesh(dolfin.Point(0, 0),dolfin.Point(1, 1), 10, 10)`
MeshFunction	Function defined over a given mesh	`dolfin.MeshFunction("size_t", mesh, mesh.topology().dim()-1)`
FunctionSpace	Object for representing a function space	`V = dolfin.FunctionSpace(mesh, 'Lagrange', 1)`
TrialFunction	Object for representing a trial function defined in a given function space	`u = dolfin.TrialFunction(V)`

(continued)

Table 11-1. (*continued*)

Function/Class	Description	Example
TestFunction	Object for representing a test function defined in a given function space	`v = dolfin.TestFunction(V)`
Function	Object for representing unknown functions appearing in the weak form of a PDE	`u_sol = dolfin.Function(V)`
Constant	Object for representing a fixed constant	`c = dolfin.Constant(1.0)`
Expression	Representation of a mathematical expression in terms of the spatial coordinates	`dolfin.Expression("x[0]*x[0] + x[1]*x[1]")`
DirichletBC	Object for representing Dirichlet-type boundary conditions	`dolfin.DirichletBC(V, u0, u0_boundary)`
Equation	Object for representing an equation, for example, generated by using the == operator with other FEniCS objects	`a == L`
inner	Symbolic representation of the inner product	`dolfin.inner(u, v)`
nabla_grad	Symbolic representation of the gradient operator	`dolfin.nabla_grad(u)`
dx	Symbolic representation of the volume measure for integration	`f*v*dx`
ds	Symbolic representation of a line measure for integration	`g_v * v * dolfin.ds(0, domain=mesh, subdomain_data=boundary_parts)`
assemble	Assemble the algebraic equations by carrying out the integrations over the basis functions.	`A = dolfin.assemble(a)`
solve	Solve an algebraic equation.	`dolfin.solve(A, u_sol.vector(), b)`
plot	Plot a function or expression.	`dolfin.plot(u_sol)`
File	Write a function to a file that can be opened with visualization software such as ParaView.	`dolfin.File('u_sol.pvd') << u_sol`
refine	Refine a mesh by splitting a selection of the existing mesh elements into smaller pieces.	`mesh = dolfin.refine(mesh, cell_markers)`
AutoSubDomain	Representation of a subset of a domain, selected from all elements by the indicator function passed to it as argument.	`dolfin.AutoSubDomain(v_boundary_func)`

Table 11-2. *Summary of Selected Functions and Classes in the mshr and dolfin Library*

Function/Class	Description
dolfin.Point	Representation of a coordinate point
mshr.Circle	Representation of a geometrical object with the shape of a circle, which can be used to compose 2D domain
mshr.Ellipse	Representation of a geometrical object with the shape of an ellipse
mshr.Rectangle	Representation of a domain defined by a rectangle in 2D
mshr.Box	Representation of a domain defined by a box in 3D
mshr.Sphere	Representation of a domain defined by a sphere in 3D
mshr.generate_ mesh	Generates a mesh from a domain composed of geometrical objects, such as those listed in the preceding section

Before using FEniCS and the dolfin Python library, we must set two configuration parameters via the dolfin.parameters dictionary to obtain the behavior that we need in the following examples.

```
In [37]: dolfin.parameters["reorder_dofs_serial"] = False
In [38]: dolfin.parameters["allow_extrapolation"] = True
```

Let's begin with FEniCS by reconsidering the steady-state heat equation in two dimensions solved earlier in this chapter using the FDM. Here, consider the problem $u_{xx} + u_{yy} = f$, where f is a source function. To begin, let's assume that the boundary conditions are $u(x = 0, y) = u(x = 1, y) = 0$ and $u(x, y = 0) = u(x, y = 1) = 0$. Later examples demonstrate how to define Dirichlet and Neumann boundary conditions.

The first step in the solution of a PDE with FEM is to define a mesh that describes the discretization of the problem domain. In the current example, the problem domain is the unit square $x, y \in [0, 1]$. For simple geometries like this, there are functions in the dolfin library for generating the mesh. Here, we use the RectangleMesh function, which, as the first two arguments, takes the coordinate points (x_0, y_0) and (x_1, y_1), represented as dolfin.Point instances, where (x_0, y_0) is the coordinates of the lower-left corner of the rectangle and (x_1, y_1) of the upper-right corner. The fifth and sixth arguments are the numbers of elements along the x and y directions, respectively. The resulting mesh object is viewed in a Jupyter Notebook via its rich display system (here, we generate a less fine mesh to display the mesh structure), as shown in Figure 11-4.

```
In [39]: N1 = N2 = 75
In [40]: mesh = dolfin.RectangleMesh(
    ...:     dolfin.Point(0, 0), dolfin.Point (1, 1), N1, N2)
In [41]: dolfin.RectangleMesh(
    ...:     dolfin.Point(0, 0), dolfin.Point(1, 1), 10, 10) # for display
```

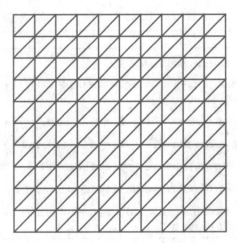

Figure 11-4. *A rectangular mesh generated using dolfin.RectangleMesh*

This mesh for the problem domain is the key to discretizing the problem into a form that can be treated using numerical methods. The next step is to define a representation of the function space for the trial and the test functions, utilizing the `dolfin.FunctionSpace` class. The constructor of this class takes at least three arguments: a mesh object, the name of the type of basis function, and the degree of the basis function. Let's use the Lagrange type of basis functions of degree one (linear basis functions).

```
In [42]: V = dolfin.FunctionSpace(mesh, 'Lagrange', 1)
```

Once the mesh and the function space objects are created, we need to create objects for the trial function u_h and the test function v_h, which we can use to define the weak form of the PDE of interest. In FEniCS, let's use the `dolfin.TrialFunction` and `dolfin.TestFunction` classes for this purpose. They both require a function space object as the first argument to their constructors.

```
In [43]: u = dolfin.TrialFunction(V)
In [44]: v = dolfin.TestFunction(V)
```

The purpose of defining representations of the function space V and the trial and test functions u and v is to be able to construct a representation of a generic PDE on the weak form. For the steady-state heat equation that we are studying here, the weak form was shown in the previous section to be (in the absence of Neumann boundary conditions)

$$\int_\Omega \nabla u . \nabla v \ dx = \int_\Omega fv \ dx.$$

Arriving at this form usually requires rewriting and transforming the direct integrals over the PDE by hand, typically by performing integration by parts. In FEniCS, the PDE itself is defined using the integrands that appear in the weak form, including the integral measure (i.e., the dx). To this end, the `dolfin` library provides several functions acting on the trial and test function objects v and u that represent operations on these functions that commonly occur in the weak form of a PDE. For example, in the present case, the integrand of the left-hand side integral is $\nabla u \cdot \nabla v$ dx. To represent this expression, we need a symbolic representation of the inner product, the gradients of u and v, and the integration measure dx. The names for these functions in the `dolfin` library are `inner`, `nabla_grad`, and `dx`, respectively, and using these functions, we can create a representation of $a(u, v) = \nabla u \cdot \nabla v$ dx that the FEniCS framework understands and can work with.

270

```
In [45]: a = dolfin.inner(dolfin.nabla_grad(u), dolfin.nabla_grad(v)) * dolfin.dx
```

Likewise, we need a representation of $b(v) = fv$ dx for the right-hand side. At this point, we need to specify an explicit form of f (the source term in the original PDF) to proceed with the solution of the problem. Here, let's look at two types of functions: $f = 1$ (a constant) and $f = x^2 + y^2$ (a function of x and y). To represent $f = 1$, we can use the dolfin.Constant object. It takes the value of the constant that it represents as its only argument.

```
In [46]: f1 = dolfin.Constant(1.0)
In [47]: L1 = f1 * v * dolfin.dx
```

If f is a function of x and y, we need to use the dolfin.Expression object to represent f. The constructor of this object takes a string as the first argument containing an expression corresponding to the function. This expression must be defined in C++ syntax, since the FEniCS framework automatically generates and compiles a C++ function to efficiently evaluate the expression. In the expression, we have access to a variable x, which is an array of coordinates at a specific point, where x is accessed as x[0], y as x[1], and so on. For example, to write the expression for $f(x, y) = x^2 + y^2$, we can use "x[0]*x[0] + x[1]*x[1]". Note that because we need to use C++ syntax in this expression, we can*not* use the Python syntax x[0]**2. The Expression class also takes the keyword argument degree that specifies the degree of the basis function or the keyword argument element that describes the finite elements, which, for example, can be obtained using the ufl_element method of the function space object V.

```
In [48]: f2 = dolfin.Expression("x[0]*x[0] + x[1]*x[1]", degree=1)
In [49]: L2 = f2 * v * dolfin.dx
```

At this point, we have defined symbolic representations of the terms that occur in the weak form of the PDE. The next step is to define the boundary conditions. Let's begin with a simple uniform Dirichlet-type boundary condition. The dolfin library contains a class DirichletBC for representing this type of boundary conditions. We can use this class to represent arbitrary functions along the boundaries of the problem domain, but in this first example, consider the simple boundary condition $u = 0$ on the entire boundary. To represent the constant value on the boundary (zero in this case), we can again use the dolfin.Constant class.

```
In [50]: u0 = dolfin.Constant(0)
```

In addition to the boundary condition value, we also need to define a function (here called u0_boundary) used to select different parts of the boundary when creating an instance of the DirichletBC class. This function takes two arguments: a coordinate array x and a flag on_boundary that indicates if a point is on the physical boundary of the mesh, and it should return True if the point x belongs to the boundary and False otherwise. Since this function is evaluated for every vertex in the mesh, by customizing the function, we could pin down the function value at arbitrary parts of the problem domain to specific values or expressions. However, here, we only need to select all the points on the physical boundary, so we can simply let the u0_boundary function return the on_boundary argument.

```
In [51]: def u0_boundary(x, on_boundary):
   ...:         return on_boundary
```

Once we have an expression for the value on the boundary, u0, and a function for selecting the boundary from the mesh vertices, u0_boundary, we can, with the function space object V, finally create the DirichletBC object.

```
In [52]: bc = dolfin.DirichletBC(V, u0, u0_boundary)
```

This completes the specification of the PDE problem, and our next step is to convert the problem into an algebraic form by assembling the matrix and vector from the weak form representations of the PDE. We can do this explicitly using the dolfin.assemble function.

```
In [53]: A = dolfin.assemble(a)
In [54]: b = dolfin.assemble(L1)
In [55]: bc.apply(A, b)
```

This results in a matrix *A* and vector *b* that define the algebraic equation system for the unknown function. The apply method of the DirichletBC class instance bc, which modifies the A and b objects so that the boundary condition is accounted for in the equations, was used here.

To finally solve the problem, we need to create a function object for storing the unknown solution and call the dolfin.solve function, providing the A matrix and the b vector, as well as the underlying data array of a Function object. We can obtain the data array for a Function instance by calling the vector method on the object.

```
In [56]: u_sol1 = dolfin.Function(V)
In [57]: dolfin.solve(A, u_sol1.vector(), b)
```

Here, the Function object for the solution u_sol1, and the call to dolfin.solve function solves the equation system and fills in the values in the data array of the u_sol1 object. Here, the PDE problem was solved by explicitly assembling the A and b matrices and passing the results to the dolfin.solve function. These steps can also be carried out automatically by the dolfin.solve function, by passing a dolfin. Equation object as the first argument to the function, the Function object for the solution as the second argument, and a boundary condition (or list of boundary conditions) as the third argument. We can create the Equation object using, for example, a == L2.

```
In [58]: u_sol2 = dolfin.Function(V)
In [59]: dolfin.solve(a == L2, u_sol2, bc)
```

This is slightly more concise than the method used to find u_sol1 using the equivalence of a == L1. But in some cases, when a problem needs to be solved for multiple situations, it can be useful to use explicit assembling of the matrix A and/or the vector b, so it is worthwhile to be familiar with both methods.

With the solution available as a FEniCS Function object, we can proceed with post-processing and visualizing the solution in many ways. A straightforward way to plot the solution is to use the built-in dolfin.plot function, which can be used to plot mesh objects, function objects, and other objects (see the docstring for dolfin.plot for more information). For example, to plot the solution u_sol2, call dolfin. plot(u_sol2). The resulting graph window is shown in Figure 11-5.

```
In [60]: dolfin.plot(u_sol2)
```

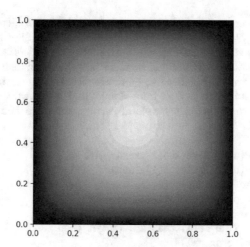

Figure 11-5. *A graph of the mesh function u_sol2, produced by the plot function in the dolfin library*

Using dolfin.plot is a good way of quickly visualizing a solution or a grid. But for better control of the visualization, it is often necessary to export the data and plot it in dedicated visualization software, such as ParaView.[1] To save the solutions u_sol1 and u_sol2 in a format that can be opened with ParaView, we can use the dolfin.File object to generate PVD files (collections of VTK files) and append objects to the file using the << operator in a C++ stream-like fashion.

```
In [61]: dolfin.File('u_sol1.pvd') << u_sol1
```

We can also add multiple objects to a PVD file using this method.

```
In [62]: f = dolfin.File('u_sol_and_mesh.pvd')
    ...: f << mesh
    ...: f << u_sol1
    ...: f << u_sol2
```

Exporting data for FEniCS objects to files that can be loaded and visualized with external visualization software is a method that benefits from the many advantages of powerful visualization software, such as interactivity, parallel processing, and high level of control of the visualizations, to mention a few. However, in many cases, it might be preferable to work within, for example, the Jupyter Notebook for visualization of the solutions and the mesh. For relatively simple problems in one, two, and, to some extent, three dimensions, we can use Matplotlib to visualize meshes and solution functions directly. To be able to use Matplotlib, we need to obtain a NumPy array with data corresponding to the FEniCS function object. There are several ways to construct such arrays. The FEniCS function object can be called like a function with an array (list) of coordinate values.

```
In [63]: u_sol1([0.21, 0.67])
Out[63]: 0.0466076997781351
```

---

[1]http://www.paraview.org

This allows us to evaluate the solution at arbitrary points within the problem domain. We can also obtain the values of a function object like u_sol1 at the mesh vertices as a FEniCS vector using the vector method, which can be converted to a NumPy array using the np.array function. The resulting NumPy arrays are flat (one-dimensional). For a two-dimensional rectangular mesh (like in the current example), it is sufficient to reshape the flat array to obtain a two-dimensional array that can be plotted with, for example, the pcolor, contour, or plot_surface functions from Matplotlib. The following steps convert the underlying data of the u_sol1 and u_sol2 function objects to NumPy arrays, which are then plotted using Matplotlib. The result is shown in Figure 11-6.

```
In [64]: u_mat1 = np.array(u_sol1.vector()).reshape(N1+1, N2+1)
In [65]: u_mat2 = np.array(u_sol2.vector()).reshape(N1+1, N2+1)
In [66]: X, Y = np.meshgrid(np.linspace(0, 1, N1+2), np.linspace(0, 1, N2+2))
In [67]: fig, (ax1, ax2) = plt.subplots(1, 2, figsize=(12, 5))
    ...:
    ...: c = ax1.pcolor(X, Y, u_mat1, cmap=mpl.cm.get_cmap('Reds'))
    ...: cb = plt.colorbar(c, ax=ax1)
    ...: ax1.set_xlabel(r"$x$", fontsize=18)
    ...: ax1.set_ylabel(r"$y$", fontsize=18)
    ...: cb.set_label(r"$u(x, y)$", fontsize=18)
    ...: cb.set_ticks([0.0, 0.02, 0.04, 0.06])
    ...:
    ...: c = ax2.pcolor(X, Y, u_mat2, cmap=mpl.cm.get_cmap('Reds'))
    ...: cb = plt.colorbar(c, ax=ax2)
    ...: ax1.set_xlabel(r"$x$", fontsize=18)
    ...: ax1.set_ylabel(r"$y$", fontsize=18)
    ...: cb.set_label(r"$u(x, x)$", fontsize=18)
    ...: cb.set_ticks([0.0, 0.02, 0.04])
```

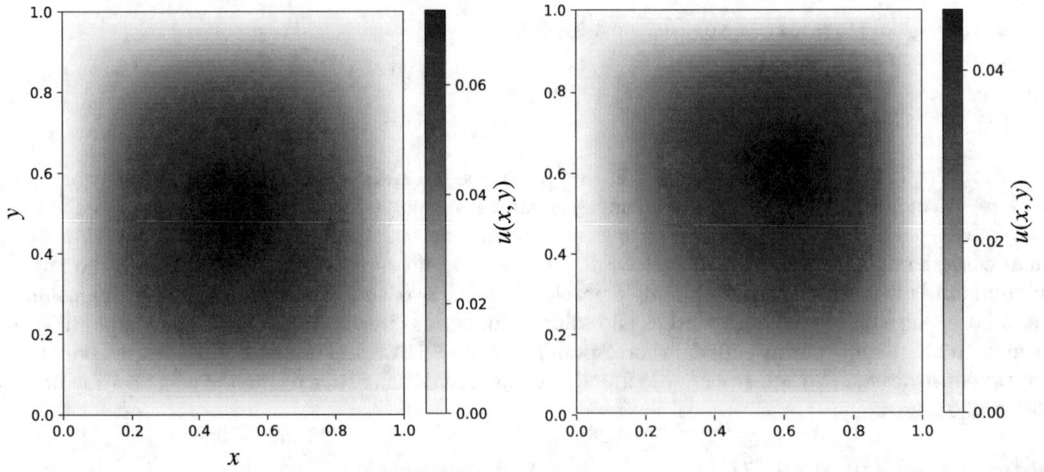

Figure 11-6. *The solution of the steady-state heat equation on the unit square, with source terms f = 1 (left) and f = x² + y² (right), subject to the condition that the u(x, y) function is zero on the boundary*

The method used to produce Figure 11-6 is simple and convenient but only works for rectangular meshes. For more complicated meshes, the vertex coordinates are not organized structurally, and a simple reshaping of the flat array data is not sufficient. However, the Mesh object representing the mesh for the problem domain contains a list of the coordinates for each vertex. Together with values from a Function object, these can be combined into a form plotted with Matplotlib `triplot` and `tripcolor` functions. To use these plot functions, we first need to create a `Triangulation` object from the vertex coordinates for the mesh.

```
In [68]: coordinates = mesh.coordinates()
    ...: triangles = mesh.cells()
    ...: triangulation = mpl.tri.Triangulation(
    ...:     coordinates[:, 0], coordinates[:, 1], triangles)
```

With the triangulation object defined, we can directly plot the array data for FEniCS functions using `triplot` and `tripcolor`, as shown in the following code. The resulting graph is shown in Figure 11-7.

```
In [69]: fig, (ax1, ax2) = plt.subplots(1, 2, figsize=(10, 4))
    ...: ax1.triplot(triangulation)
    ...: ax1.set_xlabel(r"$x$", fontsize=18)
    ...: ax1.set_ylabel(r"$y$", fontsize=18)
    ...: cmap = mpl.cm.get_cmap('Reds')
    ...: c = ax2.tripcolor(triangulation, np.array(u_sol2.vector()), cmap=cmap)
    ...: cb = plt.colorbar(c, ax=ax2)
    ...: ax2.set_xlabel(r"$x$", fontsize=18)
    ...: ax2.set_ylabel(r"$y$", fontsize=18)
    ...: cb.set_label(r"$u(x, y)$", fontsize=18)
    ...: cb.set_ticks([0.0, 0.02, 0.04])
```

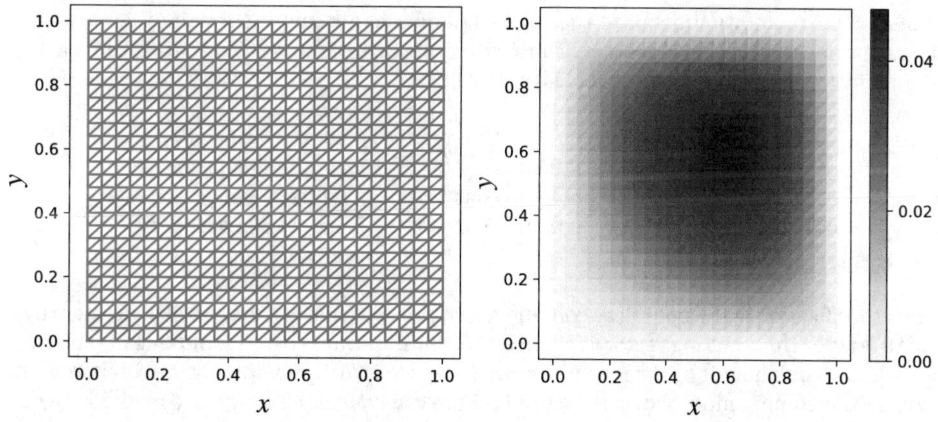

Figure 11-7. *The same as Figure 11-6, except that this graph was produced with Matplotlib's triangulation functions. The mesh is plotted to the left, and the solution of the PDE is to the right*

To see how we can work with more complicated boundary conditions, consider again the heat equation, this time without a source term $u_{xx}+u_{yy} = 0$, but with the following boundary conditions: $u(x = 0) = 3$, $u(x = 1) = -1$, $u(y = 0) = -5$, and $u(y = 1) = 5$. This problem was solved with the FDM method earlier in this chapter. Let's solve this problem again using FEM. As in the previous example, let's define a mesh for the problem domain, the function space, and trial and test function objects.

```
In [70]: V = dolfin.FunctionSpace(mesh, 'Lagrange', 1)
In [71]: u = dolfin.TrialFunction(V)
In [72]: v = dolfin.TestFunction(V)
```

Next, define the weak form of the PDE. Here, set $f = 0$ using a dolfin.Constant object to represent f.

```
In [73]: a = dolfin.inner(dolfin.nabla_grad(u), dolfin.nabla_grad(v)) * dolfin.dx
In [74]: f = dolfin.Constant(0.0)
In [75]: L = f * v * dolfin.dx
```

It remains to define the boundary conditions according to the given specification. In this example, we do not want a uniform boundary condition that applies to the entire boundary, so we need to use the first argument to the boundary selection function passed to the DirichletBC class to single out different parts of the boundary. To this end, define four functions that select the top, bottom, left, and right boundaries.

```
In [76]: def u0_top_boundary(x, on_boundary):
    ...:     # on boundary and y == 1 -> top boundary
    ...:     return on_boundary and abs(x[1]-1) < 1e-5
In [77]: def u0_bottom_boundary(x, on_boundary):
    ...:     # on boundary and y == 0 -> bottom boundary
    ...:     return on_boundary and abs(x[1]) < 1e-5
In [78]: def u0_left_boundary(x, on_boundary):
    ...:     # on boundary and x == 0 -> left boundary
    ...:     return on_boundary and abs(x[0]) < 1e-5
In [79]: def u0_right_boundary(x, on_boundary):
    ...:     # on boundary and x == 1 -> left boundary
    ...:     return on_boundary and abs(x[0]-1) < 1e-5
```

The values of the unknown function at each of the boundaries are simple constants that we can represent with instances of dolfin.Constant. Thus, we can create instances of DirichletBC for each boundary, and the resulting objects are collected in a list bcs.

```
In [80]: bc_t = dolfin.DirichletBC(V, dolfin.Constant(5), u0_top_boundary)
    ...: bc_b = dolfin.DirichletBC(V, dolfin.Constant(-5), u0_bottom_boundary)
    ...: bc_l = dolfin.DirichletBC(V, dolfin.Constant(3), u0_left_boundary)
    ...: bc_r = dolfin.DirichletBC(V, dolfin.Constant(-1), u0_right_boundary)
In [81]: bcs = [bc_t, bc_b, bc_r, bc_l]
```

With this specification of the boundary conditions, we can continue to solve the PDE problem by calling dolfin.solve. The resulting vector converted to a NumPy array is used for plotting the solution using Matplotlib's pcolor function. The result is shown in Figure 11-8. By comparing the result from the corresponding FDM computation, shown in Figure 11-2, we can conclude that the two methods give the same results.

```
In [82]: u_sol = dolfin.Function(V)
In [83]: dolfin.solve(a == L, u_sol, bcs)
In [84]: u_mat = np.array(u_sol.vector()).reshape(N1+1, N2+1)
In [85]: x = np.linspace(0, 1, N1+2)
    ...: y = np.linspace(0, 1, N1+2)
    ...: X, Y = np.meshgrid(x, y)
In [86]: fig, ax = plt.subplots(1, 1, figsize=(8, 6))
    ...: c = ax.pcolor(X, Y, u_mat, vmin=-5, vmax=5, cmap=mpl.cm.get_cmap('RdBu_r'))
    ...: cb = plt.colorbar(c, ax=ax)
    ...: ax.set_xlabel(r"$x_1$", fontsize=18)
    ...: ax.set_ylabel(r"$x_2$", fontsize=18)
    ...: cb.set_label(r"$u(x_1, x_2)$", fontsize=18)
```

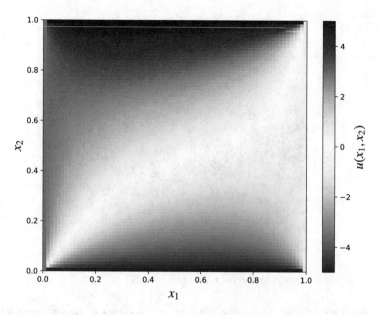

Figure 11-8. *The steady-state solution to the heat equation with different Dirichlet boundary condition on each of the sides of the unit square*

So far, we have used FEM to solve the same kind of problems that we also solved with FDM, but the true strength of FEM becomes apparent first when a PDE problem with more complicated problem geometries is considered. As an illustration, consider the heat equation on a unit circle perforated by five smaller circles, one centered at the origin and the other four smaller circles, as shown in Figure 11-9. We can use the mshr library distributed with FEniCS to generate meshes for geometries like this one. It provides geometric primitives (Point, Circle, Rectangle, etc.) that can be used in algebraic (set) operations to compose a mesh for the problem domain of interest. First, create a unit circle centered at (0, 0) using mshr.Circle, and subtract other Circle objects corresponding to the part of the mesh that should be removed. The resulting mesh is shown in Figure 11-9.

```
In [87]: r_outer = 1
    ...: r_inner = 0.25
    ...: r_middle = 0.1
    ...: x0, y0 = 0.4, 0.4
In [88]: segments = 100
    ...: domain = mshr.Circle(dolfin.Point(.0, .0), r_outer, segments) \
    ...:     - mshr.Circle(dolfin.Point(.0, .0), r_inner, segments) \
    ...:     - mshr.Circle(dolfin.Point( x0,  y0), r_middle, segments) \
    ...:     - mshr.Circle(dolfin.Point( x0, -y0), r_middl, segments) \
    ...:     - mshr.Circle(dolfin.Point(-x0,  y0), r_middle, segments) \
    ...:     - mshr.Circle(dolfin.Point(-x0, -y0), r_middle, segments)
In [89]: mesh = mshr.generate_mesh(domain, 10)
```

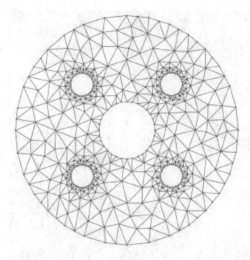

Figure 11-9. *A mesh object generated by the mshr library*

A physical interpretation of this mesh is that the geometry is a cross-section of five pipes through a block of material, where, for example, the inner pipe carries a hot fluid and the middle pipes a cold fluid for cooling the material block (e.g., an engine cylinder surrounded by cooling pipes). With this interpretation in mind, set the boundary condition of the inner pipe to a high value, $u_0(x,y)\big|_{x^2+y^2=r_{inner}^2} = 10$, and the smaller surrounding pipes to a lower value, $u_0(x,y)\big|_{(x-x_0)^2+(y-y_0)^2=r_{middle}^2} = 0$, where (x_0, y_0) is the center of each smaller pipe. Leave the outer boundary unspecified, which is equivalent to the special case of a Neumann boundary condition: $\dfrac{\partial u(x)}{\partial n} = 0$. As before, define functions for singling out vertices on the boundary. Since there are different boundary conditions, we must also use the coordinate argument x to determine which vertices belong to which boundary.

```
In [90]: def u0_inner_boundary(x, on_boundary):
    ...:     x, y = x[0], x[1]
    ...:     return on_boundary and abs(np.sqrt(x**2 + y**2) - r_inner) < 5e-2
In [91]: def u0_middle_boundary(x, on_boundary):
    ...:     x, y = x[0], x[1]
    ...:     if on_boundary:
    ...:         for _x0 in [-x0, x0]:
    ...:             for _y0 in [-y0, y0]:
    ...:                 if abs(np.sqrt((x-_x0)**2+(y-_y0)**2) - r_middle) < 5e-2:
    ...:                     return True
    ...:     return False
In [92]: bc_inner = dolfin.DirichletBC(V, dolfin.Constant(10), u0_inner_boundary)
    ...: bc_middle = dolfin.DirichletBC(V, dolfin.Constant(0), u0_middle_boundary)
In [93]: bcs = [bc_inner, bc_middle]
```

Once the mesh and boundary conditions are specified, we can proceed as usual with defining the function space and the trial and test functions and constructing the weak form representation of the PDE problem.

278

```
In [94]: V = dolfin.FunctionSpace(mesh, 'Lagrange', 1)
In [95]: u = dolfin.TrialFunction(V)
In [96]: v = dolfin.TestFunction(V)
In [97]: a = dolfin.inner(dolfin.nabla_grad(u), dolfin.nabla_grad(v)) * dolfin.dx
In [98]: f = dolfin.Constant(0.0)
In [99]: L = f * v * dolfin.dx
In [100]: u_sol = dolfin.Function(V)
```

Solving and visualizing the problem also follows the same pattern as before. The result of plotting the solution is shown in Figure 11-10.

```
In [101]: dolfin.solve(a == L, u_sol, bcs)
In [102]: coordinates = mesh.coordinates()
    ...: triangles = mesh.cells()
    ...: triangulation = mpl.tri.Triangulation(
    ...:     coordinates[:, 0], coordinates[:, 1], triangles)
In [103]: fig, (ax1, ax2) = plt.subplots(1, 2, figsize=(10, 4))
    ...: ax1.triplot(triangulation)
    ...: ax1.set_xlabel(r"$x$", fontsize=18)
    ...: ax1.set_ylabel(r"$y$", fontsize=18)
    ...: c = ax2.tripcolor(
    ...:     triangulation, np.array(u_sol.vector()), cmap=mpl.cm.get_cmap("Reds"))
    ...: cb = plt.colorbar(c, ax=ax2)
    ...: ax2.set_xlabel(r"$x$", fontsize=18)
    ...: ax2.set_ylabel(r"$y$", fontsize=18)
    ...: cb.set_label(r"$u(x, y)$", fontsize=18)
    ...: cb.set_ticks([0.0, 5, 10, 15])
```

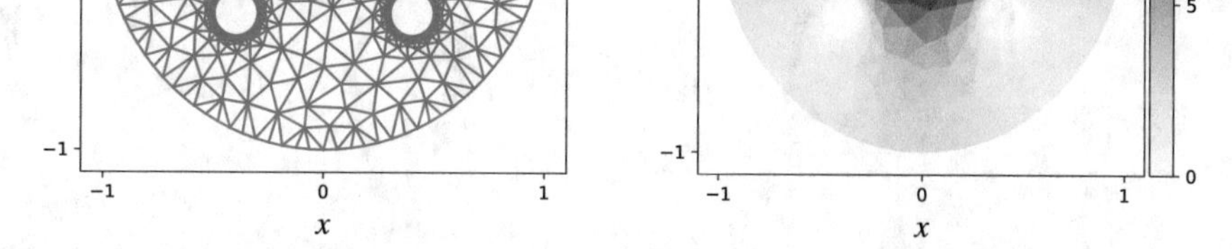

Figure 11-10. *The solution to the heat equation on a perforated unit circle*

Problems with this kind of geometry are difficult to treat with FDM methods but can be handled with relative ease using FEM. Once we obtain a solution for a FEM problem, even for intricate problem boundaries, we can also, with relative ease, post-process the solution function in other ways than plotting it. For example, we might be interested in the value of the function along one of the boundaries. For instance, in the current problem, it is natural to look at the temperature along the outer radius of the problem domain, for example, to see how much the exterior temperature of the body decreases due to the four cooling pipes.

To do this kind of analysis, we need a way of singling out the boundary values from the u_sol object. We can do this by defining an object that describes the boundary (here using dolfin.AutoSubDomain) and applying it to a new Function object used as a mask for selecting the desired elements from the u_sol and mesh. coordinates(). The following calls this mask function mask_outer.

```
In [104]: outer_boundary = dolfin.AutoSubDomain(
     ...:         lambda x, on_bnd: on_bnd and
     ...:             abs(np.sqrt(x[0]**2 + x[1]**2) - r_outer) < 5e-2)
In [105]: bc_outer = dolfin.DirichletBC(V, 1, outer_boundary)
In [106]: mask_outer = dolfin.Function(V)
In [107]: bc_outer.apply(mask_outer.vector())
In [108]: u_outer = u_sol.vector()[mask_outer.vector() == 1]
In [109]: x_outer = mesh.coordinates()[mask_outer.vector() == 1]
```

These steps created the mask for the outer boundary condition and applied it to u_sol.vector() and mesh.coordinates(), thereby obtaining the function values and the coordinates for the outer boundary points. Next, plot the boundary data as a function of the angle between the (x, y) point and the x axis. The result is shown in Figure 11-11.

```
In [110]: phi = np.angle(x_outer[:, 0] + 1j * x_outer[:, 1])
In [111]: order = np.argsort(phi)
In [112]: fig, ax = plt.subplots(1, 1, figsize=(8, 4))
     ...: ax.plot(phi[order], u_outer[order], 's-', lw=2)
     ...: ax.set_ylabel(r"$u(x,y)$ at $x^2+y^2=1$", fontsize=18)
     ...: ax.set_xlabel(r"$\phi$", fontsize=18)
     ...: ax.set_xlim(-np.pi, np.pi)
```

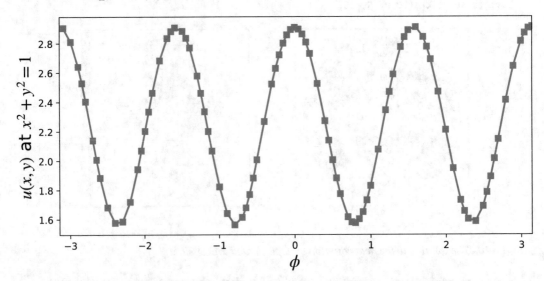

Figure 11-11. *Temperature distribution along the outer boundary of the perforated unit circle*

The accuracy of the solution to a PDE computed with FEM is intimately connected to the element sizes in the mesh that represent the problem domain: a finer mesh gives a more accurate solution. However, increasing the number of elements in the mesh also makes the problem more computationally demanding to solve. Thus, there is a trade-off between the accuracy of the mesh and the available computational resources that must be considered. A mesh with nonuniformly distributed elements is an important tool for dealing with this trade-off. With such a mesh, we can use smaller elements where the unknown function is expected to change in value quickly and fewer elements in less interesting regions. The dolfin library provides a simple way to refine a mesh using the dolfin.refine function. It takes a mesh as the first argument, and if no other arguments are given, it uniformly refines the mesh and returns a new mesh. However, the dolfin.refine function also accepts an optional second argument describing which mesh parts should be refined. This argument should be an instance of a Boolean-valued dolfin.MeshFunction, which acts as a mask that flags which elements (cells) should be divided. For example, consider a mesh for the unit circle less the part in the quadrant where $x > 0$ and $y < 0$. We can construct a mesh for this geometry using mshr.Circle and mshr.Rectangle.

```
In [113]: domain = mshr.Circle(dolfin.Point(.0, .0), 1.0) \
     ...:         - mshr.Rectangle(dolfin.Point(0.0, -1.0), dolfin.Point(1.0, 0.0))
In [114]: mesh = mshr.generate_mesh(domain, 10)
```

The resulting mesh is shown in the left part of Figure 11-12. Using meshes with finer structures near sharp corners in the geometry is often desirable. For this example, it is reasonable to attempt to refine the mesh around the edge near the origin. To do this, we need to create an instance of dolfin.MeshFunction; initialize all its elements to False, using the set_all method; iterate through the elements and mark those near the origin as True; and finally call the dolfin.refine function with the mesh and the MeshFunction instance as arguments. We can do this repeatedly until a sufficiently fine mesh is obtained. The following iteratively calls dolfin.refine with a decreasing number of cells marked for splitting.

```
In [115]: refined_mesh = mesh
     ...: for r in [0.5, 0.25]:
     ...:     cell_markers = dolfin.MeshFunction("bool", refined_mesh, dim=2)
     ...:     cell_markers.set_all(False)
     ...:     for cell in dolfin.cells(refined_mesh):
     ...:         if cell.distance(dolfin.Point(.0, .0)) < r:
     ...:             # mark cells within a radius r from the origin to be split
     ...:             cell_markers[cell] = True
     ...:     refined_mesh = dolfin.refine(refined_mesh, cell_markers)
```

Figure 11-12. *The original and the refined meshes for three-quarters of the unit circle*

The resulting mesh `refined_mesh` is a version of the original mesh with finer element partitioning near the origin. The following code plots the two meshes for comparison, and the result is shown in Figure 11-12.

```
In [116]: def mesh_triangulation(mesh):
     ...:     coordinates = mesh.coordinates()
     ...:     triangles = mesh.cells()
     ...:     triangulation = mpl.tri.Triangulation(
     ...:         coordinates[:, 0], coordinates[:, 1],
     ...:     return triangulation
In [117]: fig, (ax1, ax2) = plt.subplots(1, 2, figsize=(8, 4))
     ...:
     ...: ax1.triplot(mesh_triangulation(mesh))
     ...: ax2.triplot(mesh_triangulation(refined_mesh))
     ...:
     ...: # hide axes and ticks
     ...: for ax in [ax1, ax2]:
     ...:     for side in ['bottom','right','top','left']:
     ...:         ax.spines[side].set_visible(False)
     ...:         ax.set_xticks([])
     ...:         ax.set_yticks([])
     ...:         ax.xaxis.set_ticks_position('none')
     ...:         ax.yaxis.set_ticks_position('none')
     ...:
     ...: ax.set_xlabel(r"$x$", fontsize=18)
     ...: ax.set_ylabel(r"$y$", fontsize=18)
```

Refining a mesh using `dolfin.refine` is a practical technique for improving simple meshes that are constructed using expressions of geometrical primitives, like the one used in this chapter. As a final example of using FEniCS, let's consider another example of the steady-state heat equation, using this refined mesh for the three-quarters of the unit circle, where we impose Neumann boundary conditions on the vertical and horizontal boundaries along the missing quarter of the unit circle: for the vertical edge, assume an outflux of heat described by $\nabla u \cdot \boldsymbol{n} = -2$, $x = 0$, $y < 0$, and through the horizontal edge, assume an influx of heat described by $\nabla u \cdot \boldsymbol{n} = 1$, $x > 0$, $y = 0$, while the outer radial boundary is assumed to be described by the Dirichlet boundary condition $u(x, y) = 0$, $x^2 + y^2 = 1$.

Let's begin, as usual, by creating objects for the function space, the test function, and the trial function.

```
In [118]: mesh = refined_mesh
In [119]: V = dolfin.FunctionSpace(mesh, 'Lagrange', 1)
In [120]: u = dolfin.TrialFunction(V)
In [121]: v = dolfin.TestFunction(V)
```

For problems with Neumann boundary conditions, we need to include the boundary condition in the weak form of the PDE. Recall that the weak form for the Poisson equation is $\int_\Omega \nabla u . \nabla v \ \mathrm{d}x = \int_\Omega fv \ \mathrm{d}x + \int_{\Gamma_N} g \ v \ d\Gamma$,

so compared to the earlier examples, we need to account for the additional term $\int_{\Gamma_N} g \ v \ d\Gamma$, which is an

integral over the boundary with Neumann boundary condition. To represent the integral measure for this integral in the weak form specification, we can use `dolfin.ds`, but to be able to distinguish different parts of the boundary, we first need to mark the boundary parts. One way to do this in FEniCS is to use a `dolfin.MeshFunction` object and assign to it a unique integer value for each distinct part of the boundary. First, create a `dolfin.MeshFunction` instance.

```
In [122]: boundary_parts = dolfin.MeshFunction(
     ...:         "size_t", mesh, mesh.topology().dim()-1)
```

Next, define a function for selecting boundary points and a dolfin.AutoSubDomain instance that is initialized from the boundary selection function. The AutoSubDomain instance can then mark the corresponding cells in the MeshFunction (here called boundary_parts) with an identifying integer value. The following lines of code perform these steps for the vertical edge of the mesh, where $x = 0$ and $y < 0$.

```
In [121]: def v_boundary_func(x, on_boundary):
     ...:         """ the vertical edge of the mesh, where x = 0 and y < 0 """
     ...:         x, y = x[0], x[1]
     ...:         return on_boundary and abs(x) < 1e-4 and y < 0.0
In [122]: v_boundary = dolfin.AutoSubDomain(v_boundary_func)
In [123]: v_boundary.mark(boundary_parts, 0)
```

Repeat the same procedure for the horizontal edge of the mesh, where $y = 0$ and $x > 0$.

```
In [124]: def h_boundary_func(x, on_boundary):
     ...:         """ the horizontal edge of the mesh, where y = 0 and x > 0 """
     ...:         x, y = x[0], x[1]
     ...:         return on_boundary and abs(y) < 1e-4 and x > 0.0
In [125]: h_boundary = dolfin.AutoSubDomain(h_boundary_func)
In [126]: h_boundary.mark(boundary_parts, 1)
```

We can also use the same method to define Dirichlet boundary conditions. Mark the part of the boundary described by the Dirichlet boundary condition and then use it to create the dolfin. DirichletBC object.

```
In [127]: def outer_boundary_func(x, on_boundary):
     ...:         x, y = x[0], x[1]
     ...:         return on_boundary and abs(x**2 + y**2-1) < 1e-2
In [128]: outer_boundary = dolfin.AutoSubDomain(outer_boundary_func)
In [129]: outer_boundary.mark(boundary_parts, 2)
In [130]: bc = dolfin.DirichletBC(V, dolfin.Constant(0.0), boundary_parts, 2)
```

Once the boundaries are marked, we can create the weak form of the PDE. Since a partitioned boundary is used here, we must specify the domain and subdomain arguments to the integral measures dolfin.dx and dolfin.ds, using the mesh and boundary_parts objects.

```
In [131]: dx = dolfin.dx(domain=mesh, subdomain_data=boundary_parts)
In [132]: a = dolfin.inner(dolfin.nabla_grad(u), dolfin.nabla_grad(v)) * dx
In [133]: f = dolfin.Constant(0.0)
In [134]: g_v = dolfin.Constant(-2.0)
In [135]: g_h = dolfin.Constant(1.0)
In [136]: L = f * v * dolfin.dx(domain=mesh, subdomain_data=boundary_parts)
In [137]: L += g_v * v * dolfin.ds(0, domain=mesh, subdomain_data=boundary_parts)
In [138]: L += g_h * v * dolfin.ds(1, domain=mesh, subdomain_data=boundary_parts)
```

The last two code cells added new terms for the Neumann boundary conditions for the mesh's vertical and horizontal edges. These parts of the boundary are denoted by integers 0 and 1, respectively, as defined in the preceding section, and these integers are passed as an argument to the dolfin.ds to select integration over different parts of the boundaries.

```
In [139]: u_sol = dolfin.Function(V)
In [140]: dolfin.solve(a == L, u_sol, bc)
```

Once the representation of the weak form of the PDE is defined, we can go ahead and solve the problem using dolfin.solve, as done in earlier examples. Finally, plot the solution using Matplotlib's triangulation plot functions. The results are shown in Figure 11-13. The graph shows that, as expected, the solution has more structure near the edge at the origin. Using a mesh with smaller elements in this region is a good way to obtain sufficient resolution without inflicting excessive computational cost by using a uniformly fine-structured mesh.

```
In [141]: fig, (ax1, ax2) = plt.subplots(1, 2, figsize=(10, 4))
     ...: triangulation = mesh_triangulation(mesh)
     ...: ax1.triplot(triangulation)
     ...: ax1.set_xlabel(r"$x$", fontsize=18)
     ...: ax1.set_ylabel(r"$y$", fontsize=18)
     ...:
     ...: data = np.array(u_sol.vector())
     ...: norm = mpl.colors.Normalize(-abs(data).max(), abs(data).max())
     ...: c = ax2.tripcolor(triangulation, data, norm=norm,
     ...:     cmap=mpl.cm.get_cmap("RdBu_r"))
     ...: cb = plt.colorbar(c, ax=ax2)
     ...: ax2.set_xlabel(r"$x$", fontsize=18)
     ...: ax2.set_ylabel(r"$y$", fontsize=18)
     ...: cb.set_label(r"$u(x, y)$", fontsize=18)
     ...: cb.set_ticks([-.5, 0, .5])
```

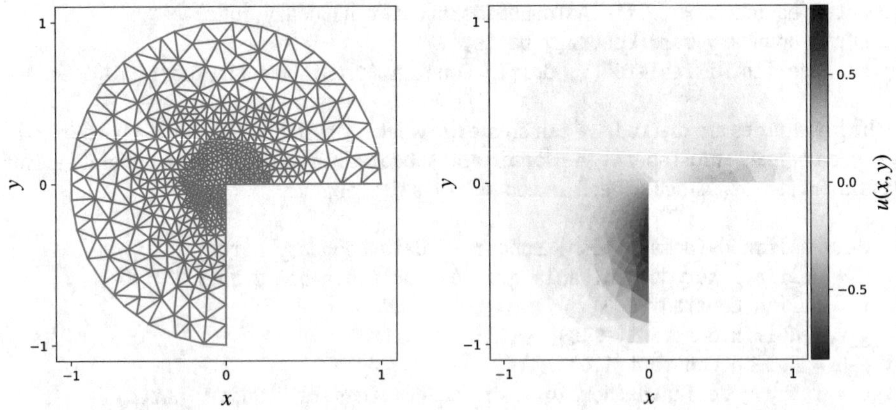

Figure 11-13. *Solution to the heat equation on a quarter of the unit circle with Neumann and Dirichlet boundary conditions*

The examples explored in this section are merely a few simple demonstrations of the types of problems for which the FEniCS framework can be used. There are many features in FEniCS that I have not even been able to mention here. For the reader who is particularly interested in solving PDE problems, I recommend studying the FEniCS book (Anders Logg, 2012) and its many example applications. In particular, important aspects of solving PDEs with FEM that I have not been able to discuss here are nontrivial Neumann boundary conditions (which need to be included in the formulation of the weak form of the PDE), PDEs for vector-valued functions, higher-dimensional PDE problems (e.g., the heat equation in three dimensions), and time-dependent PDE problems. These topics, and many others, are well supported in the FEniCS framework.

Summary

This chapter examined methods for solving partial differential equations (PDEs) and how these methods can be employed within the scientific Python environment. Specifically, it introduced the finite-difference method (FDM) and the finite-element method (FEM) for solving PDE problems and used these methods to solve several example problems. The advantage of FDM is its simplicity, and it is a very practical method for problems where it is easily applicable (simple problem domains, uniform discretization, etc.). For more complicated PDE problems, for example, where the problem domain is more complex, FEM is generally more suitable. However, the mathematical theory of the FEM is more involved, and the implementation is far more technical. While several advanced FEM frameworks can be used in Python, this chapter focused on one prominent example: the FEniCS framework. FEniCS is a full-featured FEM software that can be used for a wide range of PDE problems. With the examples considered here, we have only scraped the surface of what can be achieved with the software. However, the hope is that the examples studied in this chapter give a general sense of the workflow when solving PDE problems with FEM and when using the FEniCS software, in particular.

Further Reading

Although this chapter discussed FDM and FEM, there are other successful and useful methods for numerically solving PDEs. For instance, the finite-volume method (FVM) is a variant of the FEM method that is often used for fluid dynamics calculations, as well as in other fields. The Python library FiPy provides a framework for solving PDE problems using this method, and a theoretical introduction to the method is given in *Principles of Computational Fluid Dynamics* by P. Wesseling (Springer, 2009). The theoretical background information about the FDM and FEM that is given in this chapter is very brief indeed, and it merely serves to introduce the terminology and notation used here. For serious work with the FDM, and particularly the FEM method, it is important to understand the fundamentals of these methods thoroughly.

Good introductions to FDM and FEM are given in M. Gockenbach's *Partial Differential Equations* (SIAM, 2011) and *Understanding and Implementing the Finite Element Method* (SIAM, 2006), *Numerical Solution of Partial Differential Equations by the Finite Element Method* by C. Johnson (Dover, 2009), and *Finite Difference Methods for Ordinary and Partial Differential Equations: Steady-State and Time-Dependent Problems* by R. LeVeque (SIAM, 2007).

The FEniCS book, *Automated Solution of Differential Equations by the Finite Element Method* by A. Logg et al., (Springer, 2012), is available for free online from the FEniCS project's website (http://fenicsproject.org), also contains a nice introduction to the FEM method, in addition to detailed documentation of the FEniCS software itself.

CHAPTER 12

■ ■ ■

Data Processing and Analysis

The last several chapters covered the main topics of traditional scientific computing. These topics provide a foundation for most computational work. Starting with this chapter, let's move on to explore data processing and analysis, statistics, and statistical modeling. First, we look at the Pandas data analysis library. This library provides convenient data structures for representing series and tables of data and makes it easy to transform, split, merge, and convert data. These are important steps in the process[1] of cleansing raw data into a tidy form suitable for analysis. The Pandas library builds on top of NumPy. It complements it with features that are particularly useful when handling data, such as labeled indexing, hierarchical indices, data alignment for comparison and merging of datasets, handling of missing data, and much more. As such, the pandas library has become a de facto standard library for high-level data processing in Python, especially for statistics applications. The Pandas library contains only limited support for statistical modeling (namely, linear regression). Other packages are available for more involved statistical analysis and modeling, such as statsmodels, patsy, and scikit-learn, which are covered in later chapters. However, for statistical modeling with these packages, Pandas can still be used for data representation and preparation. The Pandas library is, therefore, a key component in the software stack for data analysis with Python.

■ **Pandas** The Pandas library is a framework for data processing and analysis in Python. At the time of writing, the most recent version of Pandas is 2.3.1. For more information about the Pandas library, and its official documentation, see the project's website at http://pandas.pydata.org.

The primary focus of this chapter is to introduce basic features and usage of the Pandas library. Toward the end of the chapter, we briefly explore the statistical visualization library Seaborn, which is built on top of Matplotlib. This library provides quick and convenient visualization of data represented as a Pandas data structure (or NumPy arrays). Visualization is an essential part of exploratory data analysis, and the Pandas library itself also provides functions for basic data visualization (which also builds on top of Matplotlib). The Seaborn library takes this further by providing additional statistical graphing capabilities and improved styling: the Seaborn library is notable for generating good-looking graphics using default settings.

■ **Seaborn** The Seaborn library is a visualization library for statistical graphics. It builds on Matplotlib and provides easy-to-use functions for common statistical graphs. At the time of writing, the most recent version of Seaborn is 0.12.0. For more information about Seaborn and its official documentation, see https://seaborn.pydata.org/index.html.

---

[1] Also known as data munging or data wrangling

© Robert Johansson 2024
R. Johansson, *Numerical Python*, https://doi.org/10.1007/979-8-8688-0413-7_12

Importing Modules

This chapter mainly works with the pandas module, which is imported under the name pd.

```
In [1]: import pandas as pd
```

We also require NumPy and Matplotlib, which are imported in the following way.

```
In [2]: import numpy as np
In [3]: import matplotlib.pyplot as plt
```

For a more aesthetically pleasing appearance of Matplotlib figures produced by the pandas library, select a suitable style for statistical graphs using the mpl.style.use function.

```
In [4]: import matplotlib as mpl
   ...: mpl.style.use('ggplot')
```

Later in this chapter, we use the seaborn module, which we import under the name sns.

```
In [5]: import seaborn as sns
```

Introduction to Pandas

The main focus of this chapter is the pandas library for data analysis. The pandas library mainly provides data structures and methods for representing and manipulating data. The two primary data structures in Pandas are the Series and DataFrame objects, which represent data series and tabular data, respectively. Both objects have an index for accessing elements or rows in the data the object represents. By default, the indices are integers starting from zero, like NumPy arrays, but it is also possible to use any sequence of identifiers as an index.

Series

The merit of indexing a data series with labels rather than integers is apparent even in the simplest examples. Consider the following construction of a Series object. Give the constructor a list of integers to create a Series object representing the provided data. Displaying the object in IPython reveals the data of the Series object together with the corresponding indices.

```
In [6]: s = pd.Series([909976, 8615246, 2872086, 2273305])
In [7]: s
Out[7]: 0    909976
        1    8615246
        2    2872086
        3    2273305
        dtype: int64
```

The resulting object is a Series instance with the data type (dtype) int64, and the elements are indexed by the integers 0, 1, 2, and 3. Using the index and values attributes, we can extract the underlying data for the index and the values stored in the series.

```
In [8]: s.index
Out[8]: RangeIndex(start=0, stop=4, step=1)
In [9]: s.values
Out[9]: array([ 909976, 8615246, 2872086, 2273305], dtype=int64)
```

While using integer-indexed arrays or data series is a fully functional representation of the data, it is not descriptive. For example, if the data represents the population of four European capitals, it is convenient and descriptive to use the city names as indices rather than integers. With a Series object, this is possible, and we can assign the index attribute of a Series object to a list with new indices to accomplish this. We can also set the name attribute of the Series object to give it a descriptive name.

```
In [10]: s.index = ["Stockholm", "London", "Rome", "Paris"]
In [11]: s.name = "Population"
In [12]: s
Out[12]: Stockholm       909976
         London         8615246
         Rome           2872086
         Paris          2273305
         Name: Population, dtype: int64
```

It is now immediately obvious what the data represents. Alternatively, we can set the index and name attributes through keyword arguments to the Series object when created.

```
In [13]: s = pd.Series([909976, 8615246, 2872086, 2273305], name="Population",
    ...:               index=["Stockholm", "London", "Rome", "Paris"])
```

While it is perfectly possible to store the data for the populations of these cities directly in a NumPy array, even in this simple example, it is much clearer what the data represent when the data points are indexed with meaningful labels. The benefits of bringing the description of the data closer to the data are even more significant when the complexity of the dataset increases.

We can access elements in a Series by indexing with the corresponding index (label) or directly through an attribute with the same name as the index (if the index label is a valid Python symbol name).

```
In [14]: s["London"]
Out[14]: 8615246
In [15]: s.Stockholm
Out[15]: 909976
```

Indexing a Series object with a list of indices gives a new Series object with a subset of the original data (corresponding to the provided list of indices).

```
In [16]: s[["Paris", "Rome"]]
Out[16]: Paris       2273305
         Rome        2872086
         Name: Population, dtype: int64
```

With a data series represented as a Series object, we can easily compute its descriptive statistics using the Series methods count (the number of data points), median (calculate the median), mean (calculate the mean value), std (calculate the standard deviation), min and max (minimum and maximum values), and the quantile (for calculating quantiles).

```
In [17]: s.median(), s.mean(), s.std()
Out[17]: (2572695.5, 3667653.25, 3399048.5005155364)
In [18]: s.min(), s.max()
Out[18]: (909976, 8615246)
In [19]: s.quantile(q=0.25), s.quantile(q=0.5), s.quantile(q=0.75)
Out[19]: (1932472.75, 2572695.5, 4307876.0)
```

All the preceding data are combined in the output of the describe method, which provides a summary of the data represented by a Series object.

```
In [20]: s.describe()
Out[20]: count         4.000000
         mean    3667653.250000
         std     3399048.500516
         min      909976.000000
         25%     1932472.750000
         50%     2572695.500000
         75%     4307876.000000
         max     8615246.000000
         Name: Population, dtype: float64
```

Using the plot method, we can quickly and easily produce graphs that visualize the data in a Series object (see Figure 12-1). The pandas library uses Matplotlib for plotting, and we can optionally pass a Matplotlib Axes instance to the plot method via the ax argument. The type of the graph is specified using the kind argument (valid options are line, hist, bar, barh, box, kde, density, area, and pie).

```
In [21]: fig, axes = plt.subplots(1, 4, figsize=(12, 3))
    ...: s.plot(ax=axes[0], kind='line', title='line')
    ...: s.plot(ax=axes[1], kind='bar', title='bar')
    ...: s.plot(ax=axes[2], kind='box', title='box')
    ...: s.plot(ax=axes[3], kind='pie', title='pie')
```

Figure 12-1. Examples of plot styles that can be produced with Pandas using the Series.plot method

DataFrame

As shown in previous examples, a pandas Series object provides a convenient container for one-dimensional arrays, which can use descriptive labels for the elements and offers quick access to descriptive statistics and visualization. For higher-dimensional arrays (mainly two-dimensional arrays or tables), the corresponding data structure is the Pandas DataFrame object. It can be viewed as a collection of Series objects with a common index.

There are numerous ways to initialize a DataFrame. For simple examples, the easiest way is to pass a nested Python list or dictionary to the constructor of the DataFrame object. For example, consider an extension of the dataset used in the previous section, where, in addition to the population of each city, we also include a column that specifies which state each city belongs to. We can create the corresponding DataFrame object in the following way.

```
In [22]: df = pd.DataFrame([[909976, "Sweden"],
    ...:                    [8615246, "United Kingdom"],
    ...:                    [2872086, "Italy"],
    ...:                    [2273305, "France"]])
In [23]: df
Out[23]:
```

	0	1
0	909976	Sweden
1	8615246	United Kingdom
2	2872086	Italy
3	2273305	France

The result is a tabular data structure with rows and columns. Like with a Series object, we can use labeled indexing for rows by assigning a sequence of labels to the index attribute, and, in addition, we can set the columns attribute to a sequence of labels for the columns.

```
In [24]: df.index = ["Stockholm", "London", "Rome", "Paris"]
In [25]: df.columns = ["Population", "State"]
In [26]: df
Out[26]:
```

	Population	State
Stockholm	909976	Sweden
London	8615246	United Kingdom
Rome	2872086	Italy
Paris	2273305	France

The index and columns attributes can also be set using the corresponding keyword arguments to the DataFrame object when it is created.

```
In [27]: df = pd.DataFrame([[909976, "Sweden"],
    ...:                    [8615246, "United Kingdom"],
    ...:                    [2872086, "Italy"],
    ...:                    [2273305, "France"]],
    ...:                    index=["Stockholm", "London", "Rome", "Paris"],
    ...:                    columns=["Population", "State"])
```

An alternative way to create the same data frame, which is sometimes more convenient, is to pass a dictionary with column titles as keys and column data as values.

```
In [28]: df = pd.DataFrame(
    ...:     {"Population": [909976, 8615246, 2872086, 2273305],
    ...:      "State": ["Sweden", "United Kingdom", "Italy", "France"]},
    ...:     index=["Stockholm", "London", "Rome", "Paris"])
```

As before, the underlying data in a DataFrame can be obtained as a NumPy array using the values attribute and the index and column arrays through the index and columns attributes, respectively. Each column in a data frame can be accessed using the column name as an attribute (or by indexing with the column label, e.g., df["Population"]).

```
In [29]: df.Population
Out[29]: Stockholm     909976
         London       8615246
         Rome         2872086
         Paris        2273305
         Name: Population, dtype: int64
```

The result of extracting a column from a DataFrame is a new Series object, which we can process and manipulate with the methods discussed in the previous section. Rows of a DataFrame instance can be accessed using the loc indexer attribute. Indexing this attribute also results in a Series object, which corresponds to a row of the original data frame.

```
In [30]: df.loc["Stockholm"]
Out[30]: Population     909976
         State          Sweden
         Name: Stockholm, dtype: object
```

Passing a list of row labels to the loc indexer results in a new DataFrame that is a subset of the original DataFrame, containing only the selected rows.

```
In [31]: df.loc[["Paris", "Rome"]]
Out[31]:
```

	Population	State
Paris	2273305	France
Rome	2872086	Italy

The loc indexer can also select both rows and columns simultaneously by passing a row label (or a list thereof) and a column label (or a list thereof). The result is a DataFrame, a Series, or an element value, depending on the number of selected columns and rows.

```
In [32]: df.loc[["Paris", "Rome"], "Population"]
Out[32]: Paris     2273305
         Rome      2872086
         Name: Population, dtype: int64
```

We can compute descriptive statistics using the same methods used for Series objects. When invoking those methods (mean, std, median, min, max, etc.) for a DataFrame, the calculation is performed for each column with numerical data types.

```
In [33]: df.mean()
Out[33]: Population    3667653.25
         dtype: float64
```

In this case, only one of the two columns has a numerical data type (the one named Population). Using the DataFrame info method and the dtypes attribute, we can obtain a summary of the content in a DataFrame and the data types of each column.

```
In [34]: df.info()
<class 'pandas.core.frame.DataFrame'>
Index: 4 entries, Stockholm to Paris
Data columns (total 2 columns):
Population    4 non-null int64
State         4 non-null object
dtypes: int64(1), object(1)
memory usage: 96.0+ bytes
In [35]: df.dtypes
Out[35]: Population        int64
         State            object
         dtype: object
```

The real advantages of using pandas emerge when dealing with larger and more complex datasets than the examples presented so far. Such data can rarely be defined as explicit lists or dictionaries, which can be passed to the DataFrame initializer. A more realistic situation is that the data must be read from a file or other external sources. The pandas library supports numerous methods for reading data from files of different formats. Here, use the read_csv function to read in data and create a DataFrame object from a CSV file.[2] This function accepts many optional arguments for tuning its behavior. See the docstring help(pd.read_csv) for details. Some of the most useful arguments are header (specifies which row, if any, contains a header with column names), skiprows (number of rows to skip before starting to read data, or a list of line numbers to skip), delimiter (the character that is used as a delimiter between column values), encoding (the name of the encoding used in the file, e.g., utf-8), and nrows (number of rows to read). The first and only mandatory argument to the pd.read_csv function is a filename or a URL to the data source. For example, to read a dataset stored in a file called european_cities.csv,[3] of which the first five lines are shown in the following code, we can call pd.read_csv("european_cities.csv"), since the default delimiter is "," and the header is by default taken from the first line. However, we could also write out all these options explicitly.

```
In [36]: !head -n 5 european_cities.csv
Rank,City,State,Population,Date of census
1,London, United Kingdom,"8,615,246",1 June 2014
2,Berlin, Germany,"3,437,916",31 May 2014
3,Madrid, Spain,"3,165,235",1 January 2014
4,Rome, Italy,"2,872,086",30 September 2014
In [37]: df_pop = pd.read_csv("european_cities.csv",
   ...:                       delimiter=",", encoding="utf-8", header=0)
```

---

[2] CSV, or comma-separated values, is a common text format where rows are stored in lines and columns are separated by a comma (or some other text delimiter). See Chapter 18 for more details about this and other file formats.

[3] This dataset was obtained from the Wiki page: http://en.wikipedia.org/wiki/Largest_cities_of_the_European_Union_by_population_within_city_limits

This dataset is similar to the example data used earlier in this chapter, but now there are additional columns and many more rows for other cities. Once a dataset is read into a DataFrame object, it is useful to start by inspecting the summary given by the info method to begin forming an idea of the dataset's properties.

```
In [38]: df_pop.info()
<class 'pandas.core.frame.DataFrame'>
Int64Index: 105 entries, 0 to 104
Data columns (total 5 columns):
Rank            105 non-null int64
City            105 non-null object
State           105 non-null object
Population      105 non-null object
Date of census  105 non-null object
dtypes: int64(1), object(4) memory usage: 4.9+ KB
```

Here, there are 105 rows in this dataset and that it has five columns. Only the Rank column is of a numerical data type. In particular, the Population column is not yet of numeric data type because its values are of the format "8,615,246" and are therefore interpreted as string values by the read_csv function. It is also informative to display a tabular view of the data. However, this dataset is too large to display in full. In situations like this, the head and tail methods are handy for creating a truncated dataset containing the first few and last few rows, respectively. These functions take an optional argument specifying how many rows to include in the truncated DataFrame. Note also that df.head(n) is equivalent to df[:n], where n is an integer.

```
In [39]: df_pop.head()
Out[39]:
```

	Rank	City	State	Population	Date of census
0	1	London	United Kingdom	8,615,246	1 June 2014
1	2	Berlin	Germany	3,437,916	31 May 2014
2	3	Madrid	Spain	3,165,235	1 January 2014
3	4	Rome	Italy	2,872,086	30 September 2014
4	5	Paris	France	2,273,305	1 January 2013

Displaying a truncated DataFrame gives a good idea of what the data looks like and what remains to be done before the data is ready for analysis. It is common to transform columns in one way or another and reorder the table by sorting by a specific column or by ordering the index. The following explores some methods for modifying DataFrame objects. First, we can create new columns and update columns in a DataFrame simply by assigning a Series object to the DataFrame indexed by the column name, and we can delete columns using the Python del keyword.

The apply method is a powerful tool to transform the content in a column. It creates and returns a new Series object for which a function passed to apply has been applied to each element in the original column. For example, we can use the apply method to transform the elements in the Population column from strings to integers by passing a lambda function that removes the "," characters from the strings and casts the results to an integer. Here, we assign the transformed column to a new one named NumericPopulation. Using the same method, tidy up the State values by removing extra white spaces in its elements using the string method strip.

```
In [40]: df_pop["NumericPopulation"] = df_pop.Population.apply(
    ...:        lambda x: int(x.replace(",", "")))
In [41]: df_pop["State"].values[:3]  # contains extra white spaces
Out[41]: array([' United Kingdom', ' Germany', ' Spain'], dtype=object)
In [42]: df_pop["State"] = df_pop["State"].apply(lambda x: x.strip())
In [43]: df_pop.head()
Out[43]:
```

	Rank	City	State	Population	Date of census	NumericPopulation
0	1	London	United Kingdom	8,615,246	1 June 2014	8615246
1	2	Berlin	Germany	3,437,916	31 May 2014	3437916
2	3	Madrid	Spain	3,165,235	1 January 2014	3165235
3	4	Rome	Italy	2,872,086	30 September 2014	2872086
4	5	Paris	France	2,273,305	1 January 2013	2273305

Inspecting the data types of the columns in the updated DataFrame confirms that the new column NumericPopulation is indeed of integer type (while the Population column is unchanged).

```
In [44]: df_pop.dtypes
Out[44]: Rank                 int64
         City                object
         State               object
         Population          object
         Date of census      object
         NumericPopulation    int64
         dtype: object
```

We may also need to change the index to one of the columns of the DataFrame. For example, we could use the City column as an index in the current case. We can accomplish this using the set_index method, which takes as an argument the name of the column to use as an index. The result is a new DataFrame object, and the original DataFrame is unchanged. Furthermore, using the sort_index method, we can sort the data frame with respect to the index.

```
In [45]: df_pop2 = df_pop.set_index("City")
In [46]: df_pop2 = df_pop2.sort_index()
In [47]: df_pop2.head()
Out[47]:
```

City	Rank	State	Population	Date of census	NumericPopulation
Aarhus	92	Denmark	326,676	1 October 2014	326676
Alicante	86	Spain	334,678	1 January 2012	334678
Amsterdam	23	Netherlands	813,562	31 May 2014	813562
Antwerp	59	Belgium	510,610	1 January 2014	510610
Athens	34	Greece	664,046	24 May 2011	664046

The sort_index method also accepts a list of column names, which creates a hierarchical index. A hierarchical index uses tuples of index labels to address rows in the data frame. We can use the sort_index method with the integer-valued argument level to sort the rows in a DataFrame according to the nth level of the hierarchical index, where level=n. The following example creates a hierarchical index with State and City as indices and uses the sort_index method to sort by the first index (State).

```
In [48]: df_pop3 = df_pop.set_index(["State", "City"]).sort_index(level=0)
In [49]: df_pop3.head(7)
Out[49]:
```

State	City	Rank	Population	Date of census
Austria	Vienna	7	1794770	1 January 2015
Belgium	Antwerp	59	510610	1 January 2014
	Brussels	16	1175831	1 January 2014
Bulgaria	Plovdiv	84	341041	31 December 2013
	Sofia	14	1291895	14 December 2014
	Varna	85	335819	31 December 2013
Croatia	Zagreb	24	790017	31 March 2011

A DateFrame with a hierarchical index can be partially indexed using only its zeroth-level index (df3. loc["Sweden"]) or completely indexed using a tuple of all hierarchical indices (df3.loc[("Sweden", "Gothenburg")]).

```
In [50]: df_pop3.loc["Sweden"]
Out[50]:
In [51]: df_pop3.loc[("Sweden", "Gothenburg")]
Out[51]: Rank                              53
         Population                   528,014
         Date of census         31 March 2013
         NumericPopulation            528014
         Name: (Sweden, Gothenburg), dtype: object
```

City	Rank	Population	Date of census	NumericPopulation
Gothenburg	53	528,014	31 March 2013	528014
Malmö	102	309,105	31 March 2013	309105
Stockholm	20	909,976	31 January 2014	909976

If we want to sort by a column rather than the index, we can use the sort_values method. It takes a column name, or a list of column names, with respect to which the DataFrame is to be sorted. It also accepts the keyword argument ascending, which is a Boolean or a list of Boolean values that specifies whether the corresponding column is to be sorted in ascending or descending order.

```
In [52]: df_pop.set_index("City").sort_values(["State", "NumericPopulation"],
   ...:                                        ascending=[False, True]).head()
Out[52]:
```

City	Rank	State	Population	Date of census	NumericPopulation
Nottingham	103	United Kingdom	308,735	30 June 2012	308735
Wirral	97	United Kingdom	320,229	30 June 2012	320229
Coventry	94	United Kingdom	323,132	30 June 2012	323132
Wakefield	91	United Kingdom	327,627	30 June 2012	327627
Leicester	87	United Kingdom	331,606	30 June 2012	331606

With categorical data such as the State column, it is often interesting to summarize how many values of each category a column contains. Such counts can be computed using the value_counts method (of the Series object). For example, we can use the following to count the number of cities each country has on the list of the 105 largest cities in Europe.

```
In [53]: city_counts = df_pop.State.value_counts()
In [54]: city_counts.head()
Out[54]: Germany          19
         United Kingdom   16
         Spain            13
         Poland           10
         Italy            10
         dtype: int64
```

This example shows that the state with the largest number of cities on the list is Germany, with 19 cities, followed by the United Kingdom, with 16 cities, and so on. A related question is how large the total population of all cities within a state is. To answer this type of question, we can proceed in two ways: First, we can create a hierarchical index using State and City and use the groupby and the sum methods to reduce the DataFrame along one of the indices. In this case, we want to sum over all entries within the index level State, so we can use groupby(level="State").sum(), which eliminates the City index. For presentation, also sort the resulting DataFrame in descending order of the column NumericPopulation.

```
In [55]: df_pop3 = df_pop[["State", "City", "NumericPopulation"]].set_index(
   ...:     ["State", "City"])
In [56]: df_pop4 = df_pop3.groupby(level="State").sum().sort_values(
   ...:     "NumericPopulation", ascending=False)
In [57]: df_pop4.head()
Out[57]:
```

State	NumericPopulation
United Kingdom	16011877
Germany	15119548
Spain	10041639
Italy	8764067
Poland	6267409

Second, we can obtain the same results using the groupby method with a column name instead of index level as an argument. It allows grouping rows of a DataFrame by the values of a given column and apply a reduction function on the resulting object (e.g., sum, mean, min, max, etc.). The result is a new DataFrame with the grouped-by column as an index. Using this method, we can compute the total population of the 105 cities, grouped by state, in the following way.

```
In [58]: df_pop5 = (df_pop[["State", "NumericPopulation"]]
    ...:                 .groupby("State").sum()
    ...:                 .sort_values("NumericPopulation", ascending=False))
```

Note that the columns "State" and "NumericPopulation" were selected from the data frame by indexing with a list of column names we need in the group by aggregation. The drop method with the keyword argument axis=1 could also have removed a column (use axis=0 to drop rows) from the DataFrame. Finally, use the plot method of the Series object to plot bar graphs for the city count and the total population. The results are shown in Figure 12-2.

```
In [59]: fig, (ax1, ax2) = plt.subplots(1, 2, figsize=(12, 4))
    ...: city_counts.plot(kind='barh', ax=ax1)
    ...: ax1.set_xlabel("# cities in top 105")
    ...: df_pop5.NumericPopulation.plot(kind='barh', ax=ax2)
    ...: ax2.set_xlabel("Total pop. in top 105 cities")
```

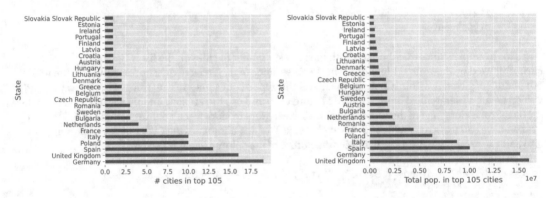

Figure 12-2. *The number of cities in the list of the top 105 most populated cities in Europe (left) and the total population in those cities (right), grouped by state*

Time Series

Time series are a common form of data in which a quantity is given, for example, at regularly or irregularly spaced timestamps or for fixed or variable time spans (periods). In pandas, there are dedicated data structures for representing these types of data. Series and DataFrame can have both columns and indices with data types describing timestamps and time spans. When dealing with temporal data, it is particularly useful to index the data with time data types. Using pandas time-series indexers, DatetimeIndex and PeriodIndex, we can carry out many common date, time, period, and calendar operations, such as selecting time ranges and shifting and resampling the data points in a time series.

To generate a sequence of dates that can be used as an index in a pandas Series or DataFrame objects, we can, for example, use the date_range function. It takes the starting point as a date and time string (or a datetime object from the Python standard library) as a first argument, and the number of elements in the range can be set using the periods keyword argument.

```
In [60]: pd.date_range("2015-1-1", periods=31)
Out[60]: <class 'pandas.tseries.index.DatetimeIndex'>
         [2015-01-01, ..., 2015-01-31]
         Length: 31, Freq: D, Timezone: None
```

To specify the frequency of the timestamps (which defaults to one day), we can use the freq keyword argument, and instead of using periods to specify the number of points, we can give both starting and ending points as date and time strings (or datetime objects) as the first and second arguments. For example, we can use the following to generate hourly timestamps between 00:00 and 12:00 on 2015-01-01.

```
In [61]: pd.date_range("2015-1-1 00:00", "2015-1-1 12:00", freq="H")
Out[61]: <class 'pandas.tseries.index.DatetimeIndex'>
         [2015-01-01 00:00:00, ..., 2015-01-01 12:00:00]
         Length: 13, Freq: H, Timezone: None
```

The date_range function returns an instance of DatetimeIndex, which can be used, for example, as an index for a Series or DataFrame object.

```
In [62]: ts1 = pd.Series(np.arange(31),
    ...:                  index=pd.date_range("2015-1-1", periods=31))
In [63]: ts1.head()
Out[63]: 2015-01-01    0
         2015-01-02    1
         2015-01-03    2
         2015-01-04    3
         2015-01-05    4
         Freq: D, dtype: int64
```

The elements of a DatetimeIndex object can, for example, be accessed using indexing with date and time strings. An element in a DatetimeIndex is of the type Timestamp, whivch is a pandas object that extends the standard Python datetime object (see the datetime module in the Python standard library).

```
In [64]: ts1["2015-1-3"]
Out[64]: 2
In [65]: ts1.index[2]
Out[65]: Timestamp('2015-01-03 00:00:00', offset='D')
```

In many aspects, a Timestamp and datetime object are interchangeable, and the Timestamp class has, like the datetime class, attributes for accessing time fields such as year, month, day, hour, minute, and so on. However, a notable difference between Timestamp and datetime is that Timestamp stores a timestamp with nanosecond resolution, while a datetime object only uses microsecond resolution.

```
In [66]: ts1.index[2].year, ts1.index[2].month, ts1.index[2].day
Out[66]: (2015, 1, 3)
In [67]: ts1.index[2].nanosecond
Out[67]: 0
```

We can convert a Timestamp object to a standard Python datetime object using the to_pydatetime method.

```
In [68]: ts1.index[2].to_pydatetime()
Out[68]: datetime.datetime(2015, 1, 3, 0, 0)
```

We can use a list of datetime objects to create a pandas time series.

```
In [69]: import datetime
In [70]: ts2 = pd.Series(
    ...:        np.random.rand(2),
    ...:        index=[datetime.datetime(2015, 1, 1), datetime.datetime(2015, 2, 1)])
In [71]: ts2
Out[71]: 2015-01-01    0.683801
         2015-02-01    0.916209
         dtype: float64
```

Data that is defined for sequences of time spans can be represented using Series and DataFrame objects that are indexed using the PeriodIndex class. We can construct an instance of the PeriodIndex class explicitly by passing a list of Period objects and then specify it as an index when creating a Series or DataFrame object.

```
In [72]: periods = pd.PeriodIndex([pd.Period('2015-01'),
    ...:                           pd.Period('2015-02'),
    ...:                           pd.Period('2015-03')])
In [73]: ts3 = pd.Series(np.random.rand(3), index=periods)
In [74]: ts3
Out[74]: 2015-01    0.969817
         2015-02    0.086097
         2015-03    0.016567
         Freq: M, dtype: float64
In [75]: ts3.index
Out[75]: <class 'pandas.tseries.period.PeriodIndex'>
         [2015-01, ..., 2015-03]
         Length: 3, Freq: M
```

We can also convert a Series or DataFrame object indexed by a DatetimeIndex object to a PeriodIndex using the to_period method (which takes an argument that specifies the period frequency, here 'M' for month).

```
In [76]: ts2.to_period('M')
Out[76]: 2015-01    0.683801
         2015-02    0.916209
         Freq: M, dtype: float64
```

The remaining part of this section explores select features of pandas time series through examples. Let's look at manipulating two-time series containing temperature measurement sequences at given timestamps. There is one dataset for an indoor temperature sensor and one for an outdoor temperature sensor, with observations approximately every 10 minutes during most of 2014. The two data files, temperature_indoor_2014.tsv and temperature_outdoor_2014.tsv, are TSV (tab-separated values, a variant of the CSV format) files with two columns: the first column contains Unix timestamps (seconds since Jan 1, 1970), and the second column is the measured temperature in degree Celsius. For example, the following are the first five lines in the outdoor dataset.

```
In [77]: !head -n 5 temperature_outdoor_2014.tsv
1388530986     4.380000
1388531586     4.250000
1388532187     4.190000
1388532787     4.060000
1388533388     4.060000
```

We can read the data files using read_csv by specifying that the delimiter between columns is the TAB character: delimiter="\t". When reading the two files, we also explicitly specify the column names using the names keyword argument since the files in this example do not have header lines with the column names.

```
In [78]: df1 = pd.read_csv('temperature_outdoor_2014.tsv', delimiter="\t",
    ...:                    names=["time", "outdoor"])
In [79]: df2 = pd.read_csv('temperature_indoor_2014.tsv', delimiter="\t",
    ...:                    names=["time", "indoor"])
```

Once we have created DataFrame objects for the time-series data, inspecting the data by displaying the first few lines is informative.

```
In [80]: df1.head()
Out[80]:
```

	time	outdoor
0	1388530986	4.38
1	1388531586	4.25
2	1388532187	4.19
3	1388532787	4.06
4	1388533388	4.06

The next step toward a meaningful representation of the time-series data is to convert the Unix timestamps to date and time objects using to_datetime with the unit="s" argument. Furthermore, we localize the timestamps (assigning a time zone) using tz_localize and convert the time zone attribute to the Europe/Stockholm time zone using tz_convert. We also set the time column as an index using set_index.

```
In [81]: df1.time = (pd.to_datetime(df1.time.values, unit="s")
    ...:                    .tz_localize('UTC').tz_convert('Europe/Stockholm'))
In [82]: df1 = df1.set_index("time")
In [83]: df2.time = (pd.to_datetime(df2.time.values, unit="s")
    ...:                    .tz_localize('UTC').tz_convert('Europe/Stockholm'))
In [84]: df2 = df2.set_index("time")
In [85]: df1.head()
Out[85]:
```

Time	outdoor
2014-01-01 00:03:06+01:00	4.38
2014-01-01 00:13:06+01:00	4.25
2014-01-01 00:23:07+01:00	4.19
2014-01-01 00:33:07+01:00	4.06
2014-01-01 00:43:08+01:00	4.06

Displaying the first few rows of the data frame for the outdoor temperature dataset shows that the index is a date and time object. A time series index represented as proper date and time objects (in contrast to using integers representing the Unix timestamps, for example) allows us to easily perform many time-oriented operations. Before exploring the data in more detail, let's plot the two time series to understand what the data looks like. We can use the DataFrame.plot method; the results are shown in Figure 12-3. Note that data is missing for a part of August. Imperfect data is a common problem, and handling missing data appropriately is an important part of the mission statement of the pandas library.

```
In [86]: fig, ax = plt.subplots(1, 1, figsize=(12, 4))
    ...: df1.plot(ax=ax)
    ...: df2.plot(ax=ax)
```

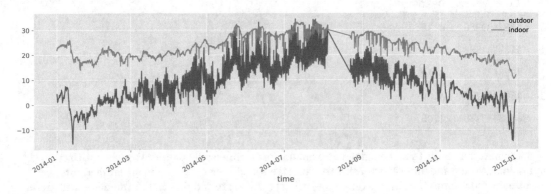

Figure 12-3. *Plot of the time series for indoor and outdoor temperature*

It is also illuminating to display the result of the info method of the DataFrame object. Doing so tells we that this dataset has nearly 50,000 data points and contains data points starting at 2014-01-01 00:03:06 and ending at 2014-12-30 23:56:35.

```
In [87]: df1.info()
<class 'pandas.core.frame.DataFrame'>
DatetimeIndex: 49548 entries, 2014-01-01 00:03:06+01:00 to 2014-12-30 23:56:35+01:00
Data columns (total 1 columns):
outdoor    49548 non-null float64
dtypes: float64(1) memory usage: 774.2 KB
```

A common operation on time series is to select and extract parts of the data. For example, from the full dataset containing all of 2014, we may be interested in selecting and analyzing only the data for January. In pandas, we can accomplish this in several ways. For example, we can use Boolean indexing of a DataFrame to create a DataFrame for a subset of the data. To create the Boolean indexing mask that selects the data for January, we can use the pandas time-series features that allow us to compare the time-series index with string representations of a date and time. In the following code, expressions like df1.index >= "2014-1-1", where df1.index is a time DateTimeIndex instance, result in a Boolean NumPy array that can be used as a mask to select the desired elements.

```
In [88]: mask_jan = (df1.index >= "2014-1-1") & (df1.index < "2014-2-1")
In [89]: df1_jan = df1[mask_jan]
In [90]: df1_jan.info()
<class 'pandas.core.frame.DataFrame'>
DatetimeIndex: 4452 entries, 2014-01-01 00:03:06+01:00 to 2014-01-31 23:56:58+01:00
Data columns (total 1 columns):
outdoor     4452 non-null float64
dtypes: float64(1) memory usage: 69.6 KB
```

Alternatively, we can use slice syntax directly with date and time strings.

```
In [91]: df2_jan = df2["2014-1-1":"2014-1-31"]
```

The results are two DataFrame objects, df1_jan and df2_jan, containing data only for January. Plotting this subset of the original data using the plot method results in the graph shown in Figure 12-4.

```
In [92]: fig, ax = plt.subplots(1, 1, figsize=(12, 4))
    ...: df1_jan.plot(ax=ax)
    ...: df2_jan.plot(ax=ax)
```

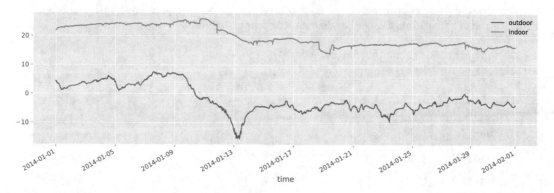

Figure 12-4. *Plot of the time series for indoor and outdoor temperature for a selected month (January)*

Like the datetime class in Python's standard library, the Timestamp class used in pandas to represent time values has attributes for accessing fields such as year, month, day, hour, minute, and so on. These fields are particularly useful when processing time series. For example, suppose we wish to calculate the average temperature for each month of the year. In this case, let's begin by creating a new column, month, which we assign to the month field of the Timestamp values of the DatetimeIndex indexer. To extract the month field from each Timestamp value, we first call reset_index to convert the index to a column in the data frame (in which case the new DataFrame object falls back to using an integer index), after which we can use the apply function on the newly created time column.[4]

---

[4]We can also directly use the month method of the DatetimeIndex index object, but for the sake of demonstration, a more explicit approach is used here.

```
In [93]: df1_month = df1.reset_index()
In [94]: df1_month["month"] = df1_month.time.apply(lambda x: x.month)
In [95]: df1_month.head()
Out[95]:
```

	time	outdoor	month
0	2014-01-01 00:03:06+01:00	4.38	1
1	2014-01-01 00:13:06+01:00	4.25	1
2	2014-01-01 00:23:07+01:00	4.19	1
3	2014-01-01 00:33:07+01:00	4.06	1
4	2014-01-01 00:43:08+01:00	4.06	1

Next, we can group the DataFrame by the new month field and aggregate the grouped values using the mean function for computing the average within each group.

```
In [96]: df1_month = df1_month[
    ...:         ["month", "outdoor"]].groupby("month").aggregate(np.mean)
In [97]: df2_month = df2.reset_index()
In [98]: df2_month["month"] = df2_month.time.apply(lambda x: x.month)
In [99]: df2_month = df2_month[
    ...:         ["month", "indoor"]].groupby("month").aggregate(np.mean)
```

After repeating the same process for the second DataFrame (indoor temperatures), we can combine df1_month and df2_month into a single DataFrame using the join method.

```
In [100]: df_month = df1_month.join(df2_month)
In [101]: df_month.head(3)
Out[101]:
```

time	outdoor	indoor
1	-1.776646	19.862590
2	2.231613	20.231507
3	4.615437	19.597748

We have leveraged Pandas' data processing capabilities in only a few lines of code to transform and compute with the data. There are often many ways to combine the tools provided by pandas to do the same or a similar analysis. For the current example, we can do the whole process in a single line of code using the to_period and groupby methods and the concat function (which, like join, combines DataFrame into a single DataFrame).

```
In [102]: df_month = pd.concat(
     ...:         [df.to_period("M").groupby(level=0).mean()
     ...:            for df in [df1, df2]], axis=1)
In [103]: df_month.head(3)
Out[103]:
```

time	outdoor	indoor
2014-01	-1.776646	19.862590
2014-02	2.231613	20.231507
2014-03	4.615437	19.597748

To visualize the results, plot the average monthly temperatures as a bar plot and a boxplot using the DataFrame method plot. The result is shown in Figure 12-5.

```
In [104]: fig, axes = plt.subplots(1, 2, figsize=(12, 4))
     ...: df_month.plot(kind='bar', ax=axes[0])
     ...: df_month.plot(kind='box', ax=axes[1])
```

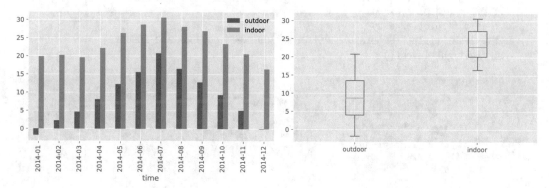

Figure 12-5. *Average indoor and outdoor temperatures per month (left) and a boxplot for monthly indoor and outdoor temperature (right)*

Finally, a handy feature of the pandas time-series objects is the ability to up- and down-sample the time series using the resample method. Resampling means changing the number of data points in a time series. It can be either increased (up-sampling) or decreased (down-sampling). For up-sampling, we need to choose a method for filling in the missing values, and for down-sampling, we need to select a method for aggregating multiple sample points between each new sample point. The resample method expects a string that specifies the new period of data points in the resampled time series as the first argument. For example, the string H represents one hour, the string D one day, the string M one month, and so on.[5] We can combine these in simple expressions, such as 7D, which denotes seven days. The resample method returns a resampler object for which we can invoke aggregation methods, such as mean and sum, to obtain the resampled data.

To illustrate using the resample method, consider the previous two-time series with temperature data. The original sampling frequency is roughly 10 minutes, which amounts to many data points over a year. For plotting purposes or to compare the two-time series sampled at slightly different timestamps, it is often necessary to downsample the original data. This can give less busy graphs and regularly spaced time series that can be readily compared to each other. The following code resamples the outdoor temperature time series to four different sampling frequencies and plots the resulting time series. We also resample the outdoor and indoor time series to daily averages that we subtract to obtain the daily average temperature difference between indoors and outdoors throughout the year (see Figure 12-6). These types of manipulations are convenient when dealing with time series, and it is one of the many areas in which the pandas library shines.

---

[5] There are a large number of available time-unit codes. For more information, see the "Offset aliases" and "Anchored offsets" sections in the Pandas reference manual.

```
In [105]: df1_hour = df1.resample("H").mean()
In [106]: df1_hour.columns = ["outdoor (hourly avg.)"]
In [107]: df1_day = df1.resample("D").mean()
In [108]: df1_day.columns = ["outdoor (daily avg.)"]
In [109]: df1_week = df1.resample("7D").mean()
In [110]: df1_week.columns = ["outdoor (weekly avg.)"]
In [111]: df1_month = df1.resample("M").mean()
In [112]: df1_month.columns = ["outdoor (monthly avg.)"]
In [113]: df_diff = (df1.resample("D").mean().outdoor -
     ...:            df2.resample("D").mean().indoor)
In [114]: fig, (ax1, ax2) = plt.subplots(2, 1, figsize=(12, 6))
     ...: df1_hour.plot(ax=ax1, alpha=0.25)
     ...: df1_day.plot(ax=ax1)
     ...: df1_week.plot(ax=ax1)
     ...: df1_month.plot(ax=ax1)
     ...: df_diff.plot(ax=ax2)
     ...: ax2.set_title("temperature difference between outdoor and indoor")
     ...: fig.tight_layout()
```

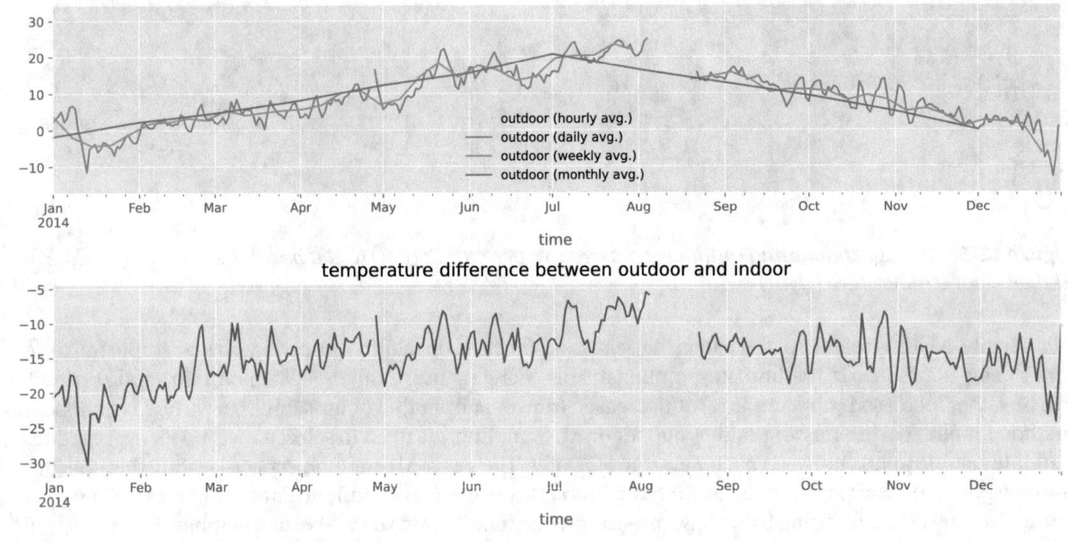

Figure 12-6. *Outdoor temperature, resampled to hourly, daily, weekly, and monthly averages (top). Daily temperature difference between outdoors and indoors (bottom)*

As an illustration of up-sampling, consider the following example, which resamples the data frame df1 to a sampling frequency of 5 minutes, using three different aggregation methods (mean, ffill for forward-fill, and bfill for back-fill). The original sample frequency is approximately 10 minutes, so this resampling is up-sampling. The result is three new data frames combined into a single DataFrame object using the concat function. The first five rows in the data frame are also shown in the following example. Note that every second data point is a new sample point, and depending on the value of the aggregation method, those values are filled (or not) according to the specified strategies. When no fill strategy is selected, the corresponding values are marked as missing using the NaN value.

```
In [115]: pd.concat(
     ...:     [df1.resample("5min").mean().rename(columns={"outdoor": 'None'}),
     ...:      df1.resample("5min").ffill().rename(columns={"outdoor": 'ffill'}),
     ...:      df1.resample("5min").bfill().rename(columns={"outdoor": 'bfill'})],
     ...:     axis=1).head()
Out[115]:
```

time	None	ffill	bfill
2014-01-01 00:00:00+01:00	4.38	4.38	4.38
2014-01-01 00:05:00+01:00	NaN	4.38	4.25
2014-01-01 00:10:00+01:00	4.25	4.25	4.25
2014-01-01 00:15:00+01:00	NaN	4.25	4.19
2014-01-01 00:20:00+01:00	4.19	4.19	4.19

The Seaborn Graphics Library

The Seaborn graphics library is built on top of Matplotlib, and it provides functions for generating graphs that are useful when working with statistics and data analysis, including distribution plots, kernel-density plots, joint distribution plots, factor plots, heatmaps, facet plots, and several ways of visualizing regressions. It also provides methods for coloring data in graphs and numerous well-crafted color palettes. The Seaborn library is created with close attention to the aesthetics of the graphs it produces, and the graphs generated by the library tend to be both good-looking and informative. The Seaborn library distinguishes itself from the underlying Matplotlib library in that it provides polished higher-level graph functions for a specific application domain, namely, statistical analysis and data visualization. The ease with which standard statistical graphs can be generated with the library makes it a valuable tool in exploratory data analysis.

To start using the Seaborn library, set a style for the graphs it produces using the sns.set function. Let's work with the style called darkgrid, which produces graphs with a gray background (also try the whitegrid style).

```
In [116]: sns.set(style="darkgrid")
```

Importing seaborn and setting a style for the library alters the default settings for how Matplotlib graphs appear, including graphs produced by the pandas library. For example, consider the following plot of the previously used indoor and outdoor temperature time series. The resulting graph is shown in Figure 12-7, and although the graph was produced using the pandas DataFrame method plot, using the sns.set function has changed the graph's appearance (compare with Figure 12-3).

```
In [117]: df1 = pd.read_csv('temperature_outdoor_2014.tsv', delimiter="\t",
     ...:                     names=["time", "outdoor"])
     ...: df1.time = (pd.to_datetime(df1.time.values, unit="s")
     ...:             .tz_localize('UTC').tz_convert('Europe/Stockholm'))
     ...: df1 = df1.set_index("time").resample("10min").mean()
In [118]: df2 = pd.read_csv('temperature_indoor_2014.tsv', delimiter="\t",
     ...:                     names=["time", "indoor"])
```

```
   ...: df2.time = (pd.to_datetime(df2.time.values, unit="s")
   ...:                 .tz_localize('UTC').tz_convert('Europe/Stockholm'))
   ...: df2 = df2.set_index("time").resample("10min").mean()
In [119]: df_temp = pd.concat([df1, df2], axis=1)
In [120]: fig, ax = plt.subplots(1, 1, figsize=(8, 4))
   ...: df_temp.resample("D").mean().plot(y=["outdoor", "indoor"], ax=ax)
```

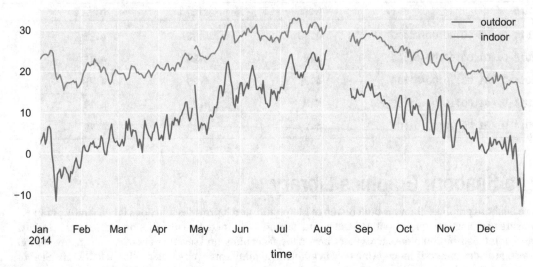

Figure 12-7. *Time-series plot produced by Matplotlib using the Pandas library, with a plot style that the Seaborn library sets up*

Apart from generating good-looking graphics, the Seaborn library's main strength is its collection of easy-to-use statistical plots. Examples are the kdeplot and histplot, which plot a kernel-density estimate plot and a histogram plot with a kernel-density estimate overlaid on top of the histogram. For example, the following two lines of code produce the graph shown in Figure 12-8. The solid blue and green lines in this figure are the kernel-density estimates that can also be graphed separately using the kdeplot function (not shown here).

```
In [121]: sns.histplot(
   ...:        df_temp.to_period("M")["outdoor"]["2014-04"].dropna().values,
   ...:        bins=50, kde=True)
   ...: sns.histplot(
   ...:        df_temp.to_period("M")["indoor"]["2014-04"].dropna().values,
   ...:        bins=50, kde=True)
```

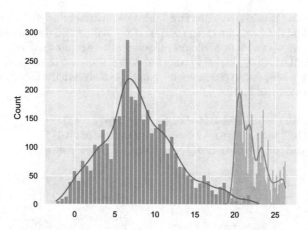

Figure 12-8. *The histogram (bars) and kernel-density plots (solid lines) for the subset of the indoor and outdoor datasets that correspond to April*

The kdeplot function can also operate on two-dimensional data, showing a contour graph of the joint kernel-density estimate. We can use the jointplot function to plot the joint distribution for two separate datasets. The following example uses the kdeplot and jointplot to show the correlation between the indoor and outdoor data series, which are resampled to hourly averages before visualized. (Missing values are dropped using the dropna method, since the functions from the seaborn module do not accept arrays with missing data.) The results are shown in Figure 12-9.

```
In [122]: sns.kdeplot(
     ...:     x=df_temp.resample("H").mean()["outdoor"].dropna().values,
     ...:     y=df_temp.resample("H").mean()["indoor"].dropna().values,
     ...:     fill=False)
In [123]: with sns.axes_style("white"):
     ...:     sns.jointplot(x=df_temp.resample("H").mean()["outdoor"].values,
     ...:                   y=df_temp.resample("H").mean()["indoor"].values,
     ...:                   kind="hex")
```

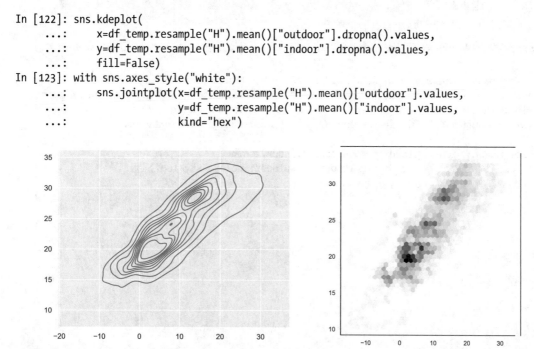

Figure 12-9. *Two-dimensional kernel-density estimate contours (left) and the joint distribution for the indoor and outdoor temperature datasets (right). The outdoor temperatures are shown on the x axis, and the indoor temperatures are on the y axis*

The seaborn library also provides functions for working with categorical data. A simple example of a graph type often useful for datasets with categorical variables is the standard boxplot for visualizing a dataset's descriptive statistics (min, max, median, and quartiles). An interesting twist on the standard boxplot is the violin plot, in which the kernel-density estimate is shown in the width of the boxplot. The boxplot and violinplot functions can produce such graphs, as shown in the following example, and the resulting graph is shown in Figure 12-10.

```
In [124]: fig, (ax1, ax2) = plt.subplots(1, 2, figsize=(8, 4))
     ...: sns.boxplot(df_temp.dropna(), ax=ax1, palette="pastel")
     ...: sns.violinplot(df_temp.dropna(), ax=ax2, palette="pastel")
```

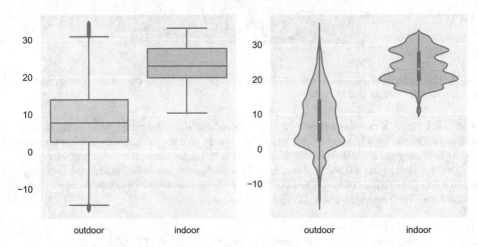

Figure 12-10. *A boxplot (left) and violin plot (right) for the indoor and outdoor temperature datasets*

As a further example of violin plots, consider the outdoor temperature dataset partitioned by the month, which can be produced by passing the month field of the data frame index as a second argument (used to group the data into categories). The resulting graph, shown in Figure 12-11, provides a compact and informative visualization of the distribution of temperatures for each month of the year.

```
In [125]: sns.violinplot(x=df_temp.dropna().index.month,
     ...:                 y=df_temp.dropna().outdoor, color="skyblue")
```

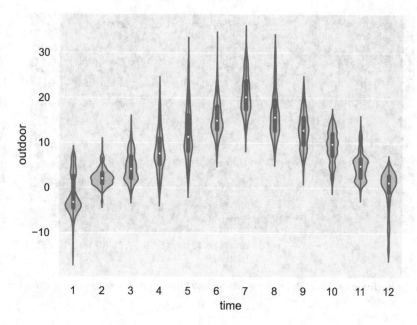

Figure 12-11. *Violin plot for the outdoor temperature grouped by month*

Heatmaps are another type of graph that is handy when dealing with categorical variables, especially for variables with many categories. The Seaborn library provides the heatmap function for generating this type of graph. For example, working with the outdoor temperature dataset, we can create two categorical columns, month and hour, by extracting those fields from the index and assigning them to new columns in the data field. Next, we can use the pivot_table function in pandas to pivot the columns into a table (matrix) where two selected categorical variables constitute the new index and columns. Here, let's pivot the temperature dataset so that the hours of the day are the columns and the months of the year are the rows (index). To aggregate the multiple data points that fall within each hour-month category, use aggfunc=np.mean argument to compute the mean of all the values.

```
In [126]: df_temp["month"] = df_temp.index.month
     ...: df_temp["hour"] = df_temp.index.hour
In [127]: table = pd.pivot_table(
     ...:         df_temp, values='outdoor', index=['month'], columns=['hour'],
     ...:         aggfunc=np.mean)
```

Once we have created a pivot table, we can visualize it as a heatmap using the heatmap function in Seaborn. The result is shown in Figure 12-12.

```
In [128]: fig, ax = plt.subplots(1, 1, figsize=(8, 4))
     ...: sns.heatmap(table, ax=ax)
```

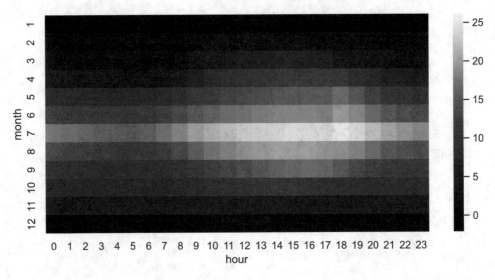

Figure 12-12. *A heatmap of the outdoor temperature data grouped by the hour of the day and month of the year*

The Seaborn library contains many more statistical visualization tools than we have been able to survey here. However, I hope that looking at a few examples of what this library can do illustrates the essence of the Seaborn library—a convenient tool for statistical analysis and exploration of data that can produce many standard statistical graphs with minimal effort. The upcoming chapters demonstrate further examples of applications of the Seaborn library.

Summary

This chapter explored data representation and processing using the Pandas library and briefly surveyed the statistical graphics tools provided by the Seaborn visualization library. The Pandas library provides the backend of much data wrangling done with Python. It achieves this by adding a higher-level abstraction layer in the data representation on top of NumPy arrays, with additional methods for operating on the underlying data. The ease with which data can be loaded, transformed, and manipulated makes it an invaluable part of the data processing workflow in Python. The Pandas library also contains essential functions for visualizing the data represented by its data structures. Visualizing data represented as the Pandas series and data frames quickly is an important tool in exploratory data analytics and for presentation. The Seaborn library takes this a step further and provides a rich collection of statistical graphs that can be produced often with a single line of code. Many functions in the Seaborn library can operate directly on Pandas data structures.

Further Reading

A great introduction to the Pandas library is given by the original creator of the library in *Python for Data Analysis* by W. McKinney (O'Reilly 2013), and it is also a rather detailed introduction to NumPy. The Pandas official documentation, available at `http://pandas.pydata.org/pandas-docs/stable`, provides an accessible and detailed description of the library's features. Another good online resource for learning Pandas is `http://github.com/jvns/pandas-cookbook`. For data visualization, we looked at the Seaborn Library, and it is well described in the documentation available on its website. With respect to higher-level visualization tools, it is also worth exploring the `ggplot` library for Python, `http://ggplot.yhathq.com`, which is an implementation based on the renowned *The Grammar of Graphics* by L. Wilkinson (Springer, 2005). This library is also closely integrated with the Pandas library, providing convenient statistical visualization tools when analyzing data. For more information about visualization in Python, see *Beginning Python Visualization* by S. Vaingast (Apress, 2014).

CHAPTER 13

Statistics

Statistics has long been a field of mathematics that is relevant to practically all applied disciplines of science and engineering, as well as business, medicine, and other fields where data is used for obtaining knowledge and making decisions. With the recent proliferation of data analytics, there has been renewed interest in statistical methods. But computer-aided statistics has a long history, and it is a field traditionally dominated by domain-specific software packages and programming environments, such as the S language and, more recently, its open source counterpart: the R language. The use of Python for statistical analysis has increased rapidly over the last several years, and there is now a mature collection of statistical libraries for Python. With these libraries, Python can match the performance and features of domain-specific languages in many areas of statistics, albeit not all, while also providing the unique advantages of the Python programming language and its environment. The Pandas library discussed in Chapter 12 is an example where traditional statistical software influenced the Python community, especially with the introduction of the data frame data structure to the Python environment. The NumPy and SciPy libraries provide computational tools for many fundamental statistical concepts, and higher-level statistical modeling and machine learning are covered by the statsmodels and scikit-learn libraries, which we will see more of in the following chapters.

This chapter focuses on fundamental statistical applications using Python, particularly the stats module in SciPy. It discusses computing descriptive statistics, random numbers, random variables, distributions, and hypothesis testing. Statistical modeling and machine-learning applications are covered in the upcoming chapters. Some fundamental statistical functions are also available through the NumPy library, such as its functions and methods for computing descriptive statistics and its module for generating random numbers. The SciPy stats module builds on top of NumPy and, for example, provides random number generators with more specialized distributions.

Importing Modules

This chapter mainly works with the stats module in SciPy. Following the convention to selectively import modules from SciPy, let's assume that this module and the optimize module are imported in the following way.

```
In [1]: from scipy import stats
   ...: from scipy import optimize
```

As usual, we also require the NumPy and the Matplotlib libraries.

```
In [2]: import numpy as np
In [3]: import matplotlib.pyplot as plt
```

R. Johansson, *Numerical Python*, https://doi.org/10.1007/979-8-8688-0413-7_13

For statistical graphs and styling, we use the Seaborn library.

```
In [4]: import seaborn as sns
In [5]: sns.set(style="whitegrid")
```

Review of Statistics and Probability

Let's begin with a brief review of statistics, introducing some of the key concepts and the notation we use in this and the following chapters. Statistics involves collecting and analyzing data to gain insights, draw conclusions, and support decision-making. Statistical methods are necessary when we have incomplete information about a phenomenon. Typically, we have incomplete information because we cannot collect data from all members of a *population* or if there is *uncertainty* in our observations (e.g., due to measurement noise). When we cannot survey an entire population, a randomly chosen *sample* can be studied instead. We can use statistical methods and compute descriptive statistics (parameters such as the mean and the variances) to make inferences about the properties of the entire population (also called *sample space*) systematically and with a controlled risk of errors.

Statistical methods are built on the foundation provided by probability theory, with which we can model uncertainty and incomplete information using probabilistic, random variables. For example, with *randomly* selected samples of a population, we can obtain representative samples whose properties can be used to infer properties of the entire population. In probability theory, each possible outcome for an observation is given a probability, and the probability for all possible outcomes constitutes the probability distribution. Given the probability distribution, we can compute the properties of the population, such as its mean and variance. However, for randomly selected samples, we only know the *expected or average* results.

In statistical analysis, it is important to distinguish between population and sample statistics. Here, we denote population parameters with Greek symbols and parameters of a sample with the corresponding population symbol with the added subscript x (or the symbol used to represent the sample). For example, a population's mean and variance are denoted with μ and σ^2, and the mean and the variance of a sample x are denoted as μ_x and σ_x^2. Furthermore, we denote variables representing a population (random variables) with capital letters, for example, X, and a set of sample elements is denoted with a lowercase letter, for example, x. A bar over a symbol denotes the average or mean, $\mu = \bar{X} = \frac{1}{N} \sum_{i=1}^{N} x_i$ and $\mu_x = \bar{x} = \frac{1}{n} \sum_{i=1}^{n} x_i$, where N is the number of elements in the population X and n is the number of elements in the sample x. The only difference between these two expressions is the number of elements in the sum ($N \geq n$). The situation is slightly more complex for the variance: the population variance is the mean of the squared distance from the mean, $\sigma^2 = \frac{1}{N} \sum_{i=1}^{N} (x_i - \mu)^2$, and the corresponding sample variance is $\sigma_x^2 = \frac{1}{n-1} \sum_{i=1}^{n} (x_i - \mu_x)^2$. In the latter expression, we have replaced the population mean μ with the sample mean μ_x and divided the sum with $n - 1$ rather than n. This is because one degree of freedom has been eliminated from the sample set when calculating the mean μ_x, so when computing the sample variance, only $n - 1$ degrees of freedom remain. Consequently, the way to compute the variance for a population and a sample is slightly different. This is reflected in functions we can use to compute these statistics in Python.

Chapter 2 demonstrated that we can compute descriptive statistics for data using NumPy functions or the corresponding ndarray methods. For example, to compute the mean and the median of a dataset, we can use the NumPy functions mean and median.

```
In [6]: x = np.array([3.5, 1.1, 3.2, 2.8, 6.7, 4.4, 0.9, 2.2])
In [7]: np.mean(x)
Out[7]: 3.1
In [8]: np.median(x)
Out[8]: 3.0
```

Similarly, we can use min and max functions or ndarray methods to compute the minimum and maximum values in the array.

```
In [9]: x.min(), x.max()
Out[9]: (0.90, 6.70)
```

To compute the variance and the standard deviation for a dataset, we use the var and std methods. By default, the population variance and standard deviation formulas are used (i.e., it is assumed that the dataset is the entire population).

```
In [10]: x.var()
Out[10]: 3.07
In [11]: x.std()
Out[11]: 1.7521415467935233
```

However, we can use the ddof argument (delta degrees of freedom) to change this behavior. The denominator in the expression for the variance is the number of elements in the array minus ddof, so to calculate the unbiased estimate of the variance and standard deviation from a sample, we need to set ddof=1.

```
In [12]: x.var(ddof=1)
Out[12]: 3.5085714285714293
In [13]: x.std(ddof=1)
Out[13]: 1.8731181032095732
```

The following sections explore how to use NumPy and SciPy's stats module to generate random numbers, represent random variables and distributions, and test hypotheses.

Random Numbers

The Python standard library contains the random module, which provides functions for generating single random numbers with a few elemental distributions. The random module in the NumPy module provides similar functionality but also offers functions that generate NumPy arrays with random numbers, and it has support for a more comprehensive selection of probability distributions. Arrays with random numbers are often practical for computational purposes, so here we focus on the random module in NumPy, and later also the higher-level functions and classes in scipy.stats, which build on top of and extend NumPy.

Earlier in this book, we used np.random.rand, which generates uniformly distributed floating-point numbers in the half-open interval [0, 1) (i.e., 0.0 is a possible outcome, but 1.0 is not). In addition to this function, the np.random module contains an extensive collection of other functions for generating random numbers that cover different intervals, have different distributions, and take values of different types (e.g., floating-point numbers and integers). For example, the randn function produces random numbers that are distributed according to the *standard normal distribution* (the normal distribution with mean 0 and standard deviation 1), and the randint function generates uniformly distributed integers between a given low (inclusive) and high (exclusive) value[1]. When the rand and randn functions are called without arguments, they produce a single random number.

---

[1] You can use the np.random.seed function to initiate the random number generator in a state that results in reproducible outcomes. Here np.random.seed(123456789) was used.

```
In [14]: np.random.rand()
Out[14]: 0.532833024789759
In [15]: np.random.randn()
Out[15]: 0.8768342101492541
```

However, passing the shape of the array as arguments to these functions produces arrays of random numbers. For example, here we generate a vector of length 5 using rand by passing a single argument 5 and a 2 × 4 array using randn by passing 2 and 4 as arguments (higher-dimensional arrays are generated by passing the length of each dimension as arguments).

```
In [16]: np.random.rand(5)
Out[16]: array([ 0.71356403,  0.25699895,  0.75269361,  0.88387918,  0.15489908])
In [17]: np.random.randn(2, 4)
Out[17]: array([[ 3.13325952,  1.15727052,  1.37591514,  0.94302846],
                [ 0.8478706 ,  0.52969142, -0.56940469,  0.83180456]])
```

To generate random integers using randint (see also random_integers), we need to either provide the upper limit for the random numbers (in which case the lower limit is implicitly zero) or provide both the lower and upper limits. The size of the generated array is specified using the size keyword arguments, and it can be an integer or a tuple that specifies the shape of a multidimensional array.

```
In [18]: np.random.randint(10, size=10)
Out[18]: array([0, 3, 8, 3, 9, 0, 6, 9, 2, 7])
In [19]: np.random.randint(low=10, high=20, size=(2, 10))
Out[19]: array([[12, 18, 18, 17, 14, 12, 14, 10, 16, 19],
                [15, 13, 15, 18, 11, 17, 17, 10, 13, 17]])
```

The randint function generates random integers in the half-open interval [low, high). To demonstrate that the random numbers produced by rand, randn, and randint, are distributed differently, we can plot the histograms of, let's say, 10,000 random numbers produced by each function. The result is shown in Figure 13-1. We note that the distributions for rand and randint appear uniform but have different ranges and types. In contrast, the distribution of the numbers produced by randn resembles a Gaussian curve centered at zero, as expected.

```
In [20]: fig, axes = plt.subplots(1, 3, figsize=(12, 3))
    ...: axes[0].hist(np.random.rand(10000))
    ...: axes[0].set_title("rand")
    ...: axes[1].hist(np.random.randn(10000))
    ...: axes[1].set_title("randn")
    ...: axes[2].hist(np.random.randint(low=1, high=10, size=10000),
    ...:              bins=9, align='left')
    ...: axes[2].set_title("randint(low=1, high=10)")
```

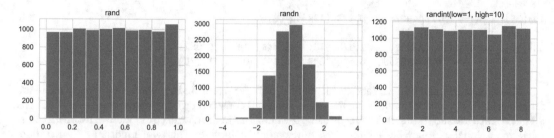

Figure 13-1. *Distributions for 10,000 random numbers generated by the rand, randn, and randint functions in NumPy's random module*

In statistical analysis, generating a unique list of integers is often necessary. This corresponds to sampling (randomly selecting) items from a set (population) without replacement (so that we do not get the same item twice). We can use the choice function from the NumPy `random` module to generate this type of random number. As the first argument, we can either provide a list (or array) with the population's values or an integer corresponding to the number of elements in the population. As the second argument, we give the number of values to be sampled. The `replace` keyword argument can specify whether the values are sampled with or without replacement, using the Boolean values `True` or `False`. For example, to sample five unique (without replacement) items from the set of integers between 0 (inclusive) and 10 (exclusive), we can use the following.

```
In [21]: np.random.choice(10, 5, replace=False)
Out[21]: array([9, 0, 5, 8, 1])
```

When working with random number generation, it can be useful to *seed* the random number generator. The seed is a number that initializes a random number generator to a specific state. Once seeded with a specific number, it always generates the same sequence of random numbers. This can be useful when testing and reproducing previous results and occasionally in applications requiring reseeding the random number generator (e.g., after having forked a process). To seed the random number generator in NumPy, we can use the seed function, which takes an integer as the argument.

```
In [22]: np.random.seed(123456789)
In [23]: np.random.rand()
Out[23]: 0.532833024789759
```

Note that after seeding the random number generator with a specific number, here 123456789, the following calls to the random number generators always produce the same results.

```
In [24]: np.random.seed(123456789); np.random.rand()
Out[24]: 0.532833024789759
```

The seed of the random number generator is a global state of the `np.random` module. A finer level of control of the state of the random number generator can be achieved by using the `RandomState` class, which optionally takes a seed integer as the argument to its initializer. The `RandomState` object keeps track of the state of the random number generator. It allows maintaining several independent random number generators in the same program (which can be useful, e.g., when working with threaded applications). Once a `RandomState` object has been created, we can use the methods of this object to generate random numbers. The `RandomState` class has methods corresponding to the functions available in the `np.random` module. For example, we can use the `randn` method of the `RandomState` class to generate standard normal distributed random numbers.

```
In [25]: prng = np.random.RandomState(123456789)
In [26]: prng.randn(2, 4)
Out[26]: array([[ 2.212902,    2.1283978,   1.8417114,   0.08238248],
                [ 0.85896368, -0.82601643,  1.15727052,  1.37591514]])
```

Similarly, there are methods, rand, randint, rand_integers, and choice, which also correspond to the functions in the np.random module with the same name. It is considered good programming practice to use a RandomState instance rather than directly using the functions in the np.random module because it avoids relying on a global state variable and improves the isolation of the code. This is an important consideration when developing library functions that use random numbers but is less critical in smaller applications and calculations.

In addition to the fundamental random number distributions we have looked at so far (discrete and continuous uniform distributions, randint and rand, and the standard normal distribution, randn), there are also functions and RandomState methods for many probability distributions that occur in statistics. These include the continuous χ^2 distribution (chisquare), the Student's t distribution (standard_t), and the F distribution (f).

```
In [27]: prng.chisquare(1, size=(2, 2))
Out[27]: array([[ 0.78631596,  0.19891367],
                [ 0.11741336,  2.8713997 ]])
In [28]: prng.standard_t(1, size=(2, 3))
Out[28]: array([[ 0.39697518, -0.19469463,  1.15544019],
                [-0.65730814, -0.55125015,  0.13578694]])
In [29]: prng.f(5, 2, size=(2, 4))
Out[29]: array([[  0.45471421,  17.64891848,   1.48620557,   2.55433261],
                [  1.21823269,   3.47619315,   0.50835525,   0.70599655]])
```

It also includes the discrete binomial distribution (binomial) and the Poisson distribution (poisson).

```
In [30]: prng.binomial(10, 0.5, size=10)
Out[30]: array([4, 5, 6, 7, 3, 5, 7, 5, 4, 5])
In [31]: prng.poisson(5, size=10)
Out[31]: array([3, 5, 5, 5, 0, 6, 5, 4, 6, 3])
```

See the docstrings for the np.random module, help(np.random), and the RandomState class for a complete list of available distribution functions. While it is possible to use the functions in np.random and methods in RandomState to draw random numbers from many different statistical distribution functions, when working with distributions, there is a higher-level interface in the scipy.stats module that combines random number sampling with many other convenient functions for probability distributions. The following section explores this in more detail.

Random Variables and Distributions

In probability theory, the set of possible outcomes of a random process is called the *sample space*. Each element in the sample space (i.e., an outcome of an experiment or an observation) can be assigned a probability, and the probabilities of all possible outcomes define the probability distribution. A *random variable* is a mapping from the sample space to the real numbers or integers. For example, the possible outcomes of a coin toss are head and tail, so the sample space is {head, tail}, and a possible random variable takes the value 0 for head and 1 for tail. There are many ways to define random variables for the possible outcomes of a given random process. Random variables are a problem-independent representation of a random process. It is easier to work with random variables because they are described by numbers instead of outcomes from problem-specific sample spaces. A common step in statistical problem-solving is to map outcomes to numerical values and figure out the probability distribution of those values.

Consequently, a random variable is characterized by its possible values and probability distribution, which assigns a probability for each possible value. Each observation of the random variable results in a random number, and the probability distribution describes the observed values. There are two main types of distributions, discrete and continuous distributions, which are integer-valued and real-valued, respectively. When working with statistics, dealing with random variables is of central importance, and in practice, this often means working with probability distributions. The SciPy `stats` module provides classes for representing random variables with many probability distributions. There are two base classes for discrete and continuous random variables: `rv_discrete` and `rv_continuous`. These classes are not used directly but as base classes for random variables with specific distributions and define a common interface for all random variable classes in SciPy `stats`. A summary of selected methods for discrete and continuous random variables is given in Table 13-1.

Table 13-1. *Selected Methods for Discrete and Continuous Random Variables in the SciPy stats Module*

Methods	Description
pdf/pmf	Probability distribution function (continuous) or probability mass function (discrete)
cdf	Cumulative distribution function
sf	Survival function (1 – cdf)
ppf	Percent-point function (inverse of cdf)
moment	Noncentral moments of nth order
stats	Statistics of the distribution (typically the mean and variance, sometimes additional statistics)
fit	Fit distribution to data using a numerical maximum likelihood optimization (for continuous distributions)
expect	Expectation value of a function with respect to the distribution
interval	The endpoints of the interval that contains a given percentage of the distribution (confidence interval)
rvs	Random variable samples Takes as arguments the size of the resulting array of samples
mean, median, std, var	Descriptive statistics: mean, median, standard deviation, and the variance of the distribution

There are many classes for the discrete and continuous random variables in the SciPy `stats` module. There are classes for 13 discrete and 98 continuous distributions at the time of writing, including the most encountered distributions (and many less common). For a complete reference, see the docstring for the `stats` module: `help(stats)`. The following explores some of the more common distributions, but the usage of all the other distributions follows the same pattern.

The random variable classes in the SciPy `stats` module have several uses. They are both representations of the distribution, which can be used to compute descriptive statistics and for graphing, and they can be used to generate random numbers following the given distribution using the `rvs` (random variable sample) method. The latter usecase is similar to the `np.random` module was used for earlier in this chapter.

To demonstrate how to use the random variable classes in SciPy `stats`, consider the following example where we create a normal distributed random variable with a mean 1.0 and standard deviation of 0.5.

```
In [32]: X = stats.norm(1, 0.5)
```

Now X is an object that represents a random variable, and we can compute descriptive statistics of this random variable using, for example, the `mean`, `median`, `std`, and `var` methods.

```
In [33]: X.mean()
Out[33]: 1.0
In [34]: X.median()
Out[34]: 1.0
In [35]: X.std()
Out[35]: 0.5
In [36]: X.var()
Out[36]: 0.25
```

Noncentral moments of arbitrary order can be computed with the `moment` method.

```
In [37]: [X.moment(n) for n in range(5)]
Out[37]: [1.0, 1.0, 1.25, 1.75, 2.6875]
```

We can obtain a distribution-dependent list of statistics using the `stats` method (here, for a normal distributed random variable, we get the mean and the variance).

```
In [38]: X.stats()
Out[38]: (1.0, 0.25)
```

We can evaluate the probability distribution function, the cumulative distribution function, the survival function, using methods like `pdf`, `cdf`, and `sf`. These all take a value, or an array of values, to evaluate the function.

```
In [39]: X.pdf([0, 1, 2])
Out[39]: array([ 0.10798193,   0.79788456,   0.10798193])
In [40]: X.cdf([0, 1, 2])
Out[40]: array([ 0.02275013,   0.5,         0.97724987])
```

The `interval` method can compute the lower and upper values of x such that a given percentage of the probability distribution falls within the interval (lower, upper). This method is helpful in computing confidence intervals and for selecting a range of x values for plotting.

```
In [41]: X.interval(0.95)
Out[41]: (0.020018007729972975, 1.979981992270027)
In [42]: X.interval(0.99)
Out[42]: (-0.28791465177445019, 2.2879146517744502)
```

To build intuition for the properties of a probability distribution, it is useful to graph it together with the corresponding cumulative probability function and the percent-point function. To make it easier to repeat this for several distributions, we first create a `plot_rv_distribution` function that plots the result of `pdf` or `pmf`, the `cdf` and `sf`, and `ppf` methods of the SciPy `stats` random variable objects over an interval that contains 99.9% of the probability distribution function. We also highlight the area that contains 95% of the probability distribution using the fill_between drawing method.

```
In [43]: def plot_rv_distribution(X, axes=None):
    ...:     """Plot the PDF or PMF, CDF, SF and PPF of a given random variable"""
    ...:     if axes is None:
    ...:         fig, axes = plt.subplots(1, 3, figsize=(12, 3))
    ...:
    ...:     x_min_999, x_max_999 = X.interval(0.999)
    ...:     x999 = np.linspace(x_min_999, x_max_999, 1000)
    ...:     x_min_95, x_max_95 = X.interval(0.95)
    ...:     x95 = np.linspace(x_min_95, x_max_95, 1000)
    ...:
    ...:     if hasattr(X.dist, "pdf"):
    ...:         axes[0].plot(x999, X.pdf(x999), label="PDF")
    ...:         axes[0].fill_between(x95, X.pdf(x95), alpha=0.25)
    ...:     else:
    ...:         # discrete random variables do not have a pdf method,
    ...:         # instead  use pmf:
    ...:         x999_int = np.unique(x999.astype(int))
    ...:         axes[0].bar(x999_int, X.pmf(x999_int), label="PMF")
    ...:     axes[1].plot(x999, X.cdf(x999), label="CDF")
    ...:     axes[1].plot(x999, X.sf(x999), label="SF")
    ...:     axes[2].plot(x999, X.ppf(x999), label="PPF")
    ...:
    ...:     for ax in axes:
    ...:         ax.legend()
```

Next, we use this function to graph a few examples of distributions: the normal distribution, the F distribution, and the discrete Poisson distribution. The result is shown in Figure 13-2.

```
In [44]: fig, axes = plt.subplots(3, 3, figsize=(12, 9))
    ...: X = stats.norm()
    ...: plot_rv_distribution(X, axes=axes[0, :])
    ...: axes[0, 0].set_ylabel("Normal dist.")
    ...: X = stats.f(2, 50)
    ...: plot_rv_distribution(X, axes=axes[1, :])
    ...: axes[1, 0].set_ylabel("F dist.")
    ...: X = stats.poisson(5)
    ...: plot_rv_distribution(X, axes=axes[2, :])
    ...: axes[2, 0].set_ylabel("Poisson dist.")
```

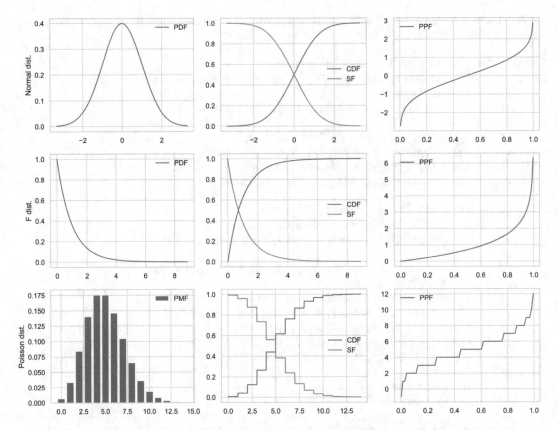

Figure 13-2. *Examples of probability distribution functions (PDF) or probability mass functions (PMFs), cumulative distribution functions (CDF), survival functions (SF), and percent-point functions (PPF) for a normal distribution (top), an F distribution (middle), and a Poisson distribution (bottom)*

The examples so far have initiated an instance of a random variable class and computed statistics and other properties using method calls. An alternative way to use the random variable classes in SciPy's `stats` module is to use class methods, for example, `stats.norm.mean`, and pass the distribution parameters as arguments (often `loc` and `scale`, as in this example for normally distributed values).

```
In [45]: stats.norm.stats(loc=2, scale=0.5)
Out[45]: (2.0, 0.25)
```

This gives the same result as creating an instance and then calling the corresponding method.

```
In [46]: stats.norm(loc=1, scale=0.5).stats()
Out[46]: (1.0, 0.25)
```

Most methods in the `rv_discrete` and `rv_continuous` classes can be used as class methods in this way.

So far, we have only looked at properties of the distribution function of random variables. Note that although a distribution function describes a random variable, the distribution itself is entirely deterministic. To draw random numbers that are distributed according to the given probability distribution, we can use the rvs (random variable sample) method. It takes as the argument the shape of the required array (can be an integer for a vector or a tuple of dimension lengths for a higher-dimensional array). Here, we use rvs(10) to generate a one-dimensional array with ten values.

```
In [47]: X = stats.norm(1, 0.5)
In [48]: X.rvs(10)
Out[48]: array([2.106451,    2.0641989,   1.9208557, 1.04119124, 1.42948184,
                0.58699179, 1.57863526,  1.68795757, 1.47151423, 1.4239353 ])
```

To see that the resulting random numbers are indeed distributed according to the corresponding probability distribution function, we can graph a histogram of many samples of a random variable and compare it to the probability distribution function. Again, to do this easily for samples of several random variables, we create a plot_dist_samples function. This function uses the interval method to obtain a suitable plot range for a random variable object.

```
In [49]: def plot_dist_samples(X, X_samples, title=None, ax=None):
    ...:     """Plot the PDF and histogram of samples of a continuous
    ...:     random variable"""
    ...:     if ax is None:
    ...:         fig, ax = plt.subplots(1, 1, figsize=(8, 4))
    ...:
    ...:     x_lim = X.interval(.99)
    ...:     x = np.linspace(*x_lim, num=100)
    ...:
    ...:     ax.plot(x, X.pdf(x), label="PDF", lw=3)
    ...:     ax.hist(X_samples, label="samples", density=1, bins=75)
    ...:     ax.set_xlim(*x_lim)
    ...:     ax.legend()
    ...:
    ...:     if title:
    ...:         ax.set_title(title)
    ...:     return ax
```

Note that in this function, we have used the tuple unpacking syntax *x_lim, which distributes the elements in the tuple x_lim to different arguments for the function. In this case, it is equivalent to np.linspace(x_lim[0], x_lim[1], num=100).

Next, we use this function to visualize 2000 samples of three random variables with different distributions: here, we use the Student's t distribution, the χ^2 distribution, and the exponential distribution, and the results are shown in Figure 13-3. Since 2000 is a reasonably large sample, the histogram graphs of the samples coincide well with the probability distribution function. The agreement can be expected to be even better with an even larger number of samples.

```
In [50]: fig, axes = plt.subplots(1, 3, figsize=(12, 3))
    ...: N = 2000
    ...: # Student's t distribution
    ...: X = stats.t(7.0)
    ...: plot_dist_samples(X, X.rvs(N), "Student's t dist.", ax=axes[0])
    ...: # The chisquared distribution
    ...: X = stats.chi2(5.0)
    ...: plot_dist_samples(X, X.rvs(N), r"$\chi^2$ dist.", ax=axes[1])
```

```
...: # The exponential distribution
...: X = stats.expon(0.5)
...: plot_dist_samples(X, X.rvs(N), "exponential dist.", ax=axes[2])
```

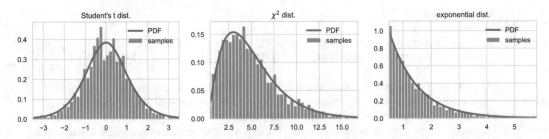

Figure 13-3. *Probability distribution function (PDF) together with histograms of 2000 random samples from the Student's t distribution (left), the χ^2 distribution (middle), and the exponential distribution (right)*

The opposite of drawing random samples from a known distribution function is to fit a given probability distribution with unknown parameters to a set of data points. In such a fit, we typically wish to optimize the unknown parameters to maximize the likelihood of observing the given data. This is called a maximum likelihood fit. Many of the random variable classes in the SciPy `stats` module implement the method `fit` that performs such a fitting to given data. As a first example, consider drawing 500 random samples from the χ^2 distribution with five degrees of freedom (`df=5`) and then refitting the random variables to the χ^2 distribution using the `fit` method.

```
In [51]: X = stats.chi2(df=5)
In [52]: X_samples = X.rvs(500)
In [53]: df, loc, scale = stats.chi2.fit(X_samples)
In [54]: df, loc, scale
Out[54]: (5.2886783664198465, 0.0077028130326141243, 0.93310362175739658)
In [55]: Y = stats.chi2(df=df, loc=loc, scale=scale)
```

The `fit` method returns the maximum likelihood parameters of the distribution for the given data. We can pass those parameters to the initializer of the `stats.chi2` to create a new random variable instance Y. The probability distribution of Y should resemble the probability distribution of the original random variable X. To verify this, we can plot the probability distribution functions for both random variables. The resulting graph is shown in Figure 13-4.

```
In [56]: fig, axes = plt.subplots(1, 2, figsize=(12, 4))
    ...: x_lim = X.interval(.99)
    ...: x = np.linspace(*x_lim, num=100)
    ...:
    ...: axes[0].plot(x, X.pdf(x), label="original")
    ...: axes[0].plot(x, Y.pdf(x), label="recreated")
    ...: axes[0].legend()
    ...:
    ...: axes[1].plot(x, X.pdf(x) - Y.pdf(x), label="error")
    ...: axes[1].legend()
```

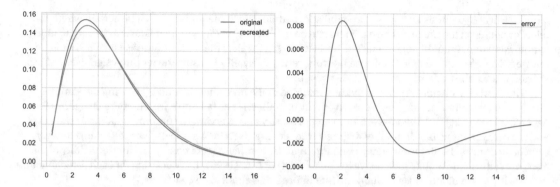

Figure 13-4. *Original and re-created probability distribution function (left) and the error (right) from a maximum likelihood fit of 500 random samples of the original distribution*

This section explored how to use random variable objects from the SciPy stats model to describe random variables with various distributions and how they can be used to compute the given distributions' properties, generate random variable samples, and perform maximum likelihood fitting. The following section explains how to use these random variable objects for hypothesis testing.

Hypothesis Testing

Hypothesis testing is a cornerstone of the scientific method, which requires that claims are investigated objectively and that a claim is rejected or accepted based on factual observations. Statistical hypothesis testing has a more specific meaning. It is a systematic methodology for evaluating whether a claim or a hypothesis is reasonable based on data. As such, it is an important application of statistics. In this methodology, we formulate the hypothesis using a null hypothesis, H_0, which represents the currently accepted state of knowledge, and an alternative hypothesis, H_A, which represents a new claim that challenges the current state of knowledge. The null hypothesis and the alternative hypothesis must be mutually exclusive and complementary so that one and only one of the hypotheses is true.

Once H_0 and H_A are defined, the data that support the test must be collected, for example, through measurements, observations, or a survey. The next step is to find a test statistic that can be computed from the data and whose probability distribution function can be found under the null hypothesis. We can evaluate the data by computing the probability (the *p-value*) of obtaining the observed value of the test statistic (or a more extreme one) using the distribution function implied by the null hypothesis. If the *p*-value is smaller than a predetermined threshold, known as the significance level, and denoted by α (typically 5% or 1%), we can conclude that the observed data is unlikely to have been described by the distribution corresponding to the null hypothesis. In that case, we can, therefore, reject the null hypothesis in favor of the alternative hypothesis. The steps for carrying out a hypothesis test are summarized in the following list.

1. Formulate the null hypothesis and the alternative hypothesis.

2. Select a test statistic such that its sampling distribution under the null hypothesis is known (exactly or approximately).

3. Collect data.

4. Compute the test statistics from the data and calculate its *p*-value under the null hypothesis.

5. If the *p*-value is smaller than the predetermined significance level α, we reject the null hypothesis. If the *p*-value is larger, we fail to reject the null hypothesis.

Statistical hypothesis testing is a probabilistic method, which means we cannot be certain whether to reject or not reject the null hypothesis. There can be two types of error: we can mistakenly reject the null hypothesis when it should not be rejected, and we can fail to reject the null hypothesis when it should be rejected. These are called type I and type II errors, respectively. By choosing the required significance level, we can balance the trade-off between these two types of error.

The most challenging step in the method outlined in the preceding section is knowing the sampling distribution of the test statistics. Fortunately, many hypothesis tests fall into a few standard categories for which the probability distributions are known. A summary and overview of common hypothesis test cases and the corresponding distribution of their test statistics are given in Table 13-2. For motivations for why each of these tests is suitable for stated situations and the complete set of conditions for the validity of the tests, see statistics textbooks such as Wasserman (2004) or Rice (1995). The docstring for each listed function in the SciPy `stats` module also contains further information about each test.

Table 13-2. *Summary of Common Hypothesis Test Cases with the Corresponding Distributions and SciPy Functions*

Null Hypothesis	Distributions	SciPy Functions for Test
Test if the mean of a population is a given value.	Normal distribution (`stats.norm`), or Student's t distribution (`stats.t`)	`stats.ttest_1samp`
Test if the means of two random variables are equal (independent or paired samples).	Student's t distribution (`stats.t`)	`stats.ttest_ind`, `stats.ttest_rel`
Test goodness of fit of a continuous distribution to data.	Kolmogorov-Smirnov distribution	`stats.kstest`
Test if categorical data occur with given frequency (sum of squared normally distributed variables).	χ^2 distribution (`stats.chi2`)	`stats.chisquare`
Test for the independence of categorical variables in a contingency table.	χ^2 distribution (`stats.chi2`)	`stats.chi2_contingency`
Test for equal variance in samples of two or more variables.	F distribution (`stats.f`)	`stats.barlett`, `stats.levene`
Test for noncorrelation between two variables.	Beta distribution (`stats.beta`, `stasts.mstats.betai`)	`stats.pearsonr`, `stats.spearmanr`
Test if two or more variables have the same population mean (ANOVA—analysis of variance).	F distribution	`stats.f_oneway`, `stats.kruskal`

The following also looks at examples of how the corresponding functions in the SciPy `stats` module can be used to carry out steps 4 and 5 in the preceding procedure: computing a test statistic and the corresponding p-value.

For example, a common null hypothesis is a claim that the mean μ of a population is a certain value μ_0. We can then sample the population and use the sample mean $\bar{x}$ to form a test statistic $z = \dfrac{\bar{x} - \mu_0}{\sigma / \sqrt{n}}$, where n is the sample size. If the population is large and the variance σ is known, then it is reasonable to assume that the test statistic is normally distributed. If the variance is unknown, we can substitute σ^2 with the sample variance σ_x^2. The test statistic then follows the Student's t distribution, which approaches the normal distribution in the limit of a large number of samples. Regardless of which distribution is used, we can compute a p-value for the test statistics using the given distribution.

As an example of how this type of hypothesis test can be carried out using the functions provided by the SciPy stats module, consider a null hypothesis that claims that a random variable X has mean $\mu_0 = 1$. Given samples of X, we then wish to test if the sampled data is compatible with the null hypothesis. Here, we simulate the samples by drawing 100 random samples from a distribution slightly different than that claimed by the null hypothesis (using $\mu = 0.8$).

```
In [57]: mu0, mu, sigma = 1.0, 0.8, 0.5
In [58]: X = stats.norm(mu, sigma)
In [59]: n = 100
In [60]: X_samples = X.rvs(n)
```

Given the sample data, X_samples, next, we need to compute a test statistic. If the population standard deviation σ is known, as in this example, we can use $z = \dfrac{\bar{x} - \mu_0}{\sigma / \sqrt{n}}$, which is normally distributed.

```
In [61]: z = (X_samples.mean() - mu0)/(sigma/np.sqrt(n))
In [62]: z
Out[62]: -2.8338979550098298
```

If the population variance is unknown, we can use the sample standard deviation instead: $t = \dfrac{\bar{x} - \mu}{\sigma_x / \sqrt{n}}$. However, in this case, the test statistics t follows the Student's t distribution instead of the normal one. To compute t in this case, we can use the NumPy method std with the ddof=1 argument to compute the sample standard deviation.

```
In [63]: t = (X_samples.mean() - mu0)/(X_samples.std(ddof=1)/np.sqrt(n))
In [64]: t
Out[64]: -2.9680338545657845
```

In either case, we get a test statistic to compare with the corresponding distribution to obtain a p-value. For example, for a normal distribution, we can use a stats.norm instance to represent a normal distributed random variable, and with its ppf method, we can look up the statistics value corresponding to a certain significance level. For a two-sided hypothesis test of significance level 5% (2.5% on each side), the statistics threshold is as follows.

```
In [65]: stats.norm().ppf(0.025)
Out[65]: -1.9599639845400545
```

Since the observed statistic is about –2.83, smaller than the threshold value of –1.96 for a two-sided test with a significance level of 5%, we have sufficient grounds to reject the null hypothesis in this case. We can explicitly compute the p-value for the observed test statistics using the cdf method (multiplied by two for a two-sided test). The resulting p-value is relatively small, which supports rejecting the null hypothesis.

```
In [66]: 2 * stats.norm().cdf(-abs(z))
Out[66]: 0.0045984013290753566
```

If we want to use the t distribution, we can use the stats.t class instead of the stats.norm. After computing the sample mean, $\bar{x}$, only $n - 1$ degrees of freedom (df) remain in the sample data. The number of degrees of freedom is an important parameter for the t distribution, which we need to specify when we create the random variable instance.

```
In [67]: 2 * stats.t(df=(n-1)).cdf(-abs(t))
Out[67]: 0.0037586479674227209
```

Again, the p-value is very small, suggesting we should reject the null hypothesis. Instead of explicitly carrying out these steps (computing the test statistics, then computing the *p*-value), there are built-in functions in SciPy's stats module for carrying out many common tests, as summarized in Table 13-2. For the test used here, we can directly compute the test statistics and the *p*-value using the stats.ttest_1samp function.

```
In [68]: t, p = stats.ttest_1samp(X_samples, mu)
In [69]: t
Out[69]: -2.9680338545657841
In [70]: p
Out[70]: 0.0037586479674227209
```

Again, we see that the *p*-value is very small (the same value as in the preceding text) and that we should reject the null hypothesis. Plotting the distribution corresponding to the null hypothesis and the sampled data is also illustrative (see Figure 13-5).

```
In [71]: fig, ax = plt.subplots(figsize=(8, 3))
    ...: sns.histplot(X_samples, kde=True, stat='density', ax=ax)
    ...: x = np.linspace(*X.interval(0.999), num=100)
    ...: ax.plot(x, stats.norm(loc=mu, scale=sigma).pdf(x))
```

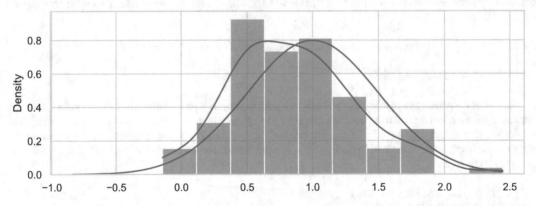

Figure 13-5. *Distribution function according to the null hypothesis (light green) and the sample estimated distribution function (dark blue)*

For another example, consider a two-variable problem where the null hypothesis states that the population means of two random variables are equal (e.g., corresponding to independent subjects with and without treatment). We can simulate this type of test by creating two random variables with normal distribution, with a randomly chosen population means. Here, we select 50 samples for each random variable.

```
In [72]: n, sigma = 50, 1.0
In [73]: mu1, mu2 = np.random.rand(2)
In [74]: X1 = stats.norm(mu1, sigma)
In [75]: X1_sample = X1.rvs(n)
In [76]: X2 = stats.norm(mu2, sigma)
In [77]: X2_sample = X2.rvs(n)
```

We want to evaluate whether the observed samples provide sufficient evidence that the two population means are unequal (rejecting the null hypothesis). For this situation, we can use the *t*-test for two independent samples, which is available in SciPy's `stats.ttext_ind`, which returns the test statistics and the corresponding *p*-value.

```
In [78]: t, p = stats.ttest_ind(X1_sample, X2_sample)
In [79]: t
Out[79]: -1.4283175246005888
In [80]: p
Out[80]: 0.15637981059673237
```

Here, the *p*-value is about 0.156, which is not small enough to support rejecting the null hypothesis that the two means are different. In this example, the two population means are different.

```
In [81]: mu1, mu2
Out[81]: (0.24764580637159606, 0.42145435527527897)
```

However, the samples drawn from these distributions did not statistically prove that these means are different (an error of type II). To increase the power of the statistical test, we would need to increase the number of samples from each random variable.

The SciPy `stats` module contains functions for common types of hypothesis testing (see the summary in Table 13-2), and their use closely follows what was shown in the examples in this section. However, some tests require additional arguments for distribution parameters. See the docstrings for each test function for details.

Nonparametric Methods

So far, we have described random variables with distributions completely determined by a few parameters, such as the mean and the variance for the normal distributions. Given the sampled data, we can fit a distribution function using maximum likelihood optimization with respect to the distribution parameters. Such distribution functions are called *parametric*, and statistical methods based on such distribution functions (e.g., a hypothesis test) are called *parametric methods*. When using those methods, we strongly assume that the given distribution describes the sampled data. An alternative approach to constructing a representation of an unknown distribution function is *kernel-density estimation* (KDE), which can be viewed as a smoothened version of the histogram of the sampled data (see, e.g., Figure 13-6). In this method, the probability distribution is estimated by a sum of the kernel function centered at each data point

$$\hat{f}(x) = \frac{1}{n.\text{bw}} \sum_{i=0}^{n} K\left(\frac{x - x_i}{\text{bw}}\right),$$ where bw is a free parameter known as the bandwidth, and K is the kernel

function (normalized to integrate to unity). The bandwidth is an important parameter that defines a scale for the influence of each term in the sum. A too-broad bandwidth gives a featureless estimate of the probability distribution, and a too-small bandwidth gives a noisy, overly structured estimate (see the middle panel in Figure 13-6). Different choices of kernel functions are also possible. A Gaussian kernel is a popular choice because of its smooth shape with local support, and it is relatively easy to perform computations.

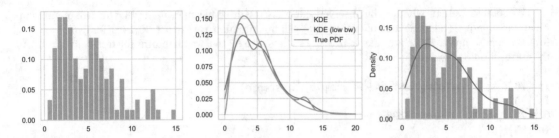

Figure 13-6. *Histogram (left), kernel-density estimation of the distribution function (middle), and both a histogram and the kernel-density estimate in the same graph (right)*

In SciPy, the KDE method using a Gaussian kernel is implemented in the `stats.gaussian_kde` function, which returns a callable object that behaves as and can be used as a probability distribution function. For example, consider a set of samples, `X_samples`, drawn from a random variable X with an unknown distribution (here simulated using the χ^2 distribution with five degrees of freedom).

```
In [82]: X = stats.chi2(df=5)
In [83]: X_samples = X.rvs(100)
```

To compute the kernel-density estimate for the given data, call the `stats.guassian_kde` function with the array of sample points as the argument.

```
In [84]: kde = stats.gaussian_kde(X_samples)
```

A standard method for computing a suitable bandwidth is used by default, often giving acceptable results. However, we could also specify a function for computing the bandwidth or directly setting the bandwidth using the `bw_method` argument. To select a smaller bandwidth, we can, for example, use the following.

```
In [85]: kde_low_bw = stats.gaussian_kde(X_samples, bw_method=0.25)
```

The `gaussian_kde` function returns an estimate of the distribution function, which we, for example, can graph or use for other applications. Here, we plot a histogram of the data and the two kernel-density estimates (with default and explicitly set bandwidth). For reference, let's also plot the true probability distribution function for the samples. The result is shown in Figure 13-6.

```
In [86]: x = np.linspace(0, 20, 100)
In [87]: fig, axes = plt.subplots(1, 3, figsize=(12, 3))
    ...: axes[0].hist(X_samples, density=True, alpha=0.5, bins=25)
    ...: axes[1].plot(x, kde(x), label="KDE")
    ...: axes[1].plot(x, kde_low_bw(x), label="KDE (low bw)")
    ...: axes[1].plot(x, X.pdf(x), label="True PDF")
    ...: axes[1].legend()
    ...: sns.distplot(X_samples, bins=25, ax=axes[2])
```

The `seaborn` statistical graphics library provides a convenient function for plotting a histogram and the kernel-density estimation for a set of data: `histplot`. A graph produced by this function is shown in the right panel of Figure 13-6.

Given the kernel-density estimate, we can also use it to generate new random numbers using the `resample` method, which takes the number of data points as arguments.

```
In [88]: kde.resample(10)
Out[88]: array([[1.75376869,  0.5812183,  8.19080268,  1.38539326,  7.56980335,
                 1.16144715,  3.07747215,  5.69498716,  1.25685068,  9.55169736]])
```

The kernel-density estimate object does not directly contain methods for computing the cumulative distribution functions (CDF) and its inverse, the percent-point function (PPF). However, several methods exist for integrating the kernel-density estimate of the probability distribution function. For example, for a one-dimensional KDE, we can use the `integrate_box_1d` to obtain the corresponding CDF.

```
In [89]: def _kde_cdf(x):
    ...:     return kde.integrate_box_1d(-np.inf, x)
In [90]: kde_cdf = np.vectorize(_kde_cdf)
```

We can use the SciPy `optimize.fsolve` function to find the inverse (the PPF).

```
In [91]: def _kde_ppf(q):
    ...:     return optimize.fsolve(
    ...:         lambda x, q: kde_cdf(x) - q, kde.dataset.mean(), args=(q,))[0]
    ...:
In [92]: kde_ppf = np.vectorize(_kde_ppf)
```

With the CDF and PPF for the kernel-density estimate, we can, for example, perform statistical hypothesis testing and compute confidence intervals. For example, using the kde_ppf function defined in the preceding section, we can compute an approximate 90% confidence interval for the mean of the population from which the sample was collected.

```
In [93]: kde_ppf([0.05, 0.95])
Out[93]: array([  0.39074674,  11.94993578])
```

As illustrated with this example, once we have a KDE that represents the probability distribution for a statistical problem, we can proceed with many of the same methods as we use in parametric statistics. The advantage of nonparametric methods is that we do not necessarily need to make assumptions about the shape of the distribution function. However, their statistical power is lower because nonparametric methods use less information (weaker assumptions) than parametric methods. Therefore, if we can justify using a parametric method, then using it is the best approach. Nonparametric methods offer a versatile generic approach that we can fall back on when parametric methods are not feasible.

Summary

This chapter explored how NumPy and the SciPy `stats` module can be used in basic statistical applications, including random number generation, for representing random variables and probability distribution functions, maximum likelihood fitting distributions to data, and using probability distributions and test statistics for hypothesis testing. We also briefly looked at kernel-density estimation of an unknown probability distribution as an example of a nonparametric method. The concepts and methods discussed in this chapter are fundamental building blocks for working with statistics, and the computational tools introduced here also provide a foundation for many statistical applications. The upcoming chapters build on what has been discussed here and explore statistical modeling and machine learning in more depth.

Further Reading

Good introductions to the fundamentals of statistics and data analysis are given in *Mathematical Statistics and Data Analysis* by J. A. Rice (Duxbury Press, 1995) and *All of Statistics* Wasserman (Springer, 2004). A computationally oriented introduction to statistics is given in *Mathematical Statistics and Data Analysis* by P. Dalgaard (Springer, 2008), which, although it uses the R language, is also relevant for statistics in Python. There are also free online resources about statistics, for example, OpenIntro Statistics, available from www.openintro.org/stat/textbook.php.

CHAPTER 14

■ ■ ■

Statistical Modeling

The previous chapter covered basic statistical concepts and methods. This chapter builds on the foundation laid out in the last chapter and explores statistical modeling, which deals with creating models that attempt to explain data. A model can have one or several parameters, and we can use a fitting procedure to find the parameter values so that the model best describes the observed data. Once a model has been fitted to data, it can be used to predict the values of new observations, given the values of the model's independent variables. We can also perform statistical analysis on the data and the fitted model and try to answer questions such as if the model accurately explains the data, which factors in the model are more relevant (predictive) than others, and if there are parameters that do not contribute significantly to the predictive power of the model.

This chapter mainly uses the statsmodels library. It provides classes and functions for defining statistical models, fitting them to observed data, calculating descriptive statistics, and carrying out statistical tests. The statsmodels library partly overlaps with the SciPy stats module we covered in the previous chapter, but it is mostly an extension of what is available in SciPy[1]. In particular, the focus of the statsmodels library is on fitting models to data rather than probability distributions and random variables, for which, in many cases, it relies on the SciPy `stats`.

■ **statsmodels**   The statsmodels library provides a rich set of functionality related to statistical tests and statistical modeling, including linear regression, logistic regression, and time-series analysis. For more information about the project and its documentation, see its web page at `www.statsmodels.org`. At the time of writing, the latest version of statsmodels is 0.14.0.

The statsmodels library is closely integrated with the Patsy library, which allows us to write statistical models as simple formulas. The Patsy library is one of the dependencies of the statsmodels library but can also be used with other statistical libraries, such as scikit-learn, which is discussed in Chapter 15. However, this chapter introduces the Patsy library in the context of using it together with the statsmodels library.

■ **patsy**   The patsy library provides features for defining statistical models with a simple formula language inspired by statistical software such as R. The patsy library is designed to be a companion library for statistical modeling packages, such as statsmodels. For more information about the project and its documentation, see the web page at `http://patsy.readthedocs.org`. At the time of writing, the most recent version of patsy is 0.5.3.

---

[1] The statsmodels library originally started as a part of the SciPy `stats` module but was later moved to a project on its own. The SciPy `stats` library remains an important dependency for statsmodels.

© Robert Johansson 2024
R. Johansson, *Numerical Python*, https://doi.org/10.1007/979-8-8688-0413-7_14

Importing Modules

This chapter works extensively with the statsmodels library. This library encourages an import convention slightly different from other libraries we have used so far: It provides api modules that collect the publically accessible symbols that the library provides. Here, we assume that the statsmodels.api is imported under the name sm, and statsmodels.formula.api is imported under the name smf. We also require the statsmodels.graphics.api module to be imported as the name smg.

```
In [1]: import statsmodels.api as sm
In [2]: import statsmodels.formula.api as smf
In [3]: import statsmodels.graphics.api as smg
```

Since the statsmodels library uses the Patsy library internally, directly accessing this library's functions is usually unnecessary. However, here we use Patsy for demonstration purposes and need to import the library explicitly.

```
In [4]: import patsy
```

As usual, we also require the Matplotlib, NumPy, and Pandas libraries to be imported as follows.

```
In [5]: import matplotlib.pyplot as plt
In [6]: import numpy as np
In [7]: import pandas as pd
```

The following is for the SciPy stats module.

```
In [8]: from scipy import stats
```

Introduction to Statistical Modeling

For a set of response (dependent) variables Y and explanatory (independent) variables X, let's find a mathematical relationship (model) between Y and X. We can generally write a mathematical model as a $Y = f(X)$ function. Knowing the $f(X)$ function would allow us to compute the value of Y for any values of X. If we do not know the $f(X)$ function. But we have access to data for observations $\{y_i, x_i\}$. We can parameterize the $f(X)$ function and fit the values of the parameters to the data. An example of a parameterization of $f(X)$ is the linear model $f(X) = \beta_0 + \beta_1 X$, where the coefficients β_0 and β_1 are the parameters of the model. Typically, we have many more data points than the number of free parameters in the model. In such cases, we can, for example, use a least square fit that minimizes the norm of the residual $r = Y - f(X)$. However, other minimization objective functions can also be used[2], for example, depending on the statistical properties of the residual r. So far, we have described a *mathematical* model. The essential component that makes a model *statistical* is that the data $\{y_i, x_i\}$ has an element of uncertainty, for example, due to measurement noise or other uncontrolled circumstances. The uncertainty in the data can be described in the model as random variables; for example, $Y = f(X) + \varepsilon$, where ε is a random variable. This is a statistical model because it includes random variables. Depending on how the random variables appear in the model and what distributions the random variables follow, we obtain different statistical models, which may require different approaches to analyze and solve.

---

[2] Examples of this are presented in Chapter 15.

A typical situation where a statistical model can be used is to describe the observations y_i in an experiment, where x_i is a vector with control knobs that are recorded together with each observation. An element in x_i may or may not be relevant for predicting the observed outcome y_i, and an important aspect of statistical modeling is determining which explanatory variables are relevant. It is, of course, also possible that relevant factors are not included in the set of explanatory variables x_i but influence the outcome of the observation y_i. In this case, it might not be possible to accurately explain the data with the model. Determining if a model accurately explains the data is another essential aspect of statistical modeling.

A widely used statistical model is $Y = \beta_0 + \beta_1 X + \varepsilon$, where β_0 and β_1 are model parameters, and ε is normally distributed with zero mean and variance σ^2: $\varepsilon \sim N(0, \sigma^2)$. This model is known as *simple linear regression* if X is a scalar, *multiple linear regression* if X is a vector, and *multivariate linear regression* if Y is a vector. Because the residual ε is normally distributed, the model can be fitted to data using *ordinary least squares* (OLS) for all these cases. Relaxing the condition that the elements in Y, in multivariate linear regression, must be independent and normally distributed with equal variance gives rise to variations of the model that can be solved with methods known as *generalized least squares* (GLS) and *weighted least squares* (WLS). All methods for solving statistical models typically have assumptions one must be mindful of when applying the models. For standard linear regression, the most critical assumption is that the residuals are independent and normally distributed.

The *generalized linear model* is an extension of the linear regression model that allows the errors in the response variable to have distributions other than the normal distribution. Particularly, the response variable is assumed to be a function of a linear predictor, where the variance of the response variable can be a function of the variable's value. This provides a broad generalization of the linear model applicable in many situations. For example, this enables modeling important problems where the response variable takes discrete values, such as binary outcomes of count values. The errors in the response variables of such models may follow different statistical distributions (e.g., the binomial and the Poisson distribution). Examples of these models include *logistic regression* for binary outcomes and *Poisson regression* for positive integer outcomes.

The following sections explore how statistical models of these types can be defined and solved using the Patsy and statsmodels libraries.

Defining Statistical Models with Patsy

Common to all statistical modeling is that we need to make assumptions about the mathematical relation between the response variable Y and explanatory variable X. In most cases, we are interested in linear models, such that Y can be written as a linear combination of the response variables X, functions of the response variables, or models with a linear component. For example, $Y = \alpha_1 X_1 + \dots + \alpha_n X_n$, $Y = \alpha_1 X + \alpha_2 X^2 \dots + \alpha_n X^n$, and $Y = \alpha_1 \sin X_1 + \alpha_2 \cos X_2$ are all examples of such linear models. Note that for the model to be linear, we only need the relation to be linear with respect to the unknown coefficients α and not necessarily in the known explanatory variables X. In contrast, an example of a nonlinear model is $Y = \exp(\beta_0 + \beta_1 X)$ since, in this case, Y is not a linear function with respect to β_0 and β_1. However, this model is *log-linear* in that taking the logarithm of the relation yields a linear model: $\tilde{Y} = \beta_0 + \beta_1 X$ for $\tilde{Y} = \log Y$. Problems that can be transformed into a linear model in this manner are the types of problems that can be handled with the generalized linear model.

Once the mathematical form of the model has been established, the next step is often to construct the design matrices y and X such that the regression problem can be written in matrix form as $y = X\beta + \varepsilon$, where y is the vector (or matrix) of observations, β is a vector of coefficients, and ε is the residual (error). The elements X_{ij} of the design matrix X are the values of the (functions of) explanatory variables corresponding to each coefficient β_j and observation y_i. Many solvers for statistical models in statsmodels and other statistical modeling libraries can take the design matrices X and y as input.

For example, if the observed values are $y = [1, 2, 3, 4, 5]$ with two independent variables with corresponding values $x_1 = [6, 7, 8, 9, 10]$ and $x_2 = [11, 12, 13, 14, 15]$, and if the linear model under consideration is $Y = \beta_0 + \beta_1 X_1 + \beta_2 X_2 + \beta_3 X_1 X_2$, then the design matrix for the right-hand side is $X = [\mathbf{1}, x_1, x_2, x_1 x_2]$. We can construct this design matrix using the NumPy vstack function.

```
In [9]: y = np.array([1, 2, 3, 4, 5])
In [10]: x1 = np.array([6, 7, 8, 9, 10])
In [11]: x2 = np.array([11, 12, 13, 14, 15])
In [12]: X = np.vstack([np.ones(5), x1, x2, x1*x2]).T
In [13]: X
Out[13]: array([[   1.,    6.,   11.,    66.],
                [   1.,    7.,   12.,    84.],
                [   1.,    8.,   13.,   104.],
                [   1.,    9.,   14.,   126.],
                [   1.,   10.,   15.,   150.]])
```

Given the design matrix X and observation vector y, we can solve for the unknown coefficient vector β, for example, using least square fit (see Chapters 5 and 6).

```
In [14]: beta, res, rank, sval = np.linalg.lstsq(X, y)
In [15]: beta
Out[15]: array([ -5.55555556e-01, 1.88888889e+00, -8.88888889e-01, -1.33226763e-15])
```

These steps are the essence of statistical modeling in its simplest form. However, variations and extensions to this basic method make statistical modeling a field in its own right and call for computational frameworks such as statsmodels for systematic analysis. For example, although constructing the design matrix X was straightforward in this simple example, it can be tedious for more involved models, especially if we wish to easily change how the model is defined. This is where the Patsy library enters the picture. It offers a convenient (although not necessarily intuitive) formula language for defining a model and automatically constructing the relevant design matrices. To construct the design matrix for a Patsy formula, we can use the patsy.dmatrices function. It takes the formula as a string as the first argument and a dictionary-like object with data arrays for the response and explanatory variables as the second argument. The basic syntax for the Patsy formula is "y ~ x1 + x2 + ...", which means that y is a linear combination of the explanatory variables x1 and x2 (implicitly including an intercept coefficient). Table 14-1 summarizes the Patsy formula syntax.

CHAPTER 14 ■ STATISTICAL MODELING

Table 14-1. *Simplified Summary of the Patsy Formula Syntax*

Syntax	Example	Description
lhs ~ rhs	y ~ x (Equivalent to y ~ 1 + x)	The ~ character separates the left-hand side (containing the dependent variables) and the right-hand side (containing the independent variables) of a model equation.
var * var	x1*x2 (Equivalent to x1+x2+x1*x2)	An interaction term that implicitly contains all its lower-order interaction terms.
var + var + ...	x1 + x2 + ... (Equivalent to y ~ 1 + x1 + x2)	The addition sign is used to denote the union of terms.
var:var	x1:x2	The colon character denotes a pure interaction term (e.g., $x_1 \cdot x_2$).
f(expr)	np.log(x), np.cos(x+y)	Arbitrary Python functions (often NumPy functions) can transform terms in the expression. The expression for the argument of a function is interpreted as an arithmetic expression rather than the set-like formula operations that are otherwise used in Patsy.
I(expr)	I(x+y)	I is a Patsy-supplied identity function that can be used to escape arithmetic expressions so that they are interpreted as arithmetic operations.
C(var)	C(x), C(x, Poly)	Treat the variable x as a categorical and expand its values into orthogonal dummy variables.

For a complete specification of the formula syntax, see the Patsy documentation at http://patsy. readthedocs.org/en/latest

As an introductory example, consider again the linear model $Y = \beta_0 + \beta_1 X_1 + \beta_2 X_2 + \beta_3 X_1 X_2$ used earlier. To define this model with Patsy, we can use the formula "y ~ 1 + x1 + x2 + x1*x2". Note that we leave out coefficients in the model formula, as each term is implicitly assumed to have a model parameter as a coefficient. In addition to specifying the formula, we also need to create a dictionary data that maps the variable names to the corresponding data arrays.

```
In [16]: data = {"y": y, "x1": x1, "x2": x2}
In [17]: y, X = patsy.dmatrices("y ~ 1 + x1 + x2 + x1*x2", data)
```

The result is two arrays y and X, which are the design matrices for the given data arrays and the specified model formula.

```
In [18]: y
Out[18]: DesignMatrix with shape (5, 1)
           y
           1
           2
           3
           4
           5
```

```
          Terms:
            'y' (column 0)
In [19]: X
Out[19]: DesignMatrix with shape (5, 4)
            Intercept  x1  x2  x1:x2
                    1   6  11     66
                    1   7  12     84
                    1   8  13    106
                    1   9  14    126
                    1  10  15    150
          Terms:
            'Intercept' (column 0)
            'x1' (column 1)
            'x2' (column 2)
            'x1:x2' (column 3)
```

These arrays are of type DesignMatrix, a Patsy-supplied subclass of the standard NumPy array, which contains additional metadata and an altered printing representation.

```
In [20]: type(X)
Out[20]: patsy.design_info.DesignMatrix
```

Note that the numerical values of the DesignMatrix array are equal to those of the explicitly constructed array we produced earlier using vstack.

As a subclass of the NumPy ndarray, the arrays of type DesignMatrix are fully compatible with code that expects NumPy arrays as input. However, we can also explicitly cast a DesignMatrix instance into a NumPy array using the np.array function, although this usually should not be necessary.

```
In [21]: np.array(X)
Out[21]: array([[    1.,     6.,    11.,     66.],
                [    1.,     7.,    12.,     84.],
                [    1.,     8.,    13.,    104.],
                [    1.,     9.,    14.,    126.],
                [    1.,    10.,    15.,    150.]])
```

Alternatively, we can set the return_type argument to "dataframe", in which case the patsy. dmatrices function returns design matrices in the form of Pandas DataFrame objects. Also, since DataFrame objects behave as dictionary-like objects, we can use data frames to specify the model data as the second argument to the patsy.dmatrices function.

```
In [22]: df_data = pd.DataFrame(data)
In [23]:
y, X = patsy.dmatrices("y ~ 1 + x1 + x2 + x1:x2", df_data, return_type="dataframe")
In [24]: X
Out[24]:
```

	Intercept	x1	x2	x1:x2
0	1	6	11	66
1	1	7	12	84
2	1	8	13	104
3	1	9	14	126
4	1	10	15	150

With the help of Patsy, we have now automatically created the design matrices required for solving a statistical model, using, for example, the np.linalg.lstsq function (as we saw an example earlier), or using one of the many statistical model solvers provided by the statsmodels library. For example, to perform an ordinary linear regression (OLS), we can use the OLS class from the statsmodels library instead of the lower-level method np.linalg.lstsq. Nearly all classes for statistical models in statsmodels take the design matrices y and X as the first and second arguments and return a class instance that represents the model. To fit the model to the data encoded in the design matrices, we need to invoke the fit method, which returns a result object that contains fitted parameters (among other attributes).

```
In [25]: model = sm.OLS(y, X)
In [26]: result = model.fit()
In [27]: result.params
Out[27]: Intercept   -5.555556e-01
         x1           1.888889e+00
         x2          -8.888889e-01
         x1:x2       -8.881784e-16
         dtype: float64
```

Note that the result is equivalent to the least square fitting computed earlier in this chapter. Using the statsmodels formula API (the module we imported as smf), we can directly pass the Patsy formula for the model when we create a model instance, eliminating the need to create the design matrices. Instead of passing y and X as arguments, we then pass the Patsy formula and the dictionary-like object (e.g., a Pandas data frame) that contains the model data.

```
In [28]: model = smf.ols("y ~ 1 + x1 + x2 + x1:x2", df_data)
In [29]: result = model.fit()
In [30]: result.params
Out[30]: Intercept   -5.555556e-01
         x1           1.888889e+00
         x2          -8.888889e-01
         x1:x2       -8.881784e-16
         dtype: float64
```

The advantage of using statsmodels instead of explicitly constructing NumPy arrays and calling the NumPy least square model is, of course, that much of the process is automated in statsmodels, which makes it possible to add and remove terms in the statistical model without any extra work. Also, when using statsmodels, we have access to a large variety of linear model solvers and statistical tests for analyzing how well the model fits the data (see Table 14-1).

Now that we have examined how a Patsy formula can be used to construct design matrices or be used directly with one of the many statistical model classes from statsmodels, let's briefly return to the syntax and notational conventions for Patsy formulae before we continue and look in more detail on different statistical models that are available in the statsmodels library. As mentioned, the basic syntax for a model formula has the form "LHS ~ RHS". The ~ character separates the left-hand side (LHS) and the right-hand side (RHS) of the model equation. The LHS specifies the terms that constitute the response variables, and the RHS specifies the terms that constitute the explanatory variables. The terms in the LHS and RHS expressions are separated by + or – signs. But these should not be interpreted as arithmetic operators but rather as set-union and set-difference operators. For example, a+b means that both *a* and *b* are included in the model, and -a means that the term *a* is *not* included. An expression of the type a*b is automatically expanded to a + b + a:b, where a:b is the pure interaction term $a \cdot b$.

As a concrete example, consider the following formula and the resulting right-hand side terms (which we can extract from the design_info attribute using the term_names attribute).

```
In [31]: from collections import defaultdict
In [32]: data = defaultdict(lambda: np.array([]))
In [33]: patsy.dmatrices("y ~ a", data=data)[1].design_info.term_names
Out[33]: ['Intercept', 'a']
```

Here, the two terms are Intercept and a, which correspond to constant and a linear dependence on *a*. By default, Patsy always includes the intercept constant, which in the Patsy formula also can be written explicitly using y ~ 1 + a. Including the 1 is optional.

```
In [34]: patsy.dmatrices("y ~ 1 + a + b", data=data)[1].design_info.term_names
Out[34]: ['Intercept', 'a', 'b']
```

In this case, we have one more explanatory variable (a and b), and here, the intercept is explicitly included in the formula. If we do not want to include the intercept in the model, we can use the notation –1 to remove this term.

```
In [35]: patsy.dmatrices("y ~ -1 + a + b", data=data)[1].design_info.term_names
Out[35]: ['a', 'b']
```

Expressions of the type a * b are automatically expanded to include all lower-order interaction terms.

```
In [36]: patsy.dmatrices("y ~ a * b", data=data)[1].design_info.term_names
Out[36]: ['Intercept', 'a', 'b', 'a:b']
```

Higher-order expansions work too.

```
In [37]: patsy.dmatrices("y ~ a * b * c", data=data)[1].design_info.term_names
Out[37]: ['Intercept', 'a', 'b', 'a:b', 'c', 'a:c', 'b:c', 'a:b:c']
```

To remove a specific term from a formula, we can write the term preceded by the minus operator. For example, to remove the pure third-order interaction term a:b:c from the automatic expansion of a*b*c, we can use the following.

```
In [38]: patsy.dmatrices(
   ...:      "y ~ a * b * c - a:b:c", data=data)[1].design_info.term_names
Out[38]: ['Intercept', 'a', 'b', 'a:b', 'c', 'a:c', 'b:c']
```

In Patsy, the + and - operators are used for set-like operations on sets of terms; if we need to represent the arithmetic operations, we must wrap the expression in a function call. For convenience, Patsy provides an identity function with the name I that can be used for this purpose. To illustrate this point, consider the following two examples, which show the resulting terms for y ~ a + b and y ~ I(a + b).

```
In [39]: data = {k: np.array([]) for k in ["y", "a", "b", "c"]}
In [40]: patsy.dmatrices("y ~ a + b", data=data)[1].design_info.term_names
Out[40]: ['Intercept', 'a', 'b']
In [41]: patsy.dmatrices("y ~ I(a + b)", data=data)[1].design_info.term_names
Out[41]: ['Intercept', 'I(a + b)']
```

Here, the column in the design matrix corresponding to the term with the name I(a+b) is the arithmetic sum of the arrays for the variables a and b. The same trick must be used if we want to include terms that are expressed as a power of a variable.

```
In [42]: patsy.dmatrices("y ~ a**2", data=data)[1].design_info.term_names
Out[42]: ['Intercept', 'a']
In [43]: patsy.dmatrices("y ~ I(a**2)", data=data)[1].design_info.term_names
Out[43]: ['Intercept', 'I(a ** 2)']
```

The notation I(...) used here is an example of a function call notation. We can apply transformations of the input data in a Patsy formula by including arbitrary Python function calls. In particular, we can transform the input data array using functions from NumPy.

```
In [44]: patsy.dmatrices("y ~ np.log(a) + b", data=data)[1].design_info.term_names
Out[44]: ['Intercept', 'np.log(a)', 'b']
```

Or we can even transform variables with arbitrary Python functions.

```
In [45]: z = lambda x1, x2: x1+x2
In [46]: patsy.dmatrices("y ~ z(a, b)", data=data)[1].design_info.term_names
Out[46]: ['Intercept', 'z(a, b)']
```

So far, we have considered models with numerical responses and explanatory variables. Statistical modeling also frequently includes categorical variables, which can take a discrete set of values that do not have a meaningful numerical order (e.g., "Female" or "Male"; type "A", "B", or "C", etc.). When using such variables in a linear model, we typically need to recode them by introducing binary dummy variables. In a patsy formula, any variable that does not have a numerical data type (float or int) is interpreted as a categorical variable and automatically encoded accordingly. For numerical variables, we can use the C(x) notation to explicitly request that a variable x should be treated as a categorical variable.

For example, compare the following two examples that show the design matrix for the formula "y ~ - 1 + a" and "y ~ - 1 + C(a)", which corresponds to models where a is a numerical and categorical explanatory variable, respectively.

```
In [48]: data = {"y": [1, 2, 3], "a": [1, 2, 3]}
In [48]: patsy.dmatrices("y ~ - 1 + a", data=data, return_type="dataframe")[1]
Out[48]:
```

	a
0	1
1	2
2	3

For a numerical variable, the corresponding column in the design matrix corresponds to the data vector. In contrast, for a categorical variable C(a) new binary-valued columns with a mask-like encoding of individual values of the original variable are added to the design matrix.

```
In [49]:
patsy.dmatrices("y ~ - 1 + C(a)", data=data, return_type="dataframe")[1]
Out[49]:
```

	C(a)[1]	C(a)[2]	C(a)[3]
0	1	0	0
1	0	1	0
2	0	0	1

Variables with nonnumerical values are automatically interpreted and treated as categorical values.

```
In [50]: data = {"y": [1, 2, 3], "a": ["type A", "type B", "type C"]}
In [51]: patsy.dmatrices("y ~ - 1 + a", data=data, return_type="dataframe")[1]
Out[51]:
```

	a[type A]	a[type B]	a[type C]
0	1	0	0
1	0	1	0
2	0	0	1

The user can change the default type of encoding of categorical variables into binary-valued treatment fields. For example, to encode the categorical variables with orthogonal polynomials instead of treatment indicators, we can use C(a, Poly).

```
In [52]: patsy.dmatrices("y ~ - 1 + C(a, Poly)",
    ...:                  data=data, return_type="dataframe")[1]
Out[52]:
```

	C(a, Poly).Constant	C(a, Poly).Linear	C(a, Poly).Quadratic
0	1	-7.071068e-01	0.408248
1	1	-5.551115e-17	-0.816497
2	1	7.071068e-01	0.408248

Patsy's automatic encoding of categorical variables is a very convenient aspect of Patsy formula, which allows the user to easily add and remove numerical and categorical variables in a model. This is arguably one of the main advantages of using the Patsy library to define model equations.

Linear Regression

The statsmodels library supports several types of statistical models that are applicable in varying situations. But nearly all follow the same usage pattern, which makes it easy to switch between different models. Statistical models in statsmodels are represented by model classes. These can be initiated given the design matrices for the response and explanatory variables of a linear model or given a Patsy formula and a data frame (or another dictionary-like object). The basic workflow when setting up and analyzing a statistical model with statsmodels includes the following steps.

1. Create an instance of a model class, for example, using `model = sm.MODEL(y, X)` or `model = smf.model(formula, data)`, where `MODEL` and `model` are the name of a particular model, such as OLS, GLS, or Logit. The convention is that uppercase names are used for classes that take design matrices as arguments and lowercase names for classes that take Patsy formulas and data frames as arguments.

2. Creating a model instance does not perform any computations. To fit the model to the data, we must invoke the `fit` method, `result = model.fit()`, which performs the fit and returns a result object with methods and attributes for further analysis.

3. Print summary statistics for the result object returned by the `fit` method. The result object varies in content slightly for each statistical model. But most models implement the `summary` method, which produces a summary text that describes the result of the fit, including several types of statistics that can be useful for judging if the statistical model successfully explains the data. Viewing the output from the `summary` method is usually a good starting point when analyzing the result of a fitting process.

4. Post-process the model fit results: in addition to the `summary` method, the result object also contains methods and attributes for obtaining the fitted parameters (`params`), the residual for the model and the data (`resid`), the fitted values (`fittedvalues`), and a method for predicting the value of the response variables for new independent variables (`predict`).

5. Finally, it may be useful to visualize the result of the fitting, for example, with the Matplotlib and Seaborn graphics libraries, using some of the many graphing routines that are directly included in the statsmodels library (see the `statsmodels.graphics` module).

To demonstrate this workflow with a simple example, the following considers fitting a model to generate data whose true value is $y = 1 + 2x_1 + 3x_2 + 4x_1x_2$. We begin with storing the data in a Pandas data frame object.

```
In [53]: N = 100
In [54]: x1 = np.random.randn(N)
In [55]: x2 = np.random.randn(N)
In [56]: data = pd.DataFrame({"x1": x1, "x2": x2})
In [57]: def y_true(x1, x2):
   ...:     return 1  + 2 * x1 + 3 * x2 + 4 * x1 * x2
In [58]: data["y_true"] = y_true(x1, x2)
```

Here we have stored the true value of *y* in the y_true column in the DataFrame object data. We simulate a noisy observation of *y* by adding a normal distributed noise to the true values and store the result in the y column.

```
In [59]: e = 0.5 * np.random.randn(N)
In [60]: data["y"] = data["y_true"] + e
```

We know from the data that we have two explanatory variables, x1 and x2, in addition to the response variable y. The simplest possible model we can start with is the linear model $Y = \beta_0 + \beta_1 x_1 + \beta_2 x_2$, which we can define with the Patsy formula "y ~ x1 + x2". Since the response variable is continuous, it is a good starting point to fit the model to the data using ordinary linear squares, for which we can use the smf. ols class.

```
In [61]: model = smf.ols("y ~ x1 + x2", data)
In [62]: result = model.fit()
```

Remember that ordinary least square regression assumes that the residuals of the fitted model and the data are normally distributed. However, before analyzing the data, we might not know if this condition is satisfied. Nonetheless, we can start by fitting the data to the model and investigate the distribution of the residual using graphical methods and statistical tests (with the null hypothesis that the residuals are normally distributed). A lot of useful information, including several types of test statistics, can be displayed using the summary method.

```
In [63]: print(result.summary())
                        OLS Regression Results
==============================================================================
Dep. Variable:                    y   R-squared:                       0.380
Model:                          OLS   Adj. R-squared:                  0.367
Method:               Least Squares   F-statistic:                     29.76
Date:              Wed, 22 Apr 2015   Prob (F-statistic):           8.36e-11
Time:                      22:40:33   Log-Likelihood:                -271.52
No. Observations:               100   AIC:                             549.0
Df Residuals:                    97   BIC:                             556.9
Df Model:                         2
Covariance Type:          nonrobust
==============================================================================
                 coef    std err          t      P>|t|      [95.0% Conf. Int.]
------------------------------------------------------------------------------
Intercept      0.9868      0.382      2.581      0.011       0.228     1.746
x1             1.0810      0.391      2.766      0.007       0.305     1.857
x2             3.0793      0.432      7.134      0.000       2.223     3.936
==============================================================================
Omnibus:                     19.951   Durbin-Watson:                   1.682
Prob(Omnibus):                0.000   Jarque-Bera (JB):               49.964
Skew:                        -0.660   Prob(JB):                     1.41e-11
Kurtosis:                     6.201   Cond. No.              1.32
==============================================================================
Warnings: [1] Standard errors assume that the covariance matrix of the errors is correctly
specified.
```

The output produced by the summary method is rather verbose, and a detailed description of all the information provided by this method is beyond the scope of this treatment. Instead, here we only focus on a few key indicators. To begin, the R-squared value is a statistic that indicates how well the model fits the data. It can take values between 0 and 1, where an R-squared statistic of 1 corresponds to a perfect fit. The R-squared value of 0.380 reported in the preceding summary method is rather poor, indicating that we need to refine our model (which is expected since we left out the interaction term $x_1 \cdot x_2$). We can also explicitly access the R-squared statistic from the result object using the rsquared attribute.

```
In [64]: result.rsquared
Out[64]: 0.38025383255132539
```

Furthermore, the coef column in the middle of the table provides the fitted model parameters. Assuming that the residuals indeed are normally distributed, the std err column provides an estimate of the standard errors for the model coefficients, and the t and P>|t| columns are the t-statistics and the corresponding p-value for the statistical test with the null hypothesis that the corresponding coefficient is zero. Therefore, while keeping in mind that this analysis assumes that the residuals are normally distributed, we can look for the columns with small p-values and judge which explanatory variables have coefficients that are very likely to be different from zero (meaning that they have a significant predictive power).

To investigate whether the assumption of normal distributed errors is justified, we need to look at how the model's residuals fit the data. The residuals are accessible via the resid attribute of the result object.

```
In [65]: result.resid.head()
Out[65]: 0    -3.370455
         1   -11.153477
         2   -11.721319
         3    -0.948410
         4     0.306215
         dtype: float64
```

Using these residuals, we can check for normality using the normaltest function from the SciPy stats module.

```
In [66]: z, p = stats.normaltest(result.fittedvalues.values)
In [67]: p
Out[67]: 4.6524990253009316e-05
```

For this example, the resulting p-value is very small, suggesting that we can reject the null hypothesis that the residuals are normally distributed (i.e., we can conclude that the assumption of normal distributed residuals is violated). A graphical method to check for the normality of a sample is to use the qqplot from the statsmodels.graphics module. The QQ-plot, which compares the sample quantiles with the theoretical quantiles, should be close to a straight line if the sampled values are normally distributed. The following function call to smg.qqplot produces the QQ-plot shown in Figure 14-1.

```
In [68]: fig, ax = plt.subplots(figsize=(8, 4))
    ...: smg.qqplot(result.resid, ax=ax)
```

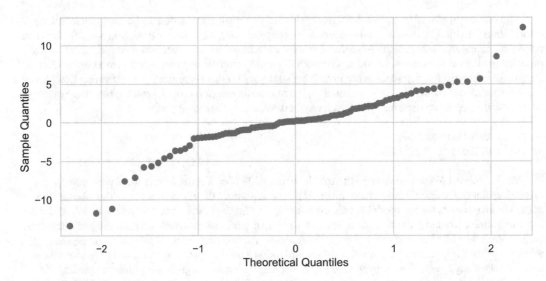

Figure 14-1. *QQ-plot of a linear model with two explanatory variables without any interaction term*

As shown in Figure 14-1, the points in the QQ-plot significantly deviate from a linear relation, suggesting that the observed residuals are unlikely to be a sample of a normal distributed random variable. In summary, these indicators provide evidence that the model used is insufficient and that we might need to refine the model. We can include the missing interaction term by adding it to the Patsy formula and repeating the steps from the previous analysis.

```
In [69]: model = smf.ols("y ~ x1 + x2 + x1*x2", data)
In [70]: result = model.fit()
In [71]: print(result.summary())
                        OLS Regression Results
===============================================================================
Dep. Variable:                      y   R-squared:                        0.963
Model:                            OLS   Adj. R-squared:                   0.961
Method:                 Least Squares   F-statistic:                      821.8
Date:                Tue, 21 Apr 2015   Prob (F-statistic):            2.69e-68
Time:                        23:52:12   Log-Likelihood:                 -138.39
No. Observations:                 100   AIC:                              284.8
Df Residuals:                      96   BIC:                              295.2
Df Model:                           3
Covariance Type:            nonrobust
===============================================================================
                 coef    std err          t      P>|t|      [95.0% Conf. Int.]
-------------------------------------------------------------------------------
Intercept      1.1023      0.100     10.996      0.000       0.903      1.301
x1             2.0102      0.110     18.262      0.000       1.792      2.229
x2             2.9085      0.095     30.565      0.000       2.720      3.097
x1:x2          4.1715      0.134     31.066      0.000       3.905      4.438
===============================================================================
```

Omnibus:	1.472	Durbin-Watson:	1.912
Prob(Omnibus):	0.479	Jarque-Bera (JB):	0.937
Skew:	0.166	Prob(JB):	0.626
Kurtosis:	3.338	Cond. No.	1.54

```
========================================================================
Warnings: [1] Standard errors assume that the covariance matrix of the errors is correctly
specified.
```

In this case, the R-squared statistic is significantly higher, 0.963, indicating a nearly perfect correspondence between the model and the data.

```
In [72]: result.rsquared
Out[72]: 0.96252198253140375
```

Note that we can always increase the R-squared statistic by introducing more variables. But we want to ensure that we do not add variables with low predictive power (small coefficient and high corresponding p-value) since it would make the model susceptible to overfitting. As usual, we require that the residuals be normally distributed. Repeating the normality test and the QQ-plot from the previous analysis with the updated model results in a relatively high p-value (0.081) and a relatively linear QQ-plot (see Figure 14-2). This suggests that, in this case, the residuals could very well be normally distributed (as we know they are, by design, in this example).

```
In [73]: z, p = stats.normaltest(result.fittedvalues.values)
In [74]: p
Out[74]: 0.081352587523644201
In [75]: fig, ax = plt.subplots(figsize=(8, 4))
    ...: smg.qqplot(result.resid, ax=ax)
```

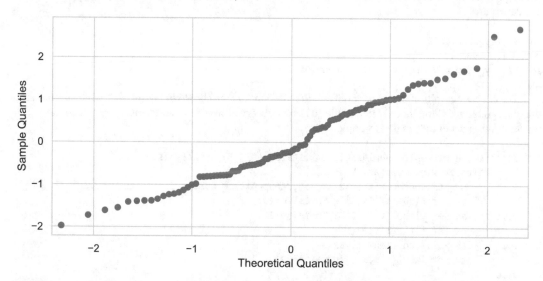

Figure 14-2. *QQ-plot of a linear model with two explanatory variables with an interaction term*

Once we are satisfied with the fit of the model, we can extract the model coefficients from the result object using the params attribute.

```
In [76]: result.params
Out[76]: Intercept    1.102297
         x1           2.010154
         x2           2.908453
         x1:x2        4.171501
         dtype: float64
```

Also, we can predict the values of new observations using the predict method, which takes as arguments a NumPy array or DataFrame object with values of the independent variables (x_1 and x_2 in this case). For example, since the current problem has only two independent variables, we can visualize the model's predictions as a contour plot. To this end, we first construct a DataFrame object with the x_1 and x_2 values for which we want to predict the y-value using the fitted model.

```
In [77]: x = np.linspace(-1, 1, 50)
In [78]: X1, X2 = np.meshgrid(x, x)
In [79]: new_data = pd.DataFrame({"x1": X1.ravel(), "x2": X2.ravel()})
```

Using the predict method of the result object obtained from the model fitting, we can compute the predicted y values for the new set of values of the response variables.

```
In [80]: y_pred = result.predict(new_data)
```

The result is a NumPy array (vector) with the same length as the data vectors X1.ravel() and X2.ravel(). To be able to plot the data using the Matplotlib contour function, we first resize the y_pred vector to a square matrix.

```
In [81]: y_pred.shape
Out[81]: (2500,)
In [82]: y_pred = y_pred.values.reshape(50, 50)
```

The contour graphs of the true model and the fitted model are shown in Figure 14-3, which demonstrate that the agreement of the model fitted to the 100 noisy observations of y is sufficient to reproduce the function rather accurately in this example.

```
In [83]: fig, axes = plt.subplots(1, 2, figsize=(12, 5), sharey=True)
    ...: def plot_y_contour(ax, Y, title):
    ...:     c = ax.contourf(X1, X2, Y, 15, cmap=plt.cm.RdBu)
    ...:     ax.set_xlabel(r"$x_1$", fontsize=20)
    ...:     ax.set_ylabel(r"$x_2$", fontsize=20)
    ...:     ax.set_title(title)
    ...:     cb = fig.colorbar(c, ax=ax)
    ...:     cb.set_label(r"$y$", fontsize=20)
    ...:
    ...: plot_y_contour(axes[0], y_true(X1, X2), "true relation")
    ...: plot_y_contour(axes[1], y_pred, "fitted model")
```

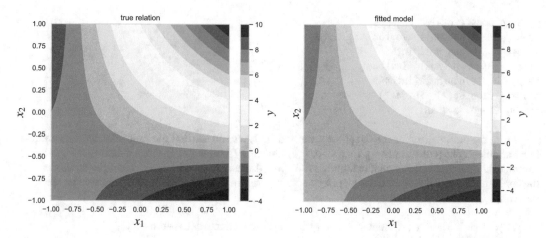

Figure 14-3. *The true relation and fit of the correct model to 100 samples from the true relation with normally distributed noise*

The example used the ordinary least square (ols) method to fit the model to the data. Several other options are also available, such as the robust linear model (rlm) that is suitable if there are significant outliers in the observations, and variants of the generalized linear model that is suitable, for example, if the response variable can take only discrete values. This is the topic of the following section. The next chapter presents examples of regularized regression, where the minimization objective is modified not only to minimize the square of the residuals but also, for example, to penalize large coefficients in the model.

Example Datasets

When working with statistical methods, having example datasets to explore is helpful. The statsmodels package provides an interface for loading example datasets from an extensive dataset repository[3] of the R statistical software. The sm.datasets module contains a get_rdataset function that can load datasets listed on the page http://vincentarelbundock.github.io/Rdatasets/datasets.html. The get_rdataset function takes the name of the dataset and optionally also the name of a package (grouping of datasets).

For example, we can use the following to load a dataset named Icecream from the package Ecdat.

```
In [84]: dataset = sm.datasets.get_rdataset("Icecream", "Ecdat")
```

The result is a data structure with the dataset and metadata describing the dataset. The title attribute gives the name of the dataset, and the __doc__ attribute contains an explanatory text describing the dataset (too long to display here).

```
In [85]: dataset.title
Out[85]: 'Ice Cream Consumption'
```

The data in the form of a Pandas DataFrame object is accessible via the data attribute.

```
In [86]: dataset.data.info()
<class 'pandas.core.frame.DataFrame'>
Int64Index: 30 entries, 0 to 29
```

---

[3] See http://vincentarelbundock.github.io/Rdatasets.

```
Data columns (total 4 columns):
cons       30 non-null float64
income     30 non-null int64
price      30 non-null float64
temp       30 non-null int64
dtypes: float64(2), int64(2)
memory usage: 1.2 KB
```

From the output given by the DataFrame method info, we can see that the Icecream dataset contains four variables: cons (consumption), income, price, and temp (temperature). Once a dataset is loaded, we can explore it and fit it to statistical models following the usual procedures. For example, to model the consumption as a linear model with price and temperature as independent variables, we can use the following.

```
In [87]: model = smf.ols("cons ~ -1 + price + temp", data=dataset.data)
In [88]: result = model.fit()
```

The result object can be analyzed using descriptive statistics and statistical tests, for example, starting with printing the output from the summary method, as we have seen before. We can also take a graphical approach and plot regression graphs, for example, using the plot_fit function in the smg module (see Figure 14-4 and also the regplot function in the seaborn library).

```
In [89]: fig, (ax1, ax2) = plt.subplots(1, 2, figsize=(12, 4))
    ...: smg.plot_fit(result, 0, ax=ax1)
    ...: smg.plot_fit(result, 1, ax=ax2)
```

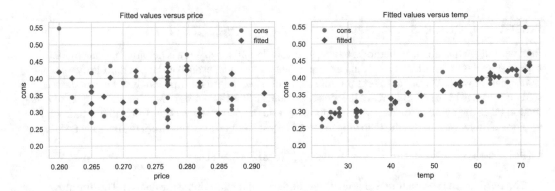

Figure 14-4. *Regression plots for the fit of the consumption vs. price and temperature in the Icecream dataset*

From the regression plots shown in Figure 14-4, we can conclude that according to this ice cream dataset, consumption seems linearly correlated to temperature but has no clear dependence on price (probably because the range of prices is relatively small). Graphical tools such as plot_fit can be a helpful tool when developing statistical models.

Discrete Regression

Regression with discrete dependent variables (e.g., binary outcomes) requires techniques different from the linear regression model we have seen. The reason is that linear regression requires that the response variable is a normally distributed *continuous* variable, which cannot be used directly for a response variable with only a few discrete possible outcomes, such as binary variables or variables taking positive integer values. However, it is possible to map a linear predictor to an interval that can be interpreted as a probability of different discrete outcomes using a suitable transformation. For example, in binary outcomes, one popular transformation is the logistic function $\log(p/(1 - p)) = \beta_0 + \beta \cdot x$, or $p = (1 + \exp(-\beta_0 - \beta_1 \cdot x))^{-1}$, which maps $x \in [-\infty, \infty]$ to $p \in [0, 1]$. In other words, the continuous or discrete feature vector x is mapped via the model parameters β_0 and β_1 and the logistic transformation onto a probability p. If $p < 0.5$, it can be taken to predict that $y = 0$, and $p \geq 0.5$ can be taken to predict $y = 1$. This procedure, known as logistic regression, is an example of a binary classifier. Classifiers are discussed more in Chapter 15.

The statsmodels library provides several methods for discrete regression, including the Logit class,[4] the related Probit class (which uses a cumulative distribution function of the normal distribution rather than the logistic function to transform the linear predictor to the [0, 1] interval), the multinomial logistic regression class MNLogit (for more than two categories), and the Poisson regression class Poisson for Poisson-distributed count variables (positive integers).

Logistic Regression

As an example of how to perform logistic regression with statsmodels, we first load a classic dataset using the sm.datasets.get_rdataset function, which contains sepal and petal lengths and widths for a sample of Iris flowers, together with a classification of the species of the flower. Here, we select a subset of the dataset corresponding to two different species and create a logistic model for predicting the species type from the petal length and width values. The info method summarizes which variables are contained in the dataset.

```
In [90]: df = sm.datasets.get_rdataset("iris").data
In [91]: df.info()
<class 'pandas.core.frame.DataFrame'>
Int64Index: 150 entries, 0 to 149
Data columns (total 5 columns):
Sepal.Length    150 non-null float64
Sepal.Width     150 non-null float64
Petal.Length    150 non-null float64
Petal.Width     150 non-null float64
Species         150 non-null object
dtypes: float64(4), object(1)
memory usage: 7.0+ KB
```

To see how many unique types of species are present in the Species column, we can use the unique method for the Pandas series that is returned when extracting the column from the data frame object.

```
In [92]: df.Species.unique()
Out[92]: array(['setosa', 'versicolor', 'virginica'], dtype=object)
```

This dataset contains three different types of species. To obtain a binary variable that we can use as the response variable in a logistic regression, here we focus only on the data for the two species *versicolor* and *virginica*. For convenience, we create a new data frame, df_subset, for the subset of the dataset corresponding to those species.

```
In [93]: df_subset = df[df.Species.isin(["versicolor", "virginica"])].copy()
```

---

[4] Logistic regression belongs to the class of model that can be viewed as a generalized linear model, with the logistic transformation as link function, so we could alternatively use sm.GLM or smf.glm.

To use logistic regression to predict the species using the other variables as independent variables, we first need to create a binary variable corresponding to the two species. Using the map method of the Pandas series object, we can map the two species names into binary values 0 and 1.

```
In [94]: df_subset.Species = df_subset.Species.map(
   ...:        {"versicolor": 1, "virginica": 0})
```

We also need to rename the columns with names that contain period characters to names that are valid symbol names in Python (e.g., by replacing the "." characters with "_"), or else Patsy formulas that include these column names are interpreted incorrectly. To rename the columns in a DataFrame object, we can use the rename method and pass a dictionary with name translations as the columns argument.

```
In [95]: df_subset.rename(columns={"Sepal.Length": "Sepal_Length",
   ...:                            "Sepal.Width": "Sepal_Width",
   ...:                            "Petal.Length": "Petal_Length",
   ...:                            "Petal.Width": "Petal_Width"}, inplace=True)
```

After these transformations, we have a DataFrame instance suitable for a logistic regression analysis.

```
In [96]: df_subset.head(3)
Out[96]:
```

	Sepal_Length	Sepal_Width	Petal_Length	Petal_Width	Species
50	7.0	3.2	4.7	1.4	1
51	6.4	3.2	4.5	1.5	1
52	6.9	3.1	4.9	1.5	1

To create a logistic model that attempts to explain the value of the Species variable with Petal_length and Petal_Width as independent variables, we can create an instance of the smf.logit class and use the Patsy formula "Species ~ Petal_Length + Petal_Width".

```
In [97]: model = smf.logit("Species ~ Petal_Length + Petal_Width", data=df_subset)
```

As usual, we need to call the resulting model instance's fit method to fit the model to the supplied data. The fit is performed with maximum likelihood optimization.

```
In [98]: result = model.fit()
Optimization terminated successfully.
        Current function value: 0.102818
        Iterations 10
```

As for regular linear regression, we can obtain a summary of the model's fit to the data by printing the output produced by the summary method of the result object. We can see the fitted model parameters with an estimate for their z-score and the corresponding p-value, which can help us judge whether an explanatory variable is significant or not in the model.

```
In [99]: print(result.summary())
                       Logit Regression Results
==============================================================================
Dep. Variable:                Species   No. Observations:                  100
Model:                          Logit   Df Residuals:                       97
Method:                           MLE   Df Model:                            2
Date:                Sun, 26 Apr 2015   Pseudo R-squ.:                  0.8517
Time:                        01:41:04   Log-Likelihood:                -10.282
converged:                       True   LL-Null:                       -69.315
LLR p-value:                 2.303e-26
==============================================================================
                 coef    std err          z      P>|z|      [95.0% Conf. Int.]
------------------------------------------------------------------------------
Intercept      45.2723     13.612      3.326      0.001      18.594     71.951
Petal_Length   -5.7545      2.306     -2.496      0.013     -10.274     -1.235
Petal_Width   -10.4467      3.756     -2.782      0.005     -17.808     -3.086
==============================================================================
```

The result object for logistic regression also provides the get_margeff method, which returns an object that implements a summary method that outputs information about the marginal effects of each explanatory variable in the model.

```
In [100]: print(result.get_margeff().summary())
         Logit Marginal Effects
=====================================
Dep. Variable:                Species
Method:                          dydx
At:                           overall
==============================================================================
                dy/dx    std err          z      P>|z|      [95.0% Conf. Int.]
------------------------------------------------------------------------------
Petal_Length   -0.1736      0.052     -3.347      0.001      -0.275     -0.072
Petal_Width    -0.3151      0.068     -4.608      0.000      -0.449     -0.181
==============================================================================
```

When we are satisfied with the fit of the model to the data, we can, for example, use it to predict the value of the response variable for new values of the explanatory variables. For this, we can use the predict method in the result object produced by the model fitting, and we need to pass a data frame object with the new values of the independent variables.

```
In [101]: df_new = pd.DataFrame({"Petal_Length": np.random.randn(20)*0.5 + 5,
     ...:                        "Petal_Width": np.random.randn(20)*0.5 + 1.7})
In [102]: df_new["P-Species"] = result.predict(df_new)
```

The result is an array with probabilities for each observation to correspond to the response $y = 1$, and by comparing this probability to the threshold value 0.5, we can generate predictions for the binary value of the response variable.

```
In [103]: df_new["P-Species"].head(3)
Out[103]: 0    0.995472
          1    0.799899
          2    0.000033
          Name: P-Species, dtype: float64
In [104]: df_new["Species"] = (df_new["P-Species"] > 0.5).astype(int)
```

The intercept and the slope of the line in the plane spanned by the coordinates Petal_Width and Petal_Length that define the boundary between a point classified as $y = 0$ and $y = 1$, respectively, can be computed from the fitted model parameters. The model parameters can be obtained using the params attribute of the result object.

```
In [105]: params = result.params
     ...: alpha0 = -params['Intercept']/params['Petal_Width']
     ...: alpha1 = -params['Petal_Length']/params['Petal_Width']
```

Finally, to access the model and its predictions for new data points, we plot a scatter plot of the fitted (squares) and predicted (circles) data where data corresponding to the species *virginica* is coded with blue (dark) color, and the species *versicolor* is coded with green (light) color. The result is shown in Figure 14-5.

```
In [106]: fig, ax = plt.subplots(1, 1, figsize=(8, 4))
     ...:  # species virginica
     ...: ax.plot(df_subset[df_subset.Species == 0].Petal_Length.values,
     ...:         df_subset[df_subset.Species == 0].Petal_Width.values,
     ...:         's', label='virginica')
     ...: ax.plot(df_new[df_new.Species == 0].Petal_Length.values,
     ...:         df_new[df_new.Species == 0].Petal_Width.values, 'o',
     ...:         markersize=10, color="steelblue", label='virginica (pred.)')
     ...:
     ...: # species versicolor
     ...: ax.plot(df_subset[df_subset.Species == 1].Petal_Length.values,
     ...:         df_subset[df_subset.Species == 1].Petal_Width.values, 's',
     ...:         label='versicolor')
     ...: ax.plot(df_new[df_new.Species == 1].Petal_Length.values,
     ...:         df_new[df_new.Species == 1].Petal_Width.values, 'o',
     ...:         markersize=10, color="green", label='versicolor (pred.)')
     ...:
     ...: # boundary line
     ...: _x = np.array([4.0, 6.1])
     ...: ax.plot(_x, alpha0 + alpha1 * _x, 'k')
     ...: ax.set_xlabel('Petal length')
     ...: ax.set_ylabel('Petal width')
     ...: ax.legend()
```

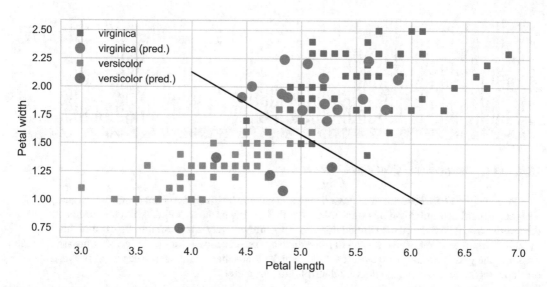

Figure 14-5. *The result of a classification of Iris species using Logit regression with petal length and width and independent variables*

Poisson Model

Another example of discrete regression is the Poisson model, which can describe a process where the response variable is a success count for many attempts, and each has a low probability of success. The Poisson model is also an example of a model that can be treated with the generalized linear model, using the natural logarithm as the link function. To see how we can fit data to a Poisson model using the statsmodels library, let's analyze another interesting dataset from the R dataset repository. The `discoveries` dataset counts the number of great discoveries between 1860 and 1959. Because of the nature of the data, it is reasonable to assume that the counts might be Poisson distributed. To explore this hypothesis, let's begin with loading the dataset using the `sm.datasets.get_rdataset` function and display the first few values to understand the data format.

```
In [107]: dataset = sm.datasets.get_rdataset("discoveries")
In [108]: df = dataset.data.set_index("time").rename(
     ...:     columns={"value": "discoveries"})
In [109]: df.head(10).T
Out[109]:
```

time	1860	1861	1862	1863	1864	1865	1866	1867	1868	1869
discoveries	5	3	0	2	0	3	2	3	6	1

Here we can see that the dataset contains integer counts in the `discoveries` series and that the first few years in the series have on average a few great discoveries. To see if this is typical data for the entire series, we can plot a bar graph of the number of discoveries per year, as shown in Figure 14-6.

```
In [109]: fig, ax = plt.subplots(1, 1, figsize=(16, 4))
     ...: df.plot(kind='bar', ax=ax)
```

357

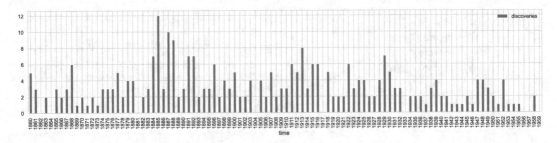

Figure 14-6. *The number of great discoveries per year*

From Figure 14-6, the number of great discoveries seems relatively constant over time, although a slightly declining trend might be noticeable. Nonetheless, the initial hypothesis that the number of discoveries might be Poisson distributed does not look immediately unreasonable. To explore this hypothesis more systematically, we can fit the data to a Poisson process, for example, using the smf.poisson class and the Patsy formula "discoveries ~ 1", which means that we model the discoveries variable with only an intercept coefficient (the Poisson distribution parameter).

```
In [110]: model = smf.poisson("discoveries ~ 1", data=df)
```

As usual, we must call the fit method to perform the fit of the model to the supplied data.

```
In [111]: result = model.fit()
Optimization terminated successfully.
         Current function value: 2.168457
         Iterations 7
```

The summary method of the result objects displays a summary of model fit and several fit statistics.

```
In [112]: print(result.summary())
                     Poisson Regression Results
==============================================================================
Dep. Variable:            discoveries   No. Observations:                  100
Model:                        Poisson   Df Residuals:                       99
Method:                           MLE   Df Model:                            0
Date:                Sun, 26 Apr 2015   Pseudo R-squ.:                   0.000
Time:                        14:51:41   Log-Likelihood:                -216.85
converged:                       True   LL-Null:                       -216.85
LLR p-value:                      nan
==============================================================================
                 coef    std err          z      P>|z|      [95.0% Conf. Int.]
------------------------------------------------------------------------------
Intercept      1.1314      0.057     19.920      0.000        1.020      1.243
==============================================================================
```

The model parameters, available via the params attribute of the result object, are related to the λ parameter of the Poisson distribution via the exponential function (the inverse of the link function).

```
In [113]: lmbda = np.exp(result.params)
```

Once we have the estimated λ parameter of the Poisson distribution, we can, for example, compare the histogram of the observed count values with the theoretical counts, which we can obtain from a Poisson-distributed random variable from the SciPy `stats` library.

```
In [114]: X = stats.poisson(lmbda)
```

In addition to the fit parameters, we can also obtain estimated confidence intervals of the parameters using the `conf_int` method.

```
In [115]: result.conf_int()
Out[115]:
```

	0	1
Intercept	1.020084	1.242721

To assess the fit of the data to the Poisson distribution, we also create random variables for the lower and upper bounds of the confidence interval for the model parameter.

```
In [116]: X_ci_l = stats.poisson(np.exp(result.conf_int().values)[0, 0])
In [117]: X_ci_u = stats.poisson(np.exp(result.conf_int().values)[0, 1])
```

Finally, we graph the histogram of the observed counts with the theoretical probability mass functions for the Poisson distributions corresponding to the fitted model parameter and its confidence intervals. The result is shown in Figure 14-7.

```
In [118]: v, k = np.histogram(df.values, bins=12, range=(0, 12), density=True)
In [119]: fig, ax = plt.subplots(1, 1, figsize=(12, 4))
     ...: ax.bar(k[:-1], v, color="steelblue",  align='center',
     ...:        label='Discoveries per year')
     ...: ax.bar(k-0.125, X_ci_l.pmf(k), color="red", alpha=0.5, align='center',
     ...:        width=0.25, label='Poisson fit (CI, lower)')
     ...: ax.bar(k, X.pmf(k), color="green", align='center', width=0.5,
     ...:        label='Poisson fit')
     ...: ax.bar(k+0.125, X_ci_u.pmf(k), color="red",  alpha=0.5, align='center',
     ...:        width=0.25, label='Poisson fit (CI, upper)')
     ...: ax.legend()
```

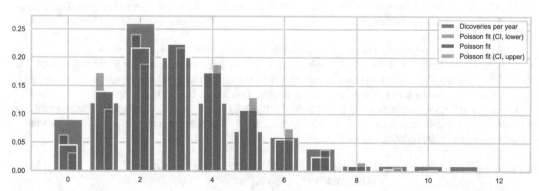

Figure 14-7. *Comparison of histograms of the number of great discoveries per year and the probability mass function for the fitted Poisson model*

359

The result shown in Figure 14-7 indicates that a Poisson process does not describe the dataset of great discoveries well since the agreement between the Poisson probability mass function and the observed counts deviates significantly. Therefore, the hypothesis that the number of great discoveries per year is a Poisson process must be rejected. A failure to fit a model to a given dataset is a natural part of the statistical modeling process. Although the dataset turned out not to be Poisson distributed (perhaps because years with a large and small number of great discovers tend to be clustered together), we still have gained insight by the failed attempt to model it as such. A time-series analysis, as discussed in the following section, could be a better approach because of the correlations between the number of discoveries in any given year and its recent past.

Time Series

Time-series analysis is an important field in statistical modeling that deals with analyzing and forecasting future values of data that are observed as a function of time. Time-series modeling differs in several aspects from regular regression models. Perhaps most importantly, a time series of observations typically cannot be considered as a series of independent random samples from a population. Instead, there is often a component of correlation between observations that are close to each other in time. Also, the independent variables in a time-series model are the past observations of the same series rather than a set of distinct factors. For example, while a regular regression can describe the demand for a product as a function of its price, in a time-series model, it is typical to attempt to predict future values from past observations. This is a reasonable approach when there are autocorrelations such as trends in the time series under consideration (e.g., daily or weekly cycles, steadily increasing trends, or inertia in the change of its value). Examples of time series include stock prices, weather and climate observations, and many other temporal processes in nature and economics.

An example of a type of statistical model for time series is the autoregressive (AR) model, in which a future value depends linearly on p previous values: $Y_t = \beta_0 + \sum_{n=1}^{p} \beta_n Y_{t-n} + \varepsilon_t$, where β_0 is a constant and $\beta n, 1$ $\leq n \leq N$ are the coefficients that define the AR model. The error εt is assumed to be white noise without autocorrelation. Within this model, all autocorrelation in the time series should be captured by the linear dependence on the p previous values. A time series that depends linearly on only one previous value (in a suitable unit of time) can be fully modeled with an AR process with $p=1$, denoted as AR(1), and a time series that depends linearly on two previous values can be modeled by an AR(2) process, and so on. The AR model is a special case of the ARMA model, a more general model that also includes a moving average (MA) of q previous residuals of the series: $Y_t = \beta_0 + \sum_{n=1}^{p} \beta_n Y_{t-n} + \sum_{n=1}^{q} \theta_n \varepsilon_{t-n} + \varepsilon_t$, where the model parameters θn are the weight factors for the moving average. This model is known as the ARMA model and is denoted by ARMA(p, q), where p is the number of autoregressive terms, and q is the number of moving-average terms. Many other time-series models exist, but the AR and ARMA capture the basic ideas fundamental to many time-series applications.

The statsmodels library has a submodule dedicated to time-series analysis: statsmodels.tsa, which implements several standard models for time-series analysis and graphical and statistical analysis tools for exploring properties of time-series data. For example, let's revisit the time series with outdoor temperature measurements used in Chapter 12 and say that we want to predict the hourly temperature for a few days into the future based on previous observations using an AR model. For concreteness, let's take the temperatures measured during March and predict the hourly temperature for the first three days of April. First, we load the dataset into a Pandas DataFrame object.

```
In [120]: df = pd.read_csv("temperature_outdoor_2014.tsv", header=None,
     ...:                   delimiter="\t", names=["time", "temp"])
     ...: df.time = pd.to_datetime(df.time, unit="s")
     ...: df = df.set_index("time").resample("H").mean()
```

For convenience, we extract the observations for March and April and store them in new DataFrame objects, df_march and df_april, respectively.

```
In [121]: df_march = df[df.index.month == 3]
In [122]: df_april = df[df.index.month == 4]
```

Here, we attempt to model the time series of the temperature observations using the AR model, and an important condition for its applicability is that it is applied to a stationary process, which does not have autocorrelation or trends other than those explained by the terms in the model. The plot_acf function in the smg.tsa model is a useful graphical tool for visualizing autocorrelation in a time series. It takes an array of time-series observations and graphs the autocorrelation with increasing time delay on the x axis. The optional lags argument can be used to determine how many time steps are to be included in the plot, which is useful for long time series and when we only wish to see the autocorrelation for a limited number of time steps. The autocorrelation functions for the temperature observations and the first-, second-, and third-order differences are generated and graphed using the plot_acf function in the following code; the resulting graph is shown in Figure 14-8.

```
In [123]: fig, axes = plt.subplots(1, 4, figsize=(12, 3))
     ...: smg.tsa.plot_acf(df_march.temp, lags=72, ax=axes[0])
     ...: smg.tsa.plot_acf(df_march.temp.diff().dropna(), lags=72, ax=axes[1])
     ...: smg.tsa.plot_acf(df_march.temp.diff().diff().dropna(),
     ...:                  lags=72, ax=axes[2])
     ...: smg.tsa.plot_acf(df_march.temp.diff().diff().diff().dropna(),
     ...:                  lags=72, ax=axes[3])
```

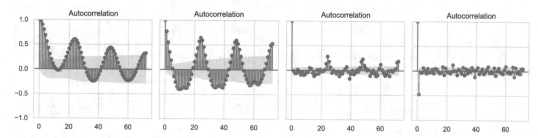

Figure 14-8. *Autocorrelation function for temperature data at increasing order of differentiation, from left to right*

We can see a clear correlation between successive values in the time series in the leftmost graph of Figure 14-8. But for increasing order, differencing of the time series reduces the autocorrelation significantly. Suggesting that while each successive temperature observation strongly correlates with its preceding value, such correlations are not as strong for the higher-order *changes* between the consecutive observations. Taking the difference of a time series is often a useful way of detrending it and eliminating correlation. Taking differences diminishes the structural autocorrelation, suggesting that a sufficiently high-order AR model might be able to model the time series.

To create an AR model for the time series under consideration, we can use the AutoReg class from the statsmodels.tsa.ar_model submodule. It can, for example, be initiated with Pandas series indexed by DatetimeIndex or PeriodIndex (see the docstring of AutoReg for alternative ways of passing time-series data to this class).

```
In [124]: from statsmodels.tsa import ar_model
     ...: model = ar_model.AutoReg(df_march.temp, lags=72)
```

We need to provide the order of the AR model, which can be set using the `lags` keyword argument to the `AutoReg` class constructor. Here, since we can see a strong autocorrelation with a lag of 24 periods (24 hours) in Figure 14-8, we must *at least* include terms for 24 previous terms in the model. To be on the safe side, and since we aim to predict the temperature for three days, or 72 hours, let's make the order of the AR model correspond to 72 hours as well. We call the `fit` method in the model class to fit the model to the data.

```
In [125]: result = model.fit()
```

An important condition for the AR process to be applicable is that the residuals of the series are stationary (no remaining autocorrelation and no trends). The Durbin-Watson statistical test can be used to test for stationary in a time series. It returns a value between 0 and 4, and values close to 2 correspond to time series that do not have remaining autocorrelation.

```
In [126]: sm.stats.durbin_watson(result.resid)
Out[126]: 1.9985623006352975
```

We can also use the `plot_acf` function to graph the autocorrelation function for the residual and verify that there is no significant autocorrelation (see Figure 14-9).

```
In [127]: fig, ax = plt.subplots(1, 1, figsize=(8, 3))
     ...: smg.tsa.plot_acf(result.resid, lags=72, ax=ax)
```

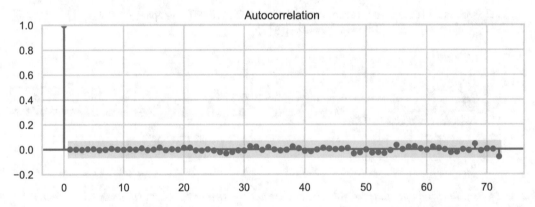

Figure 14-9. *Autocorrelation plot for the residual from the AR(72) model for the temperature observations*

The Durbin-Watson statistic close to 2 and the absence of autocorrelation in Figure 14-9 suggest that the current model successfully explains the fitted data. We can now forecast the temperature for future dates using the `predict` method in the result object returned by the model `fit` method.

```
In [128]: temp_3d_forecast = result.predict("2014-04-01", "2014-04-4")
```

Next, we graph the forecast (green) together with the previous three days of temperature observations (blue) and the actual outcome (orange), for which the result is shown in Figure 14-10.

```
In [129]: fig, ax = plt.subplots(1, 1, figsize=(12, 4))
     ...: ax.plot(df_march.index.values[-72:], df_march.temp.values[-72:],
     ...:         label="train data")
```

```
...: ax.plot(df_april.index.values[:72], df_april.temp.values[:72],
...:         label="actual outcome")
...: ax.plot(pd.date_range("2014-04-01", "2014-04-4", freq="H").values,
...:         temp_3d_forecast, label="predicted outcome")
...:
...: ax.legend()
```

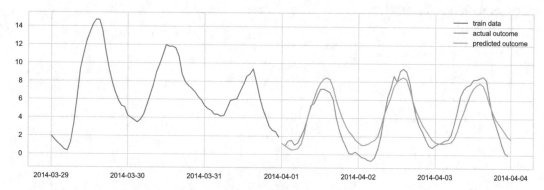

Figure 14-10. *Observed and predicted temperatures as a function of time*

The agreement of the predicted temperature and the actual outcome shown in Figure 14-10 is rather good. However, this is not always the case, as the temperature cannot be forecasted based solely on previous observations. Nonetheless, within a stable weather system, the hourly temperature of a day may be accurately forecasted with an AR model, accounting for the daily variations and other steady trends.

In addition to the basic AR model, statsmodels also provide the ARMA (autoregressive moving average) and ARIMA (autoregressive integrated moving average) models. The usage patterns for these models are similar to that of the AR model we have used here, but there are some differences in the details. Refer to the docstring statsmodels.tsa.arima.model.ARIMA class and the official statsmodels documentation for further information.

Summary

This chapter briefly surveyed statistical modeling and introduced basic statistical modeling features of the statsmodels library and model specification using Patsy formulas. Statistical modeling is a broad field, and we only scratched the surface of what the statsmodels library can be used for in this chapter. We began with an introduction to how to specify statistical models using the Patsy formula language, used in the "Linear Regression" section for response variables that are continuous (regular linear regression) and discrete (logistic and nominal regression). After covering linear regression, we briefly looked at time-series analysis, which requires slightly different methods than linear regression because of the correlations between successive observations that naturally arise in time series. We did not touch upon many aspects of statistical modeling in this introduction. However, the basics of linear regression and time-series modeling we covered here should provide a background for further exploration. Chapter 15 continues with machine learning, a topic closely related to statistical modeling in motivation and methods.

Further Reading

Excellent and thorough introductions to statistical modeling are given in *An Introduction to Statistical Learning* by G. James (Springer-Verlag, 2013), which is also available at `https://statlearning.com`, and in *Applied Predictive Modeling* by M. Kuhn (Springer, 2013). An accessible introduction to time-series analysis is given in *Forecasting: Principles and Practice* by R. J. Hyndman (OTexts, 2013), which is also available for free online at `www.otexts.org/fpp`.

CHAPTER 15

■■■

Machine Learning

This chapter explores machine learning. This topic is closely related to statistical modeling, which we considered in Chapter 14 because both use data to describe and predict outcomes of uncertain or unknown processes. The approach taken in statistical modeling emphasizes understanding how the data is generated by devising models that describe the underlying process behavior and fitting the model's parameters to the observed data. If the model fits the data well and satisfies the relevant model assumptions, then the model can be assumed to give a holistic description of the process. Such a model can, for example, be used to compute statistics with known distributions and evaluate statistical tests. However, if the data is too complex to be explained using available statistical models, this approach has reached its limits. In machine learning, on the other hand, the process that generates the data and potential models thereof is not central. Instead, the observed data and the explanatory variables are the fundamental starting point of a machine-learning application. Given data, machine-learning methods can be used to find patterns and structures in the data, which can be used to predict the outcome of new observations. Machine learning, therefore, does not provide an understanding of how data is generated. Because fewer assumptions are made regarding the distribution and statistical properties of the data, we typically cannot compute statistics and perform statistical tests regarding the significance of specific observations. Instead, machine learning strongly emphasizes the accuracy with which new observations are predicted.

Although significant differences exist in the fundamental approach taken in statistical modeling and machine learning, many mathematical methods are closely related or sometimes even the same. This chapter recognizes several methods used in Chapter 14 on statistical modeling, but they are employed with a different mindset and slightly different goals.

The chapter introduces basic machine-learning methods and surveys how such methods can be used in Python. The focus is on machine-learning methods that have broad application in many scientific and technical computing fields. The most prominent and comprehensive machine-learning library for Python is scikit-learn. However, there are several alternative and complementary libraries: TensorFlow, Keras, and PyTorch, to mention a few. This chapter uses the scikit-learn library exclusively, which implements the most common machine-learning algorithm. However, readers interested in machine learning are encouraged to explore the other libraries mentioned.

■ **scikit-learn**   The scikit-learn library contains a comprehensive collection of machine-learning-related algorithms, including regression, classification, dimensionality reduction, and clustering. For more information about the project and its documentation, see its web page at `http://scikit-learn.org`. At the time of writing, the latest version of scikit-learn is 1.3.0.

R. Johansson, *Numerical Python*, https://doi.org/10.1007/979-8-8688-0413-7_15

Importing Modules

This chapter works with the scikit-learn library, which provides the `sklearn` Python module. With the `sklearn` module, we use the same import strategy as the SciPy library to explicitly import modules from the library. Let's use the following modules from the `sklearn` library.

```
In [1]: from sklearn import datasets
In [2]: from sklearn import model_selection
In [3]: from sklearn import linear_model
In [4]: from sklearn import metrics
In [5]: from sklearn import tree
In [6]: from sklearn import neighbors
In [7]: from sklearn import svm
In [8]: from sklearn import ensemble
In [9]: from sklearn import cluster
```

For plotting and basic numerical computation, we also require the Matplotlib and NumPy libraries, which we import in the usual manner.

```
In [10]: import matplotlib.pyplot as plt
In [11]: import numpy as np
```

We also use the Seaborn library for graphics and figure styling.

```
In [12]: import seaborn as sns
```

Brief Review of Machine Learning

Machine learning is a topic in the artificial intelligence field of computer science. Machine learning can include all applications where feeding training data into a computer program enables it to perform a given task. This is a very broad definition, but machine learning is often associated with a much more specific set of techniques and methods. Here, we take a practical approach and explore several basic methods and key concepts in machine learning by example. Let's begin with a brief introduction of the terminology and core concepts.

In machine learning, the process of fitting a model or an algorithm to observed data is known as *training*. Machine-learning applications can broadly be classified into either of two types: *supervised* and *unsupervised* learning, which differ in the type of data the application is trained with. In *supervised learning*, the data includes feature variables and known response variables. Both feature and response variables can be continuous or discrete. Preparing such data typically requires manual effort and sometimes even expert domain knowledge. The application is thus trained with handcrafted data, and the training can, therefore, be viewed as supervised machine learning. Examples of applications include regression (prediction of a continuous response variable) and classification (prediction of a discrete response variable), where the value of the response variable is known for the training dataset but not for new samples.

In contrast, *unsupervised learning* corresponds to situations where machine-learning applications are trained with raw data that is not labeled or otherwise manually prepared. An example of unsupervised learning is the clustering of data into groups, or in other words, the grouping of data into initially unknown categories. In contrast to supervised classification, it is typical for unsupervised learning that the final categories are not known in advance, and the training data, therefore, cannot be labeled accordingly. It may also be the case that the manual labeling of the data is difficult or costly, for example, because the number of samples is too large. Unsupervised machine learning is more complicated and limited in what it can be used for than supervised machine learning, which should be preferred whenever possible. However, unsupervised machine learning can be a powerful tool when creating labeled training datasets is impossible or unrealistic.

Naturally, there is much more complexity to machine learning than suggested by the basic types of problems outlined in the preceding text. But these concepts are recurring themes in many machine-learning applications. This chapter looks at a few basic machine-learning techniques demonstrating several central concepts. First, let's go over some common machinelearning terminology.

- *Cross-validation* is dividing the available data into *training data* and *testing data* (also known as *validation data*), where only the training data is used to train the machine-learning model and where the test data allows the trained application to be tested on previously unseen data. This aims to measure how well the model predicts new observations and limit problems with overfitting. There are several approaches to dividing the data into training and testing datasets. For example, one extreme approach is to test all possible ways to divide the data (*exhaustive cross-validation*) and use an aggregate result (e.g., average or the minimum value, depending on the situation). However, for large datasets, the number of possible training and testing data combinations becomes extremely large, making exhaustive cross-validation impractical. Another extreme is to use all but one sample in the training set and the remaining sample in the training set (leave-one-out cross-validation) and to repeat the training-test cycle for all combinations in which one sample is chosen from the available data. A variant of this method is to divide the available data into k groups and perform a leave-one-out cross-validation with the k groups of datasets. This method is known as k-fold cross-validation and is a popular technique often used in practice. In the scikit-learn library, the `sklearn.model_selection` module contains functions for working with cross-validation.

- *Feature extraction* is an important step in the preprocessing stage of a machine-learning problem. It involves creating suitable feature variables and the corresponding feature matrices that can be passed to one of many machine-learning algorithms implemented in the scikit-learn library. The scikit-learn `sklearn.feature_extraction` module plays a similar role in many machine-learning applications as the Patsy formula library does in statistical modeling, especially for text- and image-based machine-learning problems. Using methods from the `sklearn.feature_extraction` module, we can automatically assemble feature matrices (design matrices) from various data sources.

- *Dimensionality reduction* and *feature selection* are techniques frequently used in machine-learning applications where it is common to have a large number of explanatory variables (features), many of which may not significantly contribute to the predictive power of the application. To reduce the complexity of the model, it is often desirable to eliminate less useful features, thereby reducing the problem's dimensionality. This is particularly important when the number of features is comparable to or larger than the number of observations. The scikit-learn `sklearn.decomposition` and `sklearn.feature_selection` modules contain functions for reducing the dimensionality of a machine-learning problem: for example, principal component analysis (PCA) is a popular technique for dimensionality reduction that works by performing a singular-value decomposition of the feature matrix and keeping only dimensions that correspond to the most significant singular vectors.

The following sections look at how scikit-learn can be used to solve examples of machine-learning problems using the techniques discussed in the preceding text. Here, we work with generated data and built-in datasets. Like the statsmodels library, scikit-learn comes with several built-in datasets to explore machine-learning methods. The `datasets` module in `sklearn` provides three groups of functions for loading built-in datasets (with the `load_` prefix; e.g., `load_wine`), fetching external datasets (with the `fetch_` prefix; e.g., `fetch_califorma_housing`), and generating datasets from random numbers (with the `make_` prefix; e.g., `make_regression`).

Regression

Regression is central to machine learning and statistical modeling, as demonstrated in Chapter 14. In machine learning, we are not primarily concerned with how well the regression model fits the data but instead care about how well it predicts new observations. For example, suppose we have a large number of features and a smaller number of observations. In that case, we can often fit the regression perfectly to the data without it being beneficial for predicting new values. This is an example of overfitting: A small residual between the data and the regression model does not guarantee that the model can accurately predict future observations. In machine learning, a common method to deal with this problem is to partition the available data into a training and testing datasets for validating the regression results against previously unseen data.

To see how fitting a training dataset and validating the result against a testing dataset can work out, consider a regression problem with 50 samples and 50 features, out of which only 10 features are informative (linearly correlated with the response variable). This simulates a scenario when we have 50 known features, but it turns out that only 10 of those features contribute to the predictive power of the regression model. The `make_regression` function in the `sklearn.datasets` module generates data of kind.

```
In [13]: X_all, y_all = datasets.make_regression(
    ...:        n_samples=50, n_features=50, n_informative=10)
```

The result is two arrays, `X_all` and `y_all`, of shapes `(50, 50)` and `(50,)`, corresponding to the design matrices for a regression problem with 50 samples and 50 features. Instead of performing a regression on the entire dataset (and obtaining a perfect fit because of the small number of observations), we split the dataset into two equal-sized datasets using the `train_test_split` function from `sklearn.model_selection` module. The result is a training dataset `X_train`, `y_train`, and a testing dataset `X_test`, `y_test`.

```
In [14]: X_train, X_test, y_train, y_test = \
    ...:        model_selection.train_test_split(X_all, y_all, train_size=0.5)
```

In scikit-learn, ordinary linear regression can be carried out using the `LinearRegression` class from the `sklearn.linear_model` module, which is comparable with the `statsmodels.api.OLS` from the statsmodels library. To perform a regression, we first create a `LinearRegression` instance.

```
In [15]: model = linear_model.LinearRegression()
```

To fit the model to the data, we must invoke the `fit` method, which takes the feature matrix and the response variable vector as the first and second arguments.

```
In [16]: model.fit(X_train, y_train)
Out[16]: LinearRegression(copy_X=True, fit_intercept=True, n_jobs=1,
                          normalize=False)
```

Note that compared to the `OLS` class in statsmodels, the order of the feature matrix and response variable vector is reversed, and in statsmodels the data is specified when the class instance is created instead of when calling the `fit` method. Also, in scikit-learn, calling the `fit` method does not return a new result object; instead, the result is stored directly in the model instance. These minor differences are small inconveniences when working interchangeably with the statsmodels and scikit-learn modules and are worth noting.[1]

Since the regression problem has 50 features and we only trained the model with 25 samples, we can expect complete overfitting that perfectly fits the data. This can be quantified by computing the sum of squared errors (SSEs) between the model and the data. To evaluate the model for a given set of features, we can use the `predict` method to compute the residuals and the SSE.

```
In [17]: def sse(resid):
    ...:       return np.sum(resid**2)
In [18]: resid_train = y_train - model.predict(X_train)
    ...: sse_train = sse(resid_train)
    ...: sse_train
Out[18]: 8.1172209425431673e-25
```

As expected, the residuals are all essentially zero for the training dataset due to the overfitting allowed by having twice as many features as data points. This overfitted model is, however, not at all suitable for predicting unseen data. This can be verified by computing the SSE for our test dataset.

```
In [19]: resid_test = y_test - model.predict(X_test)
    ...: sse_test = sse(resid_test)
    ...: sse_test
Out[19]: 213555.61203039082
```

The result is a very large SSE value, which indicates that the model does not predict new observations well. An alternative measure of the fit of a model to a dataset is the R-squared score (see Chapter 14), which we can compute using the `score` method. It takes a feature matrix and response variable vector as arguments and calculates the score. As expected, we obtain an r-square score of 1.0 for the training dataset, but for the testing dataset, we get a low score.

```
In [20]: model.score(X_train, y_train)
Out[20]: 1.0
In [21]: model.score(X_test, y_test)
Out[21]: 0.31407400675201746
```

The big difference between the training and testing datasets scores indicates that the model is overfitted.

Finally, we can also take a graphical approach and plot the residuals of the training and testing datasets and visually inspect the values of the coefficients and the residuals. From a `LinearRegression` object, we can extract the fitted parameters using the `coef_` attribute. To simplify repeated plotting of the training and testing residuals and the model parameters, we first create a `plot_residuals_and_coeff` function for plotting these quantities. We then call the function with the result from the ordinary linear regression model trained and tested on the training and testing datasets, respectively. The result is shown in Figure 15-1, and there is a significant difference in the magnitude of the residuals for the test and the training datasets for every sample.

---

[1] In practice it is common to work with both statsmodels and scikit-learn, as they in many respects complement each other. However, this chapter focuses solely on scikit-learn.

```
In [22]: def plot_residuals_and_coeff(resid_train, resid_test, coeff):
    ...:     fig, axes = plt.subplots(1, 3, figsize=(12, 3))
    ...:     axes[0].bar(np.arange(len(resid_train)), resid_train)
    ...:     axes[0].set_xlabel("sample number")
    ...:     axes[0].set_ylabel("residual")
    ...:     axes[0].set_title("training data")
    ...:     axes[1].bar(np.arange(len(resid_test)), resid_test)
    ...:     axes[1].set_xlabel("sample number")
    ...:     axes[1].set_ylabel("residual")
    ...:     axes[1].set_title("testing data")
    ...:     axes[2].bar(np.arange(len(coeff)), coeff)
    ...:     axes[2].set_xlabel("coefficient number")
    ...:     axes[2].set_ylabel("coefficient")
    ...:     fig.tight_layout()
    ...:     return fig, axes
In [23]: fig, ax = plot_residuals_and_coeff(resid_train, resid_test, model.coef_)
```

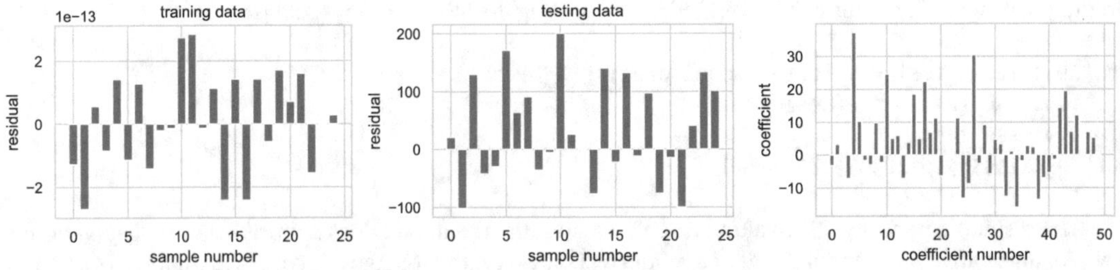

Figure 15-1. *The residual between the ordinary linear regression model and the training data (left), the model and the test data (middle), and the values of the coefficients for the 50 features (right)*

Overfitting in this example happens because we have too few samples, and one solution could be to collect more samples until overfitting is no longer a problem. However, this may not always be practical, as collecting observations may be expensive, and because in some applications, we might have a very large number of features. For such situations, it is desirable to be able to fit a regression problem in a way that avoids overfitting as much as possible (at the expense of not fitting the training data perfectly) so that the model can give meaningful predictions for new observations.

Regularized regression is one possible solution to this problem. The following looks at a few variations of regularized regression. In ordinary linear regression, the model parameters are chosen to minimize the sum of squared residuals. Viewed as an optimization problem, the objective function is, therefore, $\min_\beta \|X\beta - y\|_2^2$, where X is the feature matrix, y is the response variables, and β is the vector of model parameters and where $\|\cdot\|_2$ denotes the L2 norm. In *regularized* regression, we add a *penalty term* in the objective function of the minimization problem. Different types of penalty terms impose different types of regularization on the original regression problem. Two popular regularizations are obtained by adding the L1 or L2 norms of the parameter vector to the minimization objective function, $\min_\beta \{\|X\beta - y\|_2^2 + \alpha \|\beta\|_1\}$ and $\min_\beta \{\|X\beta - y\|_2^2 + \alpha \|\beta\|_2^2\}$. These are known as LASSO and Ridge regression, respectively. Here, α is a free parameter that determines the strength of the regularization. Adding the L2 norm $\|\beta\|_2^2$ favors model parameter vectors with smaller coefficients, and adding the L1 norm $\|\beta\|_1$ favors a model parameter vectors with as few nonzero elements as possible. Which type of regularization is more suitable depends on the problem at hand: When we wish to eliminate as many features as possible, we can use L1 regularization with LASSO regression, and when we want to limit the magnitude of the model coefficients, we can use L2 regularization with Ridge regression.

With scikit-learn, we can perform Ridge regression using the `Ridge` class from the `sklearn.linear_model` module. The usage of this class is almost the same as the `LinearRegression` class used in the preceding text, but we can also give the value of the α parameter that determines the strength of the regularization as an argument when we initialize the class. Here, we chose the value $\alpha = 2.5$. A more systematic approach to choosing α is introduced later in this chapter.

```
In [24]: model = linear_model.Ridge(alpha=2.5)
```

To fit the regression model to the data, we again use the fit method, passing the training feature matrix and response variable as arguments.

```
In [25]: model.fit(X_train, y_train)
Out[25]: Ridge(alpha=2.5, copy_X=True, fit_intercept=True, max_iter=None,
               normalize=False, solver='auto', tol=0.001)
```

Once the model has been fitted to the training data, we can compute the model predictions for the training and testing datasets and compute the corresponding SSE values.

```
In [26]: resid_train = y_train - model.predict(X_train)
    ...: sse_train = sse(resid_train)
    ...: sse_train
Out[26]: 178.50695164950841
In [27]: resid_test = y_test - model.predict(X_test)
    ...: sse_test = sse(resid_test)
    ...: sse_test
Out[27]: 212737.00160105844
```

We note that the SSE of the training data is no longer close to zero, but there is a slight decrease in the SSE for the testing data. For comparison with ordinary regression, we also plot the training and testing residuals and the model parameters using the `plot_residuals_and_coeff` function defined in the preceding text. The result is shown in Figure 15-2.

```
In [28]: fig, ax = plot_residuals_and_coeff(resid_train, resid_test, model.coef_)
```

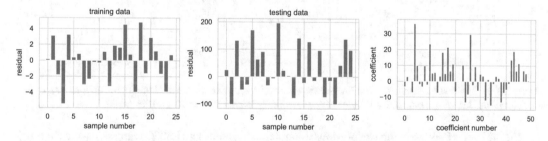

Figure 15-2. *The residual between the Ridge-regularized regression model and the training data (left), the model and the test data (middle), and the values of the coefficients for the 50 features (right)*

Similarly, we can perform the L1-regularized LASSO regression using the `Lasso` class from the `sklearn.linear_model` module. It also accepts the value of the α parameter as an argument when the class instance is initialized. Here, we choose $\alpha = 1.0$ and perform the model fitting to the training data and the computation of the SSE for the training and testing data in the same way described previously.

```
In [29]: model = linear_model.Lasso(alpha=1.0)
In [30]: model.fit(X_train, y_train)
Out[30]: Lasso(alpha=1.0, copy_X=True, fit_intercept=True, max_iter=1000,
               normalize=False, positive=False, precompute=False,
               random_state=None, selection='cyclic', tol=0.0001,
               warm_start=False)
In [31]: resid_train = y_train - model.predict(X_train)
    ...: sse_train = sse(resid_train)
    ...: sse_train
Out[31]: 309.74971389531891
In [32]: resid_test = y_test - model.predict(X_test)
    ...: sse_test = sse(resid_test)
    ...: sse_test
Out[32]: 1489.1176065002333
```

Here, we note that while the SSE of the training data increased compared to that of the ordinary regression, the SSE for the testing data decreased significantly. Thus, by paying a price for how well the regression model fits the training data, we have obtained a model with a significantly improved ability to predict the testing dataset. For comparison with the earlier methods, we again graph the residuals and the model parameters with the plot_residuals_and_coeff function. The result is shown in Figure 15-3. In the rightmost panel of this figure, we see that the coefficient profile is significantly different from those shown in Figure 15-1 and Figure 15-2, and the coefficient vector produced with the LASSO regression contains mostly zeros. This is a suitable method for the current data because, in the beginning, when we generated the dataset, we chose 50 features, out of which only 10 are informative. If we suspect we might have many features that might not contribute much to the regression model, using the L1 regularization of the LASSO regression can thus be a good approach to try.

```
In [33]: fig, ax = plot_residuals_and_coeff(resid_train, resid_test, model.coef_)
```

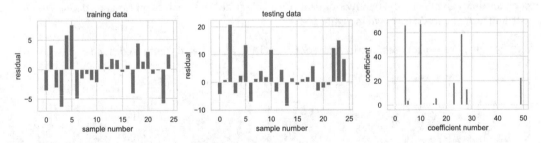

Figure 15-3. *The residual between the LASSO-regularized regression model and the training data (left), the model and the test data (middle), and the values of the coefficients for the 50 features (right)*

The values of α used in the two previous examples using Ridge and LASSO regression were chosen arbitrarily. The most suitable value of α is problem-dependent, and for every new problem, we need to find an appropriate value using trial and error. The scikit-learn library provides methods for assisting this process. But before we explore those methods, it is instructive to look at how the regression model parameters and the SSE for the training and testing datasets depend on the value of α for a specific problem. Here, we focus on LASSO regression since it was seen to work well for the current problem, and we repeatedly solve the same problem using different values for the regularization strength parameter α while storing the values of the coefficients and SSE values in NumPy arrays.

Let's begin by creating the required NumPy arrays. We use np.logspace to create a range of α values that span several orders of magnitude.

```
In [34]: alphas = np.logspace(-4, 2, 100)
In [35]: coeffs = np.zeros((len(alphas), X_train.shape[1]))
In [36]: sse_train = np.zeros_like(alphas)
In [37]: sse_test = np.zeros_like(alphas)
```

Next, we loop through the α values and perform the LASSO regression for each value.

```
In [38]: for n, alpha in enumerate(alphas):
    ...:        model = linear_model.Lasso(alpha=alpha)
    ...:        model.fit(X_train, y_train)
    ...:        coeffs[n, :] = model.coef_
    ...:        sse_train[n] = sse(y_train - model.predict(X_train))
    ...:        sse_test[n] = sse(y_test - model.predict(X_test))
```

Finally, we plot the coefficients and the SSE for the training and testing datasets using Matplotlib. The result is shown in Figure 15-4. We can see in the left panel of this figure that many coefficients are nonzero for very small values of α. This corresponds to the overfitting regime. We can also see that when α is increased above a certain threshold, many of the coefficients collapse to zero, and only a few coefficients remain nonzero. In the right panel of the figure, we see that while the SSE for the training set is steadily increasing with increasing α, there is also a sharp drop in the SSE for the testing dataset. This is the sought-after effect in LASSO regression. However, for too large values of α, all coefficients converge to zero, and the SSEs for both the training and testing datasets become large. Therefore, there is an optimal region of α that prevents overfitting and improves the model's ability to predict unseen data. While these observations are not universally true, a similar pattern can be seen for many problems.

```
In [39]: fig, axes = plt.subplots(1, 2, figsize=(12, 4), sharex=True)
    ...: for n in range(coeffs.shape[1]):
    ...:        axes[0].plot(np.log10(alphas), coeffs[:, n], color='k', lw=0.5)
    ...:
    ...: axes[1].semilogy(np.log10(alphas), sse_train, label="train")
    ...: axes[1].semilogy(np.log10(alphas), sse_test, label="test")
    ...: axes[1].legend(loc=0)
    ...:
    ...: axes[0].set_xlabel(r"${\log_{10}}\alpha$", fontsize=18)
    ...: axes[0].set_ylabel(r"coefficients", fontsize=18)
    ...: axes[1].set_xlabel(r"${\log_{10}}\alpha$", fontsize=18)
    ...: axes[1].set_ylabel(r"sse", fontsize=18)
```

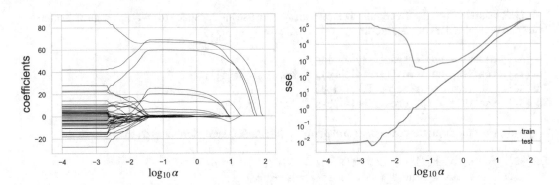

Figure 15-4. *The coefficients (left) and the sum of squared errors (SSEs) for the training and testing datasets (right), for LASSO regression as a function of the logarithm of the regularization strength parameter* α

Testing a regularized regression with several values of α can be carried out automatically using, for example, the `RidgeCV` and `LassoCV` classes. These Ridge and LASSO regression variants internally search for the optimal α using a cross-validation approach. By default, a k-fold cross-validation with $k = 3$ is used, although this can be changed using the `cv` argument to these classes. Because of the built-in cross-validation, we do not need to explicitly divide the dataset into training and testing datasets, as we have done previously.

To use the LASSO method with an automatically chosen α, we create an instance of `LassoCV` and invoke its `fit` method.

```
In [40]: model = linear_model.LassoCV()
In [41]: model.fit(X_all, y_all)
Out[41]: LassoCV(alphas=None, copy_X=True, cv=None, eps=0.001, fit_intercept=True,
                 max_iter=1000, n_alphas=100, n_jobs=1, normalize=False,
                 positive=False, precompute='auto', random_state=None,
                 selection='cyclic', tol=0.0001, verbose=False)
```

The value of regularization strength parameter α selected through the crossvalidation search is accessible through the `alpha_` attribute.

```
In [42]: model.alpha_
Out[42]: 0.13118477495069433
```

The suggested value of α agrees reasonably well with what we might have guessed from Figure 15-4. For comparison with the previous method, we also compute the SSE for the training and testing datasets (although both were used for training in the call to `LassoCV.fit`) and graph the SSE values together with the model parameters, as shown in Figure 15-5. Using the cross-validated LASSO method obtains a model that predicts both the training and testing datasets with relatively high accuracy, and we are no longer as likely to suffer from the problem of overfitting despite having few samples compared to the number of features.[2]

```
In [43]: resid_train = y_train - model.predict(X_train)
    ...: sse_train = sse(resid_train)
    ...: sse_train
```

---

[2] However, note that we can never be sure that a machine-learning application does not suffer from overfitting before we see how the application performs on new observations, and a repeated reevaluation of the application on a regular basis is a good practice.

```
Out[43]: 66.900068715063625
In [44]: resid_test = y_test - model.predict(X_test)
    ...: sse_test = sse(resid_test)
    ...: sse_test
Out[44]: 966.39293785448456
In [45]: fig, ax = plot_residuals_and_coeff(resid_train, resid_test, model.coef_)
```

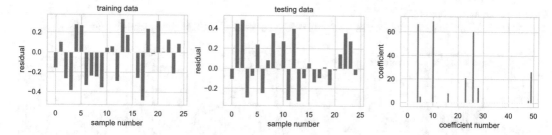

Figure 15-5. *The residuals of the LASSO-regularized regression model with cross-validation for the training data (left) and the testing data (middle). The values of the coefficients for the 50 features are also shown (right)*

Finally, another type of popular regularized regression, which combines the L1 and L2 regularization of the LASSO and Ridge methods, is called elastic-net regularization. The minimization objective function for this method is $\min_{\beta} \left\{ \| X\beta - y \|_2^2 + \alpha\rho \| \beta \|_1 + \alpha(1-\rho) \| \beta \|_2^2 \right\}$, where the parameter ρ (l1_ratio in scikit-learn) determines the relative weight of the L1 and L2 penalties and, thus, how much the method behaves like the LASSO and Ridge methods. In scikit-learn, we can perform an elastic-net regression using the ElasticNet class, to which we can give explicit values of the α (alpha) and ρ (l1_ratio) parameters, or the cross-validated version ElasticNetCV, which automatically finds suitable values of the α and ρ parameters.

```
In [46]: model = linear_model.ElasticNetCV()
In [47]: model.fit(X_train, y_train)
Out[47]: ElasticNetCV(alphas=None, copy_X=True, cv=None, eps=0.001,
                       fit_intercept=True, l1_ratio=0.5, max_iter=1000,
                       n_alphas=100, n_jobs=1, normalize=False, positive=False,
                       precompute='auto', random_state=None, selection='cyclic',
                       tol=0.0001, verbose=0)
```

The value of regularization parameters α and ρ suggested by the crossvalidation search is available through the alpha_ and l1_ratio attributes.

```
In [48]: model.alpha_
Out[48]: 0.13118477495069433
In [49]: model.l1_ratio
Out[49]: 0.5
```

For comparison with the previous method, we again compute the SSE and plot the model coefficients, as shown in Figure 15-6. As expected with $\rho = 0.5$, the result has characteristics of both LASSO regression (favoring a sparse solution vector with only a few dominating elements) and Ridge regression (suppressing the magnitude of the coefficients).

```
In [50]: resid_train = y_train - model.predict(X_train)
    ...: sse_train = sse(resid_train)
    ...: sse_train
Out[50]: 2183.8391729391255
In [51]: resid_test = y_test - model.predict(X_test)
    ...: sse_test = sse(resid_test)
    ...: sse_test
Out[51]: 2650.0504463382508
In [52]: fig, ax = plot_residuals_and_coeff(resid_train, resid_test, model.coef_)
```

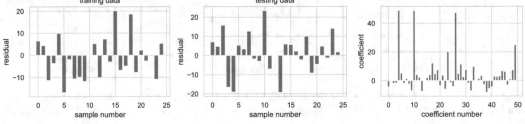

Figure 15-6. *The residuals of the elastic-net regularized regression model with cross-validation for the training data (left) and the testing data (middle). The values of the coefficients for the 50 features are also shown (right)*

Classification

Like regression, classification is a central topic in machine learning. The chapter used a logistic regression model to classify observations into discrete categories. Logistic regression is also used in machine learning for the same task. But there is also a wide variety of alternative algorithms for classification, such as nearest neighbor methods, support vector machines (SVM), decision trees, and random forest methods. The scikit-learn library provides a convenient unified API that allows all these different methods to be used interchangeably for any classification problem.

To see how we can train a classification model with a training dataset and test its performance on testing datasets, let's once again look at the Iris dataset, which provides features for Iris flower samples (sepal and petal width and height), together with the species of each sample (*setosa*, *versicolor*, and *virginica*). The Iris dataset included in the scikit-learn library (as well as in the statsmodels library) is a classic dataset commonly used for testing and demonstrating machine-learning algorithms and statistical models. Here, we revisit the problem of classifying the species of a flower sample given its sepal and petal width and height (see also Chapter 14). First, call the load_iris function in the dataset module to load the dataset. The result is a container object (called a Bunch object in the scikit-learn jargon) containing the data and metadata.

```
In [53]: iris = datasets.load_iris()
In [54]: type(iris)
Out[54]: sklearn.utils._bunch.Bunch
```

For example, descriptive names of the features and target classes are available through the feature_names and target_names attributes.

```
In [55]: iris.target_names
Out[55]: array(['setosa', 'versicolor', 'virginica'], dtype='|S10')
In [56]: iris.feature_names
Out[56]: ['sepal length (cm)', 'sepal width (cm)', 'petal length (cm)', 'petal width (cm)']
```

The actual dataset is available through the data and target attributes.

```
In [57]: iris.data.shape
Out[57]: (150, 4)
In [58]: iris.target.shape
Out[58]: (150,)
```

We begin by splitting the dataset into a training and testing part, using the train_test_split function. Here, we include 70% of the samples in the training set, leaving the remaining 30% for testing and validation.

```
In [59]: X_train, X_test, y_train, y_test = \
    ...:     model_selection.train_test_split(
    ...:         iris.data, iris.target, train_size=0.7)
```

The first step in training a classifier and performing classification tasks using scikitlearn is to create a classifier instance. As mentioned, numerous classifiers are available. Let's begin with a logistic regression classifier, which is provided by the LogisticRegression class in the linear_model module.

```
In [60]: classifier = linear_model.LogisticRegression()
```

The classifier's training is carried out by calling the fit method of the classifier instance. The arguments are the design matrices for the feature and target variables. Here, we use the training part of the Iris dataset arrays created for us when loading the dataset using the load_iris function. If the design matrices are not already available, we can use the same techniques used in Chapter 14, such as constructing the matrices by hand using NumPy functions or using the Patsy library to construct the appropriate arrays automatically. We can also use the feature extraction utilities in the feature_extraction module in the scikit-learn library.

```
In [61]: classifier.fit(X_train, y_train)
Out[61]: LogisticRegression(C=1.0, class_weight=None, dual=False,
                            fit_intercept=True, intercept_scaling=1, max_iter=100,
                            multi_class='ovr', penalty='l2', random_state=None,
                            solver='liblinear', tol=0.0001, verbose=0)
```

Once the classifier has been trained, we can immediately start using it to predict the class for new observations using the predict method. Here, we apply this method to predict the class for the samples assigned to the testing dataset to compare the predictions with the actual values.

```
In [62]: y_test_pred = classifier.predict(X_test)
```

The sklearn.metrics module contains helper functions to assist in analyzing the performance and accuracy of classifiers. For example, the classification_report function, which takes arrays of actual values and the predicted values, returns a tabular summary of the informative classification metrics related to the rate of false negatives and false positives[3].

```
In [63]: print(metrics.classification_report(y_test, y_test_pred))
             precision    recall  f1-score   support
          0       1.00      1.00      1.00        13
          1       1.00      0.92      0.96        13
          2       0.95      1.00      0.97        19
avg / total       0.98      0.98      0.98        45
```

The so-called *confusion matrix*, which can be computed using the confusion_matrix function, also presents useful classification metrics in a compact form: The diagonals correspond to the number of samples that are correctly classified for each level of the category variable, and the off-diagonal elements are the number of incorrectly classified samples. More specifically, the element C_{ij} of the confusion matrix C is the number of samples of category i that were categorized as j. For the current data, we obtain the confusion matrix.

```
In [64]: metrics.confusion_matrix(y_test, y_test_pred)
Out[64]: array([[13  0  0]
                [ 0 12  1]
                [ 0  0 19]])
```

This confusion matrix shows that all elements in the first and third classes were classified correctly, but one element of the second class was mistakenly classified as class three. Note that the elements in each row of the confusion matrix sum up the total number of samples for the corresponding category. In this testing sample, we have 13 elements each in the first and second class and 19 elements in the third class, as can be seen by counting unique values in the y_test array.

```
In [65]: np.bincount(y_test)
Out[65]: array([13, 13, 19])
```

To perform a classification using a different classifier algorithm, we only need to create an instance of the corresponding classifier class. For example, to use a decision tree instead of logistic regression, we can use the DesicisionTreeClassifier class from the sklearn.tree module. Training the classifier and predicting new observations is done in the same way for all classifiers.

```
In [66]: classifier = tree.DecisionTreeClassifier()
    ...: classifier.fit(X_train, y_train)
    ...: y_test_pred = classifier.predict(X_test)
    ...: metrics.confusion_matrix(y_test, y_test_pred)
Out[66]: array([[13,  0,  0],
                [ 0, 12,  1],
                [ 0,  1, 18]])
```

The resulting confusion matrix with the decision tree classifier is somewhat different, corresponding to one additional misclassification in the testing dataset.

---

[3] Note that we do not get the same results each time we run this process due to the randomness in the train-test data split when using the model_selection.train_test_split function. If reproducibility is required, we can use the random_state keyword argument to ensure it.

Other popular classifiers available in scikit-learn include the nearest neighbor classifier KNeighborsClassifier from the sklearn.neighbors module, the support vector classifier (SVC) from the sklearn.svm module, and the random forest classifier RandomForestClassifier from the sklearn.ensemble module. Since they all have the same usage pattern, we can programmatically apply a series of classifiers on the same problem and compare their performance (on this particular problem), for example, as a function of the training and testing sample sizes. To this end, we create a NumPy array with training size ratios ranging from 10% to 90%.

```
In [67]: train_size_vec = np.linspace(0.1, 0.9, 30)
```

Next, we create a list of classifier classes to apply.

```
In [68]: classifiers = [tree.DecisionTreeClassifier,
    ...:                neighbors.KNeighborsClassifier,
    ...:                svm.SVC,
    ...:                ensemble.RandomForestClassifier]
```

We also create an array in which we can store the diagonals of the confusion matrix as a function of training size ratio and classifier.

```
In [69]: cm_diags = np.zeros((3, len(train_size_vec), len(classifiers)), dtype=float)
```

Finally, we loop over each training size ratio and classifier. For each combination, we train the classifier, predict the values of the testing data, compute the confusion matrix, and store its diagonal divided by the ideal values in the cm_diags array.

```
In [70]: for n, train_size in enumerate(train_size_vec):
    ...:     X_train, X_test, y_train, y_test = \
    ...:         model_selection.train_test_split(iris.data, iris.target,
    ...:                                          train_size=train_size)
    ...:     for m, Classifier in enumerate(classifiers):
    ...:         classifier = Classifier()
    ...:         classifier.fit(X_train, y_train)
    ...:         y_test_p = classifier.predict(X_test)
    ...:         cm_diags[:, n, m] = metrics.confusion_matrix(
    ...:             y_test, y_test_p).diagonal()
    ...:         cm_diags[:, n, m] /= np.bincount(y_test)
```

The resulting classification accuracy for each classifier, as a function of training size ratio, is plotted and shown in Figure 15-7.

```
In [71]: fig, axes = plt.subplots(1, len(classifiers), figsize=(12, 3))
    ...: for m, Classifier in enumerate(classifiers):
    ...:     axes[m].plot(train_size_vec, cm_diags[2, :, m],
    ...:                  label=iris.target_names[2])
    ...:     axes[m].plot(train_size_vec, cm_diags[1, :, m],
    ...:                  label=iris.target_names[1])
    ...:     axes[m].plot(train_size_vec, cm_diags[0, :, m],
    ...:                  label=iris.target_names[0])
    ...:     axes[m].set_title(type(Classifier()).__name__)
    ...:     axes[m].set_ylim(0, 1.1)
    ...:     axes[m].set_ylabel("classification accuracy")
```

```
...:        axes[m].set_xlabel("training size ratio")
...:        axes[m].legend(loc=4)
```

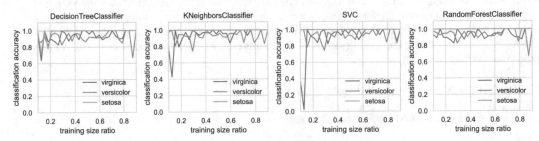

Figure 15-7. *Comparison of classification accuracy of four different classifiers*

Figure 15-7 shows that the classification error is different in each model, but for this example, they have comparable performance. Which classifier is the best depends on the problem at hand, and it is difficult to give any definite answer to which one is more suitable in general. Fortunately, it is easy to switch between different classifiers in scikit-learn and, therefore, effortless to try a few different classifiers for a given classification problem. In addition to the classification accuracy, another important aspect is the computational performance and scaling to larger problems. For large classification problems with many features, decision-tree methods such as the randomized forest method are often a good starting point.

Clustering

In the two previous sections, we explored regression and classification, both examples of supervised learning, since the response variables are given in the dataset. Clustering is a different type of problem and an important topic in machine learning. It can be considered a classification problem where the classes are unknown, making clustering an example of unsupervised learning. The training dataset for a clustering algorithm contains only the feature variables, and the algorithm's output is an array of integers that assign each sample to a cluster (or class). This output array corresponds to the response variable in a supervised classification problem.

The scikit-learn library implements a large number of clustering algorithms that are suitable for different types of clustering problems and different types of datasets. Popular general-purpose clustering methods include the *K-means algorithm*, which groups the samples into clusters such that the within-group sum of square deviation from the group center is minimized, and the *mean-shift algorithm*, which clusters the samples by fitting the data to density functions (e.g., Gaussian functions).

In scikit-learn, the sklearn.cluster module contains several clustering algorithms, including the K-means algorithm KMeans and the mean-shift algorithm MeanShift, just to mention a few. To perform a clustering task with one of these methods, we first initialize an instance of the corresponding class and train it with a feature-only dataset using the fit method, and we finally obtain the result of the clustering by calling the predict method. Many clustering algorithms require the number of clusters as an input parameter, which we can specify using the n_clusters parameter when the class instance is created.

To demonstrate clustering, let's again consider the Iris dataset. Here, we do not use the response variable, which was used in supervised classification. Instead, we attempt to automatically discover a suitable clustering of the samples using the K-means method. We begin by loading the Iris data, as before, and store the feature and target data in the variables X and y, respectively.

```
In [72]: X, y = iris.data, iris.target
```

With the K-means clustering method, we must specify how many clusters we want in the output. The most suitable number of clusters is not always apparent in advance, and trying clustering with a few different numbers of clusters is often necessary. However, here, we know that the data corresponds to three species of Iris flowers, so we use three clusters. To perform the clustering, we create an instance of the Kmeans class, using the n_clusters argument to set the number of clusters.

```
In [73]: n_clusters = 3
In [74]: clustering = cluster.KMeans(n_clusters=n_clusters)
```

To perform the computation, we call the fit method with the Iris feature matrix as an argument.

```
In [75]: clustering.fit(X)
Out[75]: KMeans(copy_x=True, init='k-means++', max_iter=300, n_clusters=3,
                n_init=10, n_jobs=1, precompute_distances='auto',
                random_state=None, tol=0.0001, verbose=0)
```

The clustering result is available through the predict method, to which we also pass a feature dataset that optionally can contain features of new samples. However, not all the clustering methods implemented in scikit-learn support predicting clusters for a new sample. In this case, the predict method is unavailable, and we need to use the fit_predict method instead. Here, we use the predict method with the training feature dataset to obtain the clustering result.

```
In [76]: y_pred = clustering.predict(X)
```

The result is an integer array of the same length and the number of samples in the training dataset. The elements in the array indicate which group (from 0 up to n_samples-1) each sample is assigned to. Since the resulting array y_pred is long, we only display every eighth element in the array using the NumPy stride indexing ::8.

```
In [77]: y_pred[::8]
Out[77]: array([1, 1, 1, 1, 1, 1, 1, 2, 2, 2, 2, 2, 2, 0, 0, 0, 0, 0, 0],
               dtype=int32)
```

We can compare the obtained clustering with the supervised classification of the Iris samples.

```
In [78]: y[::8]
Out[78]: array([0, 0, 0, 0, 0, 0, 0, 1, 1, 1, 1, 1, 1, 2, 2, 2, 2, 2, 2])
```

There is a good correlation between the two, but the clustering output has assigned different integer values to the groups than what was used in the target vector in the supervised classification. To compare the two arrays with metrics such as the confusion_matrix function, we first need to rename the elements so that the same integer values are used for the same group. We can do this operation with NumPy array manipulations.

```
In [79]: idx_0, idx_1, idx_2 = (np.where(y_pred == n) for n in range(3))
In [80]: y_pred[idx_0], y_pred[idx_1], y_pred[idx_2] = 2, 0, 1
In [81]: y_pred[::8]
Out[81]: array([0, 0, 0, 0, 0, 0, 0, 1, 1, 1, 1, 1, 1, 2, 2, 2, 2, 2, 2],
               dtype=int32)
```

Now that we represent the corresponding groups with the same integers, we can summarize the overlaps between the supervised and unsupervised classification of the Iris samples using the confusion_ matrix function.

```
In [82]: metrics.confusion_matrix(y, y_pred)
Out[82]: array([[50,  0,  0],
                 [ 0, 48,  2],
                 [ 0, 14, 36]])
```

This confusion matrix indicates that the clustering algorithm correctly identified all samples corresponding to the first species as a group of its own. But due to the overlapping samples in the second and third groups, those could not be resolved entirely as different clusters. For example, two elements from group 1 were assigned to group 2, and 14 elements from group 2 were assigned to group 1.

The clustering result can also be visualized, for example, by plotting scatter plots for each pair of features, as we do in the following. We loop over each pair of features and each cluster and plot a scatter graph for each cluster using different colors (orange, blue, and green in Figure 15-8), and we also draw a red square around each sample for which the clustering does not agree with the supervised classification. The result is shown in Figure 15-8.

```
In [83]: N = X.shape[1]
    ...: fig, axes = plt.subplots(N, N, figsize=(12, 12),
    ...:                          sharex=True, sharey=True)
    ...: colors = ["coral", "blue", "green"]
    ...: markers = ["^", "v", "o"]
    ...: for m in range(N):
    ...:     for n in range(N):
    ...:         for p in range(n_clusters):
    ...:             mask = y_pred == p
    ...:             axes[m, n].scatter(X[:, m][mask], X[:, n][mask], s=30,
    ...:                                marker=markers[p], color=colors[p],
    ...:                                alpha=0.25)
    ...:         for idx in np.where(y != y_pred):
    ...:             axes[m, n].scatter(X[idx, m], X[idx, n], s=30,
    ...:                                marker="s", edgecolor="red",
    ...:                                facecolor=(1,1,1,0))
    ...:     axes[N-1, m].set_xlabel(iris.feature_names[m], fontsize=16)
    ...:     axes[m, 0].set_ylabel(iris.feature_names[m], fontsize=16)
```

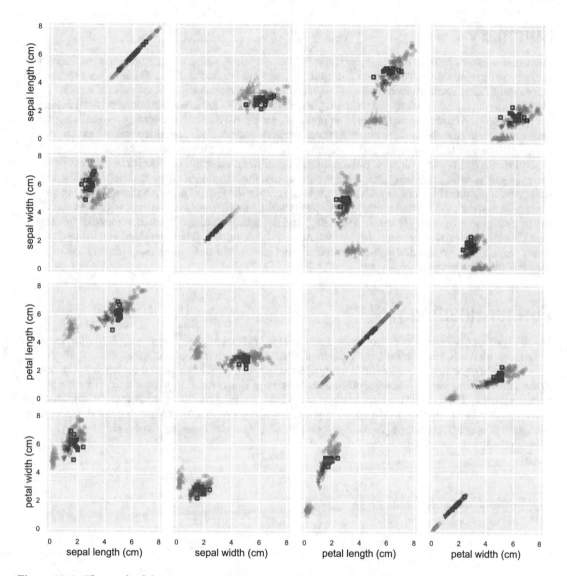

Figure 15-8. *The result of clustering, using the K-means algorithm, of the Iris dataset features*

The Iris samples' clustering result in Figure 15-8 shows that the clustering does remarkably well in recognizing which samples belong to distinct groups. Of course, because of the overlap in the features for classes shown in blue (dark gray) and green (medium gray) in the graph, we cannot expect that any unsupervised clustering algorithm can fully resolve the various groups in the dataset, and some deviation from the supervised response variable is therefore expected.

Summary

This chapter introduced machine learning using Python, beginning with a brief review and summary of the subject and its terminology. The Python library scikit-learn was applied to three different types of problems representing fundamental machine learning topics. We revisited regression from the point of view of machine learning, classification, and clustering. The first two topics are examples of supervised machine learning, while the clustering method is an example of unsupervised machine learning. Beyond what we have been able to cover here, there are many more methods and problem domains covered by machine learning. For example, an essential part of machine learning that we have not touched upon in this brief introduction is text-based problems. The scikit-learn contains an extensive module (`sklearn.text`) with tools and methods for processing text-based problems, and the Natural Language Toolkit (`www.nltk.org`) is a powerful platform for working with and processing data in the form of human language text. Image processing and computer vision are prominent problem domains in machine learning, which, for example, can be treated with OpenCV (`http://opencv.org`) and its Python bindings. Other examples of significant topics in machine learning are neural networks and deep learning, which have received much attention in recent years. Readers interested in such methods should explore the TensorFlow (`www.tensorflow.org`) and the Keras libraries (`https://keras.io`).

Further Reading

Machine learning is a part of the computer science field of artificial intelligence, a broad field with numerous techniques, methods, and applications. This chapter has only shown examples of a few basic machine-learning methods, which can be useful in many practical applications. For a more thorough introduction to machine learning, see *The Elements of Statistical Learning: Data Mining, Inference, and Prediction* by T. Hastie (Springer, 2013), and for introductions to machine learning specific to the Python environment, see *Learning scikit-learn: Machine Learning in Python* by R. Garreta (Packt, 2013), *Mastering Machine Learning With scikit-learn* by G. Hackeling (Packt, 2014), and *Building Machine Learning Systems with Python* by L. Pedro Coelho (Packt, 2015).

CHAPTER 16

■ ■ ■

Bayesian Statistics

This chapter explores an alternative interpretation of statistics—Bayesian statistics—and the methods associated with this interpretation. In contrast to the frequentist statistics used in Chapter 13 and Chapter 14, Bayesian statistics treats probability as a degree of belief rather than as a measure of the proportions of observed outcomes. This different point of view gives rise to distinct statistical methods we can use in problem-solving. While it is generally true that statistical problems can, in principle, be solved using either frequentist or Bayesian statistics, practical differences make these two approaches to statistics suitable for different types of problems.

Bayesian statistics is based on Bayes' theorem relating conditional and unconditional probabilities. Bayes' theorem is a fundamental result in probability theory, and it applies to both the frequentist and the Bayesian interpretation of statistics. In the context of Bayesian inference, unconditional probabilities are used to describe the prior knowledge of a system, and Bayes' theorem provides a rule for updating this knowledge after making new observations. The updated knowledge is described by a conditional probability conditioned on the observed data. The *prior probability distribution* describes the initial knowledge of a system, and the updated knowledge, conditioned on the observed data, is the *posterior probability distribution*. In problem-solving with Bayesian statistics, the posterior probability distribution is the unknown quantity that we seek. From it, we can compute expectation values and other statistical quantities for random variables of interest. Although Bayes' theorem describes how to compute the posterior distribution from the prior distribution, for most realistic problems, the calculations involve evaluating high-dimensional integrals that can be prohibitively difficult to compute analytically and numerically. This has, until recently, hindered Bayesian statistics from being widely used in practice. However, Bayesian methods are becoming increasingly popular with the advancement of computational statistics and efficient simulation methods that allow us to sample directly from the posterior distributions (rather than directly compute them). The methods that enable us to sample from the posterior distribution are, first and foremost, the Markov chain Monte Carlo (MCMC) methods. Several alternative implementations of MCMC methods are available. For instance, traditional MCMC methods include Gibbs sampling and the Metropolis-Hastings algorithm, and more recent methods include Hamiltonian and No-U-Turn algorithms. This chapter demonstrates how to use several of these methods.

Statistical problem-solving with Bayesian inference methods is sometimes known as *probabilistic programming*. The key steps in probabilistic programming are as follows.

1. Create a statistical model.

2. Sample from the posterior distribution for the quantity of interest using an MCMC method.

3. Use the obtained posterior distribution to compute properties of interest for the problem and make inference decisions based on the obtained results. This chapter examines how to carry out these steps from within the Python environment with the help of the PyMC library.

© Robert Johansson 2024
R. Johansson, *Numerical Python*, https://doi.org/10.1007/979-8-8688-0413-7_16

■ **PyMC** The PyMC library offers a framework for probabilistic programming—solving statistical problems using simulation with Bayesian methods. At the time of writing, the latest version is 5.6.1. For more information about the project, see the web page at www.pymc.io.

Importing Modules

This chapter mainly works with the PyMC library and a model-building library that uses PyMC called Bambi. We import these libraries in the following manner.

```
In [1]: import pymc as mc
In [2]: import bambi
```

We also require NumPy, Pandas, and Matplotlib for basic numerics, data analytics, and plotting. These libraries are imported following the usual convention.

```
In [3]: import numpy as np
In [4]: import pandas as pd
In [5]: import matplotlib.pyplot as plt
```

As a comparison to non-Bayesian statistics, we also use the stats module from SciPy, the statsmodels library, and the Seaborn library for visualization.

```
In [6]: from scipy import stats
In [7]: import statsmodels.api as sm
In [8]: import statsmodels.formula.api as smf
In [9]: import seaborn as sns
```

Introduction to Bayesian Statistics

The foundation of Bayesian statistics is Bayes' theorem, which gives a relation between unconditioned and conditional probabilities of two events A and B,

$$P(A|B)P(B) = P(B|A)P(A),$$

where $P(A)$ and $P(B)$ are the unconditional probabilities of events A and B and where P(A| B) is the conditional probability of event A given that event B is true, and $P(B|A)$ is the conditional probability of B given that A is true. Both sides of the preceding equation are equal to the probability that both A and B are true: $P(A \cap B)$. In other words, Bayes' rule states the probability that both A and B are equal to the probability of A times the probability of B given that A is true, $P(A)P(B|A)$ or, equivalently, the probability of B times the probability of A given that B is true: $P(B)P(A|B)$.

In the context of Bayesian inference, Bayes' rule is typically employed when we have a prior belief about the probability of an event A, represented by the unconditional probability $P(A)$, and wish to update this belief after observing an event B. In this language, the updated belief is represented by the conditional probability of A given the observation B: $P(A|B)$, which we can compute as follows using Bayes' rule:

$$P(A|B) = \frac{P(B|A)P(A)}{P(B)}.$$

Each factor in this expression has a distinct interpretation and a name: $P(A)$ is the *prior* probability of event A, and $P(A|B)$ is the *posterior* probability of A given the observation B. $P(B|A)$ is the *likelihood* of observing B given that A is true, and the probability of observing B regardless of A, $P(B)$, is known as *model evidence* and can be considered as a normalization constant (with respect to A).

In statistical modeling, we are typically interested in a set of random variables X characterized by probability distributions with specific parameters θ. After collecting data for the process that we are interested in modeling, we wish to infer the values of the model parameters from the data. In the frequentist's statistical approach, we can maximize the likelihood function given the observed data and obtain estimators for the model parameters. The Bayesian approach considers the unknown model parameters θ as random variables in their own right and uses Bayes' rule to derive probability distributions for the model parameters θ. If we denote the observed data as x, we can express the probability distribution for θ given the observed data x using Bayes' rule as follows:

$$p(\theta|x) = \frac{p(x|\theta)p(\theta)}{p(x)} = \frac{p(x|\theta)p(\theta)}{\int p(x|\theta)p(\theta)d\theta}.$$

The second equality in this equation follows from the law of total probability, $p(x) = \int p(x|\theta)p(\theta)d\theta$. Once we have computed the posterior probability distribution $p(\theta|x)$ for the model parameters, we can, for example, compute expectation values of the model parameters and obtain a result similar to the estimators that we can compute in a frequentist's approach. In addition, when we have an estimate of the full probability distribution for $p(\theta|x)$, we can also compute other quantities, such as credibility intervals and marginal distributions for certain model parameters when θ is multivariate. For example, if we have two model parameters, $\theta = (\theta_1, \theta_2)$, but are interested only in θ_1, we can obtain the marginal posterior probability distribution $p(\theta_1|x)$ by integrating the joint probability distribution $p(\theta_1, \theta_2|x)$ using the expression obtained from Bayes' theorem:

$$p(\theta_1|x) = \int p(\theta_1,|\theta_2,|x)d\theta_2 = \frac{\int p(x|,\theta_1|,\theta_2)p(\theta_1,\theta_2)d\theta_2}{\int\int p(x|,\theta_1|,\theta_2)p(\theta_1,\theta_2)d\theta_1 d\theta_2}.$$

Here, note that the final expression contains integrals over the known likelihood function $p(x|\theta_1, \theta_2)$ and the prior distribution $p(\theta_1, \theta_2)$, so we do not need to know the joint probability distribution $p(\theta_1, \theta_2|x)$ to compute the marginal probability distribution $p(\theta_1|x)$. This approach provides a powerful and generic methodology for computing probability distributions for model parameters and successively updating the distributions once new data becomes available. However, directly computing $p(\theta|x)$, or the marginal distributions thereof, requires that we write down the likelihood function $p(x|\theta)$ and the prior distribution $p(\theta)$ and that we can evaluate the resulting integrals. For many simple but important problems, it is possible to analytically compute these integrals and find the exact closed-form expressions for the posterior distribution. Textbooks, such as Gelman (2013), provide numerous examples of problems that are exactly solvable in this way. However, for more complicated models, with prior distributions and likelihood functions for which the resulting integrals are not easily evaluated, or for multivariate statistical models, for which the resulting integrals can be high-dimensional, both exact and numerical evaluation may be unfeasible.

It is primarily for models that cannot be solved with exact methods that we can benefit from using simulation methods, such as Markov chain Monte Carlo, which allows us to sample the posterior probability distribution for the model parameters and thereby construct an approximation of the joint or marginal posterior distributions, or directly evaluating integrals, such as expectation values. Another important advantage of simulation-based methods is that the modeling process can be automated. Here, we focus exclusively on Bayesian statistical modeling using Monte Carlo simulation methods. For a thorough review of the theory and many examples of analytically solvable problems, see the references at this chapter's end. The remaining part of this chapter discusses the definition of statistical models and sampling their posterior distribution with the PyMC library as a probabilistic programming framework.

Before moving to computational Bayesian statistics, it is worth taking a moment to summarize the key differences between the Bayesian approach and the classical frequentist approach that was used in earlier chapters. In both approaches to statistical modeling, we formulate the models in terms of random variables.

A key step in defining a statistical model is to make assumptions about the probability distributions for the random variables defined in the model. In parametric methods, each probability distribution is characterized by a small number of parameters. In the frequentist's approach, those model parameters have specific true values, and observed data is interpreted as random samples from the true distributions. In other words, the model parameters are assumed to be fixed, and the data is assumed to be stochastic. The Bayesian approach takes the opposite view: the data is interpreted as fixed, and the model parameters are described as random variables. Starting from a prior distribution for the model parameters, we can then update the distribution to account for observed data and, in the end, obtain a probability distribution for the relevant model parameters conditioned on the observed data.

Model Definition

A statistical model is defined in terms of a set of random variables. The random variables in a given model can be independent or, more interestingly, dependent on each other. The PyMC library provides classes for representing random variables for a large number of probability distributions: For example, an instance of `mc.Normal` can be used to represent a normal distributed random variable. Other examples are `mc.Bernoulli` for representing discrete Bernoulli-distributed random variables, `mc.Uniform` for uniformly distributed random variables, `mc.Gamma` for Gamma-distributed random variables, and so on. For a complete list of available distributions, see `dir(mc.distributions)` and the docstrings for each available distribution for information on how to use them. It is also possible to define custom distributions using the `mc.CustomDist` class, which takes a function that specifies the logarithm of the random variable's probability density function.

In Chapter 13, we saw that the SciPy `stats` module also contains classes for representing random variables. Like the random variable classes in SciPy `stats`, we can use the PyMC distributions to represent random variables with fixed parameters. However, the essential feature of the PyMC random variables is that the distribution parameters, such as the mean μ and variance σ^2 for a random variable following the normal distribution $\mathcal{N}(\mu, \sigma^2)$, can themselves be random variables. This allows us to chain random variables and formulate models with random variables with dependencies and hierarchical structure.

Let's start with the simplest possible example. In PyMC, models are represented by an instance of the `mc.Model` class, and random variables are added to a model using the Python context syntax: Random variable instances created within the body of a model context are automatically added to the model. Say that we are interested in a model consisting of a single random variable that follows the normal distribution with the fixed parameters $\mu = 4$ and $\sigma = 2$. We first define the fixed model parameters and then create an instance of `mc.Model` to represent our model.

```
In [10]: mu = 4.0
In [11]: sigma = 2.0
In [12]: model = mc.Model()
```

Next, we can attach random variables to the model by creating them within the model context. Here, we create a random variable X within the model context, which is activated using a `with model` statement.

```
In [13]: with model:
    ...:        mc.Normal('X', mu, tau=1/sigma**2)
```

All random variable classes in PyMC take the name of the variable as the first argument. In the case of `mc.Normal`, the second argument is the mean of the normal distribution, and the third argument `tau` is the precision $\tau = 1/\sigma^2$, where σ^2 is the variance. Alternatively, we can use the `sigma` keyword argument to specify the standard deviation rather than the precision: `mc.Normal('X', mu, sigma=sigma)`.

We can inspect which random variables exist in a model using the `continuous_value_vars` attribute. Here, we have only one random variable in the model.

```
In [14]: model.continuous_value_vars
Out[14]: [X]
```

To sample from the random variables in the model, we use the `mc.sample` function, which implements the MCMC algorithm. The `mc.sample` function accepts many arguments, but at a minimum, we need to provide the number of samples as the first argument. Optionally, we can also provide an MCMC step class instance using the `steps` keyword argument and initial values as a dictionary with parameter values from where sampling is started using the `initvals` keyword argument. For the step method, here we use an instance of the `Metropolis` class, which implements the Metropolis-Hastings step method for the MCMC sampler[1]. Note that we execute all model-related code within the model context.

```
In [15]: initvals = dict(X=2)
In [16]: with model:
    ...:     step = mc.Metropolis()
    ...:     trace = mc.sample(10000, initvals=initvals, step=step)
```

With these steps, we have sampled values from the random variable defined within the model, which in this simple case is only a normal distributed random variable. We requested 10,000 samples, but we got 4 × 10,000 samples. This is because each simulation is carried out in multiple runs, called chains in PyMC terminology. We can set the number of chains using the `chains` keyword argument to the `mc.sample` method, and the default value is four. The following continues using this recommended default value, but it influences the number of samples we get back. The individual sampling chains are visualized in graphs later in this chapter.

To access the sample data, we first define a convenience function `get_values` that extracts the data from the trace object returned by the `mc.sample` method.

```
In [17]: def get_values(trace, variable):
    ...:     return trace.posterior.stack(
    ...:         sample=['chain', 'draw']
    ...:     )[variable].values
```

Using this function, we can get hold of the samples produced in the MCMC sampling and confirm that it contains 40,000 samples as discussed.

```
In [18]: X = get_values(trace, "X")
In [19]: X.shape
Out[19]: (40000,)
```

---

[1] See also the `Slice`, `HamiltonianMC`, and `NUTS` samplers, which can be used more or less interchangeably.

The probability density function (PDF) for a normal distribution is, of course, known analytically. Using the SciPy `stats` module, we can access the PDF using the `pdf` method of the `norm` class instance for comparing to the sampled random variable. The sampled values and the true PDF for the present model are shown in Figure 16-1.

```
In [20]: x = np.linspace(-4, 12, 1000)
In [21]: y = stats.norm(mu, sigma).pdf(x)
In [22]: fig, ax = plt.subplots(figsize=(8, 3))
    ...: ax.plot(x, y, 'r', lw=2)
    ...: sns.histplot(X, ax=ax, kde=True, stat='density')
    ...: ax.set_xlim(-4, 12)
    ...: ax.set_xlabel("x")
    ...: ax.set_ylabel("Probability distribution")
```

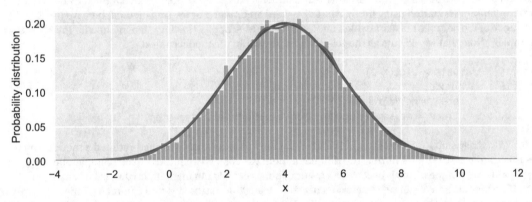

Figure 16-1. *The probability density function for the normal distributed random variable (red/thick line) and a histogram from 10,000 MCMC samples*

With the `mc.plot_trace` function, we can visualize the MCMC random walk that generated the samples, as shown in Figure 16-2. The `mc.plot_trace` function automatically plots the kernel-density estimate and the sampling trace for every random variable in the model.

```
In [23]: fig, axes = plt.subplots(1, 2, figsize=(8, 2.5), squeeze=False)
    ...: mc.plot_trace(trace, axes=axes)
    ...: axes[0, 0].plot(x, y, 'r', lw=0.5)
```

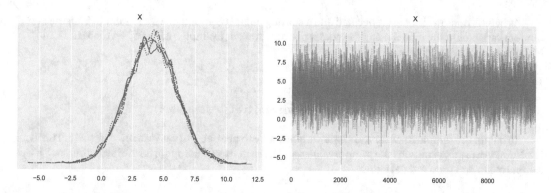

Figure 16-2. *The left panel shows the density kernel estimate (four blue lines, one for each chain) of the sampling trace and the normal probability distribution (red/thin line). The right panel shows the MCMC sampling trace*

As a next step in building more complex statistical models, consider again a model with a normal distributed random variable $X \sim \mathcal{N}\left(\mu, \sigma^2\right)$, but where parameters μ and σ are random variables. In PyMC, we can easily create dependent variables by passing them as an argument when creating other random variables. For example, with $\mu \sim \mathcal{N}(3,1)$ and $\sigma \sim |\mathcal{N}(0,1)|$, we can create the dependent random variable X using the following model specification.

```
In [24]: model = mc.Model()
In [25]: with model:
    ...:     mean = mc.Normal('mean', 3.0)
    ...:     sigma = mc.HalfNormal('sigma', sigma=1.0)
    ...:     X = mc.Normal('X', mean, sigma=sigma)
```

Here, we have used the `mc.HalfNormal` to represent the random variable $\sigma \sim |\mathcal{N}(0,1)|$, and the mean and standard deviation arguments to the `mc.Normal` class for X are random variable instances rather than fixed model parameters. As before, we can inspect which random variables a model contains using the `continuous_value_vars` attribute.

```
In [26]: model.continuous_value_vars
Out[26]: [mean, sigma_log__, X]
```

The PyMC library represents the `sigma` variable with a log-transformed variable `sigma_log__` to handle the half-normal distribution. Nonetheless, we can still directly access the `sigma` variable from the model, as shown in the following text.

When the complexity of the model increases, it may no longer be straightforward to explicitly select a suitable starting point for the sampling process. The `mc.find_MAP` function can be used to find the point in the parameter space corresponding to the maximum of the posterior distribution, which can serve as a good starting point for the sampling process.

```
In [27]: with model:
    ...:     initvals = mc.find_MAP()
In [28]: initvals
Out[28]: {'X': array(3.0), 'mean': array(3.0),
          'sigma': array(0.70710674), 'sigma_log__': array(-0.34657365)}
```

As before, once the model is specified and a starting point is computed, we can sample from the random variables in the model using the mc.sample function, for example, using mc.Metropolis as an MCMC sampling step method.

```
In [29]: with model:
    ...:         step = mc.Metropolis()
    ...:         trace = mc.sample(10000, initvals=initvals, step=step)
```

To obtain the sample trace for the sigma variable, we can use get_values(trace, 'sigma'). The result is a NumPy array that contains the sample values, and from it, we can compute further statistics, such as its sample mean and standard deviation.

```
In [30]: get_values(trace, 'sigma').mean()
Out[30]: 0.80054476153369014
```

The same approach can be used to obtain the samples of X and compute statistics from them.

```
In [31]: X = get_values(trace, 'X')
In [32]: X.mean()
Out[32]: 2.9993248663922092
In [33]: get_values(trace, 'X').std()
Out[33]: 1.4065656512676457
```

The trace plot for the current model, created using the mc.plot_trace, is shown in Figure 16-3, where we have used the var_names argument to mc.plot_trace to select which random variables to plot explicitly.

```
In [34]: fig, axes = plt.subplots(3, 2, figsize=(8, 6), squeeze=False)
    ...: mc.plot_trace(trace, var_names=['mean', 'sigma', 'X'], axes=axes)
```

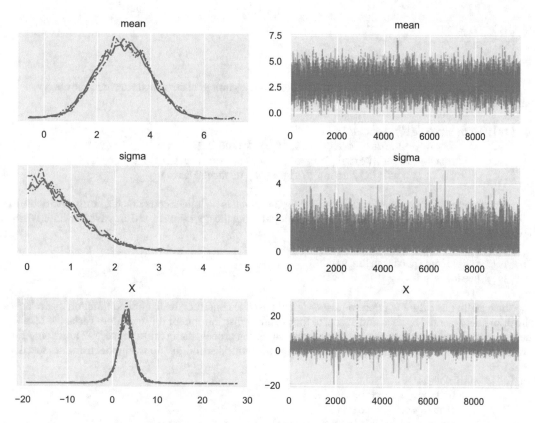

Figure 16-3. *Kernel density estimates (left) and MCMC random sampling trace (right) for the three random variables: mean, sigma, and X*

Sampling Posterior Distributions

So far, we have defined models and sampled from models that only contain random variables without any references to observed data. In the context of Bayesian models, these types of random variables represent the prior distributions of the unknown model parameters. In the previous examples, we have used the MCMC method to sample from the prior distributions of the model. However, the real advantage of the MCMC approach is that we can sample from the posterior distribution, which represents the probability distribution for the model variables after having updated the prior distribution to account for the effect of observations.

To condition the model on observed data, we only need to add the data using the observed keyword argument when the corresponding random variable is created within the model: For example, `mc.Normal('X', mean, 1/sigma**2, observed=data)` indicates that the random variable X has been observed to take the values in the array data. Adding observed random variables to a model automatically results in subsequent sampling using `mc.sample` samples the posterior distribution of the model, appropriately conditioned on the observed data according to Bayes' rule and the likelihood function implied by the distribution selected for the observed data. For example, consider the model used in the preceding text with a normal distributed random variable X whose mean and standard deviation are random variables. Here, we simulate the observations for X by drawing samples from a normally distributed random variable with $\mu = 2.5$ and $\sigma = 1.5$ using the norm class from the SciPy stats module.

```
In [35]: mu = 2.5
In [36]: s = 1.5
In [37]: data = stats.norm(mu, s).rvs(100)
```

The data is fed into the model by setting the keyword argument observed=data when the observed variable is created and added to the model.

```
In [38]: with mc.Model() as model:
   ...:        mean = mc.Normal('mean', 4.0, 1.0) # true 2.5
   ...:        sigma = mc.HalfNormal('sigma', 3.0 * np.sqrt(np.pi/2)) # true 1.5
   ...:        X = mc.Normal('X', mean, 1/sigma**2, observed=data)
```

A consequence of providing observed data for X is that it is no longer considered a random variable in the model. This can be seen from inspecting the model using the continuous_value_vars attribute, where X is now absent.

```
In [39]: model.continuous_value_vars
Out[39]: [mean, sigma_log_]
```

Instead, in this case, X is a deterministic variable used to construct the likelihood function that relates the priors, represented by mean and sigma, to the posterior distribution for these random variables. Like before, we can find a suitable starting point for the sampling process using the mc.find_MAP function. After creating an MCMC step instance, we can sample the posterior distribution for the model using mc.sample.

```
In [40]: with model:
   ...:        initvals = mc.find_MAP()
   ...:        step = mc.Metropolis()
   ...:        trace = mc.sample(10000, initvals=initvals, step=step)
```

The starting point that was calculated using mc.find_MAP maximizes the likelihood of the posterior given the observed data, and it provides an estimate of the unknown parameters of the prior distribution.

```
In [41]: initvals
Out[41]: {'mean': array(2.506494035976824), 'sigma_log': array(0.39468163345610)}
```

However, to obtain estimates of the distribution of these parameters (which are random variables in their own right), we need to carry out the MCMC sampling using the mc.sample function, as done in the preceding text. The result of the posterior distribution sampling is shown in Figure 16-4. Note that the distributions for the mean and sigma variables are closer to the true parameter values, $\mu = 2.5$ and $\sigma = 1.5$, than to the prior guesses of 4.0 and 3.0, respectively, due to the influence of the data and the corresponding likelihood function.

```
In [42]: fig, axes = plt.subplots(2, 2, figsize=(8, 4), squeeze=False)
   ...: mc.plot_trace(trace, var_names=['mean', 'sigma'], axes=axes)
```

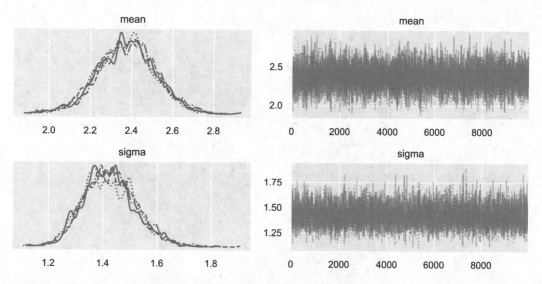

Figure 16-4. *The MCMC sampling trace of the posterior distribution for mean and sigma*

To calculate statistics and estimate quantities using the samples from the posterior distributions, we can access arrays containing the samples using the get_values function, which takes the trace object and the name of the random variable as arguments. For example, the following code computes estimates of the mean of the two random variables in the model and compares them to the corresponding true values for the distributions that the data points were drawn from.

```
In [43]: mu, get_values(trace, 'mean').mean()
Out[43]: (2.5, 2.5290001218008435)
In [44]: s, get_values(trace, 'sigma').mean()
Out[44]: (1.5, 1.5029047840092264)
```

The PyMC library also provides utilities for analyzing and summarizing the statistics of the marginal posterior distributions obtained from the mc.sample function. For example, the mc.plot_forest function visualizes the mean and credibility intervals (i.e., an interval within which the true parameter value is likely to be) for each random variable in a model. The result of visualizing the samples for the current example using the mc.plot_forest function is shown in Figure 16-5.

```
In [45]: mc.plot_forest(trace, var_names=['mean', 'sigma'])
```

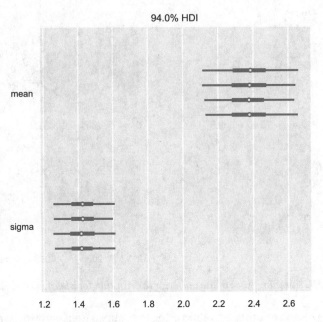

Figure 16-5. *A forest plot for the two parameters, mean and sigma, and their credibility intervals. Note that we have four lines for each variable, corresponding to the four chains of samples in the simulations*

Similar information can also be presented in text form using the mc.summary function, which, for example, includes information such as the mean, standard deviation, and posterior quantiles.

```
In [46]: mc.summary(trace, var_names=['mean', 'sigma'])
        mean     sd  hdi_3%  hdi_97%  mcse_mean  mcse_sd  ess_bulk  ess_tail  \
mean   2.637  0.143   2.367    2.902      0.002    0.001    6783.0    6542.0
sigma  1.412  0.091   1.240    1.581      0.001    0.001    4886.0    5138.0

       r_hat
mean     1.0
sigma    1.0
```

Linear Regression

Regression is one of the most essential tools in statistical modeling, and examples of linear regression within the classical statistical formalism, for example, were shown in Chapters 14 and 15. Linear regression can also be approached with Bayesian methods and treated as a modeling problem. We assign prior probability distributions to the unknown model parameters (slopes and intercept) and compute the posterior distribution given the available observations. To compare the similarities and differences between Bayesian linear regression and the frequentist's approach to the same problem, using, for example, the methods from Chapter 14, here we begin with a short analysis of a linear regression problem using the statsmodels library. The following analyzes the same problem with PyMC.

As example data for performing a linear regression analysis, here we use a dataset that contains the height and weight of 200 men and women, which we can load using the get_rdataset function from the datasets module in the statsmodels library.

```
In [47]: dataset = sm.datasets.get_rdataset("Davis", "carData")
```

For simplicity, to begin with, we work only with the subset of the dataset that corresponds to male subjects, and to avoid having to deal with outliers, we filter out all subjects with a weight that exceeds 110 kg. These operations are readily performed using Pandas methods for filtering data frames using Boolean masks.

```
In [48]: data = dataset.data[dataset.data.sex == 'M']
In [49]: data = data[data.weight < 110]
```

The resulting Pandas data frame object data contains several columns.

```
In [50]: data.head(3)
Out[50]:
```

1	sex	Weight	height	Repwt	repht
0	M	77	182	77	180
3	M	68	177	70	175
5	M	76	170	76	165

Here we focus on a linear regression model for the relationship between the weight and height columns in this dataset. Using the statsmodels library and its model for ordinary least square regression and the Patsy formula language, we create a statistical model for this relationship in a single line of code.

```
In [51]: model = smf.ols("height ~ weight", data=data)
```

To fit the specified model to the observed data, we use the fit method of the model instance.

```
In [52]: result = model.fit()
```

Once the model has been fitted and the model result object has been created, we can use the predict method to compute the predictions for new observations and plot the linear relation between the height and weight, as shown in Figure 16-6.

```
In [53]: x = np.linspace(50, 110, 25)
In [54]: y = result.predict({"weight": x})
In [55]: fig, ax = plt.subplots(1, 1, figsize=(8, 3))
    ...: ax.plot(data.weight, data.height, 'o')
    ...: ax.plot(x, y, color="blue")
    ...: ax.set_xlabel("weight")
    ...: ax.set_ylabel("height")
```

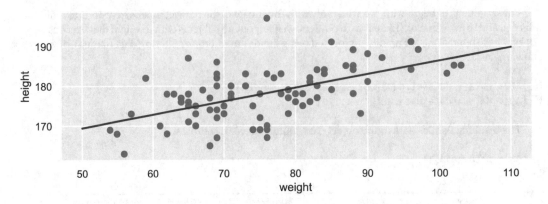

Figure 16-6. *Height vs. weight, with a linear model fitted using ordinary least square*

The linear relation shown in Figure 16-6 summarizes the main result of performing a linear regression on this dataset. It gives the best-fitting line, described by specific values of the model parameters (intercept and slope). Within the frequentist's approach to statistics, we can also compute numerous statistics, for example, *p*-values for various hypotheses, such as the hypothesis that a model parameter is zero (no effect).

A Bayesian regression analysis results in the posterior distribution for the marginal distributions for each model parameter. We can compute the mean estimates for the model parameters from such marginal distributions, which roughly correspond to the model parameters obtained from a frequentist's analysis. We can also compute other quantities, such as the credibility interval, which characterizes the uncertainty in the estimate. To model the height vs. weight using a Bayesian model, we can use a relation such as $height \sim \mathcal{N}\left(intercept + \beta \; weight, \sigma^2\right)$, where intercept, β, and σ are random variables with unknown distributions and parameters. We also need to give prior distributions to all stochastic variables in the model. Depending on the application, the exact choice of prior can be a delicate issue because the choice of priors influences the result. However, this is also an advantage of the Bayesian approach, which allows us to treat cases with little observed data if we have good reason to make assumptions about the prior distribution of the random variables in the problem. When we have many observations and data for model fitting, the exact choice of prior is not as important, and it is usually sufficient to use reasonable initial guesses. Here, we start with priors that represent broad distributions for all the model parameters.

To program the model in PyMC, we use the same methodology as earlier in this chapter. First, we create random variables for the stochastic components of the model and assign them to distributions with specific parameters that represent the prior distributions. Next, we create a deterministic variable that is a function of the stochastic variables but with observed data attached to it using the observed keyword argument, as well as in the expression for the expected value of the distribution of the heights (`height_mu`).

```
In [56]: with mc.Model() as model:
    ...:     sigma = mc.Uniform('sigma', 0, 10)
    ...:     intercept = mc.Normal('intercept', 125, sigma=30)
    ...:     beta = mc.Normal('beta', 0, sigma=5)
    ...:     height_mu = intercept + beta * data.weight
    ...:     mc.Normal('height', mu=height_mu, sigma=sigma, observed=data.height)
    ...:     predict_height = mc.Normal('predict_height', mu=intercept + beta * x,
    ...:                                sigma=sigma, shape=len(x))
```

If we want to use the model for predicting the heights at specific weight values, we can add a stochastic variable to the model. In the preceding model specification, the `predict_height` variable is an example. Here, x is the NumPy array with values between 50 and 110 created earlier. Because it is an array, we need to set the shape attribute of the `mc.Normal` class to the corresponding length of the array. If we inspect the `continuous_value_vars` attribute of the model, we now see that it contains the two model parameters (intercept and beta), the distribution of the model errors (sigma), and the `predict_height` variable for predicting the heights at specific values of weight from the x array.

```
In [57]: model.continuous_value_vars
Out[57]: [sigma_interval, intercept, beta, predict_height]
```

Once the model is fully specified, we can turn to the MCMC algorithm to sample the marginal posterior distributions for the model, given the observed data. Here, we use an alternative sampler, `mc.NUTS` (No-U-Turn Sampler), which is an efficient sampler that is suitable for many problems.

```
In [58]: with model:
    ...:     step = mc.NUTS()
    ...:     trace = mc.sample(10000, step=step)
```

The result of the sampling is stored in a `trace` object returned by `mc.sample`. We can visualize the kernel density estimate of the probability distribution and the MCMC random walk traces that generated the samples using the `mc.plot_trace` function. Here, we again use the `var_names` argument to explicitly select which stochastic variables in the model to show in the trace plot. The result is shown in Figure 16-7.

```
In [59]: fig, axes = plt.subplots(2, 2, figsize=(8, 4), squeeze=False)
    ...: mc.plot_trace(trace, var_names=['intercept', 'beta'], axes=axes)
```

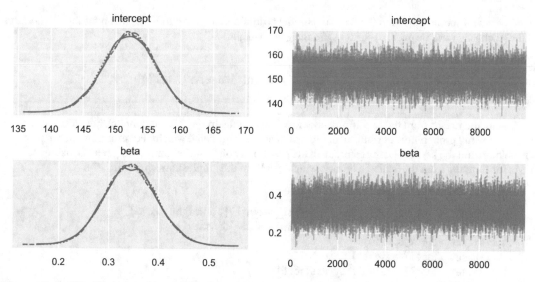

Figure 16-7. Distribution and sampling trace of the linear model intercept and beta coefficient

The values of the intercept and coefficient in the linear model that most closely correspond to the results from the statsmodels analysis are obtained by computing the mean of the traces for the stochastic variables in the Bayesian model.

```
In [60]: intercept = get_values(trace, "intercept").mean()
In [61]: intercept
Out[61]: 149.97546241676989
In [62]: beta = get_values(trace, "beta").mean()
In [63]: beta
Out[63]: 0.37077795098761318
```

The corresponding result from the statsmodels analysis is obtained by accessing the `params` attribute in the result class returned by the `fit` method (see the preceding text).

```
In [64]: result.params
Out[64]: Intercept     152.617348
         weight          0.336477
         dtype: float64
```

By comparing these values for the intercepts and the coefficients, we see that the two approaches give similar results for the maximum likelihood estimates of the unknown model parameters. In the statsmodels approach, to predict the expected height for a given weight, say 90 kg, we can use the predict method to get a specific height.

```
In [65]: result.predict({"weight": 90}).values
Out[65]: array([ 182.90030002])
```

The corresponding result in the Bayesian model is obtained by computing the mean for the distribution of the stochastic variable `predict_height`, for the given weight.

```
In [66]: weight_index = np.where(x == 90)[0][0]
In [67]: get_values(trace, "predict_height")[weight_index, :].mean()
Out[67]: 183.33943635274935
```

Again, the results from the two approaches are comparable. In the Bayesian model, however, we can estimate the full probability distribution of the height at every modeled weight. For example, we can plot a histogram and the kernel density estimate of the probability distribution for the weight 90 kg using the `histplot` function from the Seaborn library, which results in the graph shown in Figure 16-8.

```
In [68]: fig, ax = plt.subplots(figsize=(8, 3))
    ...: height = get_values(trace, "predict_height")[:, weight_index]
    ...: sns.histplot(height, ax=ax, kde=True, stat='density')
    ...: ax.set_xlim(150, 210)
    ...: ax.set_xlabel("height")
    ...: ax.set_ylabel("Probability distribution")
```

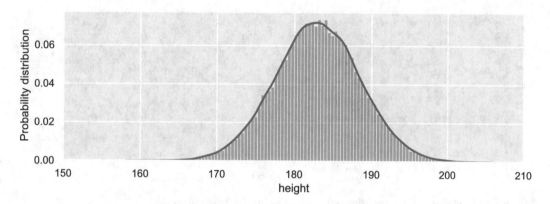

Figure 16-8. *Probability distribution for prediction of the height for the weight of 90 kg*

Every sample in the MCMC trace represents a possible value of the intercept and coefficients in the linear model that we wish to fit the observed data. To visualize the uncertainty in the mean intercept and coefficient that we can take as estimates of the final linear model parameters, it is illustrative to plot the lines corresponding to each sample point, along with the data as a scatter plot and the lines that correspond to the mean intercept and slope. This results in a graph like the one shown in Figure 16-9. The spread of the lines represents the uncertainty in estimating the height for a given weight. The spread tends to be larger toward the edges where fewer data points are available and tighter in the middle of a cloud of data points.

```
In [69]: fig, ax = plt.subplots(1, 1, figsize=(8, 3))
    ...: for n in range(500, 2000, 1):
    ...:     intercept = get_values(trace, "intercept")[n]
    ...:     beta = get_values(trace, "beta")[n]
    ...:     ax.plot(x, intercept + beta * x, color='red', lw=0.25, alpha=0.05)
    ...: intercept = get_values(trace, "intercept").mean()
    ...: beta = get_values(trace, "beta").mean()
    ...: ax.plot(x, intercept + beta * x, color='k',
    ...:         label="Mean Bayesian prediction")
    ...: ax.plot(data.weight, data.height, 'o')
    ...: ax.plot(x, y, '--', color="blue", label="OLS prediction")
    ...: ax.set_xlabel("weight")
    ...: ax.set_ylabel("height")
    ...: ax.legend(loc=0)
```

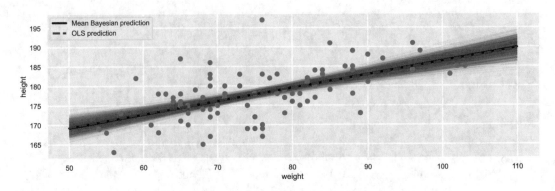

Figure 16-9. *Height vs. weight, with linear fits using OLS and a Bayesian model*

This linear regression problem explicitly defined the statistical model and the stochastic variables included in the model. This illustrates the general steps for analyzing statistical models using the Bayesian approach and the PyMC library. For model definition, however, it is more convenient to use a formula language like Patsy (see Chapter 14). The Bambi library provides such a model specification interface for PyMC models. When using the `bambi.Model` class, which takes a Patsy formula and data as arguments, this library automatically takes care of setting up the model and its variables. The `bambi.Model` class also provides a convenient and familiar `fit` method that performs the MCMC sampling, for which we can use the `draws` keyword argument to specify the number of samples to simulate.

```
In [70]: model = bambi.Model('height ~ weight', data)
    ...: trace = model.fit(draws=2000)
```

The `fit` method returns a sampling trace object with which we can analyze and visualize using the same methods. For example, we can visualize the result using the `mc.plot_trace` function, shown in Figure 16-10. In these trace plots, `height_sigma` corresponds to the `sigma` variable in the explicit model definition used earlier, and it represents the standard error of the residual of the model and the observed data.

```
In [71]: fig, axes = plt.subplots(3, 2, figsize=(8, 6), squeeze=False)
    ...: mc.plot_trace(trace, var_names=['Intercept', 'weight', 'height_sigma'],
    ...:               axes=axes)
```

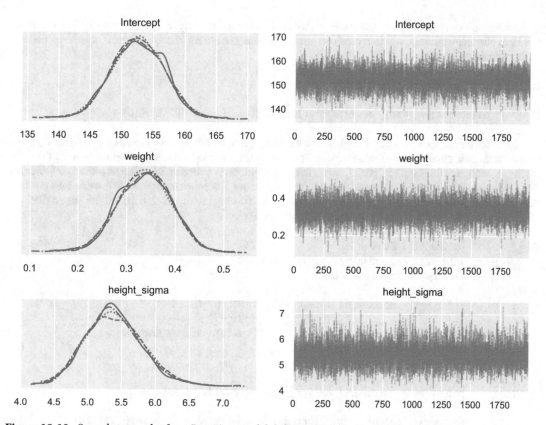

Figure 16-10. *Sample trace plot for a Bayesian model defined using the Bambi library*

With the Bambi library, we can define and analyze linear models using Bayesian statistics in almost the same way as when working with the frequentist approach with statsmodels. For the simple example studied here, the regression analysis with both statistical approaches gives similar results, and neither method is better than the other. However, there are practical differences that, depending on the situation, can favor one or the other. For example, with the Bayesian approach, we have access to estimates of the full marginal posterior distributions, which can be useful for computing statistical quantities other than the mean. However, performing MCMC on simple models like the one considered here is significantly more computationally demanding than doing ordinary least square fitting. The real advantages of the Bayesian methods arise when analyzing complicated models in high dimensions (many unknown model parameters). Defining appropriate frequentist models can be difficult in such cases, and solving the resulting models is challenging. The MCMC algorithm is very attractive in that it scales well to high-dimensional problems and can be highly competitive for complex statistical models. While the models we have considered here are simple and can easily be solved using a frequentist's approach, the general methodology used here remains unchanged, and creating more involved models is only a matter of adding more stochastic variables to the model.

The final example illustrates that the same general procedure can also be used when the Bayesian model's complexity is increased. We return to the height and weight dataset. But instead of selecting only the male subject, we consider an additional level in the model that accounts for the subject's gender so that both males and females can be modeled with potentially different slopes and intercepts. In PyMC, we can create a multilevel model using the shape argument to specify the dimension for each stochastic variable added to the model, as shown in the following example.

Let's begin by preparing the dataset. Here, we again restrict our analysis to subjects weighing less than 110 kg to eliminate outliers. We then convert the sex column to a binary variable where 0 represents male and 1 represents female.

```
In [72]: data = dataset.data.copy()
In [73]: data = data[data.weight < 110]
In [74]: data["sex"] = data["sex"].apply(lambda x: 1 if x == "F" else 0)
```

Next, we define the statistical model, which we here take to be height~N(intercept$_i$ + β_i weight, σ^2), where i is an index that takes the value 0 for male subjects and 1 for female subjects. When creating the stochastic variables for intercept$_i$ and β_i, we indicate the multilevel structure by specifying shape=2 (since, in this case, we have two levels: male and female). The only other difference compared to the previous model definition is that we also need to use an index mask when defining the expression for height_mu, so that each value in data.weight is associated with the correct level.

```
In [75]: with mc.Model() as model:
    ...:         intercept_mu, intercept_sigma = 125, 30
    ...:         beta_mu, beta_sigma = 0, 5
    ...:
    ...:         intercept = mc.Normal('intercept', intercept_mu,
    ...:                                 sigma=intercept_sigma, shape=2)
    ...:         beta = mc.Normal('beta', beta_mu, sigma=beta_sigma, shape=2)
    ...:         error = mc.Uniform('error', 0, 10)
    ...:
    ...:         sex_idx = data.sex.values
    ...:         height_mu = intercept[sex_idx] + beta[sex_idx] * data.weight
    ...:
    ...:         mc.Normal('height', mu=height_mu, sigma=error, observed=data.height)
```

Inspecting the model variables using the continuous_value_vars attribute object shows that we again have three stochastic variables in the model: intercept, beta, and error_interval. However, in contrast to the earlier model, intercept and beta both have two levels.

```
In [76]: model.continuous_value_vars
Out[76]: [intercept, beta, error_interval]
```

How we invoke the MCMC sampling algorithm is identical to the earlier examples in this chapter. Here, we use the NUTS sampler and collect 5000 samples.

```
In [77]: with model:
    ...:         initvals = mc.find_MAP()
    ...:         step = mc.NUTS()
    ...:         trace = mc.sample(5000, step=step, initvals=initvals)
```

Like before, we can use the mc.plot_trace function to visualize the result of the sampling. This allows us to quickly understand the distribution of the model parameters and verify that the MCMC sampling has produced sensible results. The trace plot for the current model is shown in Figure 16-11. Unlike earlier examples, here we have multiple curves in the panels for the intercept and beta variables, reflecting their multilevel nature: the blue lines show the results for the male subjects, and the orange lines show the results for the female subjects.

```
In [78]: mc.plot_trace(trace, figsize=(8, 6))
```

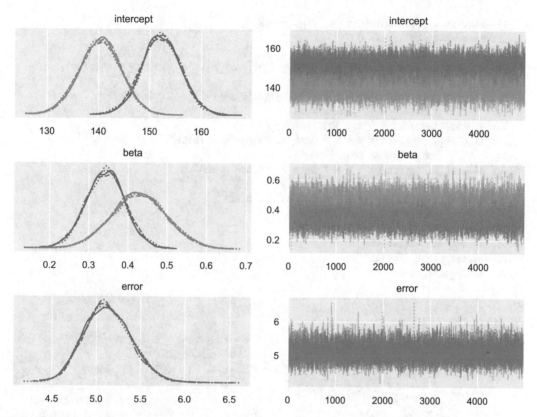

Figure 16-11. *Kernel density estimates of the probability distributions of the model parameters and the MCMC sampling traces for each variable in the multilevel model for height vs. weight*

Using the get_values function of the trace object, we can extract the sampling data for the model variables. Here, the sampling data for intercept and beta are two-dimensional arrays with shape (2, 20000): the first dimension represents the level of the variable, and the second dimension represents each sample (4 chains, 5000 samples each). Here, we are interested in the intercept and the slope for each gender, so we take the mean along the second axis (all samples).

```
In [79]: intercept_m, intercept_f = get_values(trace, 'intercept').mean(axis=1)
In [80]: beta_m, beta_f = get_values(trace, 'beta').mean(axis=1)
```

By averaging over both dimensions, we can also get the intercept and the slope that represent the entire dataset, where male and female subjects are grouped together.

```
In [81]: intercept = get_values(trace, 'intercept').mean()
In [82]: beta = get_values(trace, 'beta').mean()
```

Finally, we visualize the results by plotting the data as scatter plots and drawing the lines corresponding to the intercepts and slopes we obtained for male and female subjects and the result from grouping all subjects. The result is shown in Figure 16-12.

```
In [83]: fig, ax = plt.subplots(1, 1, figsize=(8, 3))
    ...: mask_m = data.sex == 0
    ...: mask_f = data.sex == 1
    ...: ax.plot(data.weight[mask_m], data.height[mask_m], 'o', color="steelblue",
    ...:         label="male", alpha=0.5)
    ...: ax.plot(data.weight[mask_f], data.height[mask_f], 'o', color="green",
    ...:         label="female", alpha=0.5)
    ...: x = np.linspace(35, 110, 50)
    ...: ax.plot(x, intercept_m + x * beta_m, color="steelblue",
    ...:         label="model male group")
    ...: ax.plot(x, intercept_f + x * beta_f, color="green",
    ...:         label="model female group")
    ...: ax.plot(x, intercept + x * beta, color="black",
    ...:         label="model both groups")
    ...:
    ...: ax.set_xlabel("weight")
    ...: ax.set_ylabel("height")
    ...: ax.legend(loc=0)
```

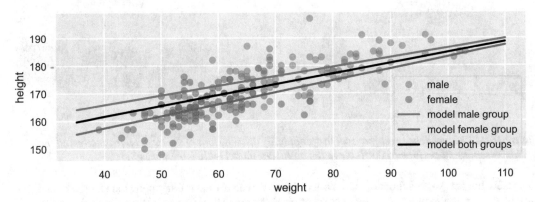

Figure 16-12. *The height vs. weight for male (dark/blue) and female (light/green) subjects*

The regression lines are shown in Figure 16-12, and the distribution plots shown in Figure 16-11 indicate that the model is improved by considering different intercepts and slopes for male and female subjects. In a Bayesian model with PyMC, changing the underlying model used in the analysis is only a matter of adding stochastic variables, defining how they are related, and assigning a prior distribution for each stochastic variable. The MCMC sampling required to solve the model is independent of the model details. This is one of the most attractive aspects of Bayesian statistical modeling. For instance, in the multilevel model considered in the preceding text, instead of specifying the priors for the intercept and slope variables as independent probability distributions, we could relate the distribution parameters of the priors to another stochastic variable and thereby obtain a hierarchical Bayesian model, where the model parameters describing the distribution of the intercept and the slope for each level are drawn from a common distribution. Hierarchical models have many uses and are one of the many applications where Bayesian statistics excel.

Summary

This chapter explored Bayesian statistics using computational methods provided by the PyMC library. The Bayesian statistical approach differs from classical frequentist statistics in several fundamental viewpoints. From a practical, computational perspective, Bayesian methods are often difficult to solve analytically. Computing the posterior distribution for a Bayesian model exactly is often prohibitively expensive. However, we can usually apply powerful and efficient sampling methods that allow us to find an approximate posterior distribution using simulations. The key role of a Bayesian statistics framework is to enable us to define statistical models and then apply sampling methods to find an approximate posterior distribution for the model. Here, we have used the PyMC library as a Bayesian modeling framework in Python. We briefly explored defining statistical models in terms of stochastic variables with given distributions and the simulation and sampling of the posterior distribution for those models using the MCMC methods implemented in the PyMC library.

Further Reading

For accessible introductions to the theory of Bayesian statistics, see *Doing Bayesian Data Analysis* by J. Kruschke (Academic Press, 2014) and *Think Bayes* by A. Downey (O'Reilly, 2013). A more technical discussion is given in *Bayesian Data Analysis* by A. Gelman (CRC Press, 2013). A computationally oriented introduction to Bayesian methods with Python is given in *Probabilistic Programming and Bayesian Methods for Hackers*, which is available for free online at `http://dataorigami.net/Probabilistic-Programming-and-Bayesian-Methods-for-Hackers`. An interesting discussion about the differences between the Bayesian and frequentist approaches to statistics, with examples written in Python, is given in *Frequentism and Bayesianism: A Python-driven Primer* by J. VanderPlas (2014), also available at `http://arxiv.org/pdf/1411.5018.pdf`.

CHAPTER 17

Signal Processing

This chapter explores signal processing, a subject with applications in diverse branches of science and engineering. A signal in this context can be a quantity that varies in time (temporal signal) or as a function of space coordinates (spatial signal). An audio signal is a typical example of a temporal signal, while an image is a typical example of a spatial signal in two dimensions. In practice, signals are often continuous functions. But it is common in computational applications to work with discretized signals, where the original continuous signal is sampled at discrete points with uniform distances. The sampling theorem gives rigorous and quantitative conditions for when a discrete sequence of samples can accurately represent a continuous signal.

Signal processing is essential in scientific computing because of its broad applicability and because there are very efficient computational methods for fundamental and highly practical problems. In particular, the fast Fourier transform (FFT) is an important algorithm for many signal-processing problems. Moreover, it is one of the most important numerical algorithms in all computing. This chapter examines how FFTs can be used in spectral analysis. But beyond this basic application, there is also broad usage of FFT both directly and indirectly as a component in other algorithms. Other signal-processing methods, such as convolution and correlation analysis and linear filters, have widespread applications, particularly in engineering fields such as control theory.

The chapter discusses spectral analysis and basic applications of linear filters using the SciPy library.

Importing Modules

This chapter mainly works with the fftpack and signal modules from the SciPy library. As usual with modules from the SciPy library, we import the modules using the following pattern.

```
In [1]: from scipy import fftpack
In [2]: from scipy import signal
```

We also use the io.wavefile module from SciPy to read and write WAV audio files in one of the examples. We import this module in the following way.

```
In [3]: import scipy.io.wavfile
In [4]: from scipy import io
```

For basic numerics and graphics, we also require the NumPy, Pandas, and Matplotlib libraries.

```
In [5]: import numpy as np
In [6]: import pandas as pd
In [7]: import matplotlib.pyplot as plt
In [8]: import matplotlib as mpl
```

© Robert Johansson 2024
R. Johansson, *Numerical Python*, https://doi.org/10.1007/979-8-8688-0413-7_17

Spectral Analysis

Let's begin this exploration of signal processing by considering spectral analysis. Spectral analysis is a fundamental application of Fourier transforms, a mathematical integral transform that allows us to take a signal from the time domain—which is described as a function of time—to the frequency domain, which is described as a function of frequency. The frequency-domain representation of a signal is useful for many purposes, for example, extracting features such as dominant frequency components of a signal, applying filters to signals, and solving differential equations (see Chapter 9), to mention a few.

Fourier Transforms

The following is the mathematical expression for the Fourier transform $F(v)$ of a continuous signal $f(t)$:[1]

$$F(v) = \int_{-\infty}^{\infty} f(t) e^{-2\pi i v t} dt,$$

and the inverse Fourier transform is given by the following:

$$f(t) = \int_{-\infty}^{\infty} F(v) e^{2\pi i v t} dv.$$

Here $F(v)$ is the complex-valued amplitude spectrum of the signal $f(t)$, and v is the frequency. From $F(v)$, we can compute other types of spectrums, such as the power spectrum $|F(v)|^2$. In this formulation, $f(t)$ is a continuous signal with infinite duration. In practical applications, we are often more interested in approximating $f(t)$ function using a finite number of samples for a finite time duration. For example, we might sample the $f(t)$ at N uniformly spaced points in the time interval $t \in [0, T]$, resulting in a sequence of samples that we denote $(x_0, x_1, ..., x_N)$. The continuous Fourier transform shown in the preceding text can be adapted to the discrete case: the discrete Fourier transform (DFT) of a sequence of uniformly spaced samples is as follows:

$$X_k = \sum_{n=0}^{N-1} x_n e^{-2\pi i n k / N}.$$

Similarly, we have the inverse DFT:

$$x_n = \frac{1}{N} \sum_{k=0}^{N-1} X_k e^{2\pi i n k / N},$$

where X_k is the discrete Fourier transform of the samples x_n, and k is a frequency bin number that can be related to an actual frequency. The DFT for a sequence of samples can be computed very efficiently using the fast Fourier transform (FFT) algorithm. The SciPy `fftpack` module[2] provides implementations of the FFT algorithm. The `fftpack` module contains FFT functions for various cases. Here, we focus on demonstrating the usage of the `fft` and `ifft` functions and several of the helper functions in the `fftpack` module. However, the general usage is similar for all FFT functions in Table 17-1.

---

[1] There are several alternative definitions of the Fourier transform, which vary in the coefficient in the exponent and the normalization of the transform integral.

[2] FFT is also implemented in the `fft` module in NumPy. It provides mostly the same functions as `scipy.fftpack`, which we use here. As a rule, when SciPy and NumPy provide the same functionality, it is generally preferable to use SciPy if available and fall back to the NumPy implementation when SciPy is unavailable.

Table 17-1. *Summary of Selected Functions from the fftpack Module in SciPy*

Function	Description
fft, ifft	General FFT and inverse FFT of a real- or complex-valued signal. The resulting frequency spectrum is complex valued.
rfft, irfft	The FFT and inverse FFT of a real-valued signal.
dct, idct	The discrete cosine transform (DCT) and its inverse.
dst, idst	The discrete sine transform (DST) and its inverse.
fft2, ifft2, fftn, ifftn	The two-dimensional and the N-dimensional FFT for complex-valued signals and their inverses.
fftshift, ifftshift, rfftshift, irfftshift	Shift the frequency bins in the result vector produced by fft and rfft, respectively, so that the spectrum is arranged such that the zero-frequency component is in the middle of the array.
fftfreq	Calculate the frequencies corresponding to the FFT bins in the result returned by fft.

For detailed usage of each function, including their arguments and return values, see their docstrings, which are available using, for example, help(fftpack.fft).

The DFT takes discrete samples as input and outputs a discrete frequency spectrum. To be able to use DFT for processes that are originally continuous, we first must reduce the signals to discrete values using sampling. According to the sampling theorem, a continuous signal with bandwidth B (i.e., the signal does not contain frequencies higher than B) can be reconstructed entirely from discrete samples with sampling frequency $f_s \geq 2B$. This is a significant result in signal processing because it tells us under what circumstances we can work with discrete instead of continuous signals. It allows us to determine a suitable sampling rate when measuring a continuous process since it is often possible to know or approximately guess the bandwidth of a process, for example, from physical arguments. While the sampling rate determines the maximum frequency we can describe with a discrete Fourier transform, the spacing of samples in frequency space is determined by the total sampling time T or equivalently from the number of sample points once the sampling frequency is determined, $T = N/f_s$.

As an introductory example, consider a simulated signal with pure sinusoidal components at 1 Hz and 22 Hz on top of a normal-distributed noise floor. We begin by defining a signal_samples function that generates noisy samples of this signal.

```
In [9]: def signal_samples(t):
   ...:         return (2 * np.sin(2 * np.pi * t) + 3 * np.sin(22 * 2 * np.pi * t) +
   ...:                 2 * np.random.randn(*np.shape(t)))
```

We can get a vector of samples by calling this function with an array with sample times as an argument. Say that we are interested in computing the frequency spectrum of this signal up to frequencies of 30 Hz. We then need to choose the sampling frequency $f_s = 60$ Hz, and if we want to obtain a frequency spectrum with a resolution of $\Delta f = 0.01$ Hz, we need to collect at least $N = f_s/\Delta f = 6000$ samples, corresponding to a sampling period of $T = N/f_s = 100$ seconds.

```
In [10]: B = 30.0
In [11]: f_s = 2 * B
In [12]: delta_f = 0.01
In [13]: N = int(f_s / delta_f); N
```

```
Out[13]: 6000
In [14]: T = N / f_s; T
Out[14]: 100.0
```

Next, we sample the signal function at N uniformly spaced points in time by creating a t array containing the sample times and then using it to evaluate the signal_samples function.

```
In [15]: t = np.linspace(0, T, N)
In [16]: f_t = signal_samples(t)
```

The resulting signal is plotted in Figure 17-1. The signal is rather noisy when viewed over the entire sampling time and for a shorter period, and the added random noise mostly masks the pure sinusoidal signals when viewed in the time domain.

```
In [17]: fig, axes = plt.subplots(1, 2, figsize=(8, 3), sharey=True)
    ...: axes[0].plot(t, f_t)
    ...: axes[0].set_xlabel("time (s)")
    ...: axes[0].set_ylabel("signal")
    ...: axes[1].plot(t, f_t)
    ...: axes[1].set_xlim(0, 5)
    ...: axes[1].set_xlabel("time (s)")
```

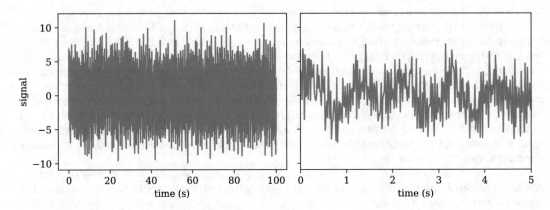

Figure 17-1. *Simulated signal with random noise. Full signal to the left and zoom into early times on the right*

To reveal the sinusoidal components in the signal, we can use the FFT to compute the signal's spectrum (or, in other words, its frequency-domain representation). We obtain the discrete Fourier transform of the signal by applying the fft function to the array of discrete samples, f_t.

```
In [18]: F = fftpack.fft(f_t)
```

The result is an F array, which contains the spectrum's frequency components at frequencies determined by the sampling rate and the number of samples. When computing these frequencies, it is convenient to use the fftfreq helper function, which takes the number of samples and the time duration between successive samples as parameters and returns an array of the same size as F that contains the frequencies corresponding to each frequency bin.

```
In [19]: f = fftpack.fftfreq(N, 1.0/f_s)
```

The frequency bins for the amplitude values returned by the fft function contain both positive and negative frequencies up to the frequency corresponding to half the sampling rate, $f_s/2$. For real-valued signals, the spectrum is symmetric at positive and negative frequencies, and we are, for this reason, often only interested in the positive-frequency components. Using the frequency array f, we can conveniently create a mask that can be used to extract the part of the spectrum that corresponds to the frequencies we are interested in. Here, we create a mask for selecting the positive-frequency components.

```
In [20]: mask = np.where(f >= 0)
```

The spectrum for the positive-frequency components is shown in Figure 17-2. The top panel contains the entire positive-frequency spectrum and is plotted on a log scale to increase the contrast between the signal and the noise. We can see sharp peaks near 1 Hz and 22 Hz, corresponding to the sinusoidal components in the signal. These peaks stand out from the noise floor in the spectrum. Despite the noise concealing the sinusoidal components in the time-domain signal, we can detect their presence in the frequency-domain representation. The two lower panels in Figure 17-2 show magnifications of the two peaks at 1 Hz and 22 Hz, respectively.

```
In [21]: fig, axes = plt.subplots(3, 1, figsize=(8, 6))
    ...: axes[0].plot(f[mask], np.log(abs(F[mask])))
    ...: axes[0].plot(B, 0, 'r*', markersize=10)
    ...: axes[0].set_ylabel("$\log(|F|)$", fontsize=14)
    ...: axes[1].plot(f[mask], abs(F[mask])/N)
    ...: axes[1].set_xlim(0, 2)
    ...: axes[1].set_ylabel("$|F|/N$", fontsize=14)
    ...: axes[2].plot(f[mask], abs(F[mask])/N)
    ...: axes[2].set_xlim(21, 23)
    ...: axes[2].set_xlabel("frequency (Hz)", fontsize=14)
    ...: axes[2].set_ylabel("$|F|/N$", fontsize=14)
```

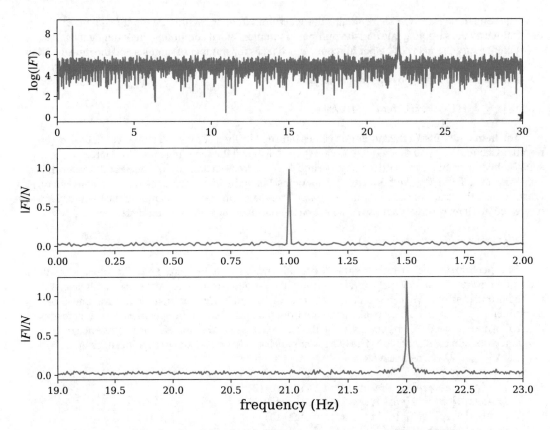

Figure 17-2. *Spectrum of the simulated signal with frequency components at 1 Hz and 22 Hz*

Frequency-Domain Filter

Just like we can compute the frequency-domain representation from the time-domain signal using the FFT function fft, we can compute the time-domain signal from the frequency-domain representation using the *inverse* FFT function ifft. For example, applying the ifft function to the F array reconstructs the f_t array. Modifying the spectrum before applying the inverse transform allows us to realize frequency-domain filters. For example, selecting only frequencies below 2 Hz in the spectrum amounts to applying a 2 Hz low-pass filter, which suppresses high-frequency components in the signal (higher than 2 Hz in this case).

```
In [22]: F_filtered = F * (abs(f) < 2)
In [23]: f_t_filtered = fftpack.ifft(F_filtered)
```

Computing the inverse FFT for the filtered signal results in a time-domain signal where the high-frequency oscillations are absent, as shown in Figure 17-3. This simple example summarizes the essence of many frequency-domain filters. Some of the many types of filters commonly used in signal-processing analysis are covered later in this chapter.

```
In [24]: fig, ax = plt.subplots(figsize=(8, 3))
    ...: ax.plot(t, f_t, label='original')
    ...: ax.plot(t, f_t_filtered.real, color="red", lw=3, label='filtered')
    ...: ax.set_xlim(0, 10)
    ...: ax.set_xlabel("time (s)")
    ...: ax.set_ylabel("signal")
    ...: ax.legend()
```

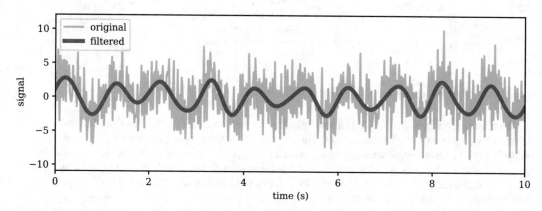

Figure 17-3. *The original time-domain signal and the reconstructed signal after applying a low-pass filter to the frequency-domain representation of the signal*

Windowing

In the previous section, we directly applied the FFT to the signal. This can give acceptable results, but it is often possible to further improve the frequency spectrum's quality and contrast by applying a *window function* to the signal before applying the FFT. A window function is a function that, when multiplied with the signal, modulates its magnitude so that it approaches zero at the beginning and the end of the sampling duration. There are many possible window functions, and the SciPy `signal` module provides implementations of many common window functions, including the Blackman function, the Hann function, the Hamming function, Gaussian window functions (with variable standard deviation), and the Kaiser window function.[3] These functions are all plotted in Figure 17-4. This graph shows that while all these window functions are slightly different, the overall shape is very similar.

```
In [25]: fig, ax = plt.subplots(1, 1, figsize=(8, 3))
    ...: N = 100
    ...: ax.plot(signal.blackman(N), label="Blackman")
    ...: ax.plot(signal.hann(N), label="Hann")
    ...: ax.plot(signal.hamming(N), label="Hamming")
    ...: ax.plot(signal.gaussian(N, N/5), label="Gaussian (std=N/5)")
    ...: ax.plot(signal.kaiser(N, 7), label="Kaiser (beta=7)")
    ...: ax.set_xlabel("n")
    ...: ax.legend(loc=0)
```

---

[3] Several other window functions are also available. See the docstring for the `scipy.signal` module for a complete list.

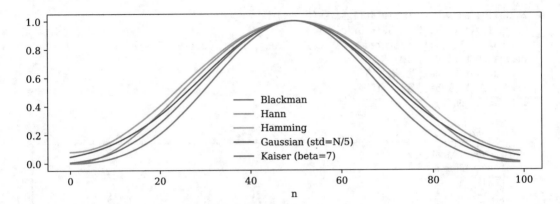

Figure 17-4. *Examples of commonly used window functions*

The alternative window functions all have slightly different properties and objectives, but for the most part, they can be used interchangeably. The main purpose of window functions is to reduce spectral leakage between nearby frequency bins, which occurs in discrete Fourier transform computation when the signal contains components with periods that are not exactly divisible with the sampling period. Signal components with such frequencies can, therefore, not fit a full number of cycles in the sampling period. Since the discrete Fourier transform assumes that the signal is periodic, the resulting discontinuity at the period boundary can give rise to spectral leakage. Multiplying the signal with a window function reduces this problem. Alternatively, we could increase the sample points (increase the sampling period) to obtain a higher frequency resolution, but this might not always be practical.

To see how we can use a window function before applying the FFT to a time-series signal, let's consider the outdoor temperature measurements that we looked at in Chapter 12. First, we use the Pandas library to load the dataset and resample it to evenly spaced hourly samples, using the `fillna` method to aggregate the elements.

```
In [26]: df = pd.read_csv('temperature_outdoor_2014.tsv', delimiter="\t",
    ...:                  names=["time", "temperature"])
In [27]: df.time = (pd.to_datetime(df.time.values, unit="s").
    ...:            tz_localize('UTC').tz_convert('Europe/Stockholm'))
In [28]: df = df.set_index("time")
In [29]: df = df.resample("H").ffill()
In [30]: df = df[(df.index >= "2014-04-01")*(df.index < "2014-06-01")].dropna()
```

Once the Pandas data frame has been created and processed, we exact the underlying NumPy arrays to be able to process the time-series data using the `fftpack` module.

```
In [31]: time = df.index.astype('int64')/1.0e9
In [32]: temperature = df.temperature.values
```

Now, we wish to apply a window function to the data in the array `temperature` before we compute the FFT. Here, we use the Blackman window function, which is available as the `blackman` function in the `signal` module in SciPy. As an argument to the window function, we need to pass the length of the sample array, and it returns an array of that same length.

```
In [33]: window = signal.blackman(len(temperature))
```

To apply the window function, we multiply it with the array containing the time-domain signal and use the result in the subsequent FFT computation. However, before we proceed with the FFT for the windowed temperature signal, we first plot the original temperature time series and the windowed version. The result is shown in Figure 17-5. The result of multiplying the time series with the window function is a signal approaching zero near the sampling period boundaries. It can be viewed as a periodic function with smooth transitions between period boundaries, and as such, the FFT of the windowed signal has better-behaved properties.

```
In [34]: temperature_windowed = temperature * window
In [35]: fig, ax = plt.subplots(figsize=(8, 3))
    ...: ax.plot(df.index, temperature, label="original")
    ...: ax.plot(df.index, temperature_windowed, label="windowed")
    ...: ax.set_ylabel("temperature", fontsize=14)
    ...: ax.legend(loc=0)
```

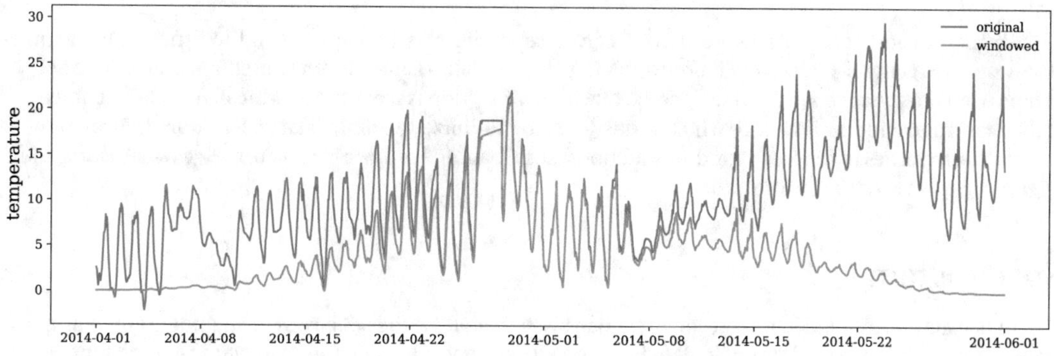

Figure 17-5. *Windowed and original temperature time-series signal*

After preparing the windowed signal, the rest of the spectral analysis proceeds as before: we can use the fft function to compute the spectrum and the fftfreq function to calculate the frequencies corresponding to each frequency bin.

```
In [36]: data_fft_windowed = fftpack.fft(temperature_windowed)
In [37]: f = fftpack.fftfreq(len(temperature), time[1]-time[0])
```

Here, we also select the positive frequencies by creating a mask array from the array f and plot the resulting positive-frequency spectrum as shown in Figure 17-6. It shows peaks at the frequency corresponding to 1 day (1/86,400 Hz) and its higher harmonics (2/86,400 Hz, 3/86,400 Hz, etc.).

```
In [38]: mask = f > 0
In [39]: fig, ax = plt.subplots(figsize=(8, 3))
    ...: ax.set_xlim(0.000005, 0.00004)
    ...: ax.axvline(1./86400, color='r', lw=0.5)
    ...: ax.axvline(2./86400, color='r', lw=0.5)
    ...: ax.axvline(3./86400, color='r', lw=0.5)
    ...: ax.plot(f[mask], np.log(abs(data_fft_windowed[mask])), lw=2)
    ...: ax.set_ylabel("$\log|F|$", fontsize=14)
    ...: ax.set_xlabel("frequency (Hz)", fontsize=14)
```

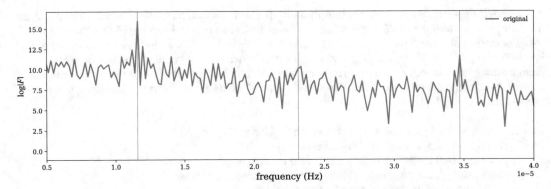

Figure 17-6. *Spectrum of the windowed temperature time series. The dominant peak occurs at the frequency corresponding to a 1-day period and its higher harmonics*

To get the most accurate spectrum from a given set of samples, applying a window function to the time-series signal is generally advisable before applying an FFT. Most of the window functions available in SciPy can be used interchangeably, and the choice of window function is usually not critical. A popular choice is the Blackman window function, which is designed to minimize spectral leakage. For more information about the properties of different window functions, see Chapter 9 of *The Scientist and Engineer's Guide to Digital Signal Processing* by S. Smith.

Spectrogram

As a final example in this section on spectral analysis, let's analyze the spectrum of an audio signal sampled from a guitar.[4] First, we load sampled data from the `guitar.wav` file using the `io.wavfile.read` function from the SciPy library.

```
In [40]: sample_rate, data = io.wavfile.read("guitar.wav")
```

The `io.wavfile.read` function returns a tuple containing the sampling rate, `sample_rate`, and a NumPy array containing the audio intensity. For this particular file, we get the sampling rate of 44.1 kHz, and the audio signal was recorded in stereo, which is represented by a data array with two channels. Each channel contains 1,181,625 samples.

```
In [41]: sample_rate
Out[41]: 44100
In [42]: data.shape
Out[42]: (1181625, 2)
```

Here, we are only concerned with analyzing a single audio channel, so we form the average of the two channels to obtain a mono-channel signal.

```
In [43]: data = data.mean(axis=1)
```

---

[4]The data used in this example was obtained from `https://www.freesound.org/people/guitarguy1985/sounds/52047`.

We can calculate the total duration of the audio recording by dividing the number of samples by the sampling rate. The result suggests that the recording is about 26.8 seconds.

```
In [44]: data.shape[0] / sample_rate
Out[44]: 26.79421768707483
```

It is often the case that we like to compute the spectrum of a signal in segments instead of the entire signal at once, for example, if the nature of the signal varies in time on a long timescale but contains nearly periodic components on a short timescale. This is particularly true for music, which can be considered nearly periodic on short timescales from the point of view of human perception (subsecond timescales) but varies on longer timescales. In the the guitar sample, we would like to apply the FFT on a sliding window in the time-domain signal. The result is a time-dependent spectrum, often visualized as an equalizer graph on music equipment and applications. Another approach is to visualize the time-dependent spectrum using a two-dimensional heatmap graph, which in this context is known as a spectrogram. The following computes the spectrogram of the guitar sample.

Before proceeding with the spectrogram visualization, we calculate the spectrum for a small part of the sample. Let's begin by determining the number of samples from the full array. If we want to analyze 0.5 seconds at the time, we can use the sampling rate to compute the number of samples to use.

```
In [45]: N = int(sample_rate/2.0) # half a second -> 22050 samples
```

Next, given the number of samples and the sampling rate, we can compute the frequencies f for the frequency bins for the result of the forthcoming FFT calculation and the sampling times t for each sample in the time-domain signal. We also create a frequency mask for selecting positive frequencies smaller than 1000 Hz, which is used later to select a subset of the computed spectrum.

```
In [46]: f = fftpack.fftfreq(N, 1.0/sample_rate)
In [47]: t = np.linspace(0, 0.5, N)
In [48]: mask = (f > 0) * (f < 1000)
```

Next, we exact the first N samples from the full sample array data and apply the fft function.

```
In [49]: subdata = data[:N]
In [50]: F = fftpack.fft(subdata)
```

The time- and frequency-domain signals are shown in Figure 17-7. The time-domain signal in the left panel is zero in the beginning before the first guitar string is plucked. The frequency-domain spectrum shows several dominant frequencies corresponding to the different tones produced by the guitar.

```
In [51]: fig, axes = plt.subplots(1, 2, figsize=(12, 3))
    ...: axes[0].plot(t, subdata)
    ...: axes[0].set_ylabel("signal", fontsize=14)
    ...: axes[0].set_xlabel("time (s)", fontsize=14)
    ...: axes[1].plot(f[mask], abs(F[mask]))
    ...: axes[1].set_ylabel("$|F|$", fontsize=14)
    ...: axes[1].set_xlabel("Frequency (Hz)", fontsize=14)
```

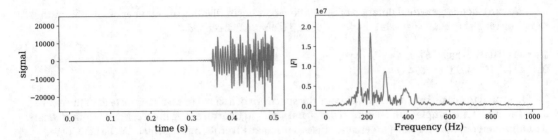

Figure 17-7. *Signal and spectrum for samples half a second duration of a guitar sound*

The next step is to repeat the analysis for successive segments from the full sample array. The time evolution of the spectrum can be visualized as a spectrogram, with frequency on the *x* axis and time on the *y* axis. To be able to plot the spectrogram with the imshow function from Matplotlib, we create a two-dimensional NumPy array spectrogram_data for storing the spectra for the successive sample segments. The shape of the spectrogram_data array is (n_max, f_values), where n_max is the number of segments of length N in the sample array data, and f_values are the number of frequency bins with frequencies that match the condition used to compute mask (positive frequencies less than 1000 Hz).

```
In [52]: n_max = int(data.shape[0] / N)
In [53]: f_values = np.sum(mask)
In [54]: spectogram_data = np.zeros((n_max, f_values))
```

To improve the resulting spectrogram's contrast, we apply a Blackman window function to each subset of the sample data before we compute the FFT. Here, we choose the Blackman window function for its spectral leakage-reducing properties, but many other window functions give similar results. The length of the window array must be the same as the length of the subdata array, so we pass its length argument to the Blackman function.

```
In [55]: window = signal.blackman(len(subdata))
```

Finally, we can compute the spectrum for each segment in the sample by looping over the array slices of size N, apply the window function, compute the FFT, and store the subset of the result for the frequencies we are interested in in the spectrogram_data array.

```
In [56]: for n in range(0, n_max):
    ...:     subdata = data[(N * n):(N * (n + 1))]
    ...:     F = fftpack.fft(subdata * window)
    ...:     spectogram_data[n, :] = np.log(abs(F[mask]))
```

When the spectrogram_data array is computed, we can visualize the spectrogram using the imshow function from Matplotlib. The result is shown in Figure 17-8.

```
In [57]: fig, ax = plt.subplots(1, 1, figsize=(8, 6))
    ...: p = ax.imshow(spectogram_data, origin='lower',
    ...:               extent=(0, 1000, 0, data.shape[0] / sample_rate),
    ...:               aspect='auto',
    ...:               cmap=mpl.cm.RdBu_r)
    ...: cb = fig.colorbar(p, ax=ax)
```

```
...: cb.set_label("$\log|F|$", fontsize=14)
...: ax.set_ylabel("time (s)", fontsize=14)
...: ax.set_xlabel("Frequency (Hz)", fontsize=14)
```

The spectrogram in Figure 17-8 contains a lot of information about the sampled signal and how it evolves. The narrow vertical stripes correspond to tones produced by the guitar, and those signals slowly decay over time. The broad horizontal bands correspond roughly to periods when strings are being plucked on the guitar, giving a very broad frequency response for a short time. Note, however, that the color axis represents a logarithmic scale, so small variations in the color represent considerable variation in the actual intensity.

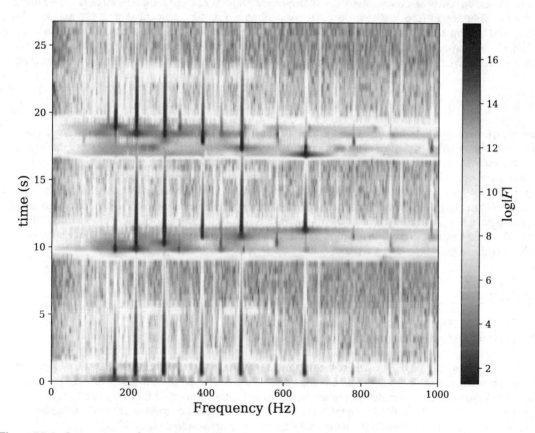

Figure 17-8. *Spectrogram of an audio sampling of a guitar sound*

Signal Filters

One of the main objectives of signal processing is to manipulate and transform temporal or spatial signals to change their characteristics. Typical applications are noise reduction, sound effects in audio signals, and effects such as blurring, sharpening, contrast enhancement, and color balance adjustments in image data. Many common transformations can be implemented as filters that act on the frequency/domain representation of the signal, for example, by suppressing specific frequency components. The previous section presented an example of a low-pass filter, which we implemented by taking the Fourier transform of the signal, removing the high-frequency components, and finally taking the inverse Fourier transform to obtain a new time-domain signal. With this approach, we can implement arbitrary frequency filters, but we cannot necessarily apply them in real-time on a streaming signal since they require buffering sufficient samples to be able to perform the discrete Fourier transform. In many applications, it is desirable to apply filters and continuously transform a signal, for example, when processing signals in transmission or live audio signals.

Convolution Filters

Certain frequency filters can be implemented directly in the time domain using a convolution of the signal with a function that characterizes the filter. An important property of Fourier transformations is that the (inverse) Fourier transform of the product of two functions (e.g., the spectrum of a signal and the filter shape function) is a convolution of the two functions (inverse) Fourier transforms. Therefore, if we want to apply a filter, H_k, to the spectrum X_k of a signal x_n, we can instead compute the convolution of x_n with h_m, the inverse Fourier transform of the H_k filter function. In general, we can write a filter on convolution form as follows:

$$y_n = \sum_{k=-\infty}^{\infty} x_k h_{n-k},$$

where x_k is the input, y_n is the output, and h_{n-k} is the convolution kernel that characterizes the filter. Note that in this general form, the signal y_n at time step n depends on both earlier and later values of the input x_k. To illustrate this point, let's return to the first example in this chapter, where we applied a low-pass filter to a simulated signal with components at 1 Hz and 22 Hz. That example Fourier transformed the signal and multiplied its spectrum with a step function that suppressed all high-frequency components. Finally, we inverse Fourier transformed the signal back into the time domain. The result was a smoothened version of the original noisy signal (see Figure 17-3). An alternative approach using convolution is to inverse Fourier transform the frequency response function for the filter H and use the result h as a kernel with which we convolve the original time-domain signal f_t.

```
In [58]: t = np.linspace(0, T, N)
In [59]: f_t = signal_samples(t)
In [60]: H = abs(f) < 2
In [61]: h = fftpack.fftshift(fftpack.ifft(H))
In [62]: f_t_filtered_conv = signal.convolve(f_t, h, mode='same')
```

To carry out the convolution, the convolve function from the signal module in SciPy was used. It takes as arguments two NumPy arrays containing the signals for which to compute the convolution. Using the optional keyword argument mode, we can set the size of the output array to be the same as the first input (mode='same'), the full convolution output after having zero-padded the arrays to account for transients (mode='full'), or to contain only elements that do not rely on zero-padding (mode='valid'). Here, we use mode='same', to easily compare and plot the result with the original signal, f_t. The result of applying this convolution filter, f_t_filtered_conv, is shown in Figure 17-9, together with the corresponding result computed using fft and ifft with a modified spectrum (f_t_filtered). As expected, the two methods give identical results.

```
In [63]: fig = plt.figure(figsize=(8, 6))
    ...: ax = plt.subplot2grid((2,2), (0,0))
    ...: ax.plot(f, H)
    ...: ax.set_xlabel("frequency (Hz)")
    ...: ax.set_ylabel("Frequency filter")
    ...: ax.set_ylim(0, 1.5)
```

```
...: ax = plt.subplot2grid((2,2), (0,1))
...: ax.plot(t - t[-1]/2.0, h.real)
...: ax.set_xlabel("time (s)")
...: ax.set_ylabel("convolution kernel")
...: ax = plt.subplot2grid((2,2), (1,0), colspan=2)
...: ax.plot(t, f_t, label='original', alpha=0.25)
...: ax.plot(t, f_t_filtered.real, 'r', lw=2,
...:         label='filtered in frequency domain')
...: ax.plot(t, f_t_filtered_conv.real, 'b--', lw=2,
...:         label='filtered with convolution')
...: ax.set_xlim(0, 10)
...: ax.set_xlabel("time (s)")
...: ax.set_ylabel("signal")
...: ax.legend(loc=2)
```

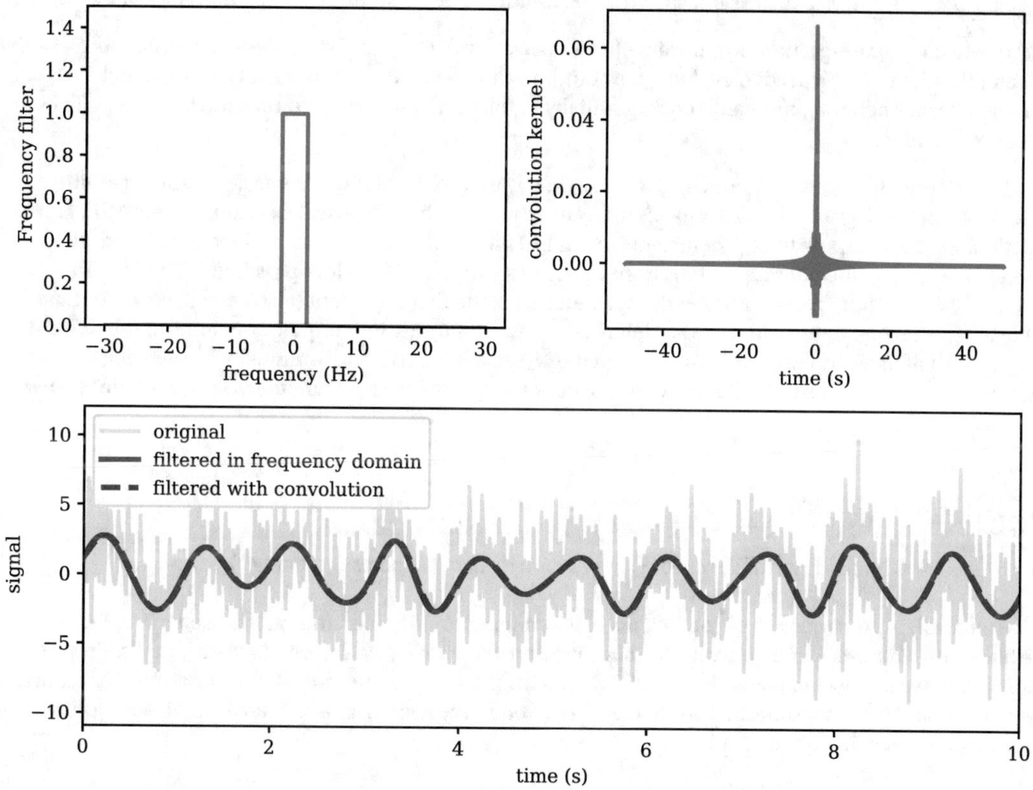

Figure 17-9. *Top left: frequency filter. Top right: convolution kernel corresponding to the frequency filter (its inverse discrete Fourier transform). Bottom: simple lowpass filter applied via convolution*

FIR and IIR Filters

In the convolution filter example, there was no computational advantage for using convolution to implement the filter rather than a sequence of a call to `fft`, spectrum modifications, followed by a call to `ifft`. The convolution here is, in general, more computationally expensive than the extra FFT transformation, and the SciPy signal module even provides a `fftconvolve` function, which implements the convolution using FFT and its inverse. Furthermore, the convolution kernel of the filter has many undesirable properties, such as being noncasual, where the output signal depends on future input values (see the upper-right panel in Figure 17-9). However, important special cases of convolution-like filters can be efficiently implemented with dedicated digital signal processors (DSPs) and general-purpose processors.

An important family of such filters is the *finite impulse response* (FIR) filters, which take the form $y_n = \sum_{k=0}^{M} b_k x_{n-k}$.

This time-domain filter is casual because the output y_n only depends on input values at earlier time steps.

Another similar type of filter is the *infinite impulse response* (IIR) filter, which can be written in the form $a_0 y_n = \sum_{k=0}^{M} b_k x_{n-k} - \sum_{k=1}^{N} a_k y_{n-k}$. This is not strictly a convolution since it additionally includes past values of the *output* when computing a new output value (a feedback term). But it is nonetheless in a similar form. Both FIR and IIR filters can be used to evaluate new output values given the recent history of the signal and the output and can, therefore, be evaluated sequentially in the time domain if we know the finite sequences of values of b_k and a_k.

Computing the values of b_k and a_k given a set of requirements on filter properties is known as filter design. The SciPy `signal` module provides many functions for this purpose. For example, using the `firwin` function, we can compute the b_k coefficients for an FIR filter given frequencies of the band boundaries, where, for example, the filter transitions from a pass to a stop filter (for a low-pass filter). The `firwin` function takes the number of values in the a_k sequence as the first argument (also known as *taps* in this context). The second argument, `cutoff`, defines the low-pass transition frequency in units of the Nyquist frequency (half the sampling rate). The scale of the Nyquist frequency can optionally be set using the `nyq` argument, which defaults to 1. Finally, we can specify the type of window function to use with the `window` argument.

```
In [64]: n = 101
In [65]: f_s = 1 / 3600
In [66]: nyq = f_s/2
In [67]: b = signal.firwin(n, cutoff=nyq/12, nyq=nyq, window="hamming")
```

The result is the sequence of coefficients b_k that defines an FIR filter and can be used to implement the filter with a time-domain convolution. Given the coefficients b_k, we can evaluate the amplitude and phase response of the filter using the `freqz` function from the `signal` module. It returns arrays containing frequencies and the corresponding complex-valued frequency response, which are suitable for plotting purposes, as shown in Figure 17-10.

```
In [68]: f, h = signal.freqz(b)
In [69]: fig, ax = plt.subplots(1, 1, figsize=(12, 3))
    ...: h_ampl = 20 * np.log10(abs(h))
    ...: h_phase = np.unwrap(np.angle(h))
    ...: ax.plot(f/max(f), h_ampl, 'b')
    ...: ax.set_ylim(-150, 5)
```

```
...: ax.set_ylabel('frequency response (dB)', color="b")
...: ax.set_xlabel(r'normalized frequency')
...: ax = ax.twinx()
...: ax.plot(f/max(f), h_phase, 'r')
...: ax.set_ylabel('phase response', color="r")
...: ax.axvline(1.0/12, color="black")
```

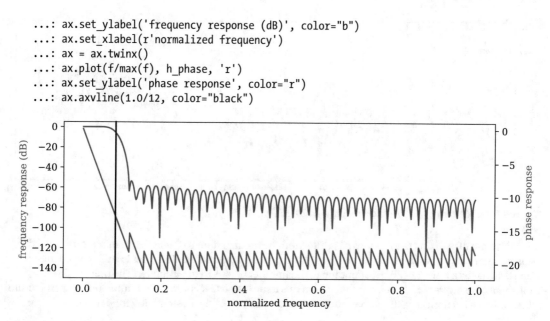

Figure 17-10. *The amplitude and phase response of a low-pass FIR filter*

The low-pass filter shown in Figure 17-10 is designed to pass through signals with frequencies less than $f_s/24$ (indicated with a vertical line) and suppress higher-frequency signal components. The finite transition region between pass and stop bands and the nonperfect suppression above the cutoff frequency is a price we have to pay to be able to represent the filter in FIR form. The accuracy of the FIR filter can be improved by increasing the number of coefficients b_k, at the expense of higher computational complexity.

The effect of an FIR filter, given the coefficients b_k, and an IIR filter, given the coefficients b_k and a_k, can be evaluated using the lfilter function from the signal module. As the first argument, this function expects the array with coefficients b_k, and as the second argument, the array with the coefficients a_k in an IIR filter or the scalar 1 in an FIR filter. The third argument to the function is the input signal array, and the return value is the filter output. For example, we can use the following to apply the FIR filter we created in the preceding text to the array with hourly temperature measurements temperature.

In [70]: temperature_filt = signal.lfilter(b, 1, temperature)

The effect of applying the low-pass FIR filter to the signal is to smoothen the function by eliminating the high-frequency oscillations, as shown in Figure 17-11. Another approach to achieve a similar result is to apply a moving average filter, in which the output is a weighted average or median of a few nearby input values. The medfilt function from the signal module applies a median filter of a given input signal, using the number of past nearby values specified with the second argument to the function.

In [71]: temperature_median_filt = signal.medfilt(temperature, 25)

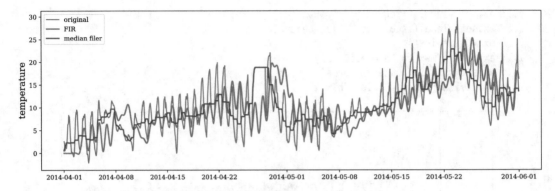

Figure 17-11. *Output of an FIR filter and a median filter*

The result of applying the FIR low-pass filter and the median filter to the hourly temperature measurement dataset is shown in Figure 17-11. Note that the output of the FIR filter is shifted from the original signal by a time delay corresponding to the number of taps in the FIR filter. The median filter implemented using medfilt does not suffer from this issue because it is computed from both past and future values, making it a noncasual filter that cannot be evaluated on the fly on streaming input data.

```
In [72]: fig, ax = plt.subplots(figsize=(8, 3))
    ...: ax.plot(df.index, temperature, label="original", alpha=0.5)
    ...: ax.plot(df.index, temperature_filt, color="red", lw=2, label="FIR")
    ...: ax.plot(df.index, temperature_median_filt, color="green", lw=2,
    ...:         label="median filer")
    ...: ax.set_ylabel("temperature", fontsize=14)
    ...: ax.legend(loc=0)
```

To design an IIR filter, we can use the iirdesign function from the signal module or use one of the many predefined IIR filter types, including the Butterworth filter (signal.butter), Chebyshev filters of types I and II (signal.cheby1 and signal.cheby2), and elliptic filter (signal.ellip). For example, we can use the following to create a Butterworth high-pass filter that allows frequencies above the critical frequency 7/365 Hz to pass while lower frequencies are suppressed.

```
In [73]: b, a = signal.butter(2, 7/365.0, btype='high')
```

The first argument to this function is the order of the Butterworth filter, and the second argument is the critical frequency of the filter (where it goes from the bandstop to the bandpass function). For this example, the optional argument btype can be used to specify if the filter is a low-pass filter (low) or high-pass filter (high). More options are described in the function's docstring: see, for example, help(signal.butter). The outputs a and b are the a_k and b_k coefficients that define the IIR filter, respectively. Here, we have computed a Butterworth filter of second order, so a and b each have three elements.

```
In [74]: b
Out[74]: array([ 0.95829139, -1.91658277,  0.95829139])
In [75]: a
Out[75]: array([ 1.        , -1.91484241,  0.91832314])
```

Like before, we can apply the filter to an input signal (here, we again use the hourly temperature dataset as an example).

```
In [76]: temperature_iir = signal.lfilter(b, a, temperature)
```

Alternatively, we can apply the filter using the filtfilt function, which applies the filter forward and backward in time, resulting in a noncasual filter.

```
In [77]: temperature_filtfilt = signal.filtfilt(b, a, temperature)
```

The results of both types of filters are shown in Figure 17-12. Eliminating the lowfrequency components detrends the time series and only retains the high-frequency oscillations and fluctuations. The filtered signal can, therefore, be viewed as measuring the volatility of the original signal. This example highlights that the daily variations are greater during the spring months of March, April, and May compared to the winter months of January and February.

```
In [78]: fig, ax = plt.subplots(figsize=(8, 3))
    ...: ax.plot(df.index, temperature, label="original", alpha=0.5)
    ...: ax.plot(df.index, temperature_iir, color="red", label="IIR filter")
    ...: ax.plot(df.index, temperature_filtfilt, color="green",
    ...:         label="filtfilt filtered")
    ...: ax.set_ylabel("temperature", fontsize=14)
    ...: ax.legend(loc=0)
```

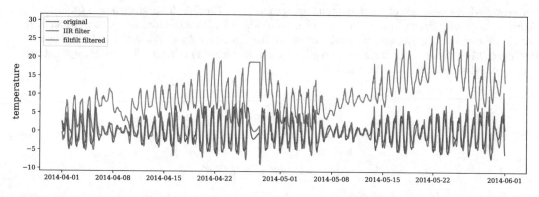

Figure 17-12. *Output from an IIR high-pass filter and the corresponding filtfilt filter (applied both forward and backward)*

These techniques can be directly applied to audio and image data. For example, to apply a filter to the audio signal of the guitar samples, we can use the lfilter functions. The coefficients b_k for the FIR filter can sometimes be constructed manually. For example, to apply a naive echo sound effect, we can create an FIR filter that repeats past signals with some time delay: $y_n = x_n + x_{n-N}$, where N is a time delay in units of time steps. The corresponding coefficients b_k are easily constructed and can be applied to the audio signal data.

```
In [79]: b = np.zeros(10000)
    ...: b[0] = b[-1] = 1
    ...: b /= b.sum()
In [80]: data_filt = signal.lfilter(b, 1, data)
```

To listen to the modified audio signal, we can write it to a WAV file using the `write` function from the `io.wavefile` module in SciPy.

```
In [81]: io.wavfile.write("guitar-echo.wav", sample_rate,
    ...:                    np.vstack([data_filt, data_filt]).T.astype(np.int16))
```

Similarly, we can implement many types of image processing filters using the tools from the `signal` module. SciPy also provides a module `ndimage`, which contains many common image manipulation functions and filters specially adapted for applying two-dimensional image data. The Scikit-Image library[5] provides a more advanced framework for working with image processing in Python.

Summary

Signal processing is a vast field with applications in most fields of science and engineering. As such, we have only covered a few fundamental applications of signal processing in this chapter, focusing on introducing methods for approaching this type of problem with computational methods using Python and the libraries and tools available within the Python ecosystem for scientific computing. In particular, the chapter explored spectral analysis of time-dependent signals using fast Fourier transform and the design and application of linear filters to signals using the `signal` module in the SciPy library.

Further Reading

For a comprehensive review of the signal processing theory, see *The Scientist and Engineer's Guide to Digital Signal Processing* by Steven Smith (1999), which can be viewed online at `www.dspguide.com/pdfbook.htm`. For a Python-oriented discussion of signal processing, see *Python for Signal Processing* by J. Unpingco (Springer, 2014), from which content is available as IPython notebooks at `http://nbviewer.org/github/unpingco/Python-for-Signal-Processing`.

---

[5] See the project's web page at `http://scikit-image.org` for more information.

CHAPTER 18

▪ ▪ ▪

Data Input and Output

In nearly all scientific computing and data analysis applications, there is a need for data input and output. This includes loading datasets and persistently storing results to files on disk or databases. Getting data in and out of programs is a critical step in the computational workflow. There are many standardized formats for storing structured and unstructured data. The benefits of using standardized formats are obvious: You can use existing libraries for reading and writing data, saving time and effort. In the course of working with scientific and technical computing, it is likely that you will face a variety of data formats through interaction with colleagues and peers or when acquiring data from sources such as equipment and databases. As a computational practitioner, it is essential to handle data efficiently and seamlessly, regardless of its format. This motivates why this entire chapter is devoted to data input and output.

Python has good support for many file formats. Multiple options exist for dealing with the most common formats. This chapter surveys data storage formats with applications in computing and discusses typical situations where each format is suitable. We also introduce Python libraries and tools for handling common data formats in computing.

Data can be classified into several categories and types. Important categories are structured and unstructured data, and values can, for example, be categorical (a finite set of values), ordinal (values with meaningful ordering), or numerical (continuous or discrete). Values also have types like string, integer, floating-point number, and so on. A data format for storing or transmitting data should ideally account for these concepts to avoid loss of data or metadata, and we frequently need to have fine-grained control of how data is represented.

In computing applications, we mainly deal with structured data, for example, arrays and tabular data. Examples of unstructured datasets include free-form texts or nested lists with nonhomogeneous types. This chapter focuses on the CSV family of formats and the HDF5 format for structured data. Toward the end of the chapter, we discuss the JSON format as a lightweight and flexible format that can store both simple and complex datasets, with a bias toward storing lists and dictionaries. This format is well-suited for storing unstructured data. We also briefly discuss methods of serializing objects into storable data using the msgpack format and Python's built-in pickle format.

Because of the importance of data input and output in many data-centric computational applications, several Python libraries have emerged to simplify and assist in handling data in different formats and moving and converting data. For example, the Blaze library (https://blaze.pydata.org) provides a high-level interface for accessing data from different formats and sources. Here, we focus mainly on lower-level libraries for reading specific types of file formats that are useful for storing numerical data and unstructured datasets. However, the interested reader is encouraged also to explore higher-level libraries such as Blaze.

© Robert Johansson 2024
R. Johansson, *Numerical Python*, https://doi.org/10.1007/979-8-8688-0413-7_18

Importing Modules

This chapter uses several different libraries for handling different types of data. For starters, we require NumPy and Pandas, which, as usual, we import as np and pd, respectively.

```
In [1]: import numpy as np
In [2]: import pandas as pd
```

We also use the csv and json modules from the Python standard library.

```
In [3]: import csv
In [4]: import json
```

For working with the HDF5 format for numerical data, we use the h5py and the pytables libraries.

```
In [5]: import h5py
In [6]: import tables
```

Parquet is another format for numerical data. We use the parquet submodule of the pyarrow library, which is imported as follows.

```
In [7]: import pyarrow.parquet as pq
```

Finally, in the context of serializing objects to storable data, we explore the pickle and msgpack libraries.

```
In [8]: import pickle
In [9]: import msgpack
```

Comma-Separated Values

Comma-separated values (CSV) is an intuitive and loosely defined[1] plain-text file format that is simple yet effective and prevalent for storing tabular data. Each record is stored as a line in this format, and each record field is separated with a delimiter character (e.g., a comma). Optionally, each field can be enclosed in quotation marks to allow for string-valued fields that contain the delimiter character. Also, the first line is sometimes used to store column names, and comment lines are common. An example of a CSV file is shown in Listing 18-1.

Listing 18-1. Example of a CSV File with a Comment Line, a Header Line, and Mixed Numerical and String-Valued Data Fields (Data source: www.nhl.com)

```
# 2013-2014 / Regular Season / All Skaters / Summary / Points
Rank,Player,Team,Pos,GP,G,A,P,+/-,PIM,PPG,PPP,SHG,SHP,GW,OT,S,S%,TOI/GP,Shift/GP,FO%
1,Sidney Crosby,PIT,C,80,36,68,104,+18,46,11,38,0,0,5,1,259,13.9,21:58,24.0,52.5
2,Ryan Getzlaf,ANA,C,77,31,56,87,+28,31,5,23,0,0,7,1,204,15.2,21:17,25.2,49.0
3,Claude Giroux,PHI,C,82,28,58,86,+7,46,7,37,0,0,7,1,223,12.6,20:26,25.1,52.9
4,Tyler Seguin,DAL,C,80,37,47,84,+16,18,11,25,0,0,8,0,294,12.6,19:20,23.4,41.5
5,Corey Perry,ANA,R,81,43,39,82,+32,65,8,18,0,0,9,1,280,15.4,19:28,23.2,36.0
```

---

[1] Although RFC 4180, http://tools.ietf.org/html/rfc4180, is sometimes taken as an unofficial specification, in practice there exist many varieties and dialects of CSV.

CSV is occasionally also considered an acronym for character-separated value, reflecting that the CSV format commonly refers to a family of formats using different delimiters between the fields. For example, instead of the comma, the Tab character is often used, in which case the format is sometimes called TSV instead of CSV. The term delimiter-separated values (DSV) is also occasionally used to refer to these formats.

Python has several ways to read and write data in the CSV format, each with different uses and advantages. To begin with, the standard Python library contains a module called `csv` for reading CSV data. To use this module, we can call the `csv.reader` function with a file handle as an argument. It returns a class instance that can be used as an iterator that parses lines from the given CSV file into Python lists of strings. For example, to read the file `playerstats-2013-2014.csv` (shown in Listing 18-1) into a nested list of strings, we can use the following.

```
In [10]: with open("playerstats-2013-2014.csv") as f:
    ...:     csvreader = csv.reader(f)
    ...:     rows = [fields for fields in csvreader]
In [11]: rows[1][1:6]
Out[11]: ['Player', 'Team', 'Pos', 'GP', 'G']
In [12]: rows[2][1:6]
Out[12]: ['Sidney Crosby', 'PIT', 'C', '80', '36']
```

Note that by default, each field in the parsed rows is string-valued, even if the field represents a numerical value, such as 80 (games played) or 36 (goals) in the preceding example. While the `csv` module provides a flexible way of defining custom CSV reader classes, this module is most convenient for reading CSV files with string-valued fields.

Storing and loading arrays with numerical values, such as vectors and matrices, is common in computational work. The NumPy library provides the `np.loadtxt` and `np.savetxt` for this purpose. These functions take several arguments to fine-tune the type of CSV format to read or write: for example, with the `delimiter` argument, we can select which character to use to separate fields, and the `header` and `comments` arguments can be used to specify a header row and comment rows that are prepended to the header, respectively.

As an example, consider saving an array with random numbers and of shape (100, 3) to a file `data.csv` using `np.savetxt`. To give the data some context, we add a header and a comment line to the file, and we explicitly request using the comma character as field delimiter with the `delimiter=","` argument (the default delimiter is the space character).

```
In [13]: data = np.random.randn(100, 3)
In [14]: np.savetxt("data.csv", data, delimiter=",", header="x,y,z",
    ...:            comments="# Random x, y, z coordinates\n")
In [15]: !head -n 5 data.csv
# Random x, y, z coordinates
x,y,z
1.652276634254504772e-01,9.522165919962696234e-01,4.659850998659530452e-01
8.699729536125471174e-01,1.187589118344758443e+00,1.788104702180680405e+00
-8.106725710122602013e-01,2.765616277935758482e-01,4.456864674903074919e-01
```

To read data on this format back into a NumPy array, we can use the `np.loadtxt` function. It takes arguments similar to those of `np.savetxt`: in particular, we again set the `delimiter` argument to `","`, to indicate the fields separated by a comma character. We also need to use the `skiprows` argument to skip over the first two lines in the file (the comment and header line) since they do not contain numerical data.

```
In [16]: data_load = np.loadtxt("data.csv", skiprows=2, delimiter=",")
```

The result is a new NumPy array equivalent to the original one written to the data.csv file using np. savetxt.

```
In [17]: (data == data_load).all()
Out[17]: True
```

Note that in contrast to the CSV reader in the csv module in the Python standard library, by default, the loadtxt function in NumPy converts all fields into numerical values, and the result is a NumPy with numerical dtype (float64).

```
In [18]: data_load[1,:]
Out[18]: array([ 0.86997295,  1.18758912,  1.7881047 ])
In [19]: data_load.dtype
Out[19]: dtype('float64')
```

To read CSV files that contain nonnumerical data using np.loadtxt, such as the playerstats-2013-2014.csv file that we read using the Python standard library in the preceding text, we must explicitly set the data type of the resulting array using the dtype argument. We get an error if we attempt to read a CSV file with nonnumerical values without setting dtype.

```
In [20]: np.loadtxt("playerstats-2013-2014.csv", skiprows=2, delimiter=",")
--------------------------------------------------------------------------
ValueError: could not convert string to float: b'Sidney Crosby'
```

Using dtype=bytes (or str or object), we get a NumPy array with unparsed values.

```
In [21]: data = np.loadtxt("playerstats-2013-2014.csv", skiprows=2, delimiter=",",
   ...:                     dtype=bytes)
In [22]: data[0][1:6]
Out[22]: array([b'Sidney Crosby', b'PIT', b'C', b'80', b'36'], dtype='|S13')
```

Alternatively, if we want to read only columns with numerical types, we can select to read a subset of columns using the usecols argument.

```
In [23]: np.loadtxt("playerstats-2013-2014.csv", skiprows=2, delimiter=",",
   ...:             usecols=[6,7,8])
Out[23]: array([[ 68., 104.,  18.],
               [ 56.,  87.,  28.],
               [ 58.,  86.,   7.],
               [ 47.,  84.,  16.],
               [ 39.,  82.,  32.]])
```

While the NumPy savetxt and loadtxt functions are configurable and flexible CSV writers and readers, they are most convenient for all-numerical data. The Python standard library module csv, on the other hand, is most convenient for CSV files with string-valued data. A third method to read CSV files in Python is to use the Pandas read_csv function. Examples of this function were presented in Chapter 12, where we used it to create Pandas data frames from TSV-formatted data files. The read_csv function in Pandas is very handy when reading CSV files with numerical and string-valued fields. In most cases, it automatically determines which type a field has and converts it accordingly. For example, when reading the playerstats-2013-2014.csv file using read_csv, we obtain a Pandas data frame with all the fields parsed into columns with suitable type.

```
In [24]: df = pd.read_csv("playerstats-2013-2014.csv", skiprows=1)
In [25]: df = df.set_index("Rank")
In [26]: df[["Player", "GP", "G", "A", "P"]]
Out[26]:
```

Rank	Player	GP	G	A	P
1	Sidney Crosby	80	36	68	104
2	Ryan Getzlaf	77	31	56	87
3	Claude Giroux	82	28	58	86
4	Tyler Seguin	80	37	47	84
5	Corey Perry	81	43	39	82

Using the `info` method of the `DataFrame` instance `df`, we can see explicitly which type each column has been converted to (here the output is truncated for brevity).

```
In [27]: df.info()
<class 'pandas.core.frame.DataFrame'>
Int64Index: 5 entries, 1 to 5
Data columns (total 20 columns):
Player      5 non-null object
Team        5 non-null object
Pos         5 non-null object
GP          5 non-null int64
G           5 non-null int64
...
S           5 non-null int64
S%          5 non-null float64
TOI/GP      5 non-null object
Shift/GP    5 non-null float64
FO%         5 non-null float64
dtypes: float64(3), int64(13), object(4)
memory usage: 840.0+ bytes
```

Data frames can also be written to CSV files using the `to_csv` method of the `DataFrame` object.

```
In [28]: df[["Player", "GP", "G", "A", "P"]].to_csv(
    ...:        "playerstats-2013-2014subset.csv")
In [29]: !head -n 5 playerstats-2013-2014-subset.csv
Rank,Player,GP,G,A,P
1,Sidney Crosby,80,36,68,104
2,Ryan Getzlaf,77,31,56,87
3,Claude Giroux,82,28,58,86
4,Tyler Seguin,80,37,47,84
```

The combination of the Python standard library, NumPy, and Pandas, provides a powerful toolbox for both reading and writing CSV files of various flavors. However, although CSV files are convenient and effective for tabular data, the format has obvious shortcomings. For starters, it can only be used to store one- or two-dimensional arrays, and it does not contain metadata that can help interpret the data. Also, it is not very efficient in either storage or reading and writing, and it cannot be used to store more than one array per file, requiring multiple files for multiple arrays even if they are closely related. The use of CSV should, therefore, be limited to simple datasets. The following section discusses the HDF5 file format, which is designed to store numerical data efficiently and overcome all the shortcomings of simple data formats such as CSV and related formats.

HDF5

The Hierarchical Data Format 5 (HDF5) is a format for storing numerical data. It is developed by The HDF Group[2], a nonprofit organization, and is available under the BSD open source license. The HDF5 format, released in 1998, is designed and implemented to handle large datasets efficiently, including support for high-performance parallel I/O. The HDF5 format is, therefore, suitable for distributed high-performance supercomputers and clusters and can be used to store and operate on terabyte-scale datasets or even larger. However, the beauty of HDF5 is that it is equally suitable for small datasets. As such it is a truly versatile format and an invaluable tool for a computational practitioner.

The hierarchical aspect of the format allows for organizing datasets within a file, using a hierarchical structure that resembles a file system. The terminology used for entities in an HDF5 file is *groups* and *datasets*, which correspond to directories and files in the file system analogy. Groups in an HDF5 file can be nested to create a tree structure, hence *hierarchical* in the format's name. A dataset in an HDF5 file is a homogenous array of certain dimensions and elements of a certain type. The HDF5 type system supports all standard basic data types and allows the defining of custom compound data types. Both groups and datasets in an HDF5 file can also have *attributes* which can be used to store metadata about groups and datasets. Attributes can themselves have different types, such as numeric or string-valued.

In addition to the file format itself, The HDF Group also provides a library and a reference implementation of the format. The main library is written in C, and wrappers to its C API are available for many programming languages. The HDF5 library for accessing data from an HDF5 file has sophisticated support for partial read and write operations, which can be used to access a small segment of the entire dataset. This powerful feature enables computations on datasets that are larger than what can fit a computer's memory.[3] The HDF5 format is a mature file format with widespread support on different platforms and computational environments. This also makes HDF5 a suitable choice for long-term storage of data. As a data storage platform, HDF5 provides a solution to several problems: cross-platform storage, efficient I/O, and storage that scales up to very large data files, a metadata system (attributes) that can be used to annotate and describe the groups and datasets in a file to make the data self-describing. Altogether, these features make HDF5 an excellent tool for computational work.

For Python, there are two libraries for using HDF5 files: h5py and PyTables. These two libraries take different approaches to using HDF5, and it is well worth being familiar with them. The h5py library provides an API relatively close to the basic HDF5 concepts, focusing on groups and datasets. It provides a NumPy-inspired API for accessing datasets, making it intuitive for someone familiar with NumPy.

■ **h5py**   The h5py library provides a Pythonic interface to the HDF5 file format and a NumPy-like interface to its datasets. For more information about the project, including its documentation, see its web page at www.h5py.org. At the time of writing, the most recent version of the library is 3.9.0.

---

[2] https://www.hdfgroup.org
[3] This is also known as out-of-core computing. For another recent project that also provides out-of-core computing capabilities in Python, see the dask library (https://dask.pydata.org/en/latest).

The PyTables library provides a higher-level data abstraction based on the HDF5 format, providing database-like features, such as tables with easily customizable data types. It also allows querying datasets as a database and using advanced indexing features.

■ **PyTables**   The PyTables library provides a database-like data model on top of HDF5. For more information about the project and its documentation, see the web page at `www.pytables.org`. At the time of writing, the latest version of PyTables is 3.8.0.

The following two sections explore how the h5py and PyTables libraries can read and write numerical data with HDF5 files.

h5py

Let's begin with a tour of the h5py library. The API for h5py is surprisingly simple and pleasant to work with, yet at the same time full-featured. This is accomplished through the thoughtful use of Pythonic idioms such as dictionary and NumPy's array semantics. A summary of basic objects and methods in the h5py library is shown in Table 18-1. The following explores using these methods through a series of examples.

Table 18-1. *Summary of the Main Objects and Methods in the h5py API*

Object	Method/Attribute	Description
h5py.File	__init__(name, mode, ...)	Open an existing HDF5, or create a new one, with filename name. Depending on the value of the mode argument, the file can be opened in read-only or read-write mode (see main text).
	flush()	Write buffers to file.
	close()	Close an open HDF5 file.
h5py.File, h5py.Group	create_group(name)	Create a new group with name name (can be a path) within the current group.
	create_dataset(name, data=..., shape=..., dtype=..., ...)	Create a new dataset.
	[] dictionary syntax	Access items (groups and datasets) within a group.
h5py.Dataset	dtype	Data type.
	shape	Shape (dimensions) of the dataset.
	value	The full array of the underlying data of the dataset.
	[] array syntax	Access elements or subsets of the data in a dataset.
h5py.File, h5py.Group, h5py.Dataset	name	Name (path) of the object in the HDF5 file hierarchy.
	attrs	Dictionary-like attribute access.

Files

Let's begin by looking at how to open existing and create new HDF5 files using the h5py.File object. The initializer for this object only takes a filename as a required argument. But we typically also need to specify the mode argument, with which we can choose to open a file in read-only or read-write mode and if a file should be truncated or not when opened. The mode argument takes string values similar to the built-in Python function open: "r" is used for read-only (file must exist), "r+" for read-write (file must exist), "w" for creating a new file (truncate if file exists), "w-" for creating a new file (error if file exists), and "a" for read-write (if file exists, otherwise create). To create a new file in read-write mode, we can therefore use the following.

```
In [30]: f = h5py.File("data.h5", mode="w")
```

Here, the result is a file handle assigned to the variable f, which we can use to access and add content to the file. Given a file handle, we can see which mode it is opened in using the mode attribute.

```
In [31]: f.mode
Out[31]: 'r+'
```

Note that even though we opened the file in mode "w", once the file has been opened it is either read-only ("r") or read-write ("r+"). Other file-level operations that can be performed using the HDF5 file object are flushing buffers containing data that has not yet been written to the file using the flush method and closing the file using the close method.

```
In [32]: f.flush()
In [33]: f.close()
```

Groups

While representing an HDF5 file handle, the File object also represents the HDF5 group object known as the *root group*. The name of a group is accessible through the name attribute of the group object. The name takes the form of a path, similar to a path in a file system, which specifies where the group is stored in the hierarchical structure of the file. The name of the root group is "/".

```
In [34]: f = h5py.File("data.h5", "w")
In [35]: f.name
Out[35]: '/'
```

A group object has the create_group method for creating a new group within an existing group. A new group created with this method becomes a subgroup of the group instance for which the create_group method is invoked.

```
In [36]: grp1 = f.create_group("experiment1")
In [37]: grp1.name
Out[37]: '/experiment1'
```

Here, the group experiment1 is a subgroup of the root group, and its name and path in the hierarchical structure is, therefore, /experiment1. When creating a new group, its immediate parent group does not necessarily have to exist beforehand. For example, to create a new group /experiment2/measurement, we can directly use the create_group method of the root group *without* first explicitly creating the experiment2 group. Intermediate groups are created automatically.

```
In [38]: grp2_meas = f.create_group("experiment2/measurement")
In [39]: grp2_meas.name
Out[39]: '/experiment2/measurement'
In [40]: grp2_sim = f.create_group("experiment2/simulation")
In [41]: grp2_sim.name
Out[41]: '/experiment2/simulation'
```

The group hierarchy of an HDF5 file can be explored using a dictionary-style interface. To retrieve a group with a given path name, we can perform a dictionary-like lookup from one of its ancestor groups (typically the root node).

```
In [42]: f["/experiment1"]
Out[42]: <HDF5 group "/experiment1" (0 members)>
In [43]: f["/experiment2/simulation"]
Out[43]: <HDF5 group "/experiment2/simulation" (0 members)>
```

The same type of dictionary lookup works for subgroups too (not only the root node).

```
In [44]: grp_experiment2 = f["/experiment2"]
In [45]: grp_experiment2['simulation']
Out[45]: <HDF5 group "/experiment2/simulation" (0 members)>
```

The keys method returns an iterator over the names of subgroups and datasets within a group, and the items method returns an iterator over (name, value) tuples for each entity in the group. These can be used to traverse the hierarchy of groups programmatically.

```
In [46]: list(f.keys())
Out[46]: ['experiment1', 'experiment2']
In [47]: list(f.items())
Out[47]: [('experiment1', <HDF5 group "/experiment1" (0 members)>),
          ('experiment2', <HDF5 group "/experiment2" (2 members)>)]
```

To traverse the hierarchy of groups in an HDF5 file, we can also use the visit method, which takes a function as an argument and calls that function with the name for each entity in the file hierarchy.

```
In [48]: f.visit(lambda x: print(x))
experiment1
experiment2
experiment2/measurement
experiment2/simulation
```

The visititems method does the same thing except that it calls the function with both the item name and the item itself as arguments.

```
In [49]: f.visititems(lambda name, item: print(name, item))
experiment1 <HDF5 group "/experiment1" (0 members)>
experiment2 <HDF5 group "/experiment2" (2 members)>
experiment2/experiment <HDF5 group "/experiment2/measurement" (0 members)>
experiment2/simulation <HDF5 group "/experiment2/simulation" (0 members)>
```

In keeping with the semantics of Python dictionaries, we can also operate on Group objects using the set membership testing with the Python keyword in.

```
In [50]: "experiment1" in f
Out[50]: True
In [51]: "simulation" in f["experiment2"]
Out[51]: True
In [52]: "experiment3" in f
Out[52]: False
```

Using the visit and visititems methods, together with the dictionary-style methods keys and items, we can easily explore the structure and content of an HDF5 file, even if we have yet to have prior information on what it contains and how the data is organized within it. The ability to conveniently explore HDF5 is an important aspect of the usability of the format. There are also external non-Python tools for exploring the content of HDF5 files that often are useful when working with this type of files. The h5ls command-line tool is particularly handy for quickly inspecting the content of an HDF5 file.

```
In [53]: f.flush()
In [54]: !h5ls -r data.h5
/                          Group
/experiment1               Group
/experiment2               Group
/experiment2/measurement   Group
/experiment2/simulation    Group
```

Here, we used the -r flag to the h5ls program to recursively show all items in the file. The h5ls program is part of a series of HDF5 utility programs provided by a package called hdf5-tools (see also h5stat, h5copy, h5diff, etc.). Even though these are not Python tools, they are useful when working with HDF5 files in general, also from within Python.

Datasets

Now that we have explored how to create and access groups within an HDF5 file, it is time to look at how to store datasets. Storing numerical data is, after all, the main purpose of the HDF5 format. There are two main methods to create a dataset in an HDF5 file using h5py. The easiest way to create a dataset is to assign a NumPy array to an item within an HDF5 group using the dictionary index syntax. The second method is to create an empty dataset using the create_dataset method, as shown in examples later in this section.

For example, we can use the following to store two NumPy arrays, array1 and meas1, into the root group and the experiment2/measurement groups, respectively.

```
In [55]: array1 = np.arange(10)
In [56]: meas1 = np.random.randn(100, 100)
In [57]: f["array1"] = array1
In [58]: f["/experiment2/measurement/meas1"] = meas1
```

To verify that the datasets for the assigned NumPy arrays were added to the file, let's traverse through the file hierarchy using the visititems method.

```
In [59]: f.visititems(lambda name, value: print(name, value))
array1 <HDF5 dataset "array1": shape (10,), type "<i8">
experiment1 <HDF5 group "/experiment1" (0 members)>
```

```
experiment2 <HDF5 group "/experiment2" (2 members)>
experiment2/measurement <HDF5 group "/experiment2/measurement" (1 members)>
experiment2/measurement/meas1 <HDF5 dataset "meas1": shape (100, 100), type "<f8">
experiment2/simulation <HDF5 group "/experiment2/simulation" (0 members)>
```

The array1 and meas1 datasets are now added to the file. Note that the paths used as dictionary keys in the assignments determine the locations of the datasets within the file. To retrieve a dataset, we can use the same dictionary-like syntax used to retrieve a group. For example, to retrieve the array1 dataset, which is stored in the root group, we can use f["array1"].

```
In [60]: ds = f["array1"]
In [61]: ds
Out[61]: <HDF5 dataset "array1": shape (10,), type "<i8">
```

The result is a Dataset object, not a NumPy array like the one we assigned to the array1 item. The Dataset object is a proxy for the underlying data within the HDF5. Like a NumPy array, a Dataset object has several attributes that describe the dataset, including name, dtype, and shape. It also has the len method that returns the length of the dataset.

```
In [62]: ds.name
Out[62]: '/array1'
In [63]: ds.dtype
Out[63]: dtype('int64')
In [64]: ds.shape
Out[64]: (10,)
In [65]: ds.len()
Out[65]: 10
```

The actual data for the dataset can be extracted by casting the dataset back to a NumPy array using np.array. This returns the entire dataset as a NumPy array, which is equivalent to the array we assigned to the array1 dataset.

```
In [66]: np.array(ds)
Out[66]: array([0, 1, 2, 3, 4, 5, 6, 7, 8, 9])
```

We can use the following to access a dataset deeper down the group hierarchy, we can use a file system-like path name. For example, to retrieve the meas1 dataset in the group experiment2/measurement.

```
In [67]: ds = f["experiment2/measurement/meas1"]
In [68]: ds
Out[68]: <HDF5 dataset "meas1": shape (100, 100), type "<f8">
```

Again, we get a Dataset object whose basic properties can be inspected using the object attributes we introduced earlier.

```
In [69]: ds.dtype
Out[69]: dtype('float64')
In [70]: ds.shape
Out[70]: (100, 100)
```

Note that the data type of this dataset is float64, while for the dataset array1, the data type is int64. This type-information was derived from the original NumPy arrays assigned to the two datasets. We could again use the np.array to retrieve the data as a NumPy array. An alternative syntax for the same operation is bracket indexing with the ellipsis notation: ds[...].

```
In [71]: data_full = ds[...]
In [72]: type(data_full)
Out[72]: numpy.ndarray
In [73]: data_full.shape
Out[73]: (100, 100)
```

This is an example of NumPy-like array indexing. The Dataset object supports most indexing and slicing types used in NumPy, providing a powerful and flexible method for partially reading data from a file. For example, we can use the following to retrieve only the first column from the meas1 dataset.

```
In [74]: data_col = ds[:, 0]
In [75]: data_col.shape
Out[75]: (100,)
```

The result is a 100-element array corresponding to the first column in the dataset. Note that this slicing is performed within the HDF5 library, not in NumPy. In this example, only 100 elements were read from the file and stored in the resulting NumPy array, without fully loading the dataset into memory. This is an important feature when working with large datasets that do not fit in memory.

For example, the Dataset object also supports stride indexing.

```
In [76]: ds[10:20:3, 10:20:3] # 3 stride
Out[76]: array([[-0.22321057, -0.61989199,  0.78215645,  0.73774187],
                [-1.03331515,  2.54190817, -0.24812478, -2.49677693],
                [ 0.17010011,  1.88589248,  1.91401249, -0.63430569],
                [ 0.4600099 , -1.3242449 ,  0.41821078,  1.47514922]])
```

as well as "fancy indexing", where a list of indices is given for one of the array's dimensions (does not work for more than one index).

```
In [77]: ds[[1,2,3], :].shape
Out[77]: (3, 100)
```

We can also use Boolean indexing, where a Boolean-valued NumPy array is used to index a Dataset. For example, to single out the first five columns (index :5 on the second axis) for each row whose value in the first column (ds[:, 0]) is larger than 2, we can index the dataset with the Boolean mask ds[:, 0] > 2.

```
In [78]: mask = ds[:, 0] > 2
In [79]: mask.shape, mask.dtype
Out[79]: ((100,), dtype('bool'))
In [80]: ds[mask, :5]
Out[80]: array([[2.1224865 ,  0.70447132, -1.71659513,  1.43759445, -0.61080907],
                [2.11780508, -0.2100993 ,  1.06262836, -0.46637199,  0.02769476],
                [2.41192679, -0.30818179, -0.31518842, -1.78274309, -0.80931757],
                [2.10030227,  0.14629889,  0.78511191, -0.19338282,  0.28372485]])
```

Since the `Dataset` object uses NumPy's indexing and slicing syntax to select subsets of the underlying data, working with large HDF5 datasets in Python using h5py comes naturally to someone familiar with NumPy. Also, remember that for large files, there is a big difference in index slicing on the Dataset object rather than on the NumPy array that can be accessed by casting the dataset using the `np.array` function since the former avoids loading the entire dataset into memory.

We have covered how to create datasets in an HDF5 file by explicitly assigning data to an item in a group object. We can also create datasets explicitly using the `create_dataset` method. It takes the name of the new dataset as the first argument, and we can either set the data for the new dataset using the data argument or create an empty array by setting the shape argument. For example, instead of the assignment `f["array2"] = np.random.randint(10, size=10)`, we can also use the `create_dataset` method.

```
In [81]: ds = f.create_dataset("array2", data=np.random.randint(10, size=10))
In [82]: ds
Out[82]: <HDF5 dataset "array2": shape (10,), type "<i8">
In [83]: np.array(value)
Out[83]: array([2, 2, 3, 3, 6, 6, 4, 8, 0, 0])
```

When explicitly calling the `create_dataset` method, we have a finer level of control of the properties of the resulting dataset. For example, we can explicitly set the data type for the dataset using the dtype argument. We can choose a compression method using the `compress` argument, specifying the chunk size using the chunks argument and setting the maximum allowed array size for resizable datasets using the maxsize argument. There are also many other advanced features related to the Dataset object. See the docstring for `create_dataset` for details.

When creating an empty array by specifying the shape argument instead of providing an array for initializing a dataset, we can also use the fillvalue argument to set the default value for the dataset. For example, we can use the following to create an empty dataset of shape (5, 5) and default value –1.

```
In [84]: ds = f.create_dataset("/experiment2/simulation/data1", shape=(5, 5),
    ...:                          fillvalue=-1)
In [85]: ds
Out[85]: <HDF5 dataset "data1": shape (5, 5), type "<f4">
In [86]: np.array(value)
Out[86]: array([[-1., -1., -1., -1., -1.],
               [-1., -1., -1., -1., -1.],
               [-1., -1., -1., -1., -1.],
               [-1., -1., -1., -1., -1.],
               [-1., -1., -1., -1., -1.]], dtype=float32)
```

HDF5 is clever about disk usage for an empty dataset and does not store more data than necessary, particularly if we select a compression method using the `compression` argument. There are several compression methods available, for example, `'gzip'`. Using dataset compression, we can create a very large dataset and gradually fill it with data, for example, when measurement results or results of computations become available, without initially wasting a lot of storage space. For example, let's create a large dataset with shape (5000, 5000, 5000) with the data1 in the group experiment1/simulation.

```
In [87]: ds = f.create_dataset("/experiment1/simulation/data1",
    ...:                          shape=(5000, 5000, 5000), fillvalue=0,
    ...:                          compression='gzip')
In [88]: ds
Out[88]: <HDF5 dataset "data1": shape (5000, 5000, 5000), type "<f4">
```

To begin, this dataset uses neither memory nor disk space until we start filling it with data. To assign values to the dataset, we can again use the NumPy-like indexing syntax and assign values to specific elements in the dataset or to subsets selected using slicing syntax.

```
In [89]: ds[:, 0, 0] = np.random.rand(5000)
In [90]: ds[1, :, 0] += np.random.rand(5000)
In [91]: ds[:2, :5, 0]
Out[91]: array([[ 0.67240328, 0.        , 0.        , 0.        , 0.        ],
                [0.99613971, 0.48227152, 0.48904559, 0.78807044, 0.62100351]],
                dtype=float32)
```

Note that the elements that have not been assigned values are set to the fillvalue value specified when the array was created. If we do not know what fill value a dataset has, we can find out by looking at the fillvalue attribute of the Dataset object.

```
In [92]: ds.fi2llvalue
Out[92]: 0.0
```

To see that the newly created dataset is indeed stored in the group where we intended to assign it, we can again use the visititems method to list the content of the experiment1 group.

```
In [93]: f["experiment1"].visititems(lambda name, value: print(name, value))
simulation <HDF5 group "/experiment1/simulation" (1 members)>
simulation/data1 <HDF5 dataset "data1": shape (5000, 5000, 5000), type "<f4">
```

Although the dataset experiment1/simulation/data1 is very large (4×5000^3 bytes ~ 465 Gb), since we have not yet filled it with much data, the HDF5 file still does not take a lot of disk space (only about 357 Kb).

```
In [94]: f.flush()
In [95]: f.filename
Out[95]: 'data.h5'
In [96]: !ls -lh data.h5
-rw-r--r--@ 1 rob  staff   357K Apr  5 18:48 data.h5
```

We have seen how to create groups and datasets within an HDF5 file. It is, of course, sometimes also necessary to delete items from a file. With h5py, we can delete items from a group using the Python del keyword, again complying with the semantics of Python dictionaries.

```
In [97]: del f["/experiment1/simulation/data1"]
In [98]: f["experiment1"].visititems(lambda name, value: print(name, value))
simulation <HDF5 group "/experiment1/simulation" (0 members)>
```

Attributes

Attributes are a component of the HDF5 format, making it an excellent format for annotating data and providing self-describing data through metadata. For example, external parameters and conditions should often be recorded together with the observed data when storing experimental data. Likewise, in a computer simulation, it is usually necessary to store additional model or simulation parameters together with the generated simulation results. In all these cases, the best solution is to ensure the required additional information is stored as metadata with the primary datasets.

The HDF5 format supports this type of metadata through attributes. An arbitrary number of attributes can be attached to each group and dataset within an HDF5 file. With the h5py library, attributes are accessed using a dictionary-like interface, just like groups are. The Python attribute attrs of Group and Dataset objects are used to access the HDF5 attributes.

```
In [99]: f.attrs
Out[99]: <Attributes of HDF5 object at 4462179384>
```

To create an attribute, we assign the value to the attrs dictionary for the target object. For example, we can use the following to create a description attribute for the root group.

```
In [100]: f.attrs["description"] = "Result sets for experiments and simulations"
```

Similarly, to add date attributes to the experiment1 and experiment2 groups.

```
In [101]: f["experiment1"].attrs["date"] = "2015-1-1"
In [102]: f["experiment2"].attrs["date"] = "2015-1-2"
```

We can also add attributes directly to datasets (not only groups).

```
In [103]: f["experiment2/simulation/data1"].attrs["k"] = 1.5
In [104]: f["experiment2/simulation/data1"].attrs["T"] = 1000
```

Like for groups, we can use the keys and items methods of the Attribute object to retrieve iterators over the attributes it contains.

```
In [105]: list(f["experiment1"].attrs.keys())
Out[105]: ['date']
In [106]: list(f["experiment2/simulation/data1"].attrs.items())
Out[106]: [('k', 1.5), ('T', 1000)]
```

The existence of an attribute can be tested with the Python in operator in keeping with the Python dictionary semantics.

```
In [107]: "T" in f["experiment2/simulation/data1"].attrs
Out[107]: True
```

To delete existing attributes, we can use the del keyword.

```
In [108]: del f["experiment2/simulation/data1"].attrs["T"]
In [109]: "T" in f["experiment2/simulation"].attrs
Out[109]: False
```

The attributes of HDF5 groups and datasets are suitable for storing metadata together with the actual datasets. Using attributes generously can help to provide context to the data, which often must be available for the data to be useful.

PyTables

The PyTables library offers an alternative interface to HDF5 for Python. This library focuses on a higher-level table-based data model implemented using the HDF5 format. However, PyTables can also be used to create and read generic HDF5 groups and datasets, like the h5py library. Here, we use the table data model, which complements the h5py library discussed in the previous section. To demonstrate the use of PyTables table objects, let's use the NHL player statistics dataset to construct a PyTables table from a Pandas data frame. We begin with reading the dataset into a `DataFrame` object using the `read_csv` function.

```
In [110]: df = pd.read_csv("playerstats-2013-2014.csv", skiprows=1)
     ...: df = df.set_index("Rank")
```

Next, we create a new PyTables HDF5 file handle using the `tables.open_file` function[4]. This function takes a filename as the first argument and the file mode as an optional second argument. The result is a PyTables HDF5 file handle (here assigned to the variable f).

```
In [111]: f = tables.open_file("playerstats-2013-2014.h5", mode="w")
```

Like the h5py library, we can create HDF5 groups with the `create_group` method of the file handle object. It takes the path to the parent group as the first argument, the group name as the second argument, and the `title` argument, with which a descriptive HDF5 attribute can be set on the group.

```
In [112]: grp = f.create_group("/", "season_2013_2014",
     ...:                       title="NHL player statistics for the 2013/2014 season")
In [113]: grp
Out[113]: /season_2013_2014 (Group) 'NHL player statistics for the 2013/2014 season'
          children := []
```

Unlike the h5py library, the file handle object in PyTables does not represent the root group in the HDF5 file. To access the root node, we must use the `root` attribute of the file handle object.

```
In [114]: f.root
Out[114]: / (RootGroup) "
          children := ['season_2013_2014' (Group)]
```

A nice feature of the PyTables library is that it is easy to create tables with mixed column types, using the struct-like compound data type of HDF5. The simplest way to define such a table data structure with PyTables is to create a class that inherits from the `tables.IsDescription` class. It should contain fields composed of data-type representations from the `tables` library. For example, we can use the following to create a specification of the table structure for the player statistics dataset.

```
In [115]: class PlayerStat(tables.IsDescription):
     ...:     player = tables.StringCol(20, dflt="")
     ...:     position = tables.StringCol(1, dflt="C")
     ...:     games_played = tables.UInt8Col(dflt=0)
     ...:     points = tables.UInt16Col(dflt=0)
     ...:     goals = tables.UInt16Col(dflt=0)

     ...:     assists = tables.UInt16Col(dflt=0)
```

---

[4] Note that the Python module provided by the PyTables library is named `tables`. Therefore, `tables.open_file` refers to `open_file` function in the `tables` module provided by the PyTables library.

```
...:        shooting_percentage = tables.Float64Col(dflt=0.0)
...:        shifts_per_game_played = tables.Float64Col(dflt=0.0)
```

Here, the PlayerStat class represents the table structure of a table with eight columns, where the first two columns are fixed-length strings (tables.StringCol), where the following four columns are unsigned integers (tables.UInt8Col and tables.UInt16Col, of 8- and 16-bit size), and where the last two columns have floating-point type (tables.Float64Col). The optional dflt argument to data-type objects specifies the fields' default value. Once the table structure is defined using a class on this form, we can create the actual table in the HDF5 file using the create_table method. It takes a group object or the path to the parent node as the first argument, the table name as the second argument, the table specification class as the third argument, and optionally, a table title as the fourth argument (stored as an HDF5 attribute for the corresponding dataset).

```
In [116]: top30_table = f.create_table(grp, 'top30', PlayerStat,
     ...:                              "Top 30 point leaders")
```

To insert data into the table, we can use the row attribute of the table object to retrieve a Row accessor class that can be used as a dictionary to populate the row with values. When the row object is fully initialized, we can use the append method to insert the row into the table.

```
In [117]: playerstat = top30_table.row
In [118]: for index, row_series in df.iterrows():
     ...:        playerstat["player"] = row_series["Player"]
     ...:        playerstat["position"] = row_series["Pos"]
     ...:        playerstat["games_played"] = row_series["GP"]
     ...:        playerstat["points"] = row_series["P"]
     ...:        playerstat["goals"] = row_series["G"]
     ...:        playerstat["assists"] = row_series["A"]
     ...:        playerstat["shooting_percentage"] = row_series["S%"]
     ...:        playerstat["shifts_per_game_played"] = row_series["Shift/GP"]
     ...:        playerstat.append()
```

The flush method forces a write of the table data to the file.

```
In [119]: top30_table.flush()
```

To access data from the table, we can use the cols attribute to retrieve columns as NumPy arrays.

```
In [120]: top30_table.cols.player[:5]
Out[120]: array([b'Sidney Crosby', b'Ryan Getzlaf', b'Claude Giroux',
                  b'Tyler Seguin', b'Corey Perry'], dtype='|S20')
In [121]: top30_table.cols.points[:5]
Out[121]: array([104,  87,  86,  84,  82], dtype=uint16)
```

To access data row-wise, we can use the iterrows method to create an iterator over all the rows in the table. Here, we use this approach to loop through all the rows and print them to the standard output (here, the output is truncated for brevity).

```
In [122]: def print_playerstat(row):
     ...:        print("%20s\t%s\t%s\t%s" %
     ...:               (row["player"].decode('UTF-8'), row["points"],
```

```
    ...:                  row["goals"], row["assists"]))
In [123]: for row in top30_table.iterrows():
    ...:            print_playerstat(row)
  Sidney Crosby      104    36    68
Ryan Getzlaf          87    31    56
Claude Giroux         86    28    58
Tyler Seguin          84    37    47
...
Jaromir Jagr          67    24    43
John Tavares          66    24    42
Jason Spezza          66    23    43
Jordan Eberle         65    28    37
```

One of the most powerful features of the PyTables table interface is the ability to extract rows from the underlying HDF5 using queries selectively. For example, the where method allows us to pass an expression in terms of the table columns as a string that PyTables uses to filter rows.

```
In [124]: for row in top30_table.where("(points > 75) & (points <= 80)"):
    ...:            print_playerstat(row)
Phil Kessel          80    37    43
Taylor Hall          80    27    53
Alex Ovechkin        79    51    28
Joe Pavelski         79    41    38
Jamie Benn           79    34    45
Nicklas Backstrom    79    18    61
Patrick Sharp        78    34    44
Joe Thornton         76    11    65
```

With the where method, we can also define conditions in terms of multiple columns.

```
In [125]: for row in top30_table.where("(goals > 40) & (points < 80)"):
    ...:            print_playerstat(row)
Alex Ovechkin        79    51    28
Joe Pavelski         79    41    38
```

This feature allows us to query a table in a database-like fashion. Although for a small dataset, like the current one, we could just as well perform these kinds of operations directly in memory using a Pandas data frame, but remember that HDF5 files are stored on disk. The efficient use of I/O in the PyTables library enables us to work with very large datasets that do not fit in memory, which would prevent us from using, for example, NumPy or Pandas on the entire dataset.

Before we conclude this section, let's inspect the structure of the resulting HDF5 file that contains the PyTables table we have just created.

```
In [126]: f
Out[126]: File(filename=playerstats-2013-2014.h5, title='', mode='w', root_uep='/',
filters=Filters(complevel=0, shuffle=False, fletcher32=False, least_significant_digit=None))
        / (RootGroup) '' /season_2013_2014 (Group) 'NHL player stats for the
        2013/2014 season'
        /season_2013_2014/top30 (Table(30,)) 'Top 30 point leaders'
            description := {
            "assists": UInt16Col(shape=(), dflt=0, pos=0),
```

```
                    "games_played": UInt8Col(shape=(), dflt=0, pos=1),
                    "goals": UInt16Col(shape=(), dflt=0, pos=2),
                    "player": StringCol(itemsize=20, shape=(), dflt=b", pos=3),
                    "points": UInt16Col(shape=(), dflt=0, pos=4),
                    "position": StringCol(itemsize=1, shape=(), dflt=b'C', pos=5),
                    "shifts_per_game_played": Float64Col(shape=(), dflt=0.0, pos=6),
                    "shooting_percentage": Float64Col(shape=(), dflt=0.0, pos=7)}
          byteorder := 'little'
          chunkshape := (1489,)
```

From the string representation of the PyTables file handle and the HDF5 file hierarchy that it contains, we can see that the PyTables library has created a dataset /season_2013_2014/top30 that uses an involved compound data type that was created according to the specification in the PlayerStat object that we created earlier. Finally, when we are finished modifying a dataset in a file, we can flush its buffers and force a write to the file using the flush method, and when we are finished working with a file, we can close it using the close method.

```
In [127]: f.flush()
In [128]: f.close()
```

Although we do not cover other types of datasets here, such as regular homogenous arrays, it is worth mentioning that the PyTables library also supports these types of data structures. For example, we can use the create_array, create_carray, and create_earray to construct fixed-size arrays, chunked arrays, and enlargeable arrays, respectively. For more information on how to use these data structures, see the corresponding docstrings.

Pandas HDFStore

A third way to store data in HDF5 files using Python is to use the HDFStore object in Pandas. It can persistently store data frames and other Pandas objects in an HDF5 file. To use this feature in Pandas, the PyTables library must be installed. We can create an HDFStore object by passing a filename to its initializer. The result is an HDFStore object that can be used as a dictionary to which we can assign Pandas DataFrame instances to have them stored in the HDF5 file.

```
In [129]: store = pd.HDFStore('store.h5')
In [130]: df = pd.DataFrame(np.random.rand(5,5))
In [131]: store["df1"] = df
In [132]: df = pd.read_csv("playerstats-2013-2014-top30.csv", skiprows=1)
In [133]: store["df2"] = df
```

The HDFStore object behaves as a regular Python dictionary, and we can, for example, see what objects it contains by calling the keys method.

```
In [134]: store.keys()
Out[134]: ['/df1', '/df2']
```

We can test for the existence of an object with a given key using the Python in keyword.

```
In [135]: 'df2' in store
Out[135]: True
```

To retrieve an object from the store, we again use the dictionary-like semantic and index the object with its corresponding key.

```
In [136]: df = store["df1"]
```

From the HDFStore object, we can also access the underlying HDF5 handle using the root attribute. This is nothing more than a PyTables file handle.

```
In [137]: store.root
Out[137]: / (RootGroup) "   children := ['df1' (Group), 'df2' (Group)]
```

Once we are finished working with an HDFStore object, we should close it using the close method to ensure that all associated data is written to the file.

```
In [138]: store.close()
```

Since HDF5 is a standard file format, nothing prevents us from opening an HDF5 file created with Pandas HDFStore or PyTables with any other HDF5 compatible software, such as the h5py library. If we open the file produced with HDFStore with h5py, we can easily inspect its content and see how the HDFStore object arranges the data of the DataFrame objects assigned to it.

```
In [139]: f = h5py.File("store.h5")
In [140]: f.visititems(lambda x, y: print(x, "\t" * int(3 - len(str(x))//8), y))
df1                 <HDF5 group "/df1" (4 members)>
df1/axis0           <HDF5 dataset "axis0": shape (5,), type "<i8">
df1/axis1           <HDF5 dataset "axis1": shape (5,), type "<i8">
df1/block0_items    <HDF5 dataset "block0_items": shape (5,), type "<i8">
df1/block0_values   <HDF5 dataset "block0_values": shape (5, 5), type "<f8">
df2                 <HDF5 group "/df2" (8 members)>
df2/axis0           <HDF5 dataset "axis0": shape (21,), type "|S8">
df2/axis1           <HDF5 dataset "axis1": shape (30,), type "<i8">
df2/block0_items    <HDF5 dataset "block0_items": shape (3,), type "|S8">
df2/block0_values   <HDF5 dataset "block0_values": shape (30, 3), type "<f8">
df2/block1_items    <HDF5 dataset "block1_items": shape (14,), type "|S4">
df2/block1_values   <HDF5 dataset "block1_values": shape (30, 14), type "<i8">
df2/block2_items    <HDF5 dataset "block2_items": shape (4,), type "|S6">
df2/block2_values   <HDF5 dataset "block2_values": shape (1,), type "|O8">
```

We can see that the HDFStore object stores each DataFrame object in a group of its own and has split each data frame into several heterogeneous HDF5 datasets (blocks) where the columns are grouped by their data type. Furthermore, the column names and values are stored in separate HDF5 datasets.

```
In [141]: f["/df2/block0_items"][:]
Out[141]: array([b'S%', b'Shift/GP', b'F0%'], dtype='|S8')
In [142]: f["/df2/block0_values"][:3]
Out[142]: array([[ 13.9,  24. ,  52.5],
                 [ 15.2,  25.2,  49. ],
                 [ 12.6,  25.1,  52.9]])
In [143]: f["/df2/block1_values"][:3, :5]
Out[143]: array([[  1,  80,  36,  68, 104],
                 [  2,  77,  31,  56,  87],
                 [  3,  82,  28,  58,  86]])
```

Parquet

The Parquet columnar storage format from Apache[5] is another important format for numerical data that has gained popularity recently, especially in the big data and cloud storage industry. The emphasis of Parquet is efficient storage and retrieval for large datasets, including efficient compression and a design that allows parallel access from multiple processes. While HDF5 takes an ambitious all-inclusive approach, with a hierarchy structure for data within the file, metadata, and many other advanced functions, the approach of Parquet is a simple storage model that prioritizes scalability over complex features. A parquet dataset has a fixed schema for the columns and their types, and individual columns from the dataset can be efficiently retrieved individually or as a group of selected columns. The data is stored on disk in a folder-and-file structure based on partition columns in the dataset, which makes partial reading, appending, and purging data very efficient.

Let's start by reading the temperature dataset used in earlier chapters to a Pandas data frame. Once the data frame is created, we assign a new column dt to the date part of the measurement date and time column time to use this as a partition column when writing the dataset in Parquet format.

```
In [144]: df = pd.read_csv('temperature_outdoor_2014.tsv',
     ...:                     delimiter="\t", names=["time", "temperature"])
In [145]: df.time = pd.to_datetime(
     ...:          df.time.values, unit="s"
     ...: ).tz_localize('UTC').tz_convert('Europe/Stockholm')
In [146]: df["dt"] = df.time.dt.strftime("%Y-%m-%d")
```

Pandas have very convenient built-in functionality for reading and writing datasets in the Parquet format. The to_parquet method of the DataFrame object can be used for this purpose. It takes the path to the dataset as the first argument, and the partition_cols keyword argument can specify a set of columns to be used as the dataset's partitions. The dataset is split into parts according to the partition-column values, and each partition is written to a separate file inside the parquet dataset path directory. We can also use the index keyword argument to indicate whether the index of the Pandas data frame should be included in the output. Here, we set the partitions_cols argument to ["dt"] so that the resulting Parquet dataset contains a partition for each day, and data collected on different days is stored in separate files.

```
In [147]: df.to_parquet("temperature_outdoor_2014.parquet", index=False,
     ...:                 partition_cols=["dt"])
```

The result is a directory with the name of the dataset path, and inside the directory is a subdirectory for each partition.

```
In [148]: !ls temperature_outdoor_2014.parquet | head -n 5
dt=2014-01-01
dt=2014-01-02
dt=2014-01-03
dt=2014-01-04
dt=2014-01-05
```

The following shows that inside one of the partition directories is the data file.

```
In [149]: !ls temperature_outdoor_2014.parquet/dt=2014-01-01
1dbea7dc6ae549f399053d3852fb67a6-0.parquet
```

---

[5] See more about the Apache Parquet project at http://parquet.apache.org.

In contrast, if we omit the `partitions_cols` keyword argument, then all data is written directory to a single file.

```
In [150]: df.to_parquet("temperature_outdoor_2014_no_partitions.parquet",
     ...:                     index=None)
In [151]: !ls temperature_outdoor_2014_no_partitions.parquet
temperature_outdoor_2014_no_partitions.parquet
```

We can also read a Parquet dataset from files through Pandas using the `pd.read_parquet` function. To demonstrate that we can selectively read only part of the dataset by selecting a partition, we select to read a specific date partition's dataset back into a Pandas data frame by adding dt=2014-04-01 to the dataset path.

```
In [152]: df_20140401 = pd.read_parquet(
     ...:       "temperature_outdoor_2014.parquet/dt=2014-04-01")
```

The result is a data frame that contains only measurements from 2014-04-01, and this was efficiently loaded from disk by only reading data relevant to this partition.

```
In [153]: df_20140401.head()
Out[153]:
```

	time	temperature
0	2014-04-01 00:00:45+02:00	2.62
1	2014-04-01 00:10:45+02:00	2.62
2	2014-04-01 00:20:46+02:00	2.62
3	2014-04-01 00:30:46+02:00	2.50
4	2014-04-01 00:40:47+02:00	2.38

Data partitioning is very important for scalability when working with large datasets, and the explicit partition functionality in Parquet makes it a storage model well-suited for large datasets.

We can also read Parquet datasets with the `pyarrow` library's `parquet` module, which we import as `pq` here. The `read_table` function allows us to read a Parquet dataset into a table object. This function takes an optional `columns` keyword argument, which we can use to select only the columns we need to load.

```
In [154]: table = pq.read_table('temperature_outdoor_2014.parquet',
     ...:                     columns=["time", "temperature"])
In [155]: type(table)
Out[155]: pyarrow.lib.Table
```

The pyarrow table object has many methods and functions. For the `to_pandas` method, let's convert the table to a Pandas data frame for convenient subsequent data processing. Note that here, since we assigned the `columns` keyword argument to `["time", "temperature"]`, we only have these two columns in the resulting table and Pandas data frame objects, despite the Parquet dataset also containing the `dt` column in this example.

```
In [156]: df_table = table.to_pandas()
In [157]: df_table.head()
Out[157]:
```

	time	temperature
0	2014-01-01 00:03:06+01:00	4.38
1	2014-01-01 00:13:06+01:00	4.25
2	2014-01-01 00:23:07+01:00	4.19
3	2014-01-01 00:33:07+01:00	4.06
4	2014-01-01 00:43:08+01:00	4.06

With the Parquet data format, reading data only for selected partitions and columns is handled efficiently without reading the whole dataset to memory once. This becomes especially important for large datasets, but Parquet is a convenient and high-performance data storage format for structured tabular data of any size and is an excellent tool for computational practitioners.

JSON

The JSON[6] (JavaScript Object Notation) is a human-readable, lightweight, plain-text format suitable for storing datasets made up of lists and dictionaries. The values of such lists and dictionaries can be lists or dictionaries or must be of the following basic data types: string, integer, float, and Boolean, or the value null (like the None value in Python). This data model allows the storage of complex and versatile datasets without structural limitations, such as the tabular form required by formats such as CSV. A JSON document can, for example, be used as a key-value store, where the values for different keys can have different structures and data types.

The JSON format was primarily designed as a data interchange format for passing information between web services and JavaScript applications. In fact, JSON is a subset of JavaScript language and, as such, a valid JavaScript code. However, JSON is a language-independent data format that can be readily parsed and generated from practically every language and environment, including Python. The JSON syntax is also almost valid Python code, making it familiar and intuitive to work with Python.

An example of a JSON dataset was shown in Chapter 10, which featured the graph of the Tokyo Metro network. Before we revisit that dataset, let's begin with a brief overview of JSON basics and how to read and write JSON in Python. The Python standard library provides the json module for working with JSON-formatted data. Specifically, this module contains functions for generating JSON data from a Python data structure (list or dictionary), json.dump and json.dumps, and for parsing JSON data into a Python data structure: json.load and json.loads. The loads and dumps functions take Python strings as input and output, while the load and dump operate on a file handle and read and write data to a file.

For example, we can generate the JSON string of a Python list by calling the json.dumps function. The return value is a JSON string representation of the given Python list that closely resembles the Python code that could be used to create the list. However, a notable exception is the Python value None, which is represented as the value null in JSON.

```
In [158]: data = ["string", 1.0, 2, None]
In [159]: data_json = json.dumps(data)
In [160]: data_json
Out[160]: '["string", 1.0, 2, null]'
```

---

[6] For more information about JSON, see http://json.org.

To convert the JSON string back into a Python object, we can use `json.loads`.

```
In [161]: data = json.loads(data_json)
In [162]: data
Out[162]: ['string', 1.0, 2, None]
In [163]: data[0]
Out[163]: 'string'
```

We can use the same method to store Python dictionaries as JSON strings. Again, the resulting JSON string is almost identical to the Python code for defining the dictionary.

```
In [164]: data = {"one": 1, "two": 2.0, "three": "three"}
In [165]: data_json = json.dumps(data)
In [166]: data_json
Out[166]: '{"two": 2.0, "three": "three", "one": 1}'
```

To parse the JSON string and convert it back into a Python object, we again use `json.loads`.

```
In [167]: data = json.loads(data_json)
In [168]: data["two"]
Out[168]: 2.0
In [169]: data["three"]
Out[169]: 'three'
```

The combination of lists and dictionaries makes a versatile data structure. For example, we can store lists or dictionaries of lists with a variable number of elements. This type of data would be difficult to store directly as a tabular array, and further levels of nested lists and dictionaries would make it impractical. When generating JSON data with the `json.dump` and `json.dumps` functions, we can optionally give the `indent=True` argument, to obtain indented JSON code that can be easier to read.

```
In [170]: data = {"one": [1],
     ...:         "two": [1, 2],
     ...:         "three": [1, 2, 3]}
In [171]: data_json = json.dumps(data, indent=True)
In [172]: data_json
Out[172]: {
             "two": [
              1,
              2
             ],
             "three": [
              1,
              2,
              3
             ],
             "one": [
              1
             ]
          }
```

As an example of a more complex data structure, consider a dictionary containing a list, a dictionary, a list of tuples, and a text string. We could use the same method as in the preceding text to generate a JSON representation of the data structure using `json.dumps`, but instead, here, we write the content to a file using the `json.dump` function. Compared to `json.dumps`, it also takes a file handle as a second argument, which needs to be createdd beforehand.

```
In [173]: data = {"one": [1],
     ...:         "two": {"one": 1, "two": 2},
     ...:         "three": [(1,), (1, 2), (1, 2, 3)],
     ...:         "four": "a text string"}
In [174]: with open("data.json", "w") as f:
     ...:     json.dump(data, f)
```

The result is that the JSON representation of the Python data structure is written to the `data.json` file.

```
In [175]: !cat data.json
{"four": "a text string", "two": {"two": 2, "one": 1}, "three": [[1], [1, 2], [1, 2, 3]],
 "one": [1]}
```

To read and parse a JSON-formatted file into a Python data structure, we can use `json.load`, to which we need to pass a handle to an open file.

```
In [176]: with open("data.json", "r") as f:
     ...:     data_from_file = json.load(f)
In [177]: data_from_file["two"]
Out[177]: [1, 2]
In [178]: data_from_file["three"]
Out[178]: [[1], [1, 2], [1, 2, 3]]
```

The data structure returned by `json.load` is not always identical to the one stored with `json.dump`. JSON is stored as Unicode, so strings in the data structure returned by `json.load` are always Unicode strings. Also, as we can see from the preceding example, JSON does not distinguish between tuples and lists, and the `json.load` always produces lists rather than tuples, and the order in which keys for a dictionary are displayed is only guaranteed if using the `sorted_keys=True` argument to the `dumps` and `dump` functions.

Now that we know how Python lists and dictionaries can be converted to and from JSON representation using the `json` module, it is worthwhile to revisit the Tokyo Metro dataset in Chapter 10. This is a more realistic dataset and an example of a data structure that mixes dictionaries, lists of variable lengths, and string values. The first 20 lines of the JSON file are shown here.

```
In [179]: !head -n 20 tokyo-metro.json
{
    "C": {
        "color": "#149848",
        "transfers": [
            [
                "C3",
                "F15"
            ],
            [
                "C4",
                "Z2"
            ],
```

```
            [
                "C4",
                "G2"
            ],
             [
                "C7",
                "M14"
            ],
```

To load the JSON data into a Python data structure, we use json.load in the same way as before.

```
In [180]: with open("tokyo-metro.json", "r") as f:
    ...:     data = json.load(f)
```

The result is a dictionary with a key for each metro line.

```
In [181]: data.keys()
Out[181]: ['N', 'M', 'Z', 'T', 'H', 'C', 'G', 'F', 'Y']
```

The dictionary value for each metro line is again a dictionary that contains line color, lists of transfer points, and the travel times between stations on the line.

```
In [182]: data["C"].keys()
Out[182]: ['color', 'transfers', 'travel_times']
In [183]: data["C"]["color"]
Out[183]: '#149848'
In [184]: data["C"]["transfers"]
Out[184]: [['C3', 'F15'], ['C4', 'Z2'],   ['C4', 'G2'], ['C7', 'M14'],
           ['C7', 'N6'],
           ['C7', 'G6'], ['C8', 'M15'], ['C8', 'H6'], ['C9', 'H7'],
           ['C9', 'Y18'],
           ['C11', 'T9'], ['C11', 'M18'], ['C11', 'Z8'], ['C12', 'M19'],
           ['C18', 'H21']]
```

With the dataset loaded as a nested structure of Python dictionaries and lists, we can iterate over and easily filter items from the data structure, for example, using Python's list comprehension syntax. The following example demonstrates how to select the set of connected nodes in the graph on the C line, which has a travel time of 1 minute.

```
In [185]: [(s, e, tt) for s, e, tt in data["C"]["travel_times"] if tt == 1]
Out[185]: [('C3', 'C4', 1), ('C7', 'C8', 1), ('C9', 'C10', 1)]
```

The hierarchy of dictionaries and the variable length of the lists stored in the dictionaries make this a good example of a dataset that does not have a strict structure and is, therefore, suitable to store in a versatile format such as JSON.

Serialization

The previous section used the JSON format to generate a representation of in-memory Python objects, such as lists and dictionaries. This process is called serialization, which, in this case, results in a JSON plain-text representation of the objects. An advantage of the JSON format is that it is language-independent and can easily be read by other software. Its disadvantages are that JSON files are not space efficient and can only be used to serialize a limited type of objects (list, dictionaries, basic types, as discussed in the previous section). Many alternative serialization techniques address these issues. Here, we briefly look at two alternatives that address the space efficiency issue and the types of objects that can be serialized: the msgpack library and the Python pickle module.

msgpack is a binary protocol for efficiently storing JSON-like data. The msgpack software is available for many languages and environments. For more information about the library and its Python bindings, see the project's web page at http://msgpack.org. In analogy to the JSON module, the msgpack library provides two sets of functions that operate on byte lists (msgpack.packb and msgpack.unpackb) and file handles (msgpack.pack and msgpack.unpack), respectively. The pack and packb functions convert a Python data structure into a binary representation, and the unpack and unpackb functions perform the reverse operation. For example, the JSON file for the Tokyo Metro dataset is relatively large and takes about 27 KB on disk.

```
In [186]: !ls -lh tokyo-metro.json
-rw-r--r--@ 1 rob  staff    27K Apr  7 23:18 tokyo-metro.json
```

Packing the data structure with msgpack rather than JSON results in a smaller file of about 3 KB.

```
In [187]: data_pack = msgpack.packb(data)
In [188]: type(data_pack)
Out[188]: bytes
In [189]: len(data_pack)
Out[189]: 3021
In [190]: with open("tokyo-metro.msgpack", "wb") as f:
     ...:         f.write(data_pack)
In [191]: !ls -lh tokyo-metro.msgpack
-rw-r--r--@ 1 rob  staff   3.0K Apr  8 00:40 tokyo-metro.msgpack
```

More precisely, the byte list representation of the dataset uses only 3021 bytes. This can be a significant improvement in applications where storage space or bandwidth is essential. However, the price we have paid for this increased storage efficiency is that we must use the msgpack library to unpack the data, and it uses a binary format and, therefore, is not human-readable. Whether this is an acceptable trade-off or not depends on the application at hand. To unpack a binary msgpack byte list, we can use the msgpack.unpackb function, which recovers the original data structure.

```
In [192]: del data
In [193]: with open("tokyo-metro.msgpack", "rb") as f:
     ...:         data_msgpack = f.read()
     ...:         data = msgpack.unpackb(data_msgpack)
In [194]: list(data.keys())
Out[194]: ['T', 'M', 'Z', 'H', 'F', 'C', 'G', 'N', 'Y']
```

The other issue with JSON serialization is that only certain types of Python objects can be stored as JSON. The Python pickle module[7] can create a binary representation of nearly any Python object, including class instances and functions. The pickle module follows the same use pattern as the json module: we have the dump and dumps functions for serializing an object to a byte array and a file handle, respectively, and the load and loads for deserializing a pickled object.

---

[7] For an alternative to the pickle, see also the dill library at https://pypi.org/project/dill.

```
In [195]: with open("tokyo-metro.pickle", "wb") as f:
    ...:         pickle.dump(data, f)
In [196]: del data
In [197]: !ls -lh tokyo-metro.pickle
-rw-r--r--@ 1 rob  staff   8.5K Apr  8 00:40 tokyo-metro.pickle
```

The size of the pickled object is considerably smaller than the JSON serialization but larger than the serialization produced by msgpack. We can recover a pickled object using the `pickle.load` function, which expects a file handle as an argument.

```
In [198]: with open("tokyo-metro.pickle", "rb") as f:
    ...:         data = pickle.load(f)
In [199]: data.keys()
Out[199]: dict_keys(['T', 'M', 'Z', 'H', 'F', 'C', 'G', 'N', 'Y'])
```

The main advantage of `pickle` is that almost any type of Python object can be serialized. However, Python pickles cannot be read by software not written in Python, and it is also not a recommended format for long-term storage because compatibility between Python versions and with different versions of libraries that define the objects that are pickled cannot always be guaranteed. If possible, using JSON for serializing list- and dictionary-based data structures is generally a better approach. If the file size is an issue, the `msgpack` library provides a popular and easily accessible alternative to JSON.

Summary

This chapter reviewed standard data formats for reading and writing numerical data to files on disk. We introduced a selection of Python libraries that are available for working with these formats. We first looked at the ubiquitous CSV file format, a simple and transparent format suitable for small and simple datasets. The main advantage of this format is that it is human-readable plain text, which makes it intuitively understandable. However, it lacks many desirable features when working with numerical data, such as metadata describing the data and support for multiple datasets. The HDF5 format naturally takes over as the go-to format for numerical data when the size and complexity of the data grow beyond what is easily handled using a CSV format. HDF5 is a binary file format, so it is not a human-readable format like CSV. But there are good tools for exploring the content in an HDF5 file, both programmatically and using command-line and GUI-based user interfaces. In fact, due to the possibility of storing metadata in attributes, HDF5 is a great format for self-describing data. It is also a very efficient file format for numerical data, both in terms of I/O and storage, and it can even be used as a data model for computing very large datasets that do not fit in the computer's memory.

Overall, HDF5 is a fantastic tool for numerical computing that anyone working with computing should benefit significantly from being familiar with. We also reviewed the Parquet format, which complements HDF5 as a more straightforward and high-performance storage solution for tabular data, which is especially suitable for large accumulative datasets or when regular pruning is necessary. Toward the end of the chapter, we also briefly reviewed JSON, msgpack, and Python pickles for serializing data into text and binary format.

Further Reading

An informal specification of the CSV file is given in RFC 4180, `http://tools.ietf.org/html/rfc4180`. It outlines many of the commonly used features of the CSV format, although not all CSV readers and writers comply with every aspect of this document. An accessible and informative introduction to the HDF5 format and the h5py library is given by the creator of h5py in *Python and HDF5* by A. Collette (O'Reilly, 2013). It is also worth reading about the NetCDF (Network Common Data Format), `www.unidata.ucar.edu/software/netcdf`, another widely used numerical data format. The Pandas library also provides I/O functions beyond what we have discussed here, such as the ability to read Excel files (`pandas.io.excel.read_excel`) and the fixed-width format (`read_fwf`).

Regarding the JSON format, a concise but complete specification of the format is available at the website `https://json.org`. With the increasingly important role of data in computing, there has been a rapid diversification of formats and data storage technologies in recent years. As a computational practitioner, reading data from databases, such as SQL and NoSQL databases, is now also an important task. Python provides a common database API for standardizing database access from Python applications, as described by PEP 249 (`www.python.org/dev/peps/pep-0249`). Another notable project for reading databases from Python is SQLAlchemy (`www.sqlalchemy.org`).

CHAPTER 19

■ ■ ■

Code Optimization

This book explored various scientific and technical computing topics using Python and its ecosystem of libraries. As touched upon in the very first chapter of this book, the Python environment for scientific computing generally strikes a good balance between a high-level environment suitable for exploratory computing and rapid prototyping that minimizes development efforts and high-performance computing that minimizes application runtimes. High-performance numerical computation is achieved not using the Python language itself, but rather through leveraging external compiled libraries, often written in C or Fortran. Because of this, in computing applications that rely heavily on libraries such as NumPy and SciPy, most of the number crunching is performed by compiled code, and the performance is vastly better than if the computation were to be implemented purely in Python.

The key to high-performance Python programs is, therefore, to efficiently utilize libraries such as NumPy and SciPy for array-based computations. The vast majority of scientific and technical computations can be expressed in terms of common array operations and fundamental computational routines. Much of this book has been dedicated to exploring this style of scientific computing with Python by introducing the main Python libraries for different fields of scientific computing. However, occasionally, there is a need for computations that cannot easily be formulated as array expressions or do not fit existing computing patterns. In such cases, it may be necessary to implement the computation from the ground up, for example, using pure Python code. However, pure Python code tends to be slow compared to the equivalent code written in a compiled language, and if the performance overhead of pure Python is too large, it can be necessary to explore alternatives. The traditional solution is to write an external library in, for example, C or Fortran, which performs the time-consuming computations, and to create an interface to Python code using an extension module.

There are several methods to create extension modules for Python. The most fundamental approach is to use Python's C API to build an extension module with functions implemented in C that can be called from Python. This is typically very tedious and requires a significant effort. The Python standard library provides the `ctypes` module to simplify the interfacing between Python and C. Other alternatives include the CFFI (C foreign function interface) library[1] for interfacing Python with C and the F2PY[2] program for generating interfaces between Python and Fortran. These are all effective tools for interfacing Python with compiled code, and they all play an important role in making Python suitable for scientific computing. However, using these tools requires programming skills and efforts in languages other than Python, and they are the most useful when working with a code base already written in, let's say, C or Fortran.

For new development, there are alternatives closer to Python that are worth considering before embarking on a complete implementation of a problem directly in a compiled language. This chapter explores two such methods: Numba and Cython. These offer a middle ground between Python and low-level languages that retains many advantages of a high-level language while achieving performance comparable to compiled code.

---

[1] https://cffi.readthedocs.org
[2] https://numpy.org/doc/stable/f2py/index.html

© Robert Johansson 2024
R. Johansson, *Numerical Python*, https://doi.org/10.1007/979-8-8688-0413-7_19

Numba is a just-in-time (JIT) compiler for Python code using NumPy that produces machine code that can be executed more efficiently than the original Python code. To achieve this, Numba leverages the LLVM compiler suite (`https://llvm.org`), which is a compiler toolchain that has become very popular for its modular and reusable design and interface, enabling, for example, applications such as Numba, which is a relatively new project and is not yet widely used in many scientific computing libraries. But it is a promising project with solid backing by Continuum Analytics Inc.,[3] and it is likely to have a bright future in scientific computing with Python.

■ **Numba**   The Numba library provides a just-in-time compiler for Python and NumPy code based on the LLVM compiler. The main advantage of Numba is that it can generate machine code with minimal or no changes to the original Python code. For more information about the project and its documentation, see the project's web page at `https://numba.pydata.org`. At the time of writing, the latest version of the library is 0.58.0. Numba is an open source project created by Continuum Analytics Inc.

Cython is a superset of the Python language that can be automatically translated into C or C++ and compiled into machine code, which can run much faster than Python code. Cython is widely used in computationally-oriented Python projects for speeding up time-critical parts of a code base that is otherwise written in Python. Several of the libraries used earlier in this book heavily rely on Cython. These include NumPy, SciPy, Pandas, and scikit-learn, to mention a few.

■ **Cython**   The Cython library translates Python code, or decorated Python code, into C or C++, which can be compiled into a binary extension module. For more information about the project and its documentation, see the project's webpage at `https://cython.org`. At the time of writing, the latest version of Cython is 3.0.0.

This chapter explores how Numba and Cython can speed up code originally written in Python. These methods can be tried when a Python implementation is unacceptably slow. However, before optimizing anything written in Python, it is advisable first to profile the code, for example, using the `cProfile` module or IPython's profiling utilities (see Chapter 1) and identifying exactly which parts of a code are the bottlenecks. If clear bottlenecks can be identified, they may be good candidates for optimization efforts. The first optimization attempt should be to use existing libraries, such as NumPy and SciPy, in the most efficient way to solve the problem at hand and use the Python language itself in the most efficient manner possible.[4] Only when existing libraries do not provide functions and methods that allow us to implement a computation efficiently should we consider optimizing our code with Numba or Cython. Code optimization should only be used as a last resort since premature optimization is often fruitless and results in less maintainable code: "premature optimization is the root of all evil" (Donald Knuth).

---

[3] The producers of the Anaconda Python environment, see Chapter 1 and Appendix.
[4] For example, carefully consider which data structures to use, and make good use of iterators to avoid unnecessary memory copy operations.

Importing Modules

This chapter works with Numba and Cython. Numba is used as a regular Python module. We assume that this library is imported in its entirety using the following.

```
In [1]: import numba
```

Cython can be used in several ways, as shown later in this chapter. Typically, we are not required to explicitly import the Cython library when using Cython code from Python, but instead, we import the pyximport library provided by Cython and register an import hook using pyximport.install().

```
In [2]: import pyximport
```

This alters the way Python modules are imported, and it allows us to directly import Cython files with the file-ending pyx as if they were pure Python modules. Occasionally, it is also useful to explicitly import the Cython library, in which case we assume that it is imported in the following manner.

```
In [3]: import cython
```

Basic numerics and plotting also requires the NumPy and Matplotlib libraries.

```
In [4]: import numpy as np
In [5]: import matplotlib.pyplot as plt
```

Numba

One of the most attractive aspects of the Numba library is that it can often be used to speed up Python code that uses NumPy without changing the target code. The only thing that we need to do is decorate a function with the @numba.jit decorator, which results in the function being just-in-time (JIT) compiled into code that can be significantly faster than the pure Python code by as much as a factor of several hundred or more. The speedup is obtained mainly for functions that use NumPy arrays, for which Numba can automatically perform type interference and generate optimized code for the required type signatures.

To begin using Numba, consider the following simple problem: compute the sum of all elements in an array. A function that performs this computation is simple to implement in Python using for loops.

```
In [6]: def py_sum(data):
   ...:     s = 0
   ...:     for d in data:
   ...:         s += d
   ...:     return s
```

Although this function is nearly trivial, it nicely illustrates the potential and power of Numba. For loops in Python are notoriously slow, due to Python's flexibility and dynamic typing. To quantify this statement and benchmark the py_sum function, we generate an array with 50,000 random numbers and use the %timeit IPython command to measure the typical computation time.

```
In [7]: data = np.random.randn(50000)
In [8]: %timeit py_sum(data)
100 loops, best of 3: 8.43 ms per loop
```

The result suggests that summing the 50,000 elements in the data array using the py_sum function typically takes 8.43 milliseconds on this particular system. Compared to other methods, this is not a good performance. The usual solution is to use array operations, such as those provided by NumPy, instead of iterating over the arrays manually. Indeed, NumPy provides the sum function that does exactly what we want to do here. To verify that the py_sum function defined in the preceding text produces the same results as the NumPy sum function, we first issue an assert statement to this effect.

```
In [9]: assert abs(py_sum(data) - np.sum(data)) < 1e-10
```

Since assert does not raise an error, we conclude that the two functions produce the same result. Next, we benchmark the NumPy sum function using %timeit in the same way it was used in the preceding example.

```
In [10]: %timeit np.sum(data)
10000 loops, best of 3: 29.8 µs per loop
```

The NumPy sum function is several hundred times faster than the py_sum function, demonstrating that vectorized expressions and operations using, for example, NumPy are the key to good performance in Python. The same phenomenon is seen in other functions that use for loops. For example, consider the accumulative sum, py_cumsum, which takes an array as input and produces an array as output.

```
In [11]: def py_cumsum(data):
    ...:     out = np.zeros_like(data)
    ...:     s = 0
    ...:     for n in range(len(data)):
    ...:         s += data[n]
    ...:         out[n] = s
    ...:     return out
```

Benchmarking this function also gives a result that is much slower than the corresponding array-based NumPy function.

```
In [12]: %timeit py_cumsum(data)
100 loops, best of 3: 14.4 ms per loop
In [13]: %timeit np.cumsum(data)
10000 loops, best of 3: 147 µs per loop
```

Let's see how Numba can speed up the slow py_sum and py_cumsum functions. To activate the JIT compilation of a function, we apply the decorator @numba.jit.

```
In [14]: @numba.jit
    ...: def jit_sum(data):
    ...:     s = 0
    ...:     for d in data:
    ...:         s += d
    ...:     return s
```

Next, we verify that the JIT-compiled function produces the same result as the NumPy sum function and benchmark it using the %timeit function.

```
In [15]: assert abs(jit_sum(data) - np.sum(data)) < 1e-10
In [16]: %timeit jit_sum(data)
10000 loops, best of 3: 47.7 µs per loop
```

Compared to the pure Python function, the jit_sum function is about 300 times faster and reaches performance comparable to the NumPy sum function, despite being written in pure Python.

In addition to JIT compiling a function by applying the numba.jit decorator when the function is defined, we can apply the decorator after the fact. For example, to JIT compile the py_cumsum function that we defined earlier, we can use the following.

```
In [17]: jit_cumsum = numba.jit()(py_cumsum)
```

We verify that the resulting jit_cumsum function produces the same result as the corresponding NumPy function and benchmark it using %timeit.

```
In [18]: assert np.allclose(np.cumsum(data), jit_cumsum(data))
In [19]: %timeit jit_cumsum(data)
10000 loops, best of 3: 66.6 µs per loop
```

In this case, the jit_cumsum function outperforms the NumPy cumsum function by a factor of two. The NumPy function cumsum is more versatile than the jit_cumsum function, so the comparison is not entirely fair, but remarkably, we can reach performance that is comparable to compiled code by JIT compiling Python code with a single function decorator. This allows us to use loop-based computations in Python without performance degradation, which is particularly useful for algorithms that are not easily written in vectorized form.

An example of such an algorithm is the computation of the Julia fractal, which requires a variable number of iterations for each element of a matrix with coordinate points in the complex plane: A point z in the complex plane belongs to the Julia set if the iteration formula $z \leftarrow z^2 + c$ does not diverge after a large number of iterations. To generate a Julia fractal graph, we can, therefore, loop over a set of coordinate points and iterate $z \leftarrow z^2 + c$ and store the number of iterations required to diverge beyond some predetermined bound (absolute value larger than 2.0 in the following implementation).

```
In [20]: def py_julia_fractal(z_re, z_im, j):
    ...:     for m in range(len(z_re)):
    ...:         for n in range(len(z_im)):
    ...:             z = z_re[m] + 1j * z_im[n]
    ...:             for t in range(256):
    ...:                 z = z ** 2 - 0.05 + 0.68j
    ...:                 if np.abs(z) > 2.0:
    ...:                     j[m, n] = t
    ...:                     break
```

This implementation is straightforward when using explicit loops, but these three nested loops are prohibitively slow in pure Python. However, with JIT compilation using Numba, we can obtain a significant speedup.

By default, Numba gracefully falls back on the standard Python interpreter in cases when it fails to produce optimized code. An exception to this rule is when the nopython=True argument to numba.jit is given, in which case the JIT compilation fails if Numba is unable to generate statically typed code.

When automatic type interference fails, the resulting JIT-compiled code generated by Numba typically does not provide any speedup, so it is often advisable to use the nopython=True argument to the jit decorator so that we fail quickly when the produced JIT-compiled code is unlikely to result in a speedup. To assist Numba in the code generation, it is sometimes helpful to explicitly define types of variables that occur in a function body, which we can do using the locals keyword argument to the jit decorator that can be assigned to a dictionary that maps symbol names to explicit types. For example, locals=dict(z=numba.complex) specifies that the variable z is a complex number. However, with the current example, we do not need to explicitly specify the types of local variables, since they can all be inferred from the data types of the NumPy arrays passed to the function. We can verify that this is the case by using the nopython=True argument to numba.jit when decorating the py_julia_fractal function.

```
In [21]: jit_julia_fractal = numba.jit(nopython=True)(py_julia_fractal)
```

Next, we call the resulting jit_julia_fractal function to compute the Julia set. Note that we have written the function here so that all the involved NumPy arrays are defined outside the function. This helps Numba recognize which types are involved in the calculation and allows it to generate efficient code in the JIT compilation.

```
In [22]: N = 1024
In [23]: j = np.zeros((N, N), np.int64)
In [24]: z_real = np.linspace(-1.5, 1.5, N)
In [25]: z_imag = np.linspace(-1.5, 1.5, N)
In [26]: jit_julia_fractal(z_real, z_imag, j)
```

After the call to the jit_julia_fractal function, the computation result is stored in the j array. To visualize the result, we can plot the j array using the Matplotlib imshow function. The result is shown in Figure 19-1.

```
In [27]: fig, ax = plt.subplots(figsize=(8, 8))
    ...: ax.imshow(j, cmap=plt.cm.RdBu_r, extent=[-1.5, 1.5, -1.5, 1.5])
    ...: ax.set_xlabel("$\mathrm{Re}(z)$", fontsize=18)
    ...: ax.set_ylabel("$\mathrm{Im}(z)$", fontsize=18)
```

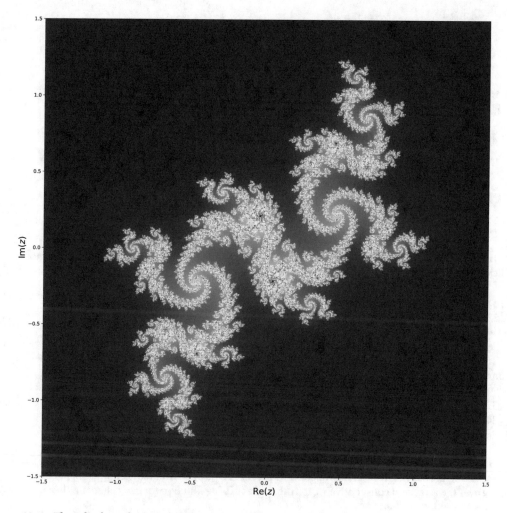

Figure 19-1. *The Julia fractal generated by a JIT-compiled Python function using Numba*

We can compare the speed of the pure Python function py_julia_fractal and the corresponding JIT-compiled function jit_julia_fractal using the %timeit command.

```
In [28]: %timeit py_julia_fractal(z_real, z_imag, j)
1 loops, best of 3: 60 s per loop
In [29]: %timeit jit_julia_fractal(z_real, z_imag, j)
10 loops, best of 3: 140 ms per loop
```

In this case, the speedup is a remarkable 430 times, again by adding a decorator to the Python function. With this type of speedup, for loops in Pythons do not really need to be avoided after all.

Another useful decorator in the Numba library is numba.vectorize. It generates and JIT compiles a vectorized function from a kernel function written for scalar input and output, much like the NumPy vectorize function. Consider, for example, the Heaviside step function:

465

$$\Theta(x) = \begin{cases} 0, x < 0 \\ \dfrac{1}{2}, x = 0. \\ 1, x > 0 \end{cases}$$

If we wanted to implement this function for scalar input x, we could use the following.

```
In [30]: def py_Heaviside(x):
    ...:     if x == 0.0:
    ...:         return 0.5
    ...:     if x < 0.0:
    ...:         return 0.0
    ...:     else:
    ...:         return 1.0
```

This function only works for scalar input, and if we want to apply it to an array or list, we have to explicitly iterate over the array and apply it to each element.

```
In [31]: x = np.linspace(-2, 2, 50001)
In [32]: %timeit [py_Heaviside(xx) for xx in x]
100 loops, best of 3: 16.7 ms per loop
```

This is inconvenient and slow. The NumPy vectorize function solves the inconvenience problem by automatically wrapping the scalar kernel function into a NumPy-array aware function.

```
In [33]: np_vec_Heaviside = np.vectorize(py_Heaviside)
In [34]: np_vec_Heaviside(x)
Out[34]: array([ 0.,   0.,   0., ...,   1.,   1.,   1.])
```

However, the NumPy vectorize function does not solve the performance problem. As we see from benchmarking the np_vec_Heaviside function with %timeit, its performance is comparable to explicitly looping over the array and consecutively calls the py_Heaviside function for each element.

```
In [35]: %timeit np_vec_Heaviside(x)
100 loops, best of 3: 13.6 ms per loop
```

Better performance can be achieved by using NumPy array expressions instead of using NumPy vectorize on a scalar kernel written in Python.

```
In [36]: def np_Heaviside(x):
    ...:     return (x > 0.0) + (x == 0.0)/2.0
In [37]: %timeit np_Heaviside(x)
1000 loops, best of 3: 268 µs per loop
```

However, even better performance can be achieved using Numba and the vectorize decorator, which takes a list of function signatures to generate JIT-compiled code. Here, we generate vectorized functions for two signatures—one that takes arrays of 32-bit floating-point numbers as input and output, defined as numba.float32(numba.float32), and one that takes arrays of 64-bit floating-point numbers as input and output, defined as numba.float64(numba.float64).

```
In [38]: @numba.vectorize([numba.float32(numba.float32),
    ...:                    numba.float64(numba.float64)])
    ...: def jit_Heaviside(x):
    ...:     if x == 0.0:
    ...:         return 0.5
    ...:     if x < 0:
    ...:         return 0.0
    ...:     else:
    ...:         return 1.0
```

Benchmarking the resulting jit_Heaviside function shows the best performance of the methods we have looked at.

```
In [39]: %timeit jit_Heaviside(x)
10000 loops, best of 3: 58.5 µs per loop
```

The jit_Heaviside function can be used as any NumPy universal function, including support for broadcasting and other NumPy features. To demonstrate that the function indeed implements the desired function, we can test it on a simple list of input values.

```
In [40]: jit_Heaviside([-1, -0.5, 0.0, 0.5, 1.0])
Out[40]: array([ 0. ,  0. ,  0.5,  1. ,  1. ])
```

This section explored speeding up Python code using JIT compilation with the Numba library. We looked at four examples: two simple examples demonstrating the basic usage of Numba, the summation and accumulative summation of an array. For a more realistic case of Numba that is not so easily defined in terms of vector expressions, we looked at the computation of the Julia set. Finally, we explored the vectorization of a scalar kernel by implementing the Heaviside step function. These examples demonstrate the typical use patterns for Numba, but there is much more to explore in the Numba library, such as code generation for GPUs. For more information about this and other topics, see the official Numba documentation at https://numba.pydata.org.

Cython

Like Numba, Cython is a solution for speeding up Python code, although Cython takes an entirely different approach to this problem. Whereas Numba is a Python library that converts pure Python code to LLVM code that is JIT-compiled into machine code, Cython is a programming language that is a superset of the Python programming language: Cython extends Python with C-like properties. Most notably, Cython allows us to use explicit and static type declarations. The purpose of the extensions to Python introduced in Cython is to make it possible to translate the code into efficient C or C++ code, which can be compiled into a Python extension module that can be imported and used from regular Python code.

There are two main usages of Cython: speeding up Python code and generating wrappers for interfacing with compiled libraries. When using Cython, we need to modify the targeted Python code, so compared to using Numba, there is a little bit more work involved, and we need to learn the syntax and behavior of Cython to use it to speed up Python code. However, Cython provides more fine-grained control of how the Python code is processed, and Cython also has features that are out of the scope of Numba, such as generating interfaces between Python and external libraries and speeding up Python code that does not use NumPy arrays.

While Numba uses transparent just-in-time compilation, Cython is mainly designed to use traditional ahead-of-time compilation. There are several ways to compile Cython code in a Python extension module, each with different uses. We begin with reviewing options for compiling Cython code and then proceed to introduce Cython features that are useful for speeding up computations written in Python. This section works with mostly the same examples examined in the previous section using Numba so that we can easily compare both the methods and the results. Next, let's look at how to speed up the py_sum and py_cumsum functions defined in the previous section.

To use Cython code from Python, it must pass through the Cython compilation pipeline: first, the Cython code must be translated into C or C++ code, after which it must be compiled into machine code using a C or C++ compiler. The translation from Cython code to C or C++ can be done using the cython command-line tool. It takes a file with Cython code, which we typically store in files using the *pyx* file extension, and produces a C or C++ file. For example, consider the cy_sum.pyx file, with the content shown in Listing 19-1. To generate a C file from this Cython file, we can run the cython cy_sum.pyx command. The result is the cy_sum.c file, which we can compile using a standard C compiler into a Python extension module. This compilation step is platform-dependent and requires the correct compiler flags and options to produce a proper Python extension.

Listing 19-1. Content of the Cython File cy_sum.pyx

```
def cy_sum(data):
    s = 0.0
    for d in data:
        s += d
    return s
```

To avoid the complications related to platform-specific compilation options for C and C++ code, we can use the distutils and Cython libraries to automate the translation of Cython code into a useful Python extension module. This requires creating a setup.py script that calls the setup function from distutils. core (which knows how to compile C code into a Python extension) and the cythonize function from Cython. Build (which knows how to translate Cython code into C code), as shown in Listing 19-2. When the setup. py file is prepared, we can compile the Cython module using the python setup.py build_ext --inplace command, which instructs distutils to build the extension module and place it in the same directory as the source code.

Listing 19-2. A setup.py Script That Can Be Used to Automatically Compile a Cython File into a Python Extension Module

```
from distutils.core import setup
from Cython.Build import cythonize
import numpy as np
setup(ext_modules=cythonize('cy_sum.pyx'),
      include_dirs=[np.get_include()],
      requires=['Cython', 'numpy'])
```

Once the Cython code has been compiled into a Python extension module, whether by hand or using the distutils library, we can import it and use it as a regular module in Python.

```
In [41]: from cy_sum import cy_sum
In [42]: cy_sum(data)
Out[42]: -189.70046227549025
```

```
In [43]: %timeit cy_sum(data)
100 loops, best of 3: 5.56 ms per loop
In [44]: %timeit py_sum(data)
100 loops, best of 3: 8.08 ms per loop
```

For this example, compiling the pure Python code in Listing 19-3 using Cython directly gives a speedup of about 30%. This is a nice speedup, but arguably not worth the trouble of going through the Cython compilation pipeline. Improving this speedup using other Cython features is discussed later.

Listing 19-3. Content of the Cython File cy_cumsum.pyx

```
cimport numpy
import numpy
def cy_cumsum(data):
    out = numpy.zeros_like(data)
    s = 0
    for n in range(len(data)):
        s += data[n]
        out[n] = s
    return out
```

The explicit compilation of Cython code into a Python extension module shown in the preceding code is useful for distributing prebuilt modules written in Cython, as the result does not require Cython to be installed to use the extension module. An alternative way to implicitly invoke the Cython compilation pipeline automatically during the import of a module is provided by the pyximport library, which is distributed with Cython. To seamlessly import a Cython file directly from Python, we can first invoke the install function from the pyximport library.

```
In [45]: pyximport.install(setup_args=dict(include_dirs=np.get_include()))
```

This modifies the behavior of the Python import statement and add support for Cython *pyx* files. When a Cython module is imported, it is first compiled in C or C++ and then to machine code in the format of a Python extension module that the Python interpreter can import. These implicit steps sometimes require additional configuration, which we can pass to the pyximport.install function via arguments. For example, to be able to import Cython code that uses NumPy-related features, we need the resulting C code to be compiled against the NumPy C header files. We can configure this by setting the include_dirs to the value given by np.get_include() in the setup_args argument to the install function, as shown in the preceding code. Several other options are also available, and we can also give custom compilation and linking arguments. See the docstring for pyximport.install for details. Once pyximport.install has been called, we can use a standard Python import statement to import a function from a Cython module.

```
In [46]: from cy_cumsum import cy_cumsum
In [47]: %timeit cy_cumsum(data)
100 loops, best of 3: 5.91 ms per loop
In [48]: %timeit py_cumsum(data)
100 loops, best of 3: 13.8 ms per loop
```

This example provides a welcome but not very impressive speedup of a factor of two for the Python code that has been passed through the Cython compilation pipeline.

Before we get into the detailed usage of Cython that allows us to improve upon this speedup factor, we quickly introduce another way of compiling and importing Cython code. When using IPython, especially the Jupyter Notebook, we can use the convenient %%cython command, which automatically compiles and loads Cython code in a code cell as a Python extension and makes it available in the IPython session. To be able to use this command, we must first activate it using the %load_ext cython command.

```
In [49]: %load_ext cython
```

With the %%cython command activated, we can write and load Cython code interactively in an IPython session.

```
In [50]: %%cython
    ...: def cy_sum(data):
    ...:     s = 0.0
    ...:     for d in data:
    ...:         s += d
    ...:     return s
In [51]: %timeit cy_sum(data)
100 loops, best of 3: 5.21 ms per loop
In [52]: %timeit py_sum(data)
100 loops, best of 3: 8.6 ms per loop
```

As before, see a direct speedup by adding the %%cython command at the first line of the IPython code cell. This is reminiscent of adding the @numba.jit decorator to a function, but the underlying mechanics of these two methods are rather different. The rest of this section uses this method for compiling and loading Cython code. When using the %%cython IPython command, it is also useful to add the -a argument. This results in Cython code annotations being displayed as the output of the code cell, as shown in Figure 19-2. The annotation shows each code line in a shade of yellow, where bright yellow indicates that the line of code is translated to C code with strong dependencies on the Python C/API and where a white line of code is directly translated into pure C code. When optimizing Cython code, we generally need to strive for Cython code that gets translated into as pure C code as possible, so it is helpful to inspect the annotation output and look for yellow lines, which typically represent the bottlenecks in the code. As a bonus, clicking a line of code in the annotation output toggles between the Cython code we provided and the C code it is being translated into.

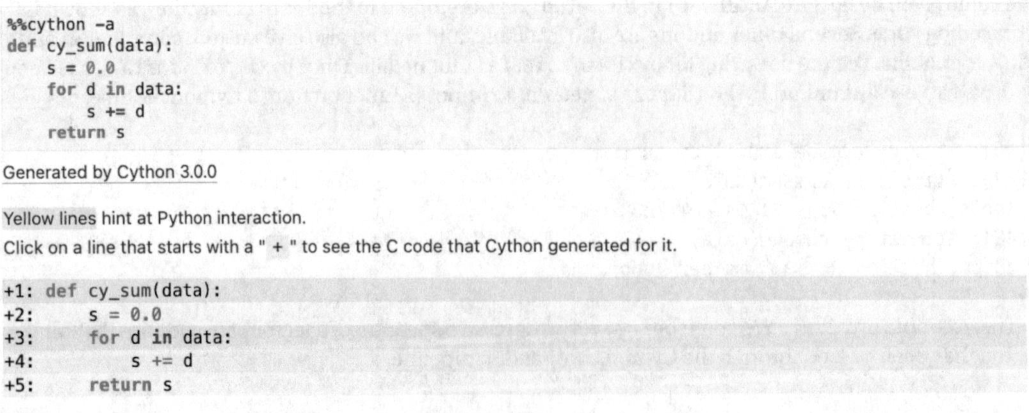

```
%%cython -a
def cy_sum(data):
    s = 0.0
    for d in data:
        s += d
    return s
```

Generated by Cython 3.0.0

Yellow lines hint at Python interaction.
Click on a line that starts with a " + " to see the C code that Cython generated for it.

```
+1: def cy_sum(data):
+2:     s = 0.0
+3:     for d in data:
+4:         s += d
+5:     return s
```

Figure 19-2. *Annotation generated by Cython using the %%cython IPython command with the -a argument*

In the rest of the section, we explore ways of speeding up Cython code using language features introduced by Cython that are particularly useful for computational problems. We first revisit the implementation of the cy_sum given in the preceding code. Our first attempt to speed up this function used the original Python code and passed it through the Cython compilation pipeline, and as a result, we saw a speedup of about 30%. The key step to see much larger speedups is to add type declarations for all the variables and arguments of the function. By explicitly declaring the types of variables, the Cython compiler can generate more efficient C code. To specify the type of a variable, we need to use the Cython keyword cdef, which we can use with any standard C type. For example, to declare the variable n of integer type, we can use cdef int n. We can also use type definitions from the NumPy library; for example, cdef numpy.float64_t s declares the variable s to be a 64-bit floating-point number. NumPy arrays can be declared using the type specification in the format numpy.ndarray[numpy.float64_t, ndim=1] data, which declares data to be an array with 64-bit floating-point number elements, with one dimension (a vector) and of unspecified length. Adding type declarations of this style to the previous cy_sum function results in the following code.

```
In [53]: %%cython
    ...: cimport numpy
    ...: cimport cython
    ...:
    ...: @cython.boundscheck(False)
    ...: @cython.wraparound(False)
    ...: def cy_sum(numpy.ndarray[numpy.float64_t, ndim=1] data):
    ...:     cdef numpy.float64_t s = 0.0
    ...:     cdef int n, N = len(data)
    ...:     for n in range(N):
    ...:         s += data[n]
    ...:     return s
```

This implementation of the cy_sum function applied two decorators, @cython.boundscheck(False) and @cython.wraparound(False), which turn off time-consuming bound checks on the indexing of NumPy arrays. This results in less safe code, but if we are confident that the NumPy arrays in this function is not indexed outside of their valid ranges, we can obtain additional speedup by turning off such checks. Now that we have explicitly declared the type of all variables and arguments of the function, Cython can generate efficient C code that, when compiled into a Python module, provides performance that is comparable to the JIT-compiled code using Numba and not far from the built-in sum function from NumPy (which also is implemented in C).

```
In [54]: %timeit cy_sum(data)
10000 loops, best of 3: 49.2 µs per loop
In [55]: %timeit jit_sum(data)
10000 loops, best of 3: 47.6 µs per loop
In [56]: %timeit np.sum(data)
10000 loops, best of 3: 29.7 µs per loop
```

Next, let's turn our attention to the cy_cumsum function. Like the cy_sum function, this function also benefits from explicit type declarations. To simplify the declarations of NumPy array types, we use the ctypedef keyword to create an alias for numpy.float64_t to the shorter FTYPE_t. Note also that in Cython code, there are two different import statements: cimport and import. The import statement can be used to import any Python module, but it results in C code that calls back into the Python interpreter and can, therefore, be slow. The cimport statement works like a regular import, but is used for importing other Cython modules. Here cimport numpy imports a Cython module named numpy that provides Cython extensions to NumPy, mostly type and function declarations. In particular, the C-like types such as numpy.float64_t are declared in this Cython module. However, the function call numpy.zeros in the function defined in the following code results in a call to the zeros function in the NumPy module, and for it, we need to include the numpy Python module using import numpy.

Adding these type declarations to the previously defined cy_cumsum function results in the implementation given in the following.

```
In [57]: %%cython
    ...: cimport numpy
    ...: import numpy
    ...: cimport cython
    ...:
    ...: ctypedef numpy.float64_t FTYPE_t
    ...:
    ...: @cython.boundscheck(False)
    ...: @cython.wraparound(False)
    ...: def cy_cumsum(numpy.ndarray[FTYPE_t, ndim=1] data):
    ...:     cdef int n, N = data.size
    ...:     cdef numpy.ndarray[FTYPE_t, ndim=1] out = \
    ...:         numpy.zeros(N, dtype=data.dtype)
    ...:     cdef numpy.float64_t s = 0.0
    ...:     for n in range(N):
    ...:         s += data[n]
    ...:         out[n] = s
    ...:     return out
```

As for cy_sum, we see a significant speedup after having declared the types of all variables in the function, and the performance of cy_cumsum is now comparable to the JIT-compiled jit_cumsum Numba function and faster than the built-in cumsum function in NumPy (which on the other hand is more versatile).

```
In [58]: %timeit cy_cumsum(data)
10000 loops, best of 3: 69.7 µs per loop
In [59]: %timeit jit_cumsum(data)
10000 loops, best of 3: 70 µs per loop
In [60]: %timeit np.cumsum(data)
10000 loops, best of 3: 148 µs per loop
```

When adding explicit type declarations, we gain performance when compiling the function with Cython, but we lose generality as the function can now not take any other type of arguments. For example, the original py_sum function and the NumPy sum function accept a much wider variety of input types. We can sum Python lists and NumPy arrays of both floating-point numbers and integers.

```
In [61]: py_sum([1.0, 2.0, 3.0, 4.0, 5.0])
Out[61]: 15.0
In [62]: py_sum([1, 2, 3, 4, 5])
Out[62]: 15
```

The Cython-compiled version with explicit type declaration, on the other hand, only works for exactly the type we declared it.

```
In [63]: cy_sum(np.array([1.0, 2.0, 3.0, 4.0, 5.0]))
Out[63]: 15.0
In [64]: cy_sum(np.array([1, 2, 3, 4, 5]))
---------------------------------------------------------------------------
ValueError: Buffer dtype mismatch, expected 'float64_t' but got 'long'
```

It is often desirable to support more than one type of input, such as the ability to sum arrays of floating-point numbers and integers with the same function. Cython provides a solution to this problem through its `ctypedef fused` keyword, with which we can define new types that are one of several provided types. For example, consider the modification to the `py_sum` function given in `py_fused_sum` here.

```
In [65]: %%cython
    ...: cimport numpy
    ...: cimport cython
    ...:
    ...: ctypedef fused I_OR_F_t:
    ...:     numpy.int64_t
    ...:     numpy.float64_t
    ...:
    ...: @cython.boundscheck(False)
    ...: @cython.wraparound(False)
    ...: def cy_fused_sum(numpy.ndarray[I_OR_F_t, ndim=1] data):
    ...:     cdef I_OR_F_t s = 0
    ...:     cdef int n, N = len(data)
    ...:     for n in range(N):
    ...:         s += data[n]
    ...:     return s
```

Here the function is defined in terms of the type `I_OR_F_t`, which is defined using `ctypedef fused` to be either `numpy.int64_t` or `numpy.float64_t`. Cython automatically generates the necessary code for both types of functions so that we can use the function on both floating-point and integer arrays (at the price of a small decrease in performance).

```
In [66]: cy_fused_sum(np.array([1.0, 2.0, 3.0, 4.0, 5.0]))
Out[66]: 15.0
In [67]: cy_fused_sum(np.array([1, 2, 3, 4, 5]))
Out[67]: 15
```

As a final example of how to speed up Python code with Cython, consider the Python code for generating the Julia set we looked at in the previous section. To implement a Cython version of this function, we take the original Python code and explicitly declare the types of all the variables used in the function, following the procedure used in the preceding text. We also add the decorators for disabling index bound checks and wraparound. We have both NumPy integer arrays and floating-point arrays as input, so we define the arguments as types `numpy.ndarray[numpy.float64_t, ndim=1]` and `numpy.ndarray[numpy.int64_t, ndim=2]`, respectively.

The implementation of `cy_julia_fractal` given in the following code also includes a Cython implementation of the square of the absolute value of a complex number. This function is declared inline using the `inline` keyword, which means that the compiler puts the function's body at every place it is called rather than creating a function that is called from those locations. This results in large code but avoid the overhead of an additional function call. We also define this function using `cdef` rather than the usual `def` keyword. In Cython, `def` defines a function that can be called from Python, while `cdef` defines a function that can be called from C. Using the cpdef keyword, we can simultaneously define a function that is callable both from C and from Python. As it is written here, using cdef, we cannot call the abs2 function from the IPython session after executing this code cell, but if we change cdef to cpdef, we can.

```
In [68]: %%cython
    ...: cimport numpy
    ...: cimport cython
    ...:
    ...: cdef inline double abs2(double complex z):
    ...:     return z.real * z.real + z.imag * z.imag
    ...:
    ...: @cython.boundscheck(False)
    ...: @cython.wraparound(False)
    ...: def cy_julia_fractal(numpy.ndarray[numpy.float64_t, ndim=1] z_re,
    ...:                      numpy.ndarray[numpy.float64_t, ndim=1] z_im,
    ...:                      numpy.ndarray[numpy.int64_t, ndim=2] j):
    ...:     cdef int m, n, t, M = z_re.size, N = z_im.size
    ...:     cdef double complex z
    ...:     for m in range(M):
    ...:         for n in range(N):
    ...:             z = z_re[m] + 1.0j * z_im[n]
    ...:             for t in range(256):
    ...:                 z = z ** 2 - 0.05 + 0.68j
    ...:                 if abs2(z) > 4.0:
    ...:                     j[m, n] = t
    ...:                     break
```

If we call the cy_julia_fractal function with the same arguments as the Python implementation that was JIT-compiled using Numba, we see that the two implementations have comparable performance.

```
In [69]: N = 1024
In [70]: j = np.zeros((N, N), dtype=np.int64)
In [71]: z_real = np.linspace(-1.5, 1.5, N)
In [72]: z_imag = np.linspace(-1.5, 1.5, N)
In [73]: %timeit cy_julia_fractal(z_real, z_imag, j)
10 loops, best of 3: 113 ms per loop
In [74]: %timeit jit_julia_fractal(z_real, z_imag, j)
10 loops, best of 3: 141 ms per loop
```

The slight edge to the cy_julia_fractal implementation is mainly due to the inline definition of the innermost loop call to the abs2 function and abs2 avoids computing the square root. Making a similar change in jit_julia_fractal improves its performance and approximately accounts for the difference shown here.

So far, we have explored Cython as a method to speed up Python code by compiling it into machine code made available as Python extension modules. Importantly, Cython can also create wrappers to compile C and C++ libraries easily. This is not explored in-depth here but we do look at a simple example that illustrates using Cython, we can call out to arbitrary C libraries in just a few lines of code. As an example, consider the math library from the C standard library. It provides mathematical functions similar to those defined in the Python standard library with the same name: math. To use these functions in a C program, we would include the math.h header file to obtain their declarations and compile and link the program against the libm library. From Cython, we can obtain function declarations using the cdef extern from keywords, after which we need to give the name of the C header file and list the declarations of the function we want to use in the following code block. For example, to make the acos function from libm available in Cython, we can use the following code.

```
In [75]: %%cython
    ...: cdef extern from "math.h":
    ...:     double acos(double)
    ...:
    ...: def cy_acos1(double x):
    ...:     return acos(x)
```

Here, we also defined the Python function cy_acos1, which we can call from Python.

```
In [76]: %timeit cy_acos1(0.5)
10000000 loops, best of 3: 83.2 ns per loop
```

Using this method, we can wrap arbitrary C functions into callable functions from regular Python code. This is a handy feature for scientific computing applications since it makes existing code written in C and C++ readily available from Python. Cython provides type declarations via the libc module for the standard libraries, so we do not need to define the functions using cdef extern from explicitly. For the acos example, we could instead directly import the function from libc.math using the cimport statement.

```
In [77]: %%cython
    ...: from libc.math cimport acos
    ...:
    ...: def cy_acos2(double x):
    ...:     return acos(x)
In [78]: %timeit cy_acos2(0.5)
10000000 loops, best of 3: 85.6 ns per loop
```

The resulting cy_acos2 function is identical to cy_acos1, which was explicitly imported from math.h earlier. It is instructive to compare the performance of these C math library functions to the corresponding functions defined in NumPy and the Python standard math library.

```
In [79]: from numpy import arccos
In [80]: %timeit arccos(0.5)
1000000 loops, best of 3: 1.07 µs per loop
In [81]: from math import acos
In [82]: %timeit acos(0.5)
10000000 loops, best of 3: 95.9 ns per loop
```

Because of the overhead related to NumPy array data structures, the NumPy version is about ten times slower than the Python math function and Cython wrappers to the C standard library function.

Summary

This chapter explored methods for speeding up Python code using Numba, which produces optimized machine code using just-in-time compilation, and Cython, which produces C code that can be compiled into machine code using ahead-of-time compilation. Numba works with pure Python code but heavily relies on type interference using NumPy arrays, while Cython works with an extension to the Python language that allows explicit type declarations. The advantage of these methods is that we can achieve performance comparable to compiled machine code while staying in a Python or Python-like programming environment. The key to speeding up Python code is using typed variables, either by using type interference from NumPy arrays, as in Numba, or by explicitly declaring the types of variables, as in Cython. Explicitly typed code can be translated into much more efficient code than the dynamically typed code in pure Python and can avoid much of the overhead involved in type lookups in Python.

Both Numba and Cython are convenient ways to obtain impressive speedups of Python code, and they often produce code with similar performance. Cython also provides an easy-to-use method for creating interfaces to external libraries to be accessed from Python. In both Numba and Cython, the common theme is using type information (from NumPy arrays or explicit declarations) to generate more efficient typed machine code. Within the Python community, there has also recently been a movement toward adding support for optional type hints to the Python language itself. For more details about type hints, see PEP 484 (`www.python.org/dev/peps/pep-0484`), which has been included in Python as of version 3.5. While type hints in Python code have been gaining popularity, they are not yet widely used in many scientific computing projects. Nonetheless, it is certainly an exciting feature and an important development to follow.

Further Reading

Thorough guides to using Cython are given in *Cython: A Guide for Python Programmers* by K. Smith (O'Reilly, 2015) and *Learning Cython Programming* by P. Herron (Packt, 2013). For more information about Numba, see its official documentation at `https://numba.pydata.org/numba-doc`. For a detailed discussion of high-performance computing with Python, also see *High Performance Python: Practical Performant Programming for Humans* by M. Gorelick (O'Reilly, 2014).

APPENDIX

■ ■ ■

Installation

This appendix covers installing and setting up a Python environment for scientific computing on commonly used platforms. As discussed in Chapter 1, the scientific computing environment for Python is not a single product but rather a diverse ecosystem of packages and libraries, and there are numerous possible ways to install and configure a Python environment on any given platform. Python is rather easy to install,[1] and on many operating systems, it is even preinstalled. All pure Python libraries hosted on the *Python Package Index*[2] are also easily installed, for example, using `pip` and a command such as `pip install PACKAGE`, where `PACKAGE` is the name of the package to install. The `pip` software then searches for the package on the Python Package Index and downloads and installs it if it is found. For example, to install IPython, we can use the following command.

```
$ pip install ipython
```

And to upgrade an already installed package, add the `--upgrade` flag to the `pip` command, as follows.

```
$ pip install --upgrade ipython
```

However, many libraries for computing with Python are not pure Python libraries, and they frequently have dependencies on system libraries written in other languages, such as C and Fortran. Pip and the Python Package Index cannot handle these dependencies, and building such libraries from source requires installing C and Fortran compilers. In other words, manually installing a full scientific computing software stack for Python can take time and effort. To solve this problem, several prepackaged Python environments with automated installers have emerged. The most prominent distribution is *Anaconda*[3] which is sponsored by Anaconda Inc., a corporation with close connections and a history of supporting the open source scientific Python community. The Anaconda environment bundles the Python interpreter, the required system libraries and tools, and many scientific-computing-oriented Python libraries in an easy-to-install distribution. The following uses this Anaconda environment to set up the software to run the code discussed in this book, particularly Miniconda, a lightweight version of Anaconda, and the conda package manager.

---

[1] Installers for all major platforms are available for download at `www.python.org/downloads`.
[2] `http://pypi.python.org`
[3] `http://anaconda.com`

© Robert Johansson 2024
R. Johansson, *Numerical Python*, https://doi.org/10.1007/979-8-8688-0413-7

Miniconda and Conda

The Anaconda environment, which comes bundled with many libraries by default, is a convenient way to get a scientific computing environment for Python up and running quickly. However, to clarify each chapter's dependencies, we start with a Miniconda environment and explicitly install the necessary packages. In this way, we control precisely which packages are included in the environment we set up. Miniconda is a minimal version of Anaconda, which only consists of the most basic components: a Python interpreter, a few fundamental libraries, and the conda package manager. The download page for the Miniconda project (https://conda.pydata.org/miniconda.html) contains installers for Linux, macOS, and Windows. Download and run the installer, and follow the on-screen instructions. When the installation has finished, you should have a directory named miniconda3 in your home directory. If you add it to your PATH variable during the installation, you should be able to invoke the conda package manager by running conda at the command prompt.

Conda[4] is a cross-platform package manager that can handle dependencies on Python packages and system tools and libraries. It is essential for installing scientific computing software, which by nature uses a diverse set of tools and libraries. Conda packages are prebuilt binaries for the target platform and are fast and convenient to install. To verify that conda is available on your system, you can use the following command.

```
$ conda --version
conda 23.10.0
```

In this case, the output tells us that conda is installed and the version is 23.10.0. To update to the latest version of conda, we can use the conda package manager itself.

```
$ conda update -n base conda
```

To update all packages installed in a particular conda environment, we can use the following command.

```
$ conda update --all
```

Once conda is installed, we can use it to install Python interpreters and libraries. We can optionally specify precise versions of the packages we want to install. The Python software ecosystem consists of many independent projects, each with their own release cycles and development goals, and there are constantly new versions of different libraries being released. This is exciting—because there is steady progress, and new features are frequently made available—but unfortunately, not all new releases of all libraries are backward compatible. It presents a dilemma for a user that requires a stable and reproducible environment over the long term and for users who simultaneously work on projects with different versions of dependencies.

The best solution in the Python ecosystem for this problem is to use a package manager such as conda to set up virtual Python environments for different projects in which different versions of the required dependencies are installed. With this approach, it is easy to maintain multiple environments with different configurations, such as separate environments for different versions of Python or environments with stable versions and development versions of required packages. I highly recommend using virtual Python environments rather than the default system-wide ones for the reasons discussed here.

With conda, new environments are created with the conda create command, to which we need to provide a name for the new environment using -n NAME or a path to where the environment is to be stored using -p PATH. When providing a name, the environment is, by default, stored in the miniconda3/envs/NAME

---

[4] http://conda.pydata.org/docs/index.html

directory. When creating a new environment, we can also list packages to install. At least one package must be specified. For example, to create two new environments based on Python 3.10 and Python 3.11, we can use the following commands.

```
$ conda create -n py3.10 python=3.10
$ conda create -n py3.11 python=3.11
```

The environments are named py3.10 and py3.11. To use one of these environments, we need to *activate* it using the conda activate py3.10 command or the conda activate py3.11 command, respectively. To *deactivate* an environment, use conda deactivate. With this method, it is easy to switch between different environments, as illustrated in the following sequence of commands.

```
$ conda activate py3.10
(py3.10)$ python --version
Python 3.10.13
(py3.10)$ conda activate py3.11
(py3.11)$ conda --version
Python 3.11.5
(py3.11)$ conda deactivate
$
```

The conda env, conda info, and conda list commands are helpful tools to manage environments. The conda info command can list available environments (same as conda env list).

```
$ conda info --envs
# conda environments:
#
base                     /Users/rob/miniconda3
py3.10                   /Users/rob/miniconda3/envs/py3.10
py3.11                   /Users/rob/miniconda3/envs/py3.11
```

The conda list command can list installed packages and their versions in a given environment.

```
$ conda list -n py3.10
# packages in environment at /Users/rob/miniconda3/envs/py3.10:
#
# Name                   Version                   Build  Channel
bzip2                    1.0.8                  h1de35cc_0
ca-certificates          2023.08.22             hecd8cb5_0
libffi                   3.4.4                  hecd8cb5_0
ncurses                  6.4                    hcec6c5f_0
openssl                  3.0.12                 hca72f7f_0
pip                      23.3.1           py310hecd8cb5_0
python                   3.10.13                h5ee71fb_0
readline                 8.2                    hca72f7f_0
setuptools               68.0.0           py310hecd8cb5_0
sqlite                   3.41.2                 h6c40b1e_0
tk                       8.6.12                 h5d9f67b_0
tzdata                   2023c                  h04d1e81_0
```

```
wheel               0.41.2          py310hecd8cb5_0
xz                  5.4.2           h6c40b1e_0
zlib                1.2.13          h4dc903c_0
```

The conda env export command produces similar information in YAML format[5].

```
(py3.10)$ conda env export

name: py3.10
channels:
  - defaults
dependencies:
  - bzip2=1.0.8=h1de35cc_0
  - ca-certificates=2023.08.22=hecd8cb5_0
  - libffi=3.4.4=hecd8cb5_0
  - ncurses=6.4=hcec6c5f_0
  - openssl=3.0.12=hca72f7f_0
  - pip=23.3.1=py310hecd8cb5_0
  - python=3.10.13=h5ee71fb_0
  - readline=8.2=hca72f7f_0
  - setuptools=68.0.0=py310hecd8cb5_0
  - sqlite=3.41.2=h6c40b1e_0
  - tk=8.6.12=h5d9f67b_0
  - tzdata=2023c=h04d1e81_0
  - wheel=0.41.2=py310hecd8cb5_0
  - xz=5.4.2=h6c40b1e_0
  - zlib=1.2.13=h4dc903c_0
prefix: /Users/rob/miniconda3/envs/py3.10
```

To install additional packages in an environment, we can specify a list of packages when the environment is created or activate the environment and use conda install or the conda install command with the -n flag to specify a target environment for the installation. For example, to create a Python 3.11 environment with NumPy version 1.24, we could use the following command.

```
$ conda create -n py3.11-np1.24 python=3.11 numpy=1.24
```

To verify that the new environment py3.11-np1.24 contains NumPy of the specified version, we can use the conda list command again.

```
$ conda list -n py3.11-np1.24
# packages in environment at /Users/rob/miniconda3/envs/py3.11-np1.24:
#
# Name                Version                 Build  Channel
blas                  1.0                       mkl
bzip2                 1.0.8               h1de35cc_0
ca-certificates       2023.08.22          hecd8cb5_0
intel-openmp          2023.1.0            ha357a0b_43548
libcxx                14.0.6              h9765a3e_0
libffi                3.4.4               hecd8cb5_0
```

---

[5] http://yaml.org

```
mkl                 2023.1.0          h8e150cf_43560
mkl-service         2.4.0             py311h6c40b1e_1
mkl_fft             1.3.8             py311h6c40b1e_0
mkl_random          1.2.4             py311ha357a0b_0
ncurses             6.4               hcec6c5f_0
numpy               1.24.3            py311h728a8a3_1
numpy-base          1.24.3            py311h53bf9ac_1
openssl             3.0.12            hca72f7f_0
pip                 23.3.1            py311hecd8cb5_0
python              3.11.5            hf27a42d_0
readline            8.2               hca72f7f_0
setuptools          68.0.0            py311hecd8cb5_0
sqlite              3.41.2            h6c40b1e_0
tbb                 2021.8.0          ha357a0b_0
tk                  8.6.12            h5d9f67b_0
tzdata              2023c             h04d1e81_0
wheel               0.41.2            py311hecd8cb5_0
xz                  5.4.2             h6c40b1e_0
zlib                1.2.13            h4dc903c_0
```

Here, we see that NumPy is indeed installed, and the precise version of the library is 1.24.3. If we do not explicitly specify the version of a library, the latest stable release is used.

To use the second method—to install additional packages in an already existing environment—we first activate the environment.

```
$ conda activate py3.11
```

Then use `conda install PACKAGE` to install the package with the name `PACKAGE`. Here, we can also give a list of package names. For example, to install the NumPy, SciPy, and Matplotlib libraries, we can use the following command.

```
(py3.11)$ conda install numpy scipy matplotlib
```

Or, we could also use the following command.

```
$ conda install -n py3.11 numpy scipy matplotlib
```

When installing packages using conda, all required dependencies are also installed automatically, and the preceding command also installed the packages `dateutil`, `freetype`, and `libpng`, among others, which are dependencies for the `matplotlib` package.

```
(py3.11)$ conda list | grep -e matplotlib -e dateutil -e freetype -e libpng
freetype            2.12.1            hd8bbffd_0
libpng              1.6.39            h6c40b1e_0
matplotlib          3.8.0             py311hecd8cb5_0
matplotlib-base     3.8.0             py311h41a4f6b_0
python-dateutil     2.8.2             pyhd3eb1b0_0
```

Note that not all the packages installed in this environment are Python libraries. For example, `libpng` and `freetype` are system libraries, but conda can handle and install them automatically as dependencies. This is one of the strengths of conda compared to, for example, the Python-centric package manager `pip`.

To update selected packages in an environment, we can use the `conda update` command. For example, to update NumPy and SciPy in the currently active environment, we can use the following command.

```
(py3.11)$ conda update numpy scipy
```

To remove a package, we can use `conda remove PACKAGE`. To completely remove an environment, we can use `conda remove -n NAME --all`. For example, to remove the environment py3.11-np1.24, we can use the following command.

```
$ conda remove -n py3.11-np1.24 --all
```

Conda locally caches packages that have once been installed. This makes it fast to reinstall a package in a new environment and quick and easy to tear down and set up new environments for testing and trying out different things without any risk of breaking environments used for other projects. To re-create a conda environment, all we need to do is to keep track of the installed packages. Using the `-e` flag with the `conda list` command gives a list of packages and their versions in a format compatible with the `pip` software. This list can be used to replicate a conda environment, for example, on another system or at a later point in time.

```
$ conda list -e > requirements.txt
```

With the `requirements.txt` file, we can now update an existing conda environment in the following manner.

```
$ conda install --file requirements.txt
```

Or create a new environment that is a replication of the environment that was used to create the `requirement.txt` file.

```
$ conda create -n NAME --file requirements.txt
```

Alternatively, we can use the YAML format dump of an environment produced by `conda env export`.

```
$ conda env export -n NAME > env.yml
```

In this case, we can reproduce the environment using the following.

```
$ conda env create --file env.yml
```

Note that here we do not need to specify the environment name since the `env.yml` file also contains this information. This method also has the advantage that packages installed using `pip` are installed when the environment is replicated or restored.

Jupyter Notebook Kernel Registration

It is a good practice to frequently create new environments for new projects, testing upgrades of external dependencies, and so on. When we work with the Jupyter Notebook environment, as introduced in Chapter 1, we need to register a new kernel for each new environment we create to be available in the kernel list in the Jupyter web application. To do this, we can activate the environment, install the `ipykernel` package, and run the command for registering the currently active.

```
$ conda activate py3.11
(py3.11)$ conda install ipykernel
(py3.11)$ python -m ipykernel install --user --name py3.11
Installed kernelspec py3.11 in /Users/rob/Library/Jupyter/kernels/py3.11
```

A Complete Environment

Now that we have explored the conda package manager and seen how it can be used to set up environments and install packages, let's go over the procedures for setting up a complete environment with all the dependencies required for the material in this book. The following uses the py3.10 environment, which was previously created using the following command.

```
$ conda create -n py3.10 python=3.10
```

This environment can be activated using the following.

```
$ conda activate py3.10
```

Once the target environment is activated, we can install the libraries used in this book with the following commands.

```
conda install ipython jupyter jupyterlab spyder pylint pyflakes pep8
conda install numpy scipy sympy matplotlib networkx pandas seaborn
conda install patsy statsmodels scikit-learn pymc
conda install h5py pytables msgpack-python cython numba cvxopt pyarrow
conda install -c conda-forge fenics mshr
conda install -c conda-forge pygraphviz
pip install scikit-monaco
pip install version_information
```

The FEniCS libraries have many intricate dependencies, making installing this standard approach on some platforms difficult.[6] For this reason, if the FEniCS installation using conda fails, it is most easily installed using the prebuilt environments available from the project's website: https://fenicsproject.org/download. Another good solution for obtaining a complete FEniCS environment can be to use a Docker[7] container with FEniCS preinstalled. See, for example, https://registry.hub.docker.com/repos/fenicsproject for more information about this method.

Table A-1 presents a breakdown of the installation commands for the dependencies on a chapter-by-chapter basis.

---

[6] There are recent efforts to create conda packages for the FEniCS libraries and their dependencies: http://fenicsproject.org/download/.

[7] For more information about software container solution Docker, see www.docker.com.

Table A-1. *Installation Instructions for Dependencies for Each Chapter*

Chapter	Used Libraries	Installation

1	IPython, Spyder, Jupyter	`conda install ipython jupyter jupyterlab`
		`conda install spyder pylint pyflakes pep8`
		
		Here, `pylint`, `pyflakes`, and `pep8` are code analysis tools that Spyder can use.
		To convert IPython notebooks to PDF, you also need a working LaTeX installation.
		To book-keep which versions of libraries were used to execute the IPython notebooks that accompany this book, the IPython extension command `%version_information` was used; it is available in the `version_information` package that can be installed with `pip`:
		
		`pip install version_information`.

2	NumPy	`conda install numpy`
3	NumPy, SymPy	`conda install numpy sympy`
4	NumPy, Matplotlib	`conda install numpy matplotlib`
5	NumPy, SymPy, SciPy, Matplotlib	`conda install numpy sympy scipy matplotlib`
6	NumPy, SymPy, SciPy, Matplotlib, cvxopt	`conda install numpy sympy scipy matplotlib cvxopt`
7	NumPy, SciPy, Matplotlib	`conda install numpy scipy matplotlib`

8	NumPy, SymPy, SciPy, Matplotlib, Scikit-Monaco	`conda install numpy sympy scipy matplotlib`
		
		There is no conda package for scikit-monaco, so we need to install this library using `pip`:
		
		`pip install scikit-monaco`
		
		At the time of writing, this library is also not compatible with the latest version of Python, and to use this library, it might be necessary to use an older version of Python, such as Python 3.6.

9	NumPy, SymPy, SciPy, Matplotlib	`conda install numpy sympy scipy matplotlib`

10	NumPy, SciPy, Matplotlib, NetworkX	`conda install numpy scipy matplotlib networkx`
		
		To visualize NetworkX graphs, we also need the Graphviz library (see `www.graphviz.org`) and its Python bindings in the `pygraphviz` library:
		
		`conda install pygraphviz`

11	NumPy, SciPy, Matplotlib, and FEniCS	`conda install numpy scipy matplotlib` `conda install -c conda-forge fenics mshr`
12	NumPy, Pandas, Matplotlib, Seaborn	`conda install numpy pandas matplotlib seaborn`

(continued)

Table A-1. (*continued*)

Chapter	Used Libraries	Installation
13	NumPy, SciPy, Matplotlib, Seaborn	`conda install numpy scipy matplotlib seaborn`
14	NumPy, Pandas, Matplotlib, Seaborn, Patsy, Statsmodels	`conda install numpy pandas matplotlib seaborn patsy statsmodels`
15	NumPy, Matplotlib, Seaborn, scikit-learn	`conda install numpy matplotlib seaborn scikit-learn`
16	NumPy, Matplotlib, PyMC	`conda install numpy matplotlib pymc`
17	NumPy, SciPy, Matplotlib	`conda install numpy scipy matplotlib`

18	NumPy, Pandas, h5py, PyTables, msgpack, PyArrow	`conda install numpy pandas h5py pytables msgpack-python pyarrow`
		
		At the time of writing, the `msgpack-python` conda package is not available for all platforms. When conda packages are not available, the msgpack library needs to be installed manually, and its Python bindings can be installed using `pip`:
		
		`pip install msgpack-python`

19	NumPy, Matplotlib, Cython, Numba	`conda install numpy matplotlib cython numba`

A list of the packages and their exact versions used to run the code included in this book is also available in the `requirements.txt` file that is available for download together with the code listing. With this file, we can create an environment with all the required dependencies with a single command.

```
$ conda create -n py3.10 --file py3.10_requirements.txt
```

Alternatively, we can re-create the py3.10 and py3.11 environments using the exports py3.10-env.yml and py3.11-env.yml. These files are also available together with the source code listings.

```
$ conda env create --file py3.10-env.yml
$ conda env create --file py3.11-env.yml
```

Summary

This appendix reviewed the installation of the various Python libraries used in this book. The Python environment for scientific computing is not a monolithic environment; rather, it consists of an ecosystem of diverse libraries maintained and developed by different groups of people, following different release cycles and development paces. Consequently, collecting all the necessary pieces of a productive setup from scratch can be difficult. In response to this problem, several solutions addressing this situation have appeared, typically in the form of prepackaged Python distributions. Anaconda is a popular example of such a solution in the Python scientific computing community. Here, I focused on the conda package manager from the Anaconda Python distribution, which, in addition to being a package manager, also allows the creation and management of virtual installation environments.

Further Reading

If we want to create Python source packages for your own projects, see, for example, `https://packaging.python.org/en/latest/index.html`. Particularly study the `setuptools` library and its documentation at `https://github.com/pypa/setuptools`. Using `setuptools`, we can create installable and distributable Python source packages. Once a source package has been created using `setuptools`, creating binary conda packages for distribution is usually straightforward. For information on creating and distributing conda packages, see `https://docs.conda.io/projects/conda-build/en/stable/user-guide/tutorials/build-pkgs.html`. See also the `conda-recipes` repository at `github.com`, which contains many examples of conda packages: `http://github.com/conda/conda-recipes`. Finally, `www.anaconda.org` is a conda package hosting service with many public channels (repositories) where custom-built conda packages can be published and installed directly using the conda package manager. Many packages unavailable in the standard Anaconda channel can be found on user-contributed channels on `anaconda.org`. Many packages are available in the `conda-forge` channel, built from conda recipes on `conda-forge.org`.

Index

A

Adams-Bashforth methods, 224
Adams-Moulton methods, 224
Application programming
    interfaces (APIs), 96
ARMA model, 360
Axes
    layouts
        GridSpec, 123
        insets, 119, 120
        subplot2grid, 122
        subplots, 120, 122
    legends, 108
    line properties, 104–107
    object, 102
    plot method, 103
    properties
        labels/titles, 111
        log plots, 116
        range, 111
        spines, 118
        ticks/tick labels/grids, 112–115
        twin axes, 117, 118
    text formatting/annotations, 109, 110

B

Backward differentiation formula (BDF), 223
Bambi, 386
Bandwidth, 331
Bayesian inference methods, 385
Bayesian statistics
    frequentist, 385
    importing modules, 386
    linear regression, 396, 398–406
    MCMC, 393
    PDF, 390
    posterior probability distribution, 385
    prior probability distribution, 385
    sample posterior distributions, 393, 394, 396
    statistical model, 388, 389, 391, 392
    unconditioned/conditional probabilities,
        386, 387
Bisection method, 141
Blackman function, 415, 416
Boolean-valued indexing, 44
Broyden-Fletcher-Goldfarb-Shanno (BFGS), 159

C

Calculus
    definition, 81
    derivatives, 82
    integrals, 83, 84
    limits, 86
    series, 85
    sums/products, 87
C and Fortran, 1, 477
cdf method, 329
C foreign function interface (CFFI), 459
classification_report function, 378
Code optimization
    Cython, 467–476
    importing modules, 461
    Numba, 461–467
Comma-separated values (CSV), 430
Composite quadrature rules, 192
Computer algebra system (CAS), 67
Conda, 478–481, 486
conda env export command, 480
confusion_matrix function, 378, 382
Coordinate list format, 239
create_dataset method, 438, 441
create_group method, 436
Cross-validation, machine-learning, 367
ctypedef fused keyword, 473
cumsum function, 472
custom_template.tplx, 25
cvxopt.solvers.lp function, 169
cy_julia_fractal function, 474
Cython, 460, 467

Printed in the United States
by Baker & Taylor Publisher Services